Pseudoscience

Pseudoscience

The Conspiracy Against Science

Edited by Allison B. Kaufman and James C. Kaufman

The MIT Press
Cambridge, Massachusetts
London, England

© 2018 Massachusetts Institute of Technology

All rights reserved. No part of this book may be reproduced in any form by any electronic or mechanical means (including photocopying, recording, or information storage and retrieval) without permission in writing from the publisher.

This book was set in ITC Stone Serif by Westchester Publishing Services. Printed and bound in the United States of America.

Library of Congress Cataloging-in-Publication Data
Names: Kaufman, Allison (Allison B.) editor. | Kaufman, James C., editor.
Title: Pseudoscience : the conspiracy against science / edited by Allison
   Kaufman and James C. Kaufman.
Description: Cambridge, MA : The MIT Press, [2017] | Includes bibliographical
   references and index.
Identifiers: LCCN 2017025528 | ISBN 9780262037426 (hardcover : alk. paper)
Subjects: LCSH: Pseudoscience.
Classification: LCC Q172.5.P77 P73 2017 | DDC 001.9—dc23
LC record available at https://lccn.loc.gov/2017025528

10 9 8 7 6 5 4 3 2 1

To my parents, Jean and Joe Katz, the original scientists in my life
—ABK

For my parents,
Drs. Alan S. and Nadeen L. Kaufman,
Who nurtured in me a lifelong love of learning and science (and the arts) and
taught me to question, wonder, and dream
—JCK

Contents

Foreword: Navigating a Post-Truth World: Ten Enduring Lessons from the Study of Pseudoscience   xi
Scott O. Lilienfeld

Acknowledgments   xix

Introduction: Pseudoscience: What It Costs and Efforts to Fight It   xxi

I   The Basics of Pseudoscience   1

1   **Pseudoscience and the Pursuit of Truth**   3
David K. Hecht

2   **The Psychology of (Pseudo)Science: Cognitive, Social, and Cultural Factors**   21
Emilio J. C. Lobato and Corinne Zimmerman

3   **The Illusion of Causality: A Cognitive Bias Underlying Pseudoscience**   45
Fernando Blanco and Helena Matute

4   **Hard Science, Soft Science, and Pseudoscience: Implications of Research on the Hierarchy of the Sciences**   77
Dean Keith Simonton

II   What Pseudoscience Costs Society   101

5   **Food-o-science Pseudoscience: The Weapons and Tactics in the War on Crop Biotechnology**   103
Kevin M. Folta

6   **An Inside Look at Naturopathic Medicine: A Whistleblower's Deconstruction of Its Core Principles**   137
Britt Marie Hermes

7   Risky Play and Growing Up: How to Understand the Overprotection
    of the Next Generation   171
    Leif Edward Ottesen Kennair, Ellen Beate Hansen Sandseter, and David Ball

8   The Anti-Vaccine Movement: A Litany of Fallacy and Errors   195
    Jonathan Howard and Dorit Rubinstein Reiss

III   Scientific (or Pseudoscientific) Soundness   221

9   Understanding Pseudoscience Vulnerability through Epistemological
    Development, Critical Thinking, and Science Literacy   223
    Arnold Kozak

10   Scientific Failure as a Public Good: Illustrating the Process of Science
    and Its Contrast with Pseudoscience   239
    Chad Orzel

11   Evidence-Based Practice as a Driver of Pseudoscience in Prevention
    Research   263
    Dennis M. Gorman

12   Scientific Soundness and the Problem of Predatory Journals   283
    Jeffrey Beall

13   Pseudoscience, Coming to a Peer-Reviewed Journal Near You   301
    Adam Marcus and Ivan Oransky

IV   Pseudoscience in the Mainstream   307

14   "Integrative" Medicine: Integrating Quackery with Science-Based Medicine   309
    David H. Gorski

15   Hypnosis: Science, Pseudoscience, and Nonsense   331
    Steven Jay Lynn, Ashwin Gautam, Stacy Ellenberg, and Scott O. Lilienfeld

16   Abuses and Misuses of Intelligence Tests: Facts and Misconceptions   351
    Mark Benisz, John O. Willis, and Ron Dumont

17   Reflections on Pseudoscience and Parapsychology: From Here to There
    and (Slightly) Back Again   375
    Christopher C. French

V Science Activism: How People Think about Pseudoscience 393

18 Using Case Studies to Combat a Pseudoscience Culture 395
Clyde Freeman Herreid

19 "HIV Does Not Cause AIDS": A Journey into AIDS Denialism 419
Seth C. Kalichman

20 Swaying Pseudoscience: The Inoculation Effect 441
Kavin Senapathy

21 The Challenges of Changing Minds: How Confirmation Bias and Pattern Recognition Affect Our Search for Meaning 451
Indre Viskontas

VI Conclusion 465

22 Truth Shall Prevail 467
Paul Joseph Barnett and James C. Kaufman

Contributor List 481
Index 483

Foreword: Navigating a Post-Truth World: Ten Enduring Lessons from the Study of Pseudoscience

Scott O. Lilienfeld

We find ourselves living increasingly in a "post-truth" world, one in which emotions and opinions count for more than well-established findings when it comes to evaluating assertions. In much of contemporary Western culture, such catchphrases as "Don't confuse me with the facts," "Everyone is entitled to my opinion," and "Trust your gut" capture a troubling reality, namely, that many citizens do not—and in some cases, apparently cannot—adequately distinguish what is true from what they wish to be true. This overreliance on the "affect heuristic," the tendency to gauge the truth value of a proposition based on our emotional reactions to it (Slovic, Finucane, Peters, and MacGregor, 2007), frequently leads us to accept dubious assertions that warm the cockles of our hearts, and to reject well-supported assertions that rub us the wrong way. We are all prone to this error, but one hallmark of an educated person is the capacity to recognize and compensate for it, at least to some degree.

We also live increasingly in an age in which intuition is prized above logical analysis. Even some prominent science writers (Horgan, 2005) and prominent scientists (Diamond, 2017) have maintained that "common sense" should be accorded more weight in the adjudication of scientific claims than it presently receives. Such endorsements of everyday intuition are surprising, not to mention ahistorical, given overwhelming evidence that commonsense reasoning often contributes to grossly mistaken conclusions (Chabris and Simons, 2011; Kahneman, 2011; Lilienfeld, 2010). As one example, the history of medicine is a much-needed reminder that excessive reliance on gut hunches and untutored behavioral observations can cause scholars and practitioners to embrace ineffective and even harmful interventions (for a horrifying example, see Scull, 2007). Indeed, according to many medical historians, prior to 1890 the lion's share of widely administered physical interventions were worthless and in some cases iatrogenic (Grove and Meehl, 1996).

These trends are deeply concerning in today's age of virtually nonstop data flow via social media, email, cable television, and the like. We live not merely in an information age, but in a *misinformation* age. In 1859 the author and preacher C. H. Spurgeon famously wrote that "A lie will go round the world while truth is pulling its boots on" (in a curious irony, this quote has itself been widely misattributed to Mark Twain). If Spurgeon's dictum was true in 1859, it is all the more true in today's world of instantaneous information transfer across the world. Not surprisingly, the levels of many pseudoscientific beliefs among the general population are high, and may well be increasing. A Chapman University poll taken in 2016 (Ledbetter, 2016) revealed that 43 percent of the American public believes that the government is concealing information regarding alien visitations to earth, and that 24 percent believes that the US government is hiding information about the NASA moon landings. This poll also found that 47 percent of Americans believe in haunted houses, 19 percent believe in psychokinesis (the ability to move objects by means of mind power alone), 14 percent believe that psychics and astrologers can foretell future events, and 14 percent think Bigfoot is real. Clearly, those of us who are committed to dispelling questionable beliefs in the general population have our work cut out for us.

It goes without saying, then, that assisting people with the task of distinguishing science from pseudoscience is more crucial than ever. This enormously informative and enlightening volume is a critical step in this direction. Its chapters remind us that pseudoscience slices across myriad domains of modern life, and that approaching the study of pseudoscience from diverse perspectives—such as the psychological, sociological, medical, philosophical, and historical—yields distinctive insights. Readers will find that each chapter brings fresh revelations regarding the causes and consequences of pseudoscience.

As I will argue later, and as many of this book's contributors observe, the distinction between science and pseudoscience is almost certainly one of degree rather than of kind. Concerted efforts by Sir Karl Popper (1962) and other philosophers of science to identify a bright line of demarcation between science and pseudoscience, such as falsifiability, have consistently fallen short of the mark. These failures notwithstanding, we must be wary of concluding that pseudoscience cannot be meaningfully differentiated from science. Doing so would be tantamount to committing what logicians term the "beard fallacy," the error of concluding that because x and y fall along a continuum, one cannot distinguish prototypical cases of x from prototypical cases of y (Walton, 1996). The fact that there is no black-and-white distinction between a beard and a five o'clock shadow does not imply that one cannot distinguish clear-cut cases of bearded from hairless faces, let alone that beards do not exist. (Even many otherwise thoughtful scientists have committed this logical error with respect to race, mental disorder, intelligence, and other intrinsically fuzzy concepts; see Bamshad and Olson, 2003, for a useful discussion.)

The present book is a superbly edited presentation of valuable lessons regarding the application of scientific thinking to the evaluation of potentially pseudoscientific and otherwise questionable claims. In the section to follow, I attempt to distill these lessons into ten user-friendly take-home messages. It may strike some readers as odd to begin a book with a list of lessons, given that such lists usually crop up in a book's afterword or postscript. But I hope to provide readers with a roadmap of sorts, pointing them to integrative principles to bear in mind while consuming this book's diverse chapters. In generating these lessons, I have drawn in part on the chapters of this volume, and in part on the extant literature on cognitive biases and scientific thinking.

(1) *We are all biased*. Yes, that includes you and me. Some evolutionary psychologists maintain that certain biases in thinking are the products of natural selection (Haselton and Buss, 2000). For example, under conditions of uncertainty we are probably evolutionarily predisposed toward certain false positive errors (Shermer, 2011). When walking through a forest, we are generally better off assuming that a moving stick-like object is a dangerous snake rather than a twig being propelled by the wind, even though the latter possibility is considerably more likely. Better safe than sorry. Whether or not these evolutionary psychologists are correct, it seems likely that many cognitive biases are deeply ingrained in the human cognitive apparatus.

(2) *We are largely unaware of our biases*. Research on *bias blind spot* (Pronin, Lin, and Ross, 2002) demonstrates that most of us can readily identify cognitive biases in just about everyone except for one person—ourselves. As a consequence of this metabias, we often believe ourselves largely immune to serious errors in thinking that afflict others. We are not merely biased; we tend to be blind to our own biases. As a consequence, we are often overconfident of our beliefs, including our false beliefs.

(3) *Science is a systematic set of safeguards against biases*. Despite what most of us learned in high school, there is probably no single "scientific method"—that is, a unitary recipe for conducting science that cuts across all research domains (McComas, 1996). Instead, what we term "science" is almost certainly an exceedingly diverse, but systematic and finely honed, set of tools that humans have developed over the centuries to compensate for our species' biases (Lilienfeld, 2010). Perhaps foremost among these biases is confirmation bias, the propensity to selectively seek out, selectively interpret, and recall evidence that supports our hypotheses, and to deny, dismiss, and distort evidence that does not (Nickerson, 1998). As social psychologists Carol Tavris and Elliott Aronson (2007) have observed, science is a method of arrogance control; it helps to keep us honest.

(4) *Scientific thinking does not come naturally to the human species*. As many authors have noted, scientific thinking is unnatural (McCauley, 2011). It needs to be acquired and practiced assiduously. Some authors (e.g., Gilbert, 1991) have even contended that our cognitive apparatus is a believing engine. We believe first, question later (see also Kahneman, 2011). In contrast, some eminent developmental psychologists and science educators have argued that human babies and young children are "natural-born scientists" (e.g.., Gopnik, 2010). True, babies are intellectually curious, seek out patterns, and even perform miniature experiments on the world. But they are not good at sorting out which patterns are real and which are illusory. Moreover, the fashionable view that babies are natural scientists is difficult to reconcile with the fact that science emerged relatively late in human history. According to some scholars, science arose only once

in the course of the human species, namely in ancient Greece, not reappearing in full-fledged form until the European enlightenment of the eighteenth century (Wolpert, 1993). Such historical realities are not easily squared with claims that science is part-and-parcel of the human cognitive apparatus.

(5) *Scientific thinking is exasperatingly domain-specific.* Findings in educational psychology suggest that scientific thinking skills generalize slowly, if at all, across different domains. This point probably helps to explain why it is so difficult to teach scientific thinking as a broad skill that can be applied to most or all fields (Willingham, 2007). This sobering truth probably also helps to explain why even many Nobel Prize winners and otherwise brilliant thinkers can easily fall prey to the seductive sway of pseudoscience. Consider Linus Pauling, the American biochemist and two-time Nobel Prize winner who became convinced that orthomolecular therapy, involving ingestion of massive doses of vitamin C, is an effective treatment for schizophrenia, cancer, and other serious maladies (see Lilienfeld and Lynn, 2016, and Offit, 2017, for other examples). We should remind ourselves that none of us is immune to the temptations of specious claims, particularly when they fall well outside of our domains of expertise.

(6) *Pseudoscience and science lie on a spectrum.* As I noted earlier, there is almost surely no bright line distinguishing pseudoscience from science. Like many pairs of interrelated concepts, such as hill versus mountain and pond versus lake, pseudoscience and science bleed into each other imperceptibly. My campus at Emory University has a modestly sized body of water that some students refer to as a large pond, others as a small lake. Who's right? Of course, there's no clear-cut answer. The pseudoscience-science distinction is probably similar. Still, as I have pointed out, the fact that there is no categorical distinction between pseudoscience and science does not mean that we cannot differentiate clear-cut exemplars of each concept. Just as no one would equate the size of a small pond in a local city park with the size of Lake Michigan, few of us would equate the scientific status of crystal healing with that of quantum mechanics.

(7) *Pseudoscience is characterized by a set of fallible, but useful, warning signs.* Some contributors to this edited volume appear to accept the view that pseudoscience is a meaningful concept, whereas others appear not to. Following the lead of the philosopher of science Larry Laudan (1983), the latter authors contend that, because the effort to demarcate pseudoscience from science has failed, there is scant substantive coherence to the pseudoscience concept. My own take, for what it is worth, is that pseudoscience is a family-resemblance concept (see also Pigliucci and Boudry, 2013) that is marked by a set of fallible, but nonetheless useful, warning signs. Such warning signs differ somewhat across authors, but often comprise an absence of self-correction, overuse of ad hoc maneuvers to immunize claims from refutation, use of scientific-sounding but vacuous language, extraordinary claims in the absence of compelling evidence, overreliance on anecdotal and testimonial assertions, avoidance of peer review, and the like

(Lilienfeld, Lynn, and Lohr, 2014). Despite their superficial differences, these warning signs all reflect a failure to compensate for confirmation bias, an overarching characteristic that sets them apart from mature sciences.

(8) *Pseudoscientific claims differ from erroneous claims.* Intuitively, we all understand that there is a fundamental difference between fake new and false news. The latter is merely incorrect, and typically results from the media getting things wrong. In contrast, the former is deceptive, often intentionally so. Similarly, many and arguably most assertions in science are surely erroneous, but that does not render them pseudoscientific. Instead, pseudoscientific claims differ from incorrect scientific claims, and in many ways are far more pernicious, because they are deceptive. Because they appear at first blush to be scientific, they can fool us. To most untrained eyes, they appear to be the real thing, but they are not.

(9) *Scientific and pseudoscientific thinking are cut from the same basic psychological cloth.* In many respects, this is one of the most profound insights imparted by contemporary psychology. Heuristics—mental shortcuts or rules of thumb—are immensely valuable in everyday life; without them, we would be psychologically paralyzed. Furthermore, in most cases, heuristics lead us to approximately correct answers. For example, if three people wearing masks and wielding handguns break into a bank and tell all of us to drop to the floor, we would be well advised to rely on the representativeness heuristic, the principle that like goes with like (Tversky and Kahneman, 1974). By doing so, we would conclude that because these individuals resemble our prototype of bank robbers, they are probably bank robbers. In fact, the invocation of the heuristic in this case and others is not only wise, but usually correct. Still, when misapplied, heuristics can lead to mistaken conclusions. For example, many unsubstantiated complementary and alternative medical remedies draw on the representativeness heuristic as a rationale for their effectiveness (Nisbett, 2015). Many companies market raw brain concentrate in pill form to enhance memory and mood (Gilovich, 1991). The reasoning, apparently, is that because psychological difficulties stem from an inadequately functioning brain, "more brain matter" will somehow help the brain to work better.

(10) *Skepticism differs from cynicism.* Skepticism has gotten a bad rap in many quarters, largely because it is commonly confused with cynicism. The term "skeptic" derives from the Greek word "skeptikos," meaning "to consider carefully" (Shermer, 2002). Skepticism requires us to keep an open mind to new claims but to insist on compelling evidence before granting them provisional acceptance. In this respect, skepticism differs from cynicism, which implies a knee-jerk dismissal of implausible claims before we have had the opportunity to investigate them carefully (Beyerstein, 1995). In fairness, some individuals in the "skeptical movement" have at times blurred this crucial distinction by rejecting assertions out of hand. Skeptics need to be on guard against their propensities toward *disconfirmation* bias, a variant of confirmation bias in which we reflexively reject assertions that challenge our preconceptions (Edwards and Smith, 1996).

If readers keep the foregoing ten lessons in mind while reading this volume, they should be well equipped to navigate their way through its stimulating chapters and their broader implications. These lessons should also remind readers that we are all susceptible to questionable claims, and that science, although hardly a panacea, is ultimately our best bulwark against our own propensities toward irrationality.

References

Bamshad, M. J., and Olson, S. E. (2003). Does Race Exist? *Scientific American, 289*(6): 78–85.

Beyerstein, B. L. (1995). *Distinguishing Science from Pseudoscience*. Victoria, Canada: Centre for Curriculum and Development.

Chabris, C., and Simons, D. (2011). *The Invisible Gorilla: And Other Ways Our Intuitions Deceive Us*. New York: Random House.

Diamond, J. (2017). Common Sense. Edge. Retrieved from https://www.edge.org/response-detail/27111

Edwards, K., & Smith, E. E. (1996). A Disconfirmation Bias in the Evaluation of Arguments. *Journal of Personality and Social Psychology, 71*, 5–24.

Gilbert, D. T. (1991). How Mental Systems Believe. *American Psychologist, 46*: 107–119.

Gilovich, T. (1991). *How We Know What Isn't So: The Fallibility of Human Reason in Everyday Life*. New York: Free Press.

Gopnik, A. (2010). How Babies Think. *Scientific American, 303*(1): 76–81.

Grove, W. M., and Meehl, P. E. (1996). Comparative Efficiency of Informal (Subjective, Impressionistic) and Formal (Mechanical, Algorithmic) Prediction Procedures: The Clinical-Statistical Controversy. *Psychology, Public Policy, and Law, 2*: 293–323.

Haselton, M. G., and Buss, D. M. (2000). Error Management Theory: A New Perspective on Biases in Cross-Sex Mind Reading. *Journal of Personality and Social Psychology, 78*(1): 81–91.

Horgan, J. (2005). In Defense of Common Sense. *New York Times*, A34.

Kahneman, D. (2011). *Thinking, Fast and Slow*. New York: Macmillan.

Laudan, L. (1983). The Demise of the Demarcation problem. In R. S. Cohen and L. Laudan (Eds.), *Physics, philosophy and psychoanalysis* (pp. 111–127). Springer: Netherlands.

Ledbetter, S. (2016). What Do Americans Fear Most? Chapman University's Third Annual Survey of American Fears Released. Chapman University Press Poll. Retrieved from https://blogs.chapman.edu/press-room/2016/10/11/what-do-americans-fear-most-chapman-universitys-third-annual-survey-of-american-fears-released/

Lilienfeld, S. O. (2010). Can Psychology Become a Science? *Personality and Individual Differences, 49*(4): 281–288.

Lilienfeld, S. O., and Lynn, S. J. (2016). You'll Never Guess Who Wrote That: 78 Surprising Authors of PsychologicalPpublications. *Perspectives on Psychological Science, 11*: 419–441.

Lilienfeld, S. O., Lynn, S. J., and Lohr, J. M. (2014). *Science and Pseudoscience in Clinical Psychology* (2nd edition). New York: Guilford.

McCauley, R. N. (2011). *Why Religion Is Natural and Science Is Not*. New York: Oxford University Press.

McComas, W. F. (1996). Ten Myths of Science: Reexamining What We Think We Know about the Nature of Science. *School Science and Mathematics, 96*: 10–16.

Nickerson, R. S. (1998). Confirmation Bias: A Ubiquitous Phenomenon in Many Guises. *Review of General Psychology, 2*: 175–220.

Nisbett, R. E. (2015). *Mindware: Tools for Smart Thinking*. New York: Macmillan.

Offit, P. A. (2017). *Pandora's Lab: Seven Stories of Science Gone Wrong*. Washington, D.C.: National Geographic Books.

Pigliucci, M., and Boudry, M. (Eds.). (2013). *Philosophy of Pseudoscience: Reconsidering the Demarcation Problem*. Chicago: University of Chicago Press.

Popper, K.R. (1962). *Conjectures and Refutations*. New York: Basic Books.

Pronin, E., Lin, D. Y., and Ross, L. (2002). The Bias Blind Spot: Perceptions of Bias in Self versus Others. *Personality and Social Psychology Bulletin, 28*: 369–381.

Scull, A. (2007). *Madhouse: A Tragic Tale of Megalomania and Modern Medicine*. New Haven, CT: Yale University Press.

Shermer, M. (2002). *Why People Believe Weird Things: Pseudoscience, Superstition, and Other Confusions of Our Time*. New York: Macmillan.

Shermer, M. (2011). *The Believing Brain: From Ghosts and Gods toPolitics and Conspiracies—How We Construct Beliefs and Reinforce Them as Truths*. New York: Macmillan.

Slovic, P., Finucane, M. L., Peters, E., and MacGregor, D. G. (2007). The Affect Heuristic. *European Journal of Operational Research, 177*: 1333–1352.

Tavris, C., and Aronson, E. (2007). *Mistakes Were Made (Not by Me): Why We Justify Foolish Beliefs, Bad Decisions, and Hurtful Actions*. Boston: Houghton-Mifflin.

Tversky, A., and Kahneman, D. (1974). Judgment Under Uncertainty: Heuristics and Biases. *Science, 185*: 1124–1130.

Walton, D. (1996). The Argument of the Beard. *Informal Logic, 18*: 235–259.

Willingham, D. T. (2007, Summer). Critical Thinking. *American Educator*, 8–19.

Wolpert, L. (1993). *The Unnatural Nature of Science: Why Science Does Not Make (Common) Sense*. Cambridge, MA: Harvard University Press.

Acknowledgments

The authors would like, first and foremost, to thank the ever-tolerant Philip Laughlin, editor extraordinaire at MIT, for his support, encouragement, and faith. We would also like to thank Anne-Marie Bono from MIT and John Hoey at Westchester Publishing Services for their help.

We would also like to thank Mark Edwards, Stephen Barrett, Steve Novella, and Paul Offit.

Finally, all of our love to Jacob and Asher.

Introduction: Pseudoscience: What It Costs and Efforts to Fight It

Allison B. Kaufman and James C. Kaufman

There is a cult of ignorance in the United States, and there always has been. The strain of anti-intellectualism has been a constant thread winding its way through our political and cultural life, nurtured by the false notion that democracy means that "my ignorance is just as good as your knowledge."
Isaac Asimov (1980)

Pseudoscience, fear-mongering, poor risk assessment, and a general misunderstanding of science have always been present in the world, but technology and social media have allowed them to permeate our everyday lives. With so many "experts" having degrees from the University of Google (as Jenny McCarthy has proudly proclaimed), how can the average person winnow out real science from fake? In some cases (such as medical treatments), these choices can mean life or death. In others (such as risk assessment), they can make people spend money on unneeded items or worry about absurd things while ignoring actual dangers. In general they prevent growth and learning.

The first step is to understand exactly what is meant by pseudoscience. We begin the book with a section on "The Basics of Pseudoscience." David K. Hecht takes us on a stroll through American history and provides an overview of the development of the public's perspective on scientific issues. On the cognitive front, Emilio J. C. Lobato and Corinne Zimmerman explore some of the reasons pseudoscience has appealed so strongly to so many people over the years, and Fernando Blanco and Helena Matute address the specific issue of cognitive bias. Finally, Dean Keith Simonton discusses how the sciences fit together into an overall hierarchy, and how and where pseudoscience weaves itself into the mix.

The next section examines the cost of pseudoscience. Kevin M. Folta details how the war on biotechnology has slowed major advances in agriculture, and relates his own personal struggles with activists who oppose genetic modification. Britt Marie Hermes, a former naturopath, explains naturopathic medicine and the dangers to society as more and more people are duped into believing it is evidence-based medicine. Leif Edward Ottesen Kennair, Ellen Beate Hansen Sandseter, and David Ball delve into how our inability to estimate risk correctly has impacted children's ability to play and learn from play. Last, Jonathan Howard and Dorit Rubinstein Reiss use a framework of logical fallacies to debunk the arguments of the anti-vaccination movement and explain how it has changed medical science in serious ways.

Scientific soundness in the public's eye is addressed next. Arnold Kozak addresses our vulnerability to accepting pseudoscience, then Chad Orzel discusses the interesting case of scientific failures that are announced to the public and then retracted. Both the existence of this vulnerability and the complications of scientific failures make the burden of evidence-based practice even heavier, as discussed by Dennis M. Gorman. The final chapters discuss the current state of peer-reviewed journals and how they can play a role in the spread of pseudoscience. Jeffrey Beall exposes the inner workings of predatory journals, and RetractionWatch editors Adam Marcus and Ivan Oransky discuss how papers are retracted and the impact of these retractions.

Although the sheer scope of pseudoscience is far too broad for a single edited book (and many areas are addressed elsewhere), the next section, on pseudoscience in the mainstream, explores examples of how pseudoscience spills into several different fields. David H. Gorski highlights how "integrative medicine" is often given as much credibility as science-based (i.e., "real") medicine. Steven Jay Lynn, Ashwin Gautam, Stacy Ellenberg, and Scott O. Lilienfeld provide an overview of hypnosis, illustrating what it can legitimately be used for and how it can been co-opted by pseudoscience. Mark Benisz, John Willis, and Ron Dumont outline the uses of intelligence tests, their limits, and a lack of understanding of the test can lead to overreaching conclusions. Lastly, Christopher C. French discusses his work in parapsychology, along with his ideas on how research in the area should be approached.

The final section offers a note of hope by highlighting varied approaches to science advocacy. Clyde Freeman Herreid presents how case studies can be used to help students think critically about pseudoscience. Seth C. Kalichman helps us understand the culture of AIDS denialism by telling his story of his own immersion in that world. We conclude with chapters by people working with new media to help people learn how to think critically: Kavin Senapathy, a blogger and community organizer, and Indre Viskontas, a television and podcast host.

Reference

Asimov, I. (1980). A Cult of Ignorance. *Newsweek*, January 21, p. 19.

I The Basics of Pseudoscience

1 Pseudoscience and the Pursuit of Truth

David K. Hecht

Pseudoscience is easy to mock. Even if one is not quite prepared to dismiss every single alternative view that has ever been called pseudoscientific, there is much cause for criticism. Like most scholars, I shake my head in disbelief at astrology and cast a skeptical eye on dowsing, séances, and crop circles. I also worry about the effect of such beliefs on acceptance of mainstream science: can a society awash in pseudoscience really take the threat of climate change seriously, or even understand it in the first place? We are faced with an enterprise that seems both ridiculous and dangerous—a tempting target for anyone committed to making a more rational world. In this essay, however, I contend that it is worth resisting that temptation. If we can avoid being critical of pseudoscientific beliefs, there is a lot that we can learn from them. They may tell us very little about the physical world, but they tell us a lot about the social one. People process information in all sorts of different ways, and understanding how societies collectively decide what is true and what is false could scarcely be a more important area of inquiry. I realize that it may seem counterintuitive to use the history of pseudoscience for such an endeavor. But if we consider only the dissemination of ideas that have proved to be correct, we may come to the intuitive but false conclusion that scientific findings get accepted (or rejected) based on how true they are. In fact, there are many factors that go into how people receive information, and accuracy plays less of a role than we might expect. People do not always interpret the world around them in the best or truest way, nor do they always go about it in the manner that scientists might hope. Trying to understand pseudoscience—and not criticize it—helps us see why this has frequently been the case.

Much energy has been spent on the "demarcation problem," the question of how to distinguish legitimate science from pretenders to that title (Pigliucci and Boudry, 2013). This is important work, with practical consequences: the easiest way, for example, to exclude intelligent design from the classroom is to show that it is not science (Nickles, 2013). But if we are looking to explain the existence of pseudoscience, we need to set aside the search for definitions. Like science itself, pseudoscience is a historical phenomenon. We can trace its emergence, its nature, its institutions, its adherents, and its changing character. And this makes it less mysterious. Pseudoscience is not a mystical, irrational, or stubborn adherence to untenable ideas—or, rather, it is not *just* that. It is also an explicable historical development. Considering it in this light offers some surprises. First, I argue that pseudoscientific beliefs are not always as random or indefensible as they might seem, and we should be careful not to dismiss this reality. Second, if pseudoscience is a historically contingent phenomenon, then so is science itself. Contrary to prevailing myth, science is neither an objective nor a detached mode of inquiry. This perhaps uncomfortable fact is a key toward understanding, and perhaps also defusing, the challenge posed by pseudoscience.

This article uses the history of pseudoscience to explore cultural attitudes about science itself. It does so without mocking pseudoscience, and in fact taking it seriously as an intellectual endeavor. I do not wish to deny the elements of error, wild speculation, and outright fraud that have characterized much of this history. But that part of the story has been well told elsewhere (Shermer, 2002). Following the lead of several historians of science who have wrestled with the subject, I contend that understanding pseudoscience is as important as debunking it. My argument begins with a brief history of science and its critics, demonstrating that the social place of science has always been more fragile than it may appear. The second section challenges the notion that it is possible to definitively demarcate "pseudo" from "real" science; there are simply too many historical instances of overlap and ambiguity for us to place any hope in solving the problem that way. The next part introduces the sociological concept of boundary work and argues that the battle against pseudoscience is not always the high-minded defense of reason that it seems to be. The article concludes by offering a few thoughts on the late–twentieth century explosion in pseudoscientific belief, setting it in context of other sorts of skepticism of science. Ultimately, I argue, ignorance has only a very limited value in explaining the persistence of pseudoscience—and if we wish to combat that persistence, we would do well to look elsewhere for explanations and solutions.

Doubting Science

Science was a marginal activity for most of human history. For centuries there were very few people who could be called scientists, and fewer still made a living out of such activity. Of course, there were exceptions to this rule, and there was no shortage of people wondering about the natural world or designing useful technology. But science as a distinct phenomenon—as something self-consciously separate from other intellectual activities and social groupings—did not emerge until the seventeenth century. Even then, it would be another two hundred years until the cultural and economic upheavals of the industrial age made science visible outside of the small circles in which it was practiced. Mid–nineteenth century American elites tried with only limited success to raise the public profile of science (Bruce, 1987), although the mechanization of industry and introduction of new modes of transportation certainly made clear how important technological innovation had become. After the Civil War, however, a decades-long series of events did succeed in making science and technology an indelible part of the American landscape. For one thing, science found stable patronage. Institutions of higher education, many of them land grant colleges with a specific mission to provide technological know-how as part of a larger program of practical learning, were one such home. And in 1876, Thomas Edison opened his laboratory at Menlo Park—one of the first of what would become many such industrial laboratories by the early twentieth century. Scientists could also count a number of visible successes in these years, including electricity, evolution, and germ theory. Moreover, spectacular advances in communication and transportation made the role of technological innovation obvious, as did the profusion of consumer products increasingly available to middle and even working-class citizens. By the middle of the twentieth century, science and technology were seen as central to the security and prosperity of the country, providing the backbone of national strength in what *Time* magazine publisher Henry Luce called "The American Century" (Bacevich, 2012).

This confidence began to erode in the 1960s. Civil rights at home and decolonization abroad took a toll on Americans' views of the moral integrity of their country. The Vietnam War undermined easy confidence in science, revealing to nonspecialists how deeply the scientific enterprise had become intertwined with the military-industrial complex. The advent of missile technology stoked nuclear fears that had been latent—and sometimes explicit—since the end of World War II (Boyer, 1985; Weart, 1988). Americans also became increasingly cognizant of the environmental consequences of technological production. Rachel Carson's 1962 bestseller *Silent Spring* exposed the toxic consequences of pesticide overuse, and the 1969 Santa Barbara oil spill was but one of many incidents that demonstrated that the natural world could not take unlimited abuse at the hands of human activity. In many respects, eroding confidence in science paralleled a general ebbing in the faith that Americans placed in institutions of all kinds, as churches and governments also came in for increased scrutiny. In 1964, for example, 76 percent of Americans answered "most of the time" or "just about always" when asked whether or not they could "trust the government in Washington to do what is right." By 1980 this had dropped to 25 percent, with the majority of respondents moving from answering "most" to answering "some" (ANES). Science had enjoyed cultural prestige through much of the twentieth century, and during the Cold War its privileged status had become ever more intertwined with the national security state. Established scientific institutions, deeply marked by the military-industrial complex of which Dwight Eisenhower had warned, were easy targets of Vietnam-era skepticism; not infrequently, scientists themselves were part of these protests (Bridger, 2015; Wolfe, 2013).

For those coming to doubt science, the historical record provided additional examples. Nineteenth-century doctors believed that women could not exercise their intellectual faculties without hurting their reproductive abilities, Progressive Era eugenicists could cite scientific backing for their advocacy of forced sterilization, and midcentury psychiatrists considered homosexuality a medical problem. These are certainly not the only such examples, and they were not simply academic mistakes. They were ideas that commanded respect and authority, and caused real harm to people. The Tuskegee syphilis study—in which researchers deliberately withheld treatment from dozens of African-American men for decades—became infamous at least in part because it was not fully anomalous, but in fact resonated with other experiences that black Americans had with the medical establishment (Reverby, 2009). It is not unreasonable to look at science's record, particularly for members of groups who have been historically disadvantaged at its hands, and conclude that something about the enterprise is awry. Massimo Pigliucci, a staunch defender of science against pseudoscience, nevertheless manages to recognize the complicated nature of why people believe what they believe. Writing about AIDS denial in South Africa, for example, he notes that "it is hardly surprising that people emerging from an apartheid regime may be inclined to suspicion of white knights in shining armor coming to their rescue, and may wish instead to emphasize their own traditions and practices" (Pigliucci, 2010; p. 60). The marriage of medicine and profit in the pharmaceutical industry is obviously a part of this as well—and the South African AIDS crisis is only one example of the ways of that biomedical practice has been problematic for many populations, largely (though not exclusively) ones that are not white, Western, or middle class (Bell and Figert, 2015; Pigliucci, 2010).

Nor need we limit ourselves to questions of obvious error and prejudice. The history of science is rife with examples of innovations that carry clear benefits and equally clear disadvantages: antibiotics save lives but also create drug-resistant diseases, and fossil fuels enable prosperity but also place ecosystems at risk. One need not be a Luddite to point out that scientific solutions to social problems are rarely as good as they seem—particularly since many of the advantages come with hidden costs, problems that are unforeseen until after the benefits are registered. There have also been deeper, more philosophical critiques of science. During World War II, the theologian Reinhold Niebuhr wrote that "the culture which venerated science in place of religion, worshipped natural causation in place of God, and which regarded the cool prudence of bourgeois man as morally more normative than Christian love, has proved itself to be less profound than it appeared to be in the seventeenth and eighteenth centuries" (Niebuhr, 1944; p. 14). This is not necessarily an antiscience sentiment, yet Niebuhr was firm in his conviction that the scientific worldview had proven "less profound than it appeared to be" two centuries earlier. He wrote these words in 1944, a year before the atomic bomb would make such sentiments increasingly widespread in American culture. In fact, one of the principal architects of that weapon—Robert Oppenheimer—would later acknowledge the limits of science in an eloquent manner, noting that its claims to truth were made possible by "an enormous renunciation of meaning and limitation of scope" (Oppenheimer, 1960; p. 20). Even this famous scientist, who retained a faith in the Enlightenment values of reason and the efficacy of human action all his life, could give voice to the way that scientific truth exists because it renounces larger claims of meaning.

At its root, science is a way of understanding the world that is extraordinarily powerful but necessarily incomplete. Only when we forget what it leaves out can we make easy condemnation of those who have chosen not to trust it. And there are a number of scholars who have reconstructed histories of science and pseudoscience in ways that let us understand a bit more about why those who believe in illogical things choose to do so. Jackson Lears, for example, talks about such pseudoscientific activities as divination and palm reading in his history of luck in American society. He places those practices alongside gambling and other games as part of what he calls a "culture of chance" (Lears, 2003). It is easy to unmask these activities as scams, and frequently (as in the case of gambling) damaging ones. But people have often had quite understandable reasons for engaging in them. Lears suggests that they were reacting to systems of control and order that had failed to deliver meaning or hope to people's lives; gambling was therefore an act of "reasserting the claims of luck against the hubris of human will" (Lears, 2003; p. 22). In *How Superstition Won and Science Lost*, John C. Burnham (1987) offers a different sort of history that nevertheless lends support to the idea that people reject science for reasons that are understandable, if unfortunate. Burnham locates a prime cause of science skepticism in the rise of media and consumer culture in the early to mid-twentieth century. His argument is counterintuitive but compelling: things like advertising and public relations work by appealing to emotions, thus helping to create a culture in which nonrational forms of decision making are valued over rational ones (Burnham, 1987). Such a society is not, to say the least, a fertile ground for the acceptance of science. These may not be positive developments. But they render a more complicated picture of why people become skeptical of science, and hence susceptible to pseudoscience.

The Demarcation Fallacy

Pseudoscience is a relational concept. It has no independent existence, but achieves meaning only through a comparison (always unfavorable) with the thing that it purports to be. This is a critical point for understanding pseudoscience, and carries at least two important corollaries. First, pseudoscience is an accusation, not an identity (Gordin, 2012). No one endeavors to do pseudoscience, but rather has that assessment of their work thrust upon them by unsympathetic observers. Nineteenth-century phrenologists and twentieth-century parapsychologists believed that they were doing legitimate work; it was mainstream scientists and later scholars who decided that they were not. A chronicle of pseudoscience through the centuries is therefore a history of rhetoric. It is not the study of a particular kind of intellectual activity, but rather of the discourse surrounding that activity. Studying pseudoscience means examining the ways that some researchers have dismissed the claims of others; it is an examination of how people have talked rather than what they have done. For it to be otherwise, we would have to identify a substantive commonality between different pseudosciences—and it seems doubtful that there is a narrative thread that goes from medieval alchemy through phrenology and quack medicine to ESP, AIDS denial, and intelligent design. There are many differences among these phenomena, and it seems an oversimplification to lump them together. Pseudosciences are related more by their shared exclusion from the category of science than by any common feature. This is part of the reason that neither side of the political spectrum—or almost any other social grouping—has a monopoly on pseudoscientific belief. Pseudoscience is not one thing but many, and people have different reasons for gravitating toward it.

The second corollary is that the history of pseudoscience cannot be understood apart from the history of science. The key word in both cases is *history*. Not only are pseudoscience and science bound together, but neither they nor their relationship to each other is static. Conceptions of science have changed over the centuries, and we have yet to find a clear and all-encompassing definition of what it is. Certainly, scientists and other scholars have devised many useful heuristics to get at this problem. But there remains something arbitrary about most definitions of science, reminiscent of Justice Potter Stewart's famous comment about pornography: "I know it when I see it." Two generations of work in the historical and sociological study of science have demonstrated that there is no timeless essence that defines the scientific enterprise as wholly distinct from other kinds of inquiry. "Science, like other productive activities," writes the geneticist R. C. Lewontin, "like the state, the family, sport, is a social institution completely integrated into and influenced by the structure of all our other social institutions" (Lewontin, 1992; p. 3). It is ultimately a social activity, and is not as fully objective or detached as popular mythology frequently assumes. The scientific method is an ideal that does not fully characterize actual scientific practice, and in fact may not even be applicable to all its subdisciplines (Cleland and Brindell, 2013). To understand pseudoscience, we must also understand the history of science—the messy, contingent, and continually changing process by which human beings have tried to understand and control the natural world.

Alchemy provides a nice illustration of this, as well as an excellent starting point for a history of pseudoscience—precisely because it is not clear that we should consider it a pseudoscience at all. Medieval and early modern alchemists experimented with a variety of techniques to transform matter; the most (in)famous such attempts involved trying to turn ordinary substances into gold. Subsequent characterizations of alchemy present it as an almost archetypal pseudoscience, with images of conjurers masquerading as scientists as they try to produce magical and fantastic transformations of matter

(Haynes, 1994). And indeed, by modern standards, there was as much philosophy and theology in alchemical practice as there was science. But the standards of today are not the relevant ones. The historian of science Lawrence Principe has demonstrated that as late as the (early) eighteenth century, audiences "did not see a clear distinction between what *we* call alchemy and chemistry" (Principe, 2011; p. 306). If we are to use hindsight to declare one of these things a pseudoscience, we must also rethink the place in the intellectual canon of people whom we *do* think of as scientists. Robert Boyle, Johannes Kepler, and Isaac Newton are all early modern scientific icons who took alchemy seriously (Principe, 2011). There is no intellectually consistent way of drawing the boundaries around science such that it includes these famous figures but excludes alchemy. Nor should we even make the attempt. The alchemists who are now so easily mocked in fact made a number of intellectual advances that were central to the scientific revolution in the sixteenth and seventeenth centuries. Newton's discoveries in optics and physics, for example, relied on a conception of particulate matter that alchemists had helped to promote (Newman, 2006; pp. 4–6). However strange this may seem in retrospect, alchemy is part of the story of how modern science formed. It was part of the same intellectual endeavor, not an unfortunate or embarrassing detour from it.

Alchemy is a suggestive but necessarily limited example, coming as it does from a time before what was called "natural philosophy" had developed into the endeavor that we now term "science." Historians generally date that transition to the nineteenth century. By the middle of the 1800s, important discoveries were being made in physics, astronomy, geology, and chemistry; these disciplines were rounding into something approaching their modern forms. Even during those years, in which scientific practice looks more identifiably modern, we can find other examples of the intertwined history of pseudoscience and science. Research into such things as phrenology, physiognomy, and mesmerism (discussed below) is easily mocked today. But like alchemy for an earlier period, the histories of these activities make much more sense when you consider the intellectual context in which they arose. Virtually no one today, and certainly no one with aspirations to a career in mainstream science, believes that the shape of a person's skull provides a key to their character. Nineteenth-century phrenologists often did, however, all the while conceiving of themselves as doing science. Their ideas have fallen into such disrepair that it may now be difficult to see how they convinced themselves of this, let alone their audiences. But it is crucial to remember that there is a difference between pseudoscience and bad science, and something can be one without being the other. And there is little to indicate that phrenology ran afoul of contemporary norms for scientific methodology or subject (Thurs, 2007). In fact, phrenology can be understood as part of an intellectual revolution: the turn toward naturalism. That it contained errors does not change this reality, nor should it blind us to the continuities it holds with modern science (Gieryn, 1999). It is only in the post-Darwin world that it has become hard to see phrenology as part of an advancing scientific worldview.

Physiognomy provides another example. The belief that facial features contain a key to character bears a number of similarities, historically and intellectually speaking, to phrenology. It is demonstrably false, even though elements of it were taken seriously well into the twentieth century; you can see its legacy, for example, in pop psychology assessments of the defendants in the sensationalized Leopold and Loeb murder trial in 1924 (Fass, 1993). But despite this falsity, physiognomy can and should be seen as part of a larger intellectual trend toward materialist explanations of the natural world. Sharrona Pearl argues that it was part of "the project of using external features to identify internal traits of specific people" (Pearl, 2010; p. 188). That the answers it provided were wrong should not obscure the fact that it was part of this project, one that grew to include such things as fingerprinting and psychoanalysis (itself a contested subject). It was also, as Pearl points out, part of an impulse for ordering and classifying that was characteristic of these decades, and is clearly scientific. Something similar can be said of mesmerism, the belief that magnetism—or something roughly akin to it—held therapeutic promise. Its specific claims were wrong. But like phrenology and physiognomy, mesmerism was part of an attempt to extend the reach of naturalist explanations for phenomena that had previously been out of the reach of such investigation (Lindauer, 2013; Winter, 1998). Today, as we take for granted that mental and emotional processes are at least somewhat amenable to scientific investigation, it may be hard to appreciate the continuities between mesmerism and subjects such as physiology or physics (Winter, 1998). But they are there. It is only with the luxury of hindsight that we can filter out the scientific aspects of mesmerism (or physiognomy, or alchemy) and misconstrue the whole enterprise as having been pseudoscientific.

A careful look at history allows us to discover reasons why particular fields that seem absurd today might justly have been considered science in centuries past. We can also do the opposite and identify the pseudoscientific origins of ideas that have since moved into the category of science. Atomic energy, for example, existed in the imaginations of science fiction novelists decades before it came to fruition underneath the University of Chicago's Stagg Field, and phrenology can be understood as early neuroscience (Anderson, 2014; Weart, 1988). Both fields of inquiry demonstrate that something can begin as either speculative or contested science, and later evolve into something legitimate. This is also true, perhaps surprisingly, of evolutionary science. We are accustomed to viewing Darwinian natural selection as a fundamental transition in the history of ideas—and with good reason. It has proven to be a durable and powerful way of explaining life's history without recourse to divinity. But it is unclear that those who accepted evolution in the mid-nineteenth century did so because of its technical solidity. Before the advent of modern genetics, neither Darwin nor anyone else could demonstrate a mechanism that made inheritance work in the way that natural selection required. As a consequence, many scientists accepted the reality of evolution while proposing frameworks other than natural selection to explain how it worked. This remained true for most of the nineteenth century—the era that Michael Ruse has called the "popular science" phase of evolutionary history (Ruse, 2013).

Evolutionary ideas had existed well before the nineteenth century, but were largely speculative. They were grounded, Ruse argues, not in evidence but in an Enlightenment ideology of progress. And it was not until the middle third of the twentieth century that evolution became fully professionalized, with university-based researchers, graduate students, dedicated journals, and mathematical rigor. The intervening years—Darwin's era—are best characterized as an era of popularization. Evolutionary theory was certainly grounded in scientific evidence during the late nineteenth century, but was most frequently "used in the public domain as a tool to push an overall metaphysical and social vision" (Ruse, 2013; p. 238). Evolution was not yet a fully professionalized science for decades after Darwin wrote, and was publicly most prominent for its use in the culture wars of the time. It was a rhetorical tool that allowed proponents to expound on the power of materialist explanations for phenomena; advancing this vision was at least as important as defending the specific details of Darwin's theory. Nor was evolution the only such tool that scientifically inclined writers had at their disposal. It was, in other words, part of the same turn toward naturalism as other intellectual propositions of the day. That the theory has proved more successful scientifically than its contemporary brethren should not obscure this fundamental similarity.

It is certainly true that many of the things that we now term pseudoscience had spiritual or other nonmaterial elements to them. But so did evolution in its nineteenth-century context; many scientists and intellectuals turned to "theistic evolution" in an attempt to reconcile the new theories with religious belief (Larson, 2004). Just because evolution is a fully professionalized science today—and has been for almost a century—does not mean that it always was. Nor can we relegate this blurriness to the past, before an imagined era of scientific maturity. Sociobiology and evolutionary psychology are more recent examples of the same phenomenon that Ruse describes; the latter, for example, has a solid if not totally secure place within psychology but has attracted much criticism from biologists (Rose and Rose, 2000; Segerstråle, 2000). It is hard to call either pseudoscientific; both use too many elements of accepted scientific theory to be so easily dismissed. Yet both are controversial and rely on conjectures to make their interpretative schemes work. It is quite possible that, one hundred years from now, they will look either like pseudoscience or accepted science. But today we can no more easily categorize them than generations past could categorize their own ambiguous areas of investigation.

Recognizing the shifting definitions of what counts as science does not mean that we need to be relativists and eschew the search for certainty altogether. There are plenty of ways to determine if a particular piece of research is faulty, bad, or fraudulent. Unlike pseudoscience, these are not relational categories. The work of Stanley Pons and Martin Fleischmann on cold fusion was not confirmed (Lewenstein, 2002). Astrologers do no better than chance when asked to discern people's zodiac signs (Carlson, 1985). Andrew

Wakefield's work on vaccines was faulty and unethical (Dyer, 2010). These claims do not depend on definitions of what science is; they can be demonstrated through analyzing what individual practitioners did, thought, and claimed. But pseudoscience is different. Invoking it entails making a corresponding assumption about the nature of science. And this is a notoriously difficult thing to do. Philosophers of science have debunked most of the usual ways of doing so: there is no common methodology to sciences that distinguishes them from nonsciences, nor does Karl Popper's falsifiability principle fully solve the problem (Pigliucci and Boudry, 2013). For most purposes, the absence of a rigorous definition of science does not matter; lacking one hardly prevents scientists from advancing knowledge, contributing to innovation, devising medical therapies, and so forth. But it does matter, I contend, for how we think about pseudoscience. As convenient as it would be to define pseudoscience precisely, and as tantalizingly close as we can sometimes get, we do not have one.

Boundary Work

What we retrospectively call the pseudoscience of the nineteenth century was characteristic of its particular moment in the history of science: an age when scientists (and others) were trying to extend the reach of what could be described with naturalist explanations and debating how best that should be done. Similarly, what we have come to call the pseudoscience of alchemy reflects its historical moment, one in which "natural philosophy" had not yet broken down into scientific and nonscientific domains. In both of these examples, we can see some surprising harmony between pseudoscience and science: not an equivalence in validity, but identifiable ways in which they were each characteristic of their particular historical moment. This is an important point about analytical method. One approach to understanding any given pseudoscience is to compare it with others, thus attempting to arrive at a normative definition of pseudoscience writ large. But it may be more fruitful to compare it with other intellectual endeavors that shared its historical moment, whether or not those were later judged to be sound or suspect. Such an approach privileges historical context over determinations of accuracy. And while that sort of analysis will not teach us anything about what is a true (or false) characterization of nature, it will let us understand how science functions as a social activity. It enables us to ask important questions such as what, sociologically speaking, is science? What have people judged to be worthy of respect, funding, licensing, or institutional support? What have they trusted to provide reliable answers to pressing personal and societal problems? A lot is at stake in the social validation of knowledge, and historically speaking it has been far from clear how this validation should be done. Invoking pseudoscience has been one way of trying to do so.

Consider, for example, the history of medicine. In the early nineteenth century, the allopathic medical tradition—what we now think of as mainstream medicine—faced a series of challenges from other visions of proper healing. In his history of alternative medicine, James C. Whorton quotes a physician who in 1850 tried to catalogue the range of unconventional medical practitioners then visible, including "homeopaths, hydropaths, eclectics, botanics, chrono-thermalists, clairvoyants, natural bone-setters, mesmerists, galvanic doctors, astrologic doctors, magnetic doctors, uriscopic doctors, blowpipe doctors, the less than a decade old plague of 'Female Physicians' (that is, women MDs), and 'etc., etc., etc.'" (Whorton, 2002; p. xii). This physician was unsympathetic to such endeavors, and the fact that he included female practitioners in his list suggests that his objections lay at least partially with *who* was doing alternative medicine, not just *what* they were doing. He had good reason to be worried. At the midpoint of the nineteenth century, the status of "regular" medicine was far from assured. The American Medical Association was only three years old and had been founded in part to help stave off the threat posed by alternative practitioners (Starr, 1982). Two decades earlier, amid the democratic ethos of the Jacksonian era, a wave of delicensing had lowered the barriers for entry into medicine. This may look strange in retrospect, as recent historical trends have been toward licensing rather than away from it. But it was not illogical in the context of the early 1800s. Few medical practices of the time could have withstood the scrutiny of modern tests of efficacy; this was true of both regular and irregular medicine. For mainstream practitioners, this was the age of "heroic medicine," which involved severe and often painful interventions such as bleeding, blistering, and purging. It should therefore come as no surprise that the gentler methods of irregular approaches could be appealing, however suspect their theoretical grounding.

Furthermore, alternative medicine changed regular medicine. Homeopathy—the belief that diluted amounts of natural substances could jump-start the body's innate ability to heal itself—has scant clinical justification. But by virtue of the simple fact that it was not heroic medicine, it helped reform the cultures of healing in nineteenth-century America. Whorton notes that Oliver Wendell Holmes—a member of the Establishment by any definition—wrote of homeopathy that "'the dealers in this preposterous system of pseudo-therapeutics have cooperated with the wiser class of practitioners in breaking up the system of over-dosing and over-drugging which has been one of the standing reproaches of medical practice" (Whorton, 2002; p. 75). Holmes was clearly no friend of homeopathy. But he was forced to acknowledge that its advocates had understood at least one thing about healing better than regular medical practitioners had. Similar things can be said about the more recent turn to holistic medicine. Many of the specific claims of alternative approaches to healing still lack clinical proof, but they have been influential nevertheless in prompting medical practitioners to think in terms of treating whole bodies rather than compartmentalized problems. It is tempting to regard such changes—both then and now—as internally driven, the result of an increasingly sophisticated medical establishment. But such a narrative would deny the almost constant pressure that this establishment has received from a host of Christian

Scientists, homeopaths, natural healers, osteopaths, and others over the more than a century and a half since the founding of the American Medical Association. After all, doctors are not just scientifically informed healers; they are also professionals who need to convince prospective clients of the legitimacy of their approaches in order to stay in business.

This is what science studies scholars call boundary work: the means by which people validate some forms of activity or knowledge and delegitimate others. Most closely associated with the work of the sociologist Thomas Gieryn, it is a powerful conceptual tool. All acts of creating categories or definitions are also acts of boundary work; they will inevitably include some ambiguous cases and exclude others. Consider, for example, the ways that definitions of "whiteness" have changed over the course of American history, or how what counts as "natural" varies across different contexts. But the application of boundary work to science is particularly interesting, since one of the great myths we have about scientific investigation is that its boundaries should be self-evident. Whether characterized by its product (knowledge) or its method (objective analysis), science seems to be a definite thing. The history of science, however, demonstrates that it is not. Every scientific endeavor involves activities—data coding, equipment management, specimen collecting—that fall into a gray area between scientific and support work. Whether or not such activities are considered science has varied across different historical contexts—and any such categorization constitutes an act of boundary work. We might also look at *who* gets considered to be a scientist: Julie DesJardins has shown how the joint work of the scientific management experts Frank and Lillian Gilbreth was seen as science when he took the lead, but not after his death when she became the public face. She also documents a wide range of instances in which the work of women, such as the mathematics underlying much of nineteenth-century astronomy, was routinely excluded from being considered science. This is yet another sort of boundary work (DesJardins, 2010). Even whole disciplines can be subject to this kind of categorization. There has never been a consensus, for example, on the status of the social sciences. Since the early twentieth century, many economists, psychologists, and sociologists have conducted their research with the ideals of scientific methodology in mind. But they have only sometimes been considered as being truly on par with the natural sciences, and these judgments have varied considerably across different practitioners, disciplines, and historical contexts (Gieryn, 1999).

Calling something pseudoscientific is another example of boundary work. It is an accusation, and there is more at stake in the term than simply calling something wrong. Whenever we invoke the idea of pseudoscience, we are revealing that we care not simply if something is wrong, but also *why* it is mistaken. There is an important difference, in this line of thinking, between different sorts of incorrect claims. And it is not immediately clear why this should be. Certainly, scientists wish to know if further investigation of a subject is likely to produce better results. But this can be analyzed without resorting to the idea of pseudoscience; one need not adopt such dismissive language in order to demonstrate that a particular methodological approach is too flawed to ever produce reliable results in the future. Instead, the primary benefits of invoking "pseudoscience" are social rather than intellectual. Being viewed as within the fold of science carries practical benefits: cultural authority, access to funding, and job opportunities. These things are scarce resources, and so scientists feel the need to police the boundaries of what counts as science. This is no different from what goes on in other disciplines and professions all the time. Academic departments routinely make decisions about what counts as sociology, economics, or literature. And professional organizations—whether of lawyers, doctors, accountants, or teachers—exist to keep some people in and some people out. The reasons for this are not always bad; after all, exclusivity is a necessary function of maintaining standards. The only unusual thing about the activity when scientists do it is that we are surprised by it; we are unaccustomed to calling the activity what it is. Such is the mythology of science in our culture that it is assumed to be a detached and objective thing above the petty squabbling of boundary work.

Confronting the fact that pseudoscience is a matter of boundary work has important implications. It locates us squarely within the social analysis of science. Of course, we might attempt to draw boundaries between science and pseudoscience based wholly on intellectual merit; this is what efforts at demarcation are all about. But even this would not suffice to explain the appeal of pseudoscience. Human beings believe many things that are not true, and empirical evidence does not always settle controversies (as any frustrated veteran of climate change debates can attest). Gieryn himself makes this point: "Those who seek essentialist demarcation criteria should not assume that these explain the epistemic authority of science" (Gieryn, 1999; p. 26). In other words, even if it were possible to demarcate science from nonscience in a definitive and final way, demonstrating this boundary would not be sufficient to bestow authority upon science and deny it to pseudoscience. Decisions about who (or what) to trust are not simple, rational affairs. They are the results of what Gieryn calls "credibility contests" in which one way of knowing is pitted against another (1999, 4–6). Even knowledge that is true must be constantly defended, because it will inevitably encroach on the interests or values of somebody in society. Examples of this phenomenon abound: climate change deniers, tobacco industry apologists, and school boards that give credence to the claims of creationists (Oreskes and Conway, 2010). What appears as obviously nonsensical to one person can seem reasonable to another. This is because the debates aren't really about science in the first place. They are about some issue of politics, ethics, or values that runs deeper than science. This is not necessarily illogical. From the perspective of an individual decision maker, it may make sense to give disproportionate weight to prior beliefs when confronted with contradictory information (Kahan, Jenkins-Smith, and Braman, 2010).

If one lesson of history is that the line between science and pseudoscience is a blurry one, then another is that belief in either one is at least as much of a social phenomenon as an intellectual one. This is true despite the reality that some beliefs are correct and others incorrect; people who accept the findings of scientists often do so for nonrational reasons as well (Kahan, Jenkins-Smith, and Braman, 2010). Authority turns out to be a much more complicated question than assessing which methodology is sound or which answer is true. It is a messy cultural question, in which a host of factors plays a role: politics, tradition, hierarchy, personal networks, and values. Adding to the confusion is that pseudoscience does not have a simple relationship with authority. As the final section illustrates, belief in pseudoscience is a way of resisting authority and appealing to it at the same time—and this may account for its persistence, as well as the reasons why scientists and their defenders become so threatened by it.

The Pseudoscientific Threat

One helpful way to think about pseudoscience is to compare it to another feature of late–twentieth century American culture, conspiracy thinking. Each is a way of challenging conventional wisdom, and each has a long history but intensified in the last third of twentieth century—just the years when mistrust in authority of all kinds began to crest (Knight 2007). They also share a similar analytical approach. The philosopher Brian Keely (1999) notes that conspiracy thinking is perhaps the only mode of thought in which evidence against a theory is taken as positive evidence for it. There is more than a trace of this disregard for evidence in pseudoscientific belief, and it is perhaps no accident that of the six examples of conspiracy thinking Keely lists to begin his article, four have to do with science (either real or pseudo.) Science and technology saturate modern society, and a lot of people are worried about it. And a lot of that worry finds its way into conspiracy thinking. Of course, there are plenty of differences between the two: pseudoscience is highly diffuse and variable, and it has no central organization or even organizing principle. (There are plenty of actual conspiracies against science—such as climate change denial or tobacco industry lobbying, as Naomi Oreskes and Erik M. Conway explain in their 2010 book, *Merchants of Doubt*.)

But perhaps the most interesting point of comparison between pseudoscientific belief and conspiracy thinking is the place where they part ways. Certainly, devotees of each make a habit of rejecting the best available evidence. But most people who are skeptical of an institution want to undermine it. Pseudoscience, by contrast, depends on science. It does not seek to disrupt the authority of science—as a conspiracy theorist might challenge the state—but instead aspires to achieve that same status for itself. Theories of intelligent design, for example, have their intellectual roots in creationism. But whereas the latter is a fundamentally nonscientific way of rejecting evolution, intelligent design attempts to appropriate the trappings and methods of science to make its case. Pseudoscientists depend on the cultural authority of science just as much as regular scientists do. They simply wish to extend its boundaries around their unorthodox practice and belief. Michael D. Gordin argues that pseudoscience may be an inevitable byproduct of a scientifically and technologically advanced society. It cannot thrive, he suggests, unless the status of science is something to which others aspire (Gordin, 2012).

We can confirm this argument in many ways. Consider, for example, the etymology of the word itself. It does not appear until 1796, well after the debut of intellectual activities (alchemy, astrology) that we now consider to be pseudoscientific. And its usage remained relatively sparse for at least a hundred years, throughout most of the nineteenth century. A Google n-gram tracking the historical usage of the word "pseudoscience" identifies a small but steady upward trend beginning around 1880—and then greatly intensifying after about 1960. This correlates well with Gordin's claim, as the scientific enterprise began coalescing into a significant social and institutional presence in the late nineteenth century. And it also correlates well with what we know about the history of public scientific authority, which begins to become more complicated and vexed in the 1960s. The explosion of pseudoscience over the past fifty years is thus a product both of the high status of science and the persistent concerns that nonscientists have frequently harbored about science.

If Gordin is right, pseudoscience cannot be eradicated—at least not without leaving science behind as well. But this does not mean that we should be complacent about pseudoscience (nor does he make that claim). Faulty ideas are damaging things, even if we can learn to see them as the inevitable byproduct of a scientific enterprise we wish to preserve. Our defense against them, however, cannot be as simple as trying to establish universally valid demarcation criteria. That is mere wishful thinking, an attempt to define the problem out of existence. Nor is it practical to refute every pretender to the throne of science one by one; there are simply too many people who are motivated to appropriate its prestige. We might start instead by reflecting on the high status of science, and whether that always serves science as well as it seems. Unrealistic expectations surround scientific activity. Among other effects, these expectations create the myth that complete certainty is possible—and skeptics have become quite good at exploiting this misunderstanding. Here again there is an instructive parallel with conspiracy thinking. Keely suggests that the issue is not about truth, but instead about "warranted belief" (Keely, 1999; p. 111). Whether or not a given conspiracy theory is true matters less, in this formulation, then whether or not the available evidence makes it reasonable to believe in the theory. This raises the intriguing possibility that there may be occasions on which dismissing something true is more logical than believing in it. But that is not so far from the ideal of scientific practice, in which all knowledge is provisional. Perhaps we could learn something from this, and start defending science as "warranted belief" instead of as objective truth.

References

American National Election Studies. (n.d.). "Trust the Federal Government, 1958–2012." Retrieved November 12, 2016, from ANES Guide to Public Opinion and Electoral Behavior website, http://www.electionstudies.org/nesguide/toptable/tab5a_1.htm

Anderson, M. (2014). *After Phrenology: Neural Reuse and the Interactive Brain*. Cambridge, Mass.: Massachusetts Institute of Technology Press.

Bacevich, A. (Ed.). (2012). *The Short American Century: A Postmortem*. Cambridge, Mass.: Harvard University Press.

Bell, S., and Figert, A. (Eds.). (2015). *Reimagining (Bio)medicalization, Pharmaceuticals, and Genetics: Old Critiques and New Engagements*. New York: Routledge.

Boyer, P. (1985). *By the Bomb's Early Light: American Thought and Culture at the Dawn of the Atomic Age*. New York: Pantheon.

Bridger, S. (2015). *Scientists at War: The Ethics of Cold War Weapons Research*. Cambridge, MA: Harvard University Press.

Bruce, R. (1987). *The Launching of Modern American Science, 1846–1876*. New York: Knopf.

Burnham, J. (1987). *How Superstition Won and Science Lost: Popularizing Science and Health in the United States*. New Brunswick, NJ: Rutgers University Press.

Carlson, S. (1985). A Double-Blind Test of Astrology. Nature 318(6045): 419–425.

Cleland, C., and Brindell, S. (2013). "The Borderlands between Science and Pseudoscience: Science and the Messy, Uncontrollable World of Nature." In M. Pigliucci and M. Boudry (Eds.), *Philosophy of Pseudoscience: Reconsidering the Demarcation Problem* (pp. 183–202). Chicago: University of Chicago Press.

DesJardins, J. (2010). *The Madame Curie Complex: The Hidden History of Women in Science*. New York: Feminist Press.

Dyer, C. (2010). *Lancet* Retracts MMR Paper after GMC Finds Andrew Wakefield Guilty of Dishonesty. *British Medical Journal* 340(7741): 281.

Fass, P. (1993). Making and Remaking an Event: The Leopold and Loeb Case in American Culture. *Journal of American History* 80(3): 919–951.

Gieryn, T. (1999). *Cultural Boundaries of Science: Credibility on the Line*. Chicago: University of Chicago Press.

Gordin, M. (2012). *The Pseudoscience Wars: Immanuel Velikovsky and the Birth of the Modern Fringe*. Chicago: University of Chicago Press.

Haynes, R. (1994). *From Faust to Strangelove: Representations of the Scientist in Western Literature*. Baltimore: Johns Hopkins University Press.

Kahan, D., Jenkins-Smith, H. and Braman, D. (2010). Cultural Cognition of Scientific Consensus. *Journal of Risk Research* 14(2): 147–174.

Keely, B. (1999). Of Conspiracy Theories. *Journal of Philosophy* 96(3): 109–126.

Knight, P. (2007). *The Kennedy Assassination.* Jackson, MS: University Press of Mississippi.

Larson, E. (2004). *Evolution: The Remarkable History of a Scientific Theory.* New York: Modern Library.

Lears, J. (2003). *Something for Nothing: Luck in America.* New York: Viking.

Lewenstein, B. (1992). "Cold Fusion and Hot History." *Osiris,* 7: 135–163.

Lewontin, R. (1992). *Biology as Ideology: The Doctrine of DNA.* New York: HarperPerennial.

Lindauer, M. (2013). *The Expressiveness of Perceptual Experience: Physiognomy Reconsidered.* Amsterdam: John Benjamins.

Newman, W. (2006). *Atoms and Alchemy: Chymistry and Experimental Origins of the Scientific Revolution.* Chicago: University of Chicago Press.

Nickles, T. (2013). "The Problem of Demarcation: History and Future." In M. Pigliucci and M. Boudry (Eds.), *Philosophy of Pseudoscience: Reconsidering the Demarcation Problem* (pp. 101–120). Chicago: University of Chicago Press.

Niebuhr, R. (1944) *The Children of Light and the Children of Darkness.* (Reprinted 2011, Chicago: University of Chicago Press.)

Oppenheimer, J. (1960). In the Keeping of Unreason. *Bulletin of the Atomic Scientists*, 16(1): 18–22.

Oreskes, N. and Conway, E. (2010). *Merchants of Doubt: How a Handful of Scientists Obscured the Truth on Issues from Tobacco Smoke to Global Warming.* New York: Bloomsbury.

Pearl, S. (2010). *About Faces: Physiognomy in Nineteenth-Century Britain.* Cambridge, MA: Harvard University Press.

Pigliucci, M. (2010). *Nonsense on Stilts: How to Tell Science from Bunk.* Chicago: University of Chicago Press.

Pigliucci, M., and Boudry, M. (Eds.) (2013). *Philosophy of Pseudoscience: Reconsidering the Demarcation Problem.* Chicago: University of Chicago Press.

Principe, L. (2011). Alchemy Restored. *Isis*, 102(2): 305–312.

Reverby, S. (2009). *Examining Tuskegee: The Infamous Syphilis Study and Its Legacy.* Chapel Hill: University of North Carolina Press.

Rose H., and Rose S. (Eds.) (2000). *Alas, Poor Darwin: Arguments Against Evolutionary Psychology.* New York: Harmony Books.

Ruse, M. (2013). Evolution: From Pseudoscience to Popular Science, from Popular Science to Professional Science. In M. Pigliucci and M. Boudry (Eds.), *Philosophy of Pseudoscience: Reconsidering the Demarcation Problem* (pp. 225–244). Chicago: University of Chicago Press.

Segerstråle, U. (2000). *Defenders of the Truth: The Battle for Science in the Sociobiology Debate and Beyond*. Oxford: Oxford University Press.

Shermer, M. (2002). *The Skeptic Encyclopedia of Pseudoscience*. Santa Barbara, CA: ABC-CLIO.

Starr, P. (1982). *The Social Transformation of American Medicine*. New York: Basic Books.

Thurs, D. (2007). *Science Talk: Changing Notions of Science in American Popular Culture*. New Brunswick, NJ: Rutgers University Press.

Weart, S. (1988). *Nuclear Fear: A History of Images*. Cambridge, MA: Harvard University Press.

Whorton, J. (2002). *Nature Cures: The History of Alternative Medicine in America*. Oxford: Oxford University Press.

Winter, A (1998). *Mesmerized: Powers of Mind in Victorian Britain*. Chicago: University of Chicago Press.

Wolfe. A. (2013). *Competing with the Soviets: Science, Technology, and the State in Cold War America*. Baltimore: Johns Hopkins University Press.

2 The Psychology of (Pseudo)Science: Cognitive, Social, and Cultural Factors

Emilio J. C. Lobato and Corinne Zimmerman

Science and technology have had profound effects on human culture, and in recent decades there has been an exponential growth in (and access to) scientific information. The key role of science education in society has been of concern in the United States since the launching of Sputnik (DeBoer, 2000; Newcombe et al., 2009). It remains a focus of educational reform (NGSS Lead States, 2013) and the topic of several reports by the National Research Council (e.g., National Research Council, 2007, 2010, 2012). As a culture, we are faced with making decisions about socio-scientific issues of local and global importance (e.g., water quality, climate change, stem cell research, space travel, genetic modification). In order to fully participate in democratic decision making, scientific knowledge and skills are required of our students (as future citizens). It has been argued that science is *the* discipline needed to promote twenty-first-century skills such as nonroutine problem solving, adaptability, complex communication skills, self-management, and systems thinking (National Research Council, 2010).

Science is clearly important; we require a workforce trained in the STEM (science, technology, engineering, and mathematics) disciplines. However, there has been a declining interest in STEM careers and a high dropout rate from STEM majors (National Research Council, 2012). Attracting and retaining individuals in science careers is a desirable goal of modern science education, but we also need to be concerned with the scientific literacy of *everybody else*. Many personal, professional, and public policy decisions require an understanding of the concepts, the processes, and the nature of science. Beyond a lack of interest in science are concerns about scientific *illiteracy*. It is alarming to scientists and educators to see the apparent "anti-science" sentiments of citizens and misinformed politicians.

We witness the consequences of scientific illiteracy in several current issues: climate change denial, anti-vaccination lobbying, end-of-world prophecies, and school boards voting to include creationism in the science curriculum. With such ready access to scientific information, there is also a corresponding access to scientific *misinformation*, which facilitates the potential for science rejection, science denialism, and support for pseudoscience (Rapp and Braasch, 2014).

The core of the argument we put forth is that much of what we already know about individual and social behavior can be used to help us understand why people have a tendency to either believe in pseudoscientific claims or to reject particular scientific claims. That is, from the principle of *parsimony* we argue that to increase our understanding of the psychology of pseudoscience we do not necessarily need to appeal to different cognitive and social mechanisms than the ones we already study. In this chapter, we define pseudoscience to mean any belief in epistemologically unwarranted claims alleged to be scientific but that use nonscientific evidentiary processes (Lobato, Mendoza, Sims, and Chin, 2014). We include (a) anti-science beliefs (e.g., anti-vaccination beliefs), (b) belief systems that have little empirical support (e.g., homeopathy, naturopathy, astrology), (c) robust science misconceptions (e.g., that heavier objects fall faster than lighter objects), and (d) ubiquitous pop-culture science myths (e.g., "learning styles," left brain vs. right brain, the fallacy that people use only 10 percent of the brain).

The Psychology of Science

For many decades, there has been an interest in the development of scientific thinking. Jean Piaget credited his meeting with Albert Einstein in 1928 for inspiring a long line of research on children's understanding of scientific concepts and their abilities to acquire, integrate, and refine their knowledge about the natural and social world (Piaget, 1946). Decades of post-Piagetian research on scientific thinking have resulted in the "psychology of science" being formalized (Feist and Gorman, 2013). The exercise of this uniquely human ability has been studied across the lifespan, from young children to professional scientists. In order to understand pseudoscientific thinking, we start with what we mean by scientific thinking.

What Is Scientific Thinking?

Scientific thinking involves the skills and abilities that allow us to coordinate our beliefs with the evidence for those beliefs (Kuhn, 2005). In this extensive literature (see Zimmerman, 2000, 2007, for reviews), scientific thinking is more precisely described as intentional information seeking (Kuhn, 2011) that reduces uncertainty by generating or confirming knowledge that is meant to be generalizable (Jirout and Zimmerman, 2015; Zimmerman and Croker, 2014). Scientific thinking develops in highly educationally mediated contexts and is scaffolded by cultural tools (Lemke, 2001). With respect to frameworks that guide this line of work, two are highly influential. The work of Klahr and his colleagues (e.g., Klahr and Dunbar, 1988) focuses on the cognitive and procedural aspects of the endeavor, whereas the work of Kuhn and her colleagues (e.g., Kuhn, Amsel, and O'Loughlin, 1988) emphasizes evidence evaluation and question generation, with particular focus on the metacognitive and metastrategic control of the process of coordinating evidence and theory.

The Cognitive Components

Klahr and Dunbar's (1988) *scientific discovery as dual search* (SDDS) model captures the complex cyclical nature of scientific thinking and includes both inquiry skills and conceptual change. SDDS is an extension of a classic model of problem solving from the field of cognitive science (Simon and Lea, 1974) and explains how people carry out problem solving in varied scientific contexts, from simulated inquiry to professional scientific practice (see Klahr, 2000, for extended discussion). There are three major cognitive components in the SDDS model: searching for hypotheses, searching for experiments, and evaluating evidence. In brief, once a hypothesis is proposed, experiments are conducted to determine the truth status of that hypothesis (or to decide among a set of competing hypotheses). Experiments may also be conducted to generate enough data to be able to propose a hypothesis (as might be the case when one has little or no prior knowledge). Evidence is then evaluated so that inferences can be made whether a hypothesis is correct or incorrect (or, in some cases, that the evidence generated is inconclusive).

The Metacognitive Components

Kuhn (2005) has argued that the defining feature of scientific thinking is the set of cognitive and metacognitive skills involved in differentiating and coordinating theory (or belief) and evidence. The effective coordination of theory and evidence depends on three metacognitive abilities: (a) the ability to encode and represent evidence and theory separately, so that relations between them can be recognized, (b) the ability to treat theories as independent objects of thought (and not as "the way things are"), and (c) the ability to recognize that theories or beliefs can be false, flawed, or incorrect. These metacognitive abilities are precursors to sophisticated scientific thinking and represent one of the ways in which children, adults, and professional scientists differ.

In the next three sections, we will outline empirical findings to support the proposal that pseudoscientific beliefs arise in much the same way as our other scientific and nonscientific beliefs do. In particular, we will focus on (a) individual factors, (b) social factors, and (c) cultural factors. Although there are many psychological findings that could be discussed under each of these levels of analysis, we review only a subset of the relevant research. At the individual level, we review the development of teleological thinking, essentialism, intuitive cognitive style, competing representations, and difficulties with belief revision. At the social level, we examine trust in testimony and evaluating expertise. At the cultural level, we explore the role of cultural identities in shaping attitudes toward or against science.

Individual Factors: Cognition and Metacognition

Although we identified many candidate individual-level factors, including a rich literature on cognitive biases and heuristics that has been described in detail elsewhere (e.g., Blanco and Matute, 2018; Kahneman, 2011; Pohl, 2004), understanding (and misunderstanding) the nature of science (e.g., Lederman, 2007), the different personal epistemological theories individuals hold (e.g., Hofer, 2001), and a substantial literature on misconceptions about both the concepts and processes of science (e.g., Lilienfeld et al., 2015; McComas, 1996; National Research Council, 2007; Posner, Strike, Hewson, and Gertzog, 1982; Treagust, 1988), our focus will be on five interrelated phenomena.

Teleological Thinking

Research has shown that children rely heavily on a naïve conceptualization of the world as intentionally and purposefully driven, a phenomenon termed *promiscuous teleology* (Kelemen, 1999). That is, it is easier to understand objects in the environment as being "for" something. An example of teleological thinking is that clouds are "for raining." Explaining that clouds exist "for rain" is easier than understanding why clouds exist at all and much easier than understanding the water cycle. Similarly, explaining that giraffes have long necks because they *need to* reach treetops for food sidesteps the naturalistic causal explanation (i.e., evolution) in favor of a purpose or intention-based explanation (Kelemen, 2012). Teleological thinking is pervasive in both biology (e.g., polar bears developed thick white fur *in order to* live in the arctic; Tamir and Zohar, 1991) and chemistry (e.g., oxygen atoms form bonds because *they want to* become more stable; Talanquer, 2013).

This bias toward teleological thinking is not confined to children. Adults also exhibit this bias, and teleological explanations are more likely to be endorsed when cognitive resources for processing information are limited, such as when judgments must be made rapidly. Teleological thinking is reduced by science education, but even educated adults exhibit a tendency to think teleologically about the natural world (Kelemen and Rosset, 2009). Professional scientists are not immune from teleological explanations when information-processing resources are limited (Kelemen, Rottman, and Seston, 2013). This pattern of results suggests that teleological thinking, which emerges at a young age, may be a default explanatory mode that can be superseded but never wholly replaced, even with education. It seems that, as a general rule, people look for agency or purpose as a causal explanation for the events around them.

Teleological thinking may help explain belief in some pseudoscientific claims, particularly those based on beliefs about *intentionality* of entities that do not possess the capacity for intentionality (e.g., alternative medical practices that argue that energy intends to cause or prevent harm). Others posit the existence of intentional agents to explain phenomena, even when a phenomenon is understood sufficiently in non-agentic terms (e.g., creationism positing the existence of a deity). A default explanatory mode focused on purpose-based or function-based explanations may be a substantial barrier to understanding natural processes (Kelemen, 2012).

Essentialism
Various forms of *essentialist thinking* also emerge in childhood and continue through adolescence into adulthood (Gelman, 2004). Essentialist thinking occurs when an individual thinks that an entity has necessary, immutable, and inherent characteristics. Essentialist thinking is not necessarily wrong, and can in fact be useful in making categorical distinctions about the world, such as distinguishing between living and nonliving entities on the basis of the source of their behavior (Gelman and Kremer, 1991). Where such thinking can take people astray, however, is when it results in assumptions or overgeneralizations about essential characteristics for entities that do not always have them.

Many social prejudices rest on assumed essential characteristics for particular groups of people (Bastian and Haslam, 2006). For example, pervasive negative stereotypes exist about the mathematical competence of women and the academic abilities of African Americans as well as positive stereotypes about the mathematical competence of Asian Americans and the athletic abilities of African Americans (Steele, 2011). These abilities (or lack thereof) are considered *inherent* characteristics of each group. There have been attempts to disguise these prejudices with the language and veneer of science, as a way of justifying and defending prejudice on both empirical grounds (i.e., claiming that there is data to support that a specific group has a particular set of characteristics) and social grounds (i.e., claiming the authority of science). *Scientific racism* is a pseudoscience that attempts to propagate and encourage racial discrimination because of essential differences between racial groups on characteristics such as intelligence, motivation, or criminality (Gillborn, 2016; Smedley and Smedley, 2005). Essentialist thinking is also prevalent in pseudoscientific beliefs about the inferiority of minority sexual orientations (Mohr, 2008), as well as attempts to use science to justify sexist beliefs about men and women (Morton, Postmes, Haslam, and Hornsey, 2009).

Pseudoscientific beliefs in domains other than the social sciences also express essentialist thinking. Anti-evolution pseudoscience proponents argue that biological evolution could not occur because one kind of life cannot produce offspring of another kind, where "kind" is a term of sufficient vagueness that anti-evolution proponents can always redefine it if need be (Scott, 2005). Psychological essentialism is an apparent impediment to accepting naturalistic explanations regarding evolution (Evans, 2001). The naturalistic fallacy expressed in some alternative medical practices is another example of essentialist thinking, where "natural" cures are inherently good or more effective and "artificial" cures are inherently bad or less effective (Gorski, 2014, 2018). Anti-scientific beliefs also exhibit essentialist thinking. Climate change denial, for instance, is sometimes justified by appealing to essential and immutable characteristics about the earth's climate, namely that climate change is natural and therefore nothing to worry about or address (Washington and Cook, 2011).

Intuitive and Analytic Thinking Styles
Teleological thinking and essentialism are specific manifestations of naïve or intuitive ways of understanding and explaining the world. A predisposition toward understanding the world "intuitively" has been the subject of much research examining differences in people's beliefs in empirically supported claims and beliefs in nonempirical claims (e.g., Lindeman, 2011; Swami et al., 2014). Here intuition refers to cognitions that arise from automatic processing that can either be holistic, inferential, or affective (Pretz et al., 2014). Intuitive thinking is often contrasted with analytic thinking, which involves focused and effortful mental activity (Kahneman, 2011). Relying heavily on intuition has been found to be related to endorsement and acceptance of various forms of epistemically unwarranted beliefs, including paranormal claims (Lindeman and Aarnio, 2007), conspiracy theories (Swami et al., 2014), endorsement of alternative medical claims (Lindeman, 2011), and credence in meaningless statements designed to sound profound (Pennycook, Cheyne, Barr, Koehler, and Fugelsang, 2015). A lower reliance on analytic thinking is associated with the acceptance of various epistemically unwarranted beliefs (Lobato et al., 2014; Pennycook, Fugelsang, and Koehler, 2015).

A predisposition toward intuitive explanations may provide a cognitively efficient way to process and react to information; however, intuitions do not necessarily best reflect how the world functions. In other words, our intuitions help some of the time, but intuitions do not typically result in accurate scientific explanations (Wolpert, 1993). Many scientific discoveries and explanations are counterintuitive. Persistent naïve physics, naïve biology, or naïve psychology are still useful for most people most of the time (Keil, 2010), but when an individual is exposed to counterintuitive scientific concepts—such as the concept of "deep time" originating from geology and necessary to understand the geological and biological history of earth—one of two things can occur. One possibility is that the counterintuitive scientific concept is rejected because it contradicts prior belief (Bloom and Weisberg, 2007). The other possibility is that the counterintuitive concept is accepted but without displacing the intuitively appealing explanation, resulting in two competing or coexisting mental models.

Competing Mental Models or Schemas
It is common for individuals to persist in holding prescientific ideas (or misconceptions) even after formal science education (National Research Council, 2007). Recent research shows that it is possible to hold mutually exclusive scientific and nonscientific beliefs, whether they are pseudoscientific or supernatural (Legare and Gelman, 2008; Shtulman and Valcarcel, 2012). The persistence and resilience of these coexisting beliefs can be shown by inducing cognitive load in participants while assessing their

knowledge about scientific topics. A simple method to induce cognitive load is by having people respond as quickly as possible (Svenson and Maule, 1993). Shtulman and Valcarcel (2012) asked participants to judge 200 statements across ten domains of knowledge (e.g., astronomy, evolution, matter, thermodynamics) as true or false as quickly as possible. These statements could be intuitively true and scientifically true (e.g., "Rocks are composed of matter"), intuitively true and scientifically false (e.g., "Fire is composed of matter"), intuitively false and scientifically true (e.g., "Air is composed of matter"), or intuitively false and scientifically false (e.g., "Numbers are composed of matter"). Participants were more accurate in their judgments of statements that were consistently true or consistently false relative to statements that were inconsistent. For inconsistent statements (e.g., intuitively true but scientifically false), response times for judging the statements were significantly slower compared to response times for consistent statements. Shtulman and Valcarcel suggest that participants experienced a *cognitive conflict* that interfered with their performance when judging statements with inconsistent truth-values. Such a conflict could only occur if both ways of mentally representing the phenomenon existed simultaneously, implying that subsequent learning of relevant scientific information only suppresses earlier naïve theories rather than replacing them.

Legare and Gelman (2008) examined the coexistence of natural and supernatural frameworks for understanding illness, disease, and treatment in children, adolescents, and adults in South African communities where modern biomedical and supernatural explanations (e.g., bewitchment) of sickness and healing were available. Participants were interviewed about their understanding of the nature of illness. Of particular interest, participants were presented with a series of vignettes describing people contracting a disease and were then asked why the person in the vignette caught the disease. Explanation choices offered were *biological* (e.g., contaminated blood, interacting with a sick person), *moral* (e.g., the person lied and the illness is a punishment), *supernatural* (e.g., the person was cursed or bewitched), or *irrelevant* (e.g., the color of the person's clothing). Participants could select more than one cause, and were asked to explain their answers. Among all age groups, there was a high rate of endorsement for both biological and supernatural explanations by the same individuals. Such a result implies that endorsing supernatural explanations for illness is not necessarily the result of a lack of knowledge about scientific explanations. Again, we see that learning about biological explanations does not necessitate the replacement of existing supernatural explanations.

Explaining the findings from Shtulman and Valcarcel (2012) and Legare and Gelman (2008) could be tricky if one assumes there must be something particularly special about either scientific beliefs or nonscientific beliefs, including pseudoscientific and supernatural beliefs. If, however, these dramatically different beliefs about the same topic codevelop and coexist in the same individual, it is unlikely that there is anything unique about either. They are all beliefs.

An additional line of evidence comes from the similar justifications people provide when asked about their belief in scientific and paranormal claims (Shtulman, 2013). When prompted to explain why people hold the beliefs they do, justifications for believing in scientific concepts (e.g., evolution, genes, electrons) and paranormal concepts (e.g., telepathy, ghosts, fate) are largely similar. In particular, Shtulman (2013) found that the most common justification that college adults provided for their beliefs was categorized as *deferential*: someone the individual perceives as trustworthy provided information to support their belief. This finding supports our argument that the processes that give rise to scientific beliefs also give rise to nonscientific beliefs, but it also highlights the importance of learning from others and issues of perceiving trustworthiness or expertise (addressed in more detail, below).

Belief Revision

A robust finding in psychology is that established beliefs are very resistant to change. It is easier to form new beliefs than it is to revise existing beliefs (Koslowski, 1996; Kuhn et al., 1988). To illustrate, one type of research asks participants to consider a topic they have strong beliefs about (e.g., capital punishment, gun control) and then evaluate evidence that either supports or refutes their position. For example, Munro, Leary, and Lasane (2004) found that participants whose beliefs were disconfirmed by research about the (fictitious) relationship between homosexuality and mental illness were more critical of the study's methodology than those whose beliefs were confirmed. In general, research is evaluated more favorably if it supports initial beliefs; this phenomenon is known as *biased assimilation* (Lord, Ross, and Lepper, 1979). Rather than becoming more moderate in the face of disconfirming evidence, beliefs often become more extreme, which is known as *attitude polarization* (see MacCoun, 1998, for a review). Scientific conclusions that contradict existing beliefs may result in a weakening of the perception of scientific inquiry as capable of investigating particular empirical questions (Munro et al., 2004). Some individuals may protect their beliefs by discounting the possibility that the phenomenon related to the belief in question could be studied scientifically. This phenomenon is known as *scientific impotence discounting* (Munro, 2010). For noncontroversial topics such as the speed of falling, rolling, or sinking objects (e.g., that heavier objects fall faster than lighter objects), children and adults also resist changing their beliefs in the face of witnessed firsthand evidence (Chinn and Malhotra, 2002; Renken and Nunez, 2010). Secondhand scientific research reports with anomalous evidence (i.e., evidence that contradicts current belief) also results in resistance to belief revision. Chinn and Brewer (1998) have documented a set of eight responses to anomalous evidence for people's beliefs, six of which do not result in belief change.

In summary, psychological research has documented a number of difficulties along the extended developmental trajectory toward mature scientific thinking. Scientific thinking, like reading and mathematics, is highly culturally and educationally mediated;

the thinking skills involved require practice and are developed slowly and deliberately with the aid of cultural tools (Lemke, 2001; Zimmerman and Croker, 2014). The finding that scientific explanations coexist with, rather than replace, intuitively appealing explanations illustrates the challenges for modern science education. Together, these limitations demonstrate how important *metacognition* is for mature scientific thinking to overcome anti-science and pseudoscientific thinking. The ability to reflect on one's own belief acquisition and revision is an example of metacognition and a hallmark of fully developed scientific thinking (Kuhn, 2005). Kuhn's research suggests we often meld our beliefs and the reasons for our beliefs into a single representation of "the way things are." In order for people to go beyond asserting the correctness of existing beliefs, metalevel competencies must be developed and practiced. Critically important is the awareness that theories or beliefs (and evidence for those beliefs) are different kinds of information and possess unique properties. With metacognitive control, people can develop and change what they believe based on evidence. In doing so, not only are they aware they are changing a belief, they also know *why* (and on the basis of which *evidence*).

Despite the critical role of metacognition in scientific thinking, developing metacognitive skills in students and lay adults is not straightforward. STEM professionals are immersed in years of disciplinary training to overcome the cognitive heuristics and biases that may work well enough for us in everyday contexts, but which lead us astray in scientific contexts (Lilienfeld, Ammirati, and David, 2012). Even then, scientists may still be prone to biased or fallacious reasoning, possibly motivated by the desire to preserve a particular theory (T. Kuhn, 1970). Fortunately, science is a social endeavor, and the limitations of the individual cognizer are mitigated through the collective and shared efforts of the culture of science.

Social Factors

Science is a cultural institution that relies upon the social transmission of knowledge. Scientific information is communicated within the scientific community among professional scientists, but it is also communicated to the public via informal media (e.g., print, television, Internet) and formal science education (Goldman and Bisanz, 2002). Teachers and textbooks communicate established science, but science journalists (and now, science bloggers) translate and transmit information about "cutting edge" or "frontier" science (Zimmerman, Bisanz, and Bisanz, 1998; Zimmerman, Bisanz, Bisanz, Klein, and Klein, 2001). Nonscientific and pseudoscientific ideas are not typically communicated through formal science education (with some exceptions, e.g., "intelligent design"); the key communication channels are the traditional media and informal sources, such as family members, friends, and social media. In the past decade or so, access to information has increased substantially because of the Internet, a proliferation of TV channels, and "both sides" science journalism. These sources vary in the reliability and accuracy of the scientific information they communicate. Two key issues that we will address are how we come to trust the testimony of others and how we evaluate the expertise of others.

Trust in Testimony
Trust is an essential component to the functioning of society in general, but it is particularly important within the scientific and educational communities. Much of what we learn in schools, in textbooks, and from other academic communication is based on verbal and written communication by others whom we must trust (Tschannen-Moran and Hoy, 2000). Independent verification of all of our knowledge would be an unwieldy endeavor. From developmental psychology, we know that this trust in testimony develops very early. Children must rely on adults in many situations: "To gain understanding of, for example, new words, future events, digestion, God, and the afterlife, children presumably depend on information from other people" (Koenig, Clément, and Harris, 2004; p. 694). Even at sixteen months of age, infants are able to differentiate adult speakers who use a false label for a known object from those who use truthful labels (Koenig and Echols, 2003). Because of the dependence on adults, by age three, children have a "highly robust bias" toward trust in the testimony of others (Jaswal, Croft, Setia, and Cole, 2010). Even though trust is a "default setting," when children do have knowledge about a particular topic, they do not necessarily accept testimonials that contradict what they know to be true (Koenig et al., 2004).

Later in life, when we consider implicit "conversational rules," adults take the default position that we should believe what another person tells us. One of the unspoken conversational rules that we follow is to assume others are telling the truth, unless we have some reason or evidence to believe someone is lying or trying to deceive (Grice, 1981). Thus, as children and as adults, our predisposition is to trust; our current methods of education and social communication are highly dependent on it. This same trust that is pervasive in all social interactions and communications makes pseudoscientific information accessible and believable, particularly when we have no reason to doubt the source. One of the skills that citizens can and should develop, therefore, is the ability to evaluate expertise.

Evaluating Expertise
Norris (1995, 1997) argues that the communal nature of science has profound implications for understanding science. Scientists rely on the expertise of other scientists in their field for aspects of their knowledge. Nonscientists also rely on reasons other than direct experience in science for believing or disbelieving the scientists who communicate scientific information (Norris, 1995). Because of our interdependence, we must develop skills to evaluate expertise and trustworthiness of sources of expertise. In an

information-dense culture, we cannot all be experts, nor can we independently verify all of what we believe:

[T]hat the moon orbits the earth, that there is a vaccine for polio, that water freezes at zero degrees Celsius and that the nature of light can be described both as a particle and as a wave—are not things we actually know, properly speaking, but mere things that we accept. (Gaon and Norris, 2001; p. 190)

Therefore, that scientists and nonscientists must defer to others for many aspects of knowledge is an important component of understanding not only science acceptance, but also science denial and pseudoscience acceptance.

New scientific information is disseminated within the scientific community via peer-reviewed outlets (Berkenkotter and Huckin, 1995). Scientists, academics, and other professionals are trained in discipline-specific practices for conducting primary research, literature searches, and how to read and critically evaluate scholarly publications. The reputation of the scientist or institutional affiliation (e.g., qualifications, conflict of interest), funding sources, and the prestige of the journals where work is published are considered alongside methodology, statistical analysis, and strength of evidence (Bazerman, 1988). In contrast, citizens, policy makers, and scientists outside of their area of expertise all rely predominantly on *second-hand reports* of scientific research (Palincsar and Magnusson, 2001). Popular press reports of scientific research often include researcher credentials, source of funding, publication outlet, and university affiliation (Mallow, 1991; Zimmerman et al., 2001).

When it comes to making judgments about the scientific information reported in secondhand sources, research shows some of our limitations. Norris and Phillips (1994) found that high school seniors had difficulty interpreting the pragmatic meaning of news articles about science. Students had difficulty differentiating statements made about evidence and conclusions drawn on the basis of evidence and assumed a higher degree of certainty in the findings than was expressed in the reports. Norris, Phillips, and Korpan (2003) found a similar pattern for undergraduate students. Zimmerman and colleagues (1998) manipulated the quality of features in research reports. Although the open-ended justifications for credibility ratings from about a third of participants included reference to features such as researcher credentials, reputation, and funding sources, manipulation of the quality of these features did not influence university students' judgments of the credibility of the research. Given the prevalence of the Internet as a source of both reliable and unreliable information, Wiley and colleagues (2009) examined undergraduates' evaluation of Internet sources. They found that some students could discriminate between reliable and unreliable sources, but also that these students had difficulty explaining why they thought a particular website was more reliable. The researchers also found that explicitly teaching students how to tell reliable and unreliable sources apart benefited learning and comprehension. Their instructional unit focused on actively attending to the source of the information, the evidence presented, how the evidence fit into an explanation, and how the explanation fit with prior knowledge.

Cultural Factors

In this section, we focus on the ways that societal values and cultural identities influence how individuals come to believe what they do about scientific and pseudoscientific claims. The influence of identity on the reception and interpretation of information has been a topic of research in social psychology for a long time. One of the earliest studies examined how groups of students from two rival universities perceived the same film of a football game. Interpretations were radically different, with penalties being judged in a way that was most favorable to their team (Hastorf and Cantril, 1954). Simply being asked to adopt the identity of a scientist ("think like a scientist") improves performance on physics problems (Amsel and Johnston, 2008). Through our different identities, whether political, religious, or as a member of a group (such as a fan of a football team or as someone who does or does not identify with science), we perceive information in a way that is consistent with the values of these identities.

Cultural Identity
In contemporary culture, it is common to hear discussion of scientific, anti-scientific, or pseudoscientific topics framed in terms of cultural identity or values that have little, if anything, to do with evidentiary support. The validity of claims about climate change does not depend on political party affiliation. Evidence (or lack thereof) for psychic powers or astrology does not depend on being open-minded or skeptical. And yet, public discourse on these topics frequently includes arguments about such identities. Cultural identity is salient in the formation of attitudes and beliefs about scientific issues (Kahan, 2015).

How an issue is *framed* affects how it is received (Levin, Schneider, and Gaeth, 1998). Marketers and advertisers take advantage of this, conducting market research to determine whether to advertise a product as, for example, "98 percent fat-free" or "contains 2 percent fat," even though these two statements mean the same thing. Framing effects also influence how empirical claims are presented to and received by the public because cultural identities act as a filter through which information is interpreted. As such, when soliciting what people think about a scientific issue, the way the topic is framed will influence how people respond. Even for noncontroversial topics, frames that prime cultural identities influence people's interpretation. For example, Shen and Gromet (2015) investigated how "neurolaw" is perceived by the general public. Neurolaw is a discipline focusing on the influence of neuroscientific advances on legal and public policy. Shen and Gromet asked participants to think of ways that neuroscience could influence law and public policy. Many thought that neuroscience could help with criminal law, such as lie detection and diagnosis of defendants' sanity or competence. In a follow-up study with a nationally representative sample, Shen and Gromet asked participants to rate whether they approved or disapproved of legal reforms in light of neuroscience advances. The researchers manipulated whether this question was presented in (a) a neutral frame, (b) a frame highlighting possible benefits to defendants and defense attorneys via reduced sentencing for extenuating circumstances, or (c) a frame highlighting possible benefits to the prosecution by justifying harsher criminal sentencing. For the neutral frame, responses showed a normal distribution. When framed in a way that indicated legal benefits, the distribution of responses changed depending on political affiliation. Responses by self-identified Democrats and independents remained roughly the same regardless of framing. For self-identified Republicans, when the framing justified longer prison sentences, responses were not different from the neutral frame; when the framing justified reduced prison sentences, there was a significantly stronger disapproval for legal reforms due to advances in neuroscience. This result highlights not only the importance of the framing of an issue, but also the importance of cultural identities on acceptance (or rejection) of certain scientific ideas.

Shen and Gromet's research shows how cultural identity affects acceptance of a scientific topic that can be framed in ideologically dissonant ways. This phenomenon is more common and more robust for scientific topics that have become divisive. For example, liberals tend to be more negative about fracking and nuclear power, and conservatives more negative about evolution and climate change (Nisbet, Cooper, and Garrett, 2015). Ideologically dissonant communications have been shown to result in people engaging in *motivated cognition*; people selectively interpret information in a way that is ideologically congruent (Kunda, 1990). Ideologically dissonant information is typically either transformed in a way that makes it ideologically consistent with prior values and beliefs or the information is dismissed as wrong or false. Interestingly, Nisbet and colleagues (2015) found that the magnitude of the negative reactions to dissonant science communications was stronger for conservatives than for liberals, which they speculate may be a result of the greater partisan politicization and media attention for topics such as evolution and climate change relative to topics such as fracking and nuclear power.

Cultural Cognition
Looking at the role of cultural identity, Kahan (2015) argues that there is a tight association between group membership and discussions on topics such as climate change. Moreover, the transmission of scientific information and the ability to accurately measure the public's understanding of an issue has become complicated because individuals resort to engaging in a specific form of motivated cognition referred to as *cultural cognition* (Kahan, Hoffman, Braman, Evans, and Rachlinski, 2012). Group membership

is a salient cue for how to interpret information so that it aligns with one's identity. This form of motivated cognition results from selectively interpreting (or dismissing) information in a way that is congruent with the values and ideals of a group to which someone belongs. Thus, people with different group allegiances can view the same information and come to opposite conclusions (as with the students watching the film of a football game; Hastorf and Cantril, 1954), possibly ending up antagonistic toward another group's interpretations.

For example, a text analysis by Medimorec and Pennycook (2015) compared a report by the Intergovernment Panel on Climate Change (IPCC) with a report by the Non-government International Panel on Climate Change (NIPCC) created by the Heartland Institute (a conservative and libertarian think tank). Both documents relied on and interpreted similar empirical research for their conclusions. However, the two reports resulted in opposite conclusions, with the NIPCC document denying the scientific consensus on climate change and the IPCC supporting the scientific consensus on climate change. This divergence in conclusions from the same evidence illustrates the influence of cultural cognition (i.e., being motivated by cultural values). Medimorec and Pennycook's text analysis revealed that the language used in the IPCC and the NIPCC reports differed in quite specific ways. The IPCC report made greater use of cautious language than the NIPCC report, possibly reflecting the tentative nature of scientific conclusions or a desire to avoid being attacked politically. The more absolute and certain language used in the NIPCC report to deny the scientific consensus about climate change has been found in other science-denial literature. Barnes and Church (2013) found that, compared to writing by proponents of evolutionary theory, creationist writing was three times more likely to use the language of certainty about answering questions regarding the origin of life.

Together, these two studies comparing texts from different sources suggest that language choice may reflect a larger cultural identity that is predisposed for or against a particular scientific conclusion. Identities that are traditionally associated with a particular anti-science viewpoint use more absolutist language when discussing science, in contrast to scientists who use more tentative and reserved language.

Relationship between Scientists and the Public
Looking forward, scientists interested in engaging the public may be better served by tailoring their arguments toward the cultural values of the intended audience (e.g., conservative economic values when discussing climate change, liberal values of care vs. harm when discussing nuclear power; Graham, Haidt, and Nosek, 2009), rather than relying on empirical evidence by itself to carry the persuasive weight. It may also be beneficial to allow the culture of scientific communities to be part of the discussion (Wynne, 2006), with scientists taking a more active role in how their culture is portrayed. Although we have mentioned research about political identities, it is important

to note that being a scientist is itself an identity that has its own cultural values and beliefs. Scientific institutions place a high regard on methodological rigor and independent replication to confirm or refute findings, and have a high tolerance for ambiguity, uncertainty, and unanswered questions. Holding this set of values is quite different from other cultural identities that have expectations of absolute and unambiguous answers that can be ascertained from epistemically poor sources of information. Scientists (and those who identify with science) are less likely to rely on personal experience, interpretation of scientific findings by nonexperts (e.g., media celebrities), or to be swayed by how intuitively appealing an explanation is. Despite some portrayals by the media, science is not in crisis when a scientific publication (e.g., Open Science Collaboration, 2015) calls some scientific findings into question. Instead this is the direct result of the values of scientific communities.

Summary and Conclusions

When we consider the range of pseudoscientific beliefs that people can hold, from the anti-scientific (e.g., Washington and Cook, 2011) to incorrect myths (e.g., Howard-Jones, 2014), it is important to recognize that many of the cognitive, metacognitive, social, and cultural processes that psychologists use to understand scientific thought and behavior can be parsimoniously recruited to explain nonscientific (pseudoscientific and anti-scientific) thought. Whether we are considering the emergence or development of beliefs that are scientific or beliefs that are pseudoscientific, it is important to recognize that both are beliefs. Therefore, the processes and mechanisms that give rise to the emergence of scientific (and prescientific) beliefs are also the processes and mechanisms that give rise to pseudoscientific and other nonscientific beliefs.

Teleological language is sometimes used to teach fundamental scientific concepts via metaphor (e.g., that atoms "want" to form bonds; Talanquer, 2013) because of the intuitive appeal of such language, but it can result in future misunderstandings that can be exploited by pseudoscience advocates. Metacognitive skills can be used to defend existing beliefs (whether they are scientific or not) as well as to refine, update, or override previously held beliefs. Trusting people we perceive to have expertise may result in trusting charlatans as well as experts. Everyone has multiple social and cultural identities that we use to make sense of the world and that bias the way we interpret information about empirical claims.

We do not argue that all beliefs are equal; we propose that when attempting to understand why individuals or groups hold a particular belief, the parsimonious strategy is to look for what is common to forming and holding all beliefs. Across the research reviewed, the factors that influence acceptance or rejection of science and acceptance or rejection of pseudoscience apply to all of us. Everyone is susceptible to these cognitive and social biases, even if it is easier to recognize bias in others (i.e., the *bias blind spot*;

Pronin, Lin, and Ross, 2002), and this susceptibility does not go away with greater cognitive sophistication or cognitive ability (West, Meserve, and Stanovich, 2012). Claims that have no epistemic warrant are filtered through the same processes as claims that have solid empirical support behind them.

Exploring the social and cognitive profiles of individuals who reject science is not likely to result in finding a unique characteristic that differentiates them from people who accept science and are guarded against believing unsubstantiated claims. Instead social scientists examine large constellations of variables, looking for trends between a given set of individual, social, or cultural factors and acceptance of scientific or nonscientific beliefs. Results from these research endeavors have yet to find the silver bullet that answers the question of why some people accept a scientific conclusion while others reject it in favor of a pseudoscientific or anti-scientific claim. As the scientific culture values parsimony, perhaps it should come as no surprise that we, as scientists, are arguing there is no need to look for such an answer, and instead accept the (perhaps uncomfortable) conclusion that all beliefs are subject to the same influences.

Author's Note

We thank Steve Croker for comments on earlier drafts of the manuscript. Correspondence should be addressed to Corinne Zimmerman (e-mail: czimmer@ilstu.edu).

References

Amsel, E., and Johnston, A. (2008). The Role of Imagination in Conceptual Change. Paper presented at the annual meeting of the American Educational Research Association (New York, NY).

Barnes, R. M., and Church, R. A. (2013). Proponents of Creationism but Not Proponents of Evolution Frame the Origins Debate in Terms of Proof. *Science & Education, 22*(3): 577–603.

Bastian, B., and Haslam, N. (2006). Psychological Essentialism and Stereotype Endorsement. *Journal of Experimental Social Psychology, 42*(2): 228–235.

Bazerman, C. (1988). *Shaping Written Knowledge: The Genre and Activity of the Experimental Article in Science.* Madison: University of Wisconsin Press.

Berkenkotter, C., and Huckin, T. N. (1995). *Genre Knowledge in Disciplinary Communication: Cognition/Culture/Power.* Hillsdale, NJ: Erlbaum.

Blanco, F., and Matute, H. (2018). The Illusion of Causality: A Cognitive Bias Underlying Pseudoscience. In Allison B. Kaufman and James C. Kaufman (Eds.), *Pseudoscience: The Conspiracy Against Science.* Cambridge, MA: MIT Press.

Bloom, P., and Weisberg, D. S. (2007). Childhood Origins of Adult Resistance to Science. *Science, 316*(5827): 996–997.

Carey, S., Evans, R., Honda, M., Jay, E., and Unger, C. (1989). "An Experiment Is When You Try It and See If It Works": A Study of Grade 7 Students' Understanding of the Construction of Scientific Knowledge. *International Journal of Science Education, 11*(5): 514–529.

Chinn, C. A., and Brewer, W. F. (1998). An Empirical Test of a Taxonomy of Responses to Anomalous Data in Science. *Journal of Research in Science Teaching, 35*(6): 623–654.

Chinn, C. A., and Malhotra, B. A. (2002). Children's Responses to Anomalous Scientific Data: How Is Conceptual Change Impeded? *Journal of Educational Psychology, 94*(2): 327–343.

DeBoer, G. E. (2000). Scientific Literacy: Another Look at Its Historical and Contemporary Meanings and Its Relationship to Science Education Reform. *Journal of Research in Science Teaching, 37*(6): 582–601.

Evans, E. M. (2001). Cognitive and Contextual Factors in the Emergence of Diverse Belief Systems: Creation versus Evolution. *Cognitive Psychology, 42*(3): 217–266.

Feist, G. J., and Gorman, M. E. (Eds.). (2013). *Handbook of the Psychology of Science*. New York: Springer.

Gelman, S. A. (2004). Psychological Essentialism in Children. *Trends in Cognitive Sciences, 8*(9): 404–409.

Gelman, S. A., and Kremer, K. E. (1991). Understanding Natural Cause: Children's Explanations of How Objects and Their Properties Originate. *Child Development, 62*(2): 396–414.

Gillborn, D. (2016). Softly, Softly: Genetics, Intelligence and the Hidden Racism of the New Geneism. *Journal of Education Policy, 31*(4): 1–24.

Goldman, S. R., and Bisanz, G. L. (2002). Toward a Functional Analysis of Scientific Genres: Implications for Understanding and Learning Processes. In J. Otero, J. A. Leon, and A. C. Graesser (Eds.), *The Psychology of Science Text Comprehension*, pp. 19–50. Mahwah, NJ: Erlbaum.

Gorski, D. (2018). "Integrative" Medicine: Integrating Quackery with Science-Based Medicine. In Allison B. Kaufman and James C. Kaufman (Eds.), *Pseudoscience: The Conspiracy Against Science*. Cambridge, MA: MIT Press.

Gorski, D. H. (2014). Integrative Oncology: Really the Best of Both Worlds? *Nature Reviews Cancer, 14*(10): 692–700.

Graham, J., Haidt, J., and Nosek, B. A. (2009). Liberals and Conservatives Rely on Different Sets of Moral Foundations. *Journal of Personality and Social Psychology, 96*(5): 1029–1046.

Grice, H. P. (1981). Presupposition and Conversational Implicature. In P. Cole. (Ed.), *Radical Pragmatics*, pp. 183–198. New York: Academic Press.

Hastorf, A. H., and Cantril, H. (1954). They Saw a Game: A Case Study. *Journal of Abnormal and Social Psychology, 49*(1): 129–134.

Hofer, B. K. (2001). Personal Epistemology Research: Implications for Learning and Teaching. *Educational Psychology Review*, *13*(4): 353–383.

Howard-Jones, P. A. (2014). Neuroscience and Education: Myths and Messages. *Nature Reviews Neuroscience*, *15*(12): 817–824.

Jaswal, V. K., Croft, C., Setia, A. R., and Cole, C. A. (2010). Young Children Have a Specific, Highly Robust Bias to Trust Testimony. *Psychological Science*, *21*(10): 1541–1547.

Jirout, J., and Zimmerman, C. (2015). Development of Science Process Skills in the Early Childhood Years. In K. C. Trundle and M. Sackes (Eds.), *Research in Early Childhood Science Education*, pp. 143–165. Dordrecht: Springer.

Kahan, D. M. (2015). Climate-Science Communication and the Measurement Problem. *Political Psychology*, *36*(S1): 1–43.

Kahan, D. M., Hoffman, D. A., Braman, D., Evans, D., and Rachlinski, J. J. (2012). They Saw a Protest: Cognitive Illiberalism and the Speech-Conduct Distinction. *Stanford Law Review*, *64*(4): 851.

Kahneman, D. (2011). *Thinking, Fast and Slow*. New York: Macmillan.

Keil, F. C. (2010). The Feasibility of Folk Science. *Cognitive Science*, *34*(5): 826–862.

Kelemen, D. (1999). Function, Goals and Intention: Children's Teleological Reasoning about Objects. *Trends in Cognitive Sciences*, *3*(12): 461–468.

Kelemen, D. (2012). Teleological Minds: How Natural Intuitions about Agency and Purpose Influence Learning about Evolution. In K. S. Rosengren, S. K. Brem, E. M. Evans, and G. M. Sinatra (Eds.), *Evolution Challenges: Integrating Research and Practice in Teaching and Learning about Evolution*, pp. 66–92. Oxford: Oxford University Press.

Kelemen, D., and Rosset, E. (2009). The Human Function Compunction: Teleological Explanation in Adults. *Cognition*, *111*(1): 138–143.

Kelemen, D., Rottman, J., and Seston, R. (2013). Professional Physical Scientists Display Tenacious Teleological Tendencies: Purpose-Based Reasoning as a Cognitive Default. *Journal of Experimental Psychology: General*, *142*(4): 1074–1083.

Klahr, D. (2000). *Exploring Science: The Cognition and Development of Discovery Processes*. Cambridge, MA: MIT Press.

Klahr, D., and Dunbar, K. (1988). Dual Space Search during Scientific Reasoning. *Cognitive Science*, *12*(1), 1–48.

Koenig, M. A., Clément, F., and Harris, P. L. (2004). Trust in Testimony: Children's Use of True and False Statements. *Psychological Science*, *15*(10): 694–698.

Koenig, M. A., and Echols, C. H. (2003). Infants' Understanding of False Labeling Events: The Referential Roles of Words and the Speakers Who Use Them. *Cognition*, *87*(3): 179–208.

Koslowski, B. (1996). *Theory and Evidence: The Development of Scientific Reasoning*. Cambridge, MA: MIT Press.

Kuhn, D. (2005). *Education for Thinking*. Cambridge, MA: Harvard University Press.

Kuhn, D. (2011). What Is Scientific Thinking and How Does It Develop? In U. Goswami (Ed.), *Handbook of Childhood Cognitive Development* (2nd ed.), pp. 497–523. Oxford, UK: Wiley-Blackwell.

Kuhn, D., Amsel, E., and O'Loughlin, M. (1988). *The Development of Scientific Thinking Skills*. Orlando, FL: Academic Press.

Kuhn, T. (1970). *The Structure of Scientific Revolutions* (2nd ed.). Chicago: University of Chicago Press.

Kunda, Z. (1990). The Case for Motivated Reasoning. *Psychological Bulletin, 108*(3): 480–498.

Lederman, N. G. (2007). Nature of Science: Past, Present, and Future. In S. K. Abell and N. G. Lederman (Eds.), *Handbook of Research on Science Education*, pp. 831–879. Mahwah, NJ: Lawrence Erlbaum.

Legare, C. H., and Gelman, S. A. (2008). Bewitchment, Biology, or Both: The Co-existence of Natural and Supernatural Explanatory Frameworks across Development. *Cognitive Science, 32*(4): 607–642.

Lemke, J. L. (2001). Articulating Communities: Sociocultural Perspectives on Science Education. *Journal of Research in Science Teaching, 38*(3): 296–316.

Levin, I. P., Schneider, S. L., and Gaeth, G. J. (1998). All Frames Are Not Created Equal: A Typology and Critical Analysis of Framing Effects. *Organizational Behavior and Human Decision Processes, 76*(2): 149–188.

Lilienfeld, S. O., Ammirati, R., and David, M. (2012). Distinguishing Science from Pseudoscience in School Psychology: Science and Scientific Thinking as Safeguards against Human Error. *Journal of School Psychology, 50*(1): 7–36.

Lilienfeld, S. O., Sauvigné, K. C., Lynn, S. J., Cautin, R. L., Latzman, R. D., and Waldman, I. D. (2015). Fifty Psychological and Psychiatric Terms to Avoid: A List of Inaccurate, Misleading, Misused, Ambiguous, and Logically Confused Words and Phrases. *Frontiers in Psychology, 6*.

Lindeman, M. (2011). Biases in Intuitive Reasoning and Belief in Complementary and Alternative Medicine. *Psychology and Health, 26*(3): 371–382.

Lindeman, M., and Aarnio, K. (2007). Superstitious, Magical, and Paranormal Beliefs: An Integrative Model. *Journal of Research in Personality, 41*(4): 731–744.

Lobato, E., Mendoza, J., Sims, V., and Chin, M. (2014). Examining the Relationship between Conspiracy Theories, Paranormal Beliefs, and Pseudoscience Acceptance among a University Population. *Applied Cognitive Psychology, 28*(5): 617–625.

Lord, C. G., Ross, L., and Lepper, M. R. (1979). Biased Assimilation and Attitude Polarization: The Effects of Prior Theories on Subsequently Considered Evidence. *Journal of Personality and Social Psychology, 37*(11): 2098–2109.

MacCoun, R. J. (1998). Biases in the Interpretation and Use of Research Results. *Annual Review of Psychology, 49*(1): 259–287.

Mallow, J. V. (1991). Reading Science. *Journal of Reading*, *34*(5): 324–338.

McComas, W. F. (1996). Ten Myths of Science: Reexamining What We Think We Know about the Nature of Science. *School Science and Mathematics*, *96*(1): 10–16.

Medimorec, S., and Pennycook, G. (2015). The Language of Denial: Text Analysis Reveals Differences in Language Use between Climate Change Proponents and Skeptics. *Climatic Change*, *133*(4): 597–605.

Mohr, J. M. (2008). Oppression by Scientific Method: The Use of Science to "Other" Sexual Minorities. *Journal of Hate Studies*, *7*(21): 21–45.

Morton, T. A., Postmes, T., Haslam, S. A., and Hornsey, M. J. (2009). Theorizing Gender in the Face of Social Change: Is There Anything Essential about Essentialism? *Journal of Personality and Social Psychology*, *96*(3): 653.

Munro, G. D. (2010). The Scientific Impotence Excuse: Discounting Belief-Threatening Scientific Abstracts. *Journal of Applied Social Psychology*, *40*(3): 579–600.

Munro, G. D., Leary, S. P., and Lasane, T. P. (2004). Between a Rock and a Hard Place: Biased Assimilation of Scientific Information in the Face of Commitment. *North American Journal of Psychology*, *6*(3): 431–444.

National Research Council. (2007). *Taking Science to School: Learning and Teaching Science in Grades K–8*. Washington, DC: National Academies Press.

National Research Council. (2010). *Exploring the Intersection of Science Education and 21st Century Skills: A Workshop Summary*. Washington, DC: National Academies Press.

National Research Council. (2012). *A Framework for K–12 Science Education: Practices, Crosscutting Concepts, and Core Ideas*. Washington, DC: National Academies Press.

Newcombe, N. S., Ambady, N., Eccles, J., Gomez, L., Klahr, D., Linn, M., Miller, K., and Mix, K. (2009). Psychology's Role in Mathematics and Science Education. *American Psychologist*, *64*(6): 538–550.

NGSS Lead States. (2013). *Next Generation Science Standards: For States, by States*. Washington, DC: National Academies Press.

Nisbet, E. C., Cooper, K. E., and Garrett, R. K. (2015). The Partisan Brain: How Dissonant Science Messages Lead Conservatives and Liberals to (Dis)trust Science. *ANNALS of the American Academy of Political and Social Science*, *658*(1): 36–66.

Norris, S. P. (1995). Learning to Live with Scientific Expertise: Towards a Theory of Intellectual Communalism for Guiding Science Teaching. *Science Education*, *79*(2): 201–217.

Norris, S. P. (1997). Intellectual Independence for Nonscientists and Other Content-Transcendent Goals of Science Education. *Science Education*, *81*(2): 239–258.

Norris, S. P., and Phillips, L. M. (1994). Interpreting Pragmatic Meaning When Reading Popular Reports of Science. *Journal of Research in Science Teaching*, *31*(9): 947–967.

Norris, S. P., Phillips, L. M., and Korpan, C. A. (2003). University Students' Interpretation of Media Reports of Science and Its Relationship to Background Knowledge, Interest, and Reading Difficulty. *Public Understanding of Science*, 12(2): 123–145.

Open Science Collaboration. (2015). Estimating the Reproducibility of Psychological Science. *Science*, 349(6251), aac4716.

Palincsar, A. S., and Magnusson, S. J. (2001). The Interplay of First- and Second-Hand Investigations to Model and Support the Development of Scientific Knowledge and Reasoning. In S. M. Carver and D. Klahr (Eds.), *Cognition and Instruction: Twenty-Five Years of Progress*, pp. 151–193. Mahwah, NJ: Erlbaum.

Pennycook, G., Cheyne, J. A., Barr, N., Koehler, D. J., and Fugelsang, J. A. (2015). On the Reception and Detection of Pseudo-Profound Bullshit. *Judgment and Decision Making*, 10(6): 549.

Pennycook, G., Fugelsang, J. A., and Koehler, D. J. (2015). Everyday Consequences of Analytic Thinking. *Current Directions in Psychological Science*, 24(6): 425–432.

Piaget, J. (1946). *The Child's Conception of Time* (Reprinted 1969, New York: Ballantine.)

Pohl, R. (Ed.). (2004). *Cognitive Illusions: A Handbook on Fallacies and Biases in Thinking, Judgement and Memory*. New York: Psychology Press.

Posner, G. J., Strike, K. A., Hewson, P. W., and Gertzog, W. A. (1982). Accommodation of a Scientific Conception: Toward a Theory of Conceptual Change. *Science Education*, 66: 211–227.

Pretz, J. E., Brookings, J. B., Carlson, L. A., Humbert, T. K., Roy, M., Jones, M., and Memmert, D. (2014). Development and Validation of a New Measure of Intuition: The Types of Intuition Scale. *Journal of Behavioral Decision Making*, 27(5): 454–467.

Pronin, E., Lin, D. Y., and Ross, L. (2002). The Bias Blind Spot: Perceptions of Bias in Self versus Others. *Personality and Social Psychology Bulletin*, 28(3): 369–381.

Rapp, D. N., and Braasch, J. L. (Eds). (2014). *Processing Inaccurate Information: Theoretical and Applied Perspectives from Cognitive Science and the Educational Sciences*. Cambridge, MA: MIT Press.

Renken, M. D., and Nunez, N. (2010). Evidence for Improved Conclusion Accuracy After Reading about Rather Than Conducting a Belief-Inconsistent Simple Physics Experiment. *Applied Cognitive Psychology*, 24(6): 792–811.

Scott, E. C. (2005). *Evolution vs. Creationism: An Introduction*. Berkeley, CA: University of California Press.

Shen, F. X., and Gromet, D. M. (2015). Red States, Blue States, and Brain States: Issue Framing, Partisanship, and the Future of Neurolaw in the United States. *ANNALS of the American Academy of Political and Social Science*, 658(1): 86–101.

Shtulman, A. (2013). Epistemic Similarities between Students' Scientific and Supernatural Beliefs. *Journal of Educational Psychology*, 105(1): 199.

Shtulman, A., and Valcarcel, J. (2012). Scientific Knowledge Suppresses but Does Not Supplant Earlier Intuitions. *Cognition*, *124*(2): 209–215.

Simon, H. A., and Lea, G. (1974). Problem Solving and Rule Induction: A Unified View. In L. W. Gregg (Ed.), *Knowledge and Cognition*, pp. 105–128. Hillsdale, NJ: Lawrence Erlbaum.

Smedley, A., and Smedley, B. D. (2005). Race as Biology Is Fiction, Racism as a Social Problem Is Real: Anthropological and Historical Perspectives on the Social Construction of Race. *American Psychologist, 60*(1): 16.

Steele, C. M. (2011). *Whistling Vivaldi: How Stereotypes Affect Us and What We Can Do*. New York: W. W. Norton.

Svenson, O., and Maule, A. J. (Eds.). (1993). *Time Pressure and Stress in Human Judgment and Decision Making*, pp. 133–144. New York: Plenum Press.

Swami, V., Voracek, M., Stieger, S., Tran, U. S., and Furnham, A. (2014). Analytic Thinking Reduces Belief in Conspiracy Theories. *Cognition*, *133*(3): 572–585.

Talanquer, V. (2013). When Atoms Want. *Journal of Chemical Education*, *90*(11): 1419–1424.

Tamir, P., and Zohar, A. (1991). Anthropomorphism and Teleology in Reasoning about Biological Phenomena. *Science Education*, *75*(1): 57–67.

Treagust, D. F. (1988). Development and Use of Diagnostic Tests to Evaluate Students' Misconceptions in Science. *International Journal of Science Education*, *10*(2): 159–169.

Tschannen-Moran, M., and Hoy, W. K. (2000). A Multidisciplinary Analysis of the Nature, Meaning, and Measurement of Trust. *Review of Educational Research, 70*(4): 547–593.

Washington, H., and Cook, J. (2011). *Climate Change Denial: Heads in the Sand*. Oxford: Earthscan.

West, R. F., Meserve, R. J., and Stanovich, K. E. (2012). Cognitive Sophistication Does Not Attenuate the Bias Blind Spot. *Journal of Personality and Social Psychology*, *103*(3): 506.

Wiley, J., Goldman, S. R., Graesser, A. C., Sanchez, C. A., Ash, I. K., and Hemmerich, J. A. (2009). Source Evaluation, Comprehension, and Learning in Internet Science Inquiry Tasks. *American Educational Research Journal*, *46*(4): 1060–1106.

Wolpert, L. (1993). *The Unnatural Nature of Science*. London: Faber and Faber.

Wynne, B. (2006). Public Engagement as a Means of Restoring Public Trust in Science—Hitting the Notes, but Missing the Music? *Public Health Genomics*, *9*(3): 211–220.

Zimmerman, C. (2000). The Development of Scientific Reasoning Skills. *Developmental Review*, *20*(1): 99–149.

Zimmerman, C. (2007). The Development of Scientific Thinking Skills in Elementary and Middle School. *Developmental Review*, *27*(2): 172–223.

Zimmerman, C., Bisanz, G. L., and Bisanz, J. (1998). Everyday Scientific Literacy: Do Students Use Information about the Social Context and Methods of Research to Evaluate News Briefs about Science? *Alberta Journal of Educational Research, 44*(2): 188–207.

Zimmerman, C., Bisanz, G. L., Bisanz, J., Klein, J. S., and Klein, P. (2001). Science at the Supermarket: A Comparison of What Appears in the Popular Press, Experts' Advice to Readers, and What Students Want to Know. *Public Understanding of Science, 10*(1): 37–58.

Zimmerman, C., and Croker, S. (2014). A Prospective Cognition Analysis of Scientific Thinking and the Implications for Teaching and Learning Science. *Journal of Cognitive Education and Psychology, 13*(2): 245–257.

3 The Illusion of Causality: A Cognitive Bias Underlying Pseudoscience

Fernando Blanco and Helena Matute

The rise of pseudoscience in the twenty-first century is difficult to understand. In past times the scarcity of reliable information may have led us to trust our intuitions or act upon the suggestions of a shaman or figure drawn by the stars. But today we live in a (supposedly) knowledge-based society that has developed science and technology, a society that makes a vast amount of data available on many issues. In today's society, therefore, one might reasonably expect decisions to be based on facts rather than on superstitions. Nevertheless, many people resort to pseudoscientific ideas as a guide in their lives, instead of using the knowledge that is at their disposal.

Health-related decisions provide a clear example of how pseudoscientific beliefs can affect our lives for the worse. Consider the case of pseudo-medicines (which we define as any medicine that has not been proven to work beyond the placebo effect). While these are remedies that appear to be scientific, they are not supported by evidence, and yet many people today say that they prefer them over conventional remedies. This has serious implications for all areas of human well-being, at both societal and individual levels. Sadly, a well-documented example can be seen in the vaccines crisis (Carroll, 2015; Nyhan and Reifler, 2015). Diseases that were long extinct in the West managed to stage a comeback because some people adopted the idea that vaccines cause autism, and therefore choose to avoid these preventive therapies. Another example related to health is the extended use of homeopathy to treat a range of diseases in many countries, in spite of the fact that no effect beyond placebo has been reported for this practice (Ernst, 2002; National Health and Medical Research Council, 2015; see also Hermes, this volume), with severe consequences frequently being reported from the use of homeopathy and other alternative treatments when these treatments prevent people from the following evidence-based therapies (recall that they are called "alternative" because they are not supported by scientific evidence). Indeed, even if used as complementary medicine (that is, to complement an evidence-based treatment rather than to substitute it), they could end up producing undesired results. For example, in these cases people might end up attributing their recovery to the alternative practice, thus abandoning the effective treatment (Yarritu, Matute, and Luque, 2015). The vaccine crisis and the popularity of homeopathy are just two examples of phenomena that might be the product of the illusion of causality, a cognitive bias that we will discuss throughout this chapter.

The problems that arise from pseudoscience are not only related to health. A further example can be found in people making financial or romantic decisions based on intuitions, pseudoscience, and superstitions rather than on evidence-based knowledge (Malmendier and Tate, 2005). It is therefore not surprising that two in five Europeans are superstitious (European Commission, 2010), 34 percent believe that homeopathy is good science (European Commission, 2005), 54 percent of Americans believe in psychic healing (Newport and Strausberg, 2001), and more than 35 percent of Americans believe in haunted houses or extrasensory perception (Moore, 2005).

There are certainly many factors involved in the development of pseudoscience and superstition, but it now seems clear that these beliefs are not due to a lack of intelligence, or to a strange personality (Wiseman and Watt, 2006). Indeed, there have been cases of Nobel laureates who were superstitious (Hutson, 2015; Wiseman and Watt, 2006), and it is now known that Steve Jobs, the founder of Apple, although unquestionably intelligent, well educated, and extraordinarily successful, still chose to receive a pseudoscientific treatment (based on acupuncture, dietary supplements, and fruit juices) instead of an evidence-based therapy in the months that followed his cancer diagnosis, a decision he later regretted (Swaine, 2011).

We will suggest that in order to understand why people believe in strange things it is essential to find out how the human mind works and how it evolved. A very important feature of our cognitive system is that it has evolved in a very noisy environment where information is often ambiguous, incomplete, vague, and imprecise. And at the same time we needed (and still need) to react very rapidly to changes in the environment, such as needing to run without hesitation upon seeing a shadow that might be a predator, or to strike when seeing a pattern in the forest that might hide a prey. One might be incorrect in making these rapid decisions, but being able to detect those patterns and to associate potential causes to their most likely effects, even in the absence of clear and complete information, was surely an adaptive strategy that conferred our ancestors a survival advantage (Haselton and Nettle, 2006). As a consequence, we see figures in the clouds, and causal patterns in mere coincidences.

The mechanism by which these decisions are made is based on *heuristics* (Gilovich, Griffin, and Kahneman, 2002). These are shortcuts that our cognitive system uses to reach rapid conclusions and decisions when, in our daily lives, we encounter ambiguous and incomplete information. Heuristics are both useful and necessary. Throughout our evolutionary history we have repeatedly survived thanks to our heuristics, which have allowed a fast and efficient response to changes in the environment, even if this response is not always mathematically precise. Heuristics are used, for example, to help us interpret the patterns of light that our eyes capture, which are almost always incomplete and ambiguous. At times, however, these heuristics lead us to commit errors (Haselton and Nettle, 2006), such as perceiving what are familiarly called visual illusions.

The same applies to other types of heuristics and biases. In particular, we are interested in a cognitive bias called the *illusion of causality* (Matute et al., 2015), which we propose as a factor underlying the belief in pseudoscience. Thus, after discussing visual illusions as an example of how heuristics and biases may operate, we will then turn our attention to a discussion on the illusion of causality, a cognitive bias that we have investigated in our laboratory for the last two decades. We will first describe how causality is typically detected under ambiguity, how it should be detected, and what type of heuristics people typically use to facilitate and speed up the detection of causality under ambiguity. Then we will show that this same process that allows people and other animals to accurately detect causality (most of the time) is the same process that produces overestimations (illusions) of causality under well-known conditions. These cases where illusions of causality occur are common in pseudoscience. Examples include the doctor who is convinced that coincidences between a given treatment and a patient reporting that she feels better are synonymous with causal efficacy, or the farmer who believes that rain is due to his having purchased a miraculous sculpture for the living room. In most cases of pseudoscience there is an illusory perception of causality between a potential cause and a desired event that actually is occurring independently of the potential cause. A similar problem occurs when people believe that vaccines cause autism or that humidity causes joint pain. The causal-detection mechanism, aided by heuristics, is jumping too quickly to a conclusion that is not supported by evidence.

Current research allows us to predict when these illusions of causality are most likely to occur. Thus, knowledge of how and why these causality biases occur may help us understand how pseudoscience develops, and, importantly, could provide us with a starting point for attempting to combat the problem.

The Adaptive Bias in Pattern Detection

Although the brain accounts for only 2 percent of the body's mass, it requires up to 20 percent of the body's total resting energy consumption (Solokoff, 1960), and is therefore the most expensive organ of all. Why then do mammals preserve an organ with such a high metabolic demand?

The answer could be related to the fact that all animals have a tendency to obtain a balanced state (homeostasis) in which the use of resources reaches an optimum point to keep us alive and to meet all of our basic needs (safety, warmth, nutrition, and the assurance of offspring). In spite of this, we still live in constantly changing contexts that make this task difficult. Whether it was in the forest, where we first evolved, or the modern cities that contain most of the human animals living in the twenty-first century, a person with rigid behavior who does not respond to an ever-changing context would experience serious difficulties. The consequences of rigid behavior in nature are easy to imagine. In the long term, rivers do change their course, mountains are born or can erode, and some species become extinct. It is therefore not surprising that animals need to adapt to these changes by natural selection. Similarly, in the short term, within a person's life there are other changes that can challenge this balance: among other things, prey can become more difficult to find with seasonal changes, and droughts can occur. As well as being able to physically change in order to adapt to their environment, animals also need to change their behaviors to stay close to the optimal balance—this being our goal. Behavioral evolution that occurs quickly, and on an individual scale, is called *learning* (Domjan, 2014). In this context, the brain is a perfectly controlled machine that is capable of learning in order to adapt to our constantly changing environment. It is therefore unsurprising that the resources devoted to keep this machine running are vast and guaranteed.

Surprisingly, the significant use of resources involved in the learning processs—not only in terms of energy—is not always inevitable. Animals can sometimes use strategies to save energy by slightly sacrificing their adaptation capabilities, some examples of which are explained in the following paragraphs.

One particular example of this could be observed in visual illusions. Figure 3.1 represents a famous object configuration known as the Müller-Lyer illusion (Müller-Lyer, 1889). We encourage the reader to observe the picture and try to figure out which of the two horizontal segments (a or b) is the longest.

If you are like the majority of people in Western society, you would have perceived segment b as being slightly longer than segment a. Both segments have, in fact, the exact same length. The only difference between them is in terms of the orientation of the adornments (arrows) that round off the edges. A visual illusion of this sort is so

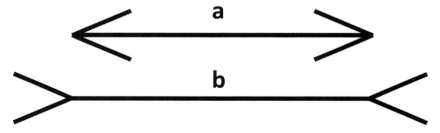

Figure 3.1
The Müller-Lyer illusion.
Figure adapted from Howe and Purves (2005).

robust that, even if we know the right answer, we cannot avoid perceiving both segments as being of a different length.

We can now see how the Müller-Lyer illusion works, and therefore we understand why the brain tends to use certain mechanisms as shortcuts and how these relate to the development of strange beliefs.

As far as psychologists are aware, visual perception, along with other sensory modalities, appears to involve some inferential or interpretational component on the part of the person who is observing the image. The traditional explanation of the Müller-Lyer illusion is as follows. When people observe lines that converge at a point, this is usually a vertex. The angle of the converging lines gives us information about the distance of the vertex with regard to the observer (see figure 3.2). If you are in a room right now, you could see several examples of this. If you move toward one of the corners of the room, you will see how both adjacent walls form an edge, creating a configuration similar to the one in figure 3.2b. If, on the contrary, you observe an

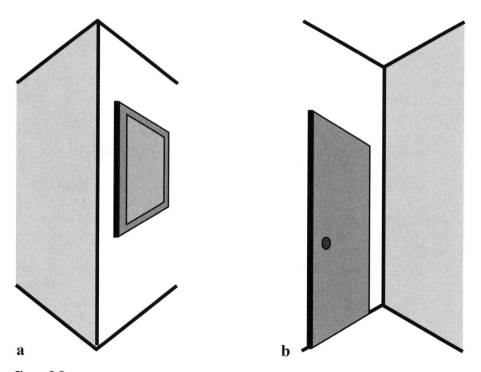

Figure 3.2
The same configuration of the Müller-Lyer illusion induces depth perception when further details (such as shades, windows) are added. (Figure elaborated by the authors.)

object such as a wall unit or a table, convergence of these two edges would indicate that the object stands out from the background (figure 3.2a). That is, the configuration of lines in figure 3.2 does actually induce perception of depth and distance in the observer.

Recent studies suggest that this type of illusion can be learned (Howe and Purves, 2005). Our cognitive system is accustomed to interpreting the visual input in such a way that it uses some invariant features of the environment in which we live. In a three-dimensional world, two segments that converge at one point usually indicate an edge, which is either incoming or outgoing (first invariant). Further, an edge far away from us will look smaller than one that is near (second invariant). In the real world in which we are constantly moving around, applying this rule is clearly an advantage for us, given that it saves us from implementing a costly learning process. We do not have to interpret each visual stimulus in detail to extract information related to depth, distance, and perspective. (For instance, the window in figure 3.2 seems closer to us than the door.) However, with some ambiguous stimuli such as those represented in figure 3.1, applying this rule leads us to commit an error. While observing figure 3.1, we process a bidimensional figure as if we were processing it in the real world (three-dimensional). It is for this reason that we interpret the convergence of the lines as a signal of perspective, concluding that the edge represented by segment a is further from us than the edge represented in segment b. Thus, we apply a second rule whereby distant objects in space are smaller, which then leads us to incorrectly believe that segment a is closer, and shorter, than segment b.

In conclusion, this and other visual illusions occur when we apply rules that work correctly most of the time, but that lead us to commit errors of judgment when the stimulus has certain characteristics (i.e., the stimulus breaks one of those rules), with the underlying reason for our error being the attempt to conserve resources and avoid a costly analysis of the visual stimulus. These rules are not inherent at birth, but are instead based on regularities or invariants that we extract from repeated exposure to stimulation (Howe and Purves, 2005). These regularities can be expressed in an intuitive way such as "distant objects seem smaller." As we will see, the logic underlying this economic cognitive strategy is the same one that could lead us to perceive causality where there is none and, consequently, to develop irrational beliefs against evidence.

Correct Estimation of Causality

Causality is a complex concept that cannot be directly perceived. Rather, it needs to be estimated or inferred from sensory input. Just as visual objects are subject to invariants, causality also obeys invariable rules that may be used by humans and other animals when attributing effects to causes. According to philosophers, the three main principles that are present in all causal relationships are priority, contiguity, and contingency

The Illusion of Causality

(Hume, 1748). Using these three principles, people, animals, and even artificial systems could draw correct inferences about causality most of the time, detecting causal patterns efficiently. At least, that is in theory.

The principle of priority simply asserts that causes always occur before their effects. Although this temporal order is an invariant (i.e., it is fixed and impossible to reverse), priority is not a perfect cue to causality. First, even if causes precede their effects, sometimes effects might be detected before causes, just as smoke can be sighted before the fire that produced it is found. The second problem is, of course, that not every event that is followed by another represents a valid cause for that event (e.g., sunrise always precedes breakfast, but does not cause it). In any case, organisms tend to use the principle that causes occur before their effects as an important cue for causality detection (Waldmann and Holyoak, 1992).

The second principle, contiguity, means that causes and their effects occur closely in time and space. A causal inference based solely on this principle would sometimes be erroneous because contiguity can be masked when a chain of intermediate causes lies between a distal cause and the effect. For instance, exercising only produces visible health benefits in the long term, because a chain of many internal changes in the organism is responsible for mediating these effects. This being said, contiguity is one of the critical cues that organisms use to detect causality; accordingly, the closer in time and space two events occur, the easier it is to infer a causal relationship between them (Lagnado and Speekenbrink, 2010).

Finally, there is the third principle, contingency, or covariation. Effects are always contingent on their causes. However, this cue to causality is, just like those previously described, far from perfect. Sometimes two stimuli or events are correlated without being causally linked ("Correlation is not causation"). Conversely, on other occasions, the actual contingency between a cause and an effect is masked by the presence of external factors, such as alternative causes of the same effect. In spite of its fallibility, people and other animals use contingency as a critical cue for helping them decide when a causal relationship exists between two events (Shanks and Dickinson, 1987; Wasserman, 1990). Interestingly, professional magicians are aware of this and set up their magic tricks so that a card seems to instantly move from their deck to a pocket, while the actual cause of the movement remains undiscovered.

Computing Contingency

Unlike perceiving priority and contiguity, assessing contingencies is a highly complex cognitive task. Assuming a simple situation in which causes and effects are binary variables (i.e., they can be either present or absent), the four possible combinations between them can be represented in a contingency matrix like the one displayed in figure 3.3.

According to the contingency matrix, a type *a* event is one in which both the potential cause and the effect are observed, and a type *b* event is one in which the potential

	Effect	¬Effect
Cause	a	b
¬Cause	c	d

Figure 3.3
Contingency matrix with four possible event types, with one cause and one effect.

cause, but not the effect, is observed. When the effect is observed but not the potential cause, this is a type c event; finally, in a type d event neither the potential cause nor the effect is observed. From this matrix, it is possible to compute a contingency index, ΔP (Allan, 1980):

$$\Delta P = P(E|C) - P(E|\neg C) = \frac{a}{a+b} - \frac{c}{c+d} \qquad \text{(Equation 1)}$$

The letters a, b, c, and d in equation 1 refer to the observed frequencies of each type of event in the contingency matrix. The ΔP rule is a contrast between two conditional probabilities: the probability of the occurrence of the effect E when cause C is present, and the probability of the occurrence of the effect E when the cause is absent, $\neg C$. In other words, the ΔP rule expresses how much the probability of the effect changes when the cause is present, as compared to when the cause is absent.

This may be better understood with an example. Imagine that you suffer from frequent episodes of back pain that you attempt to alleviate with a given drug. You tend to feel better when you take the drug, but you are uncertain that this is the real cause of your recovery. Thus, you want to discover whether there is a true causal connection between the potential cause C (taking the drug), and the effect E (feeling better), and to do so you assess the contingency between the two events. First, you observe how frequently you recover from your crises when you have not taken the drug, $P(E|\neg C)$. Bear in mind that these recoveries might be produced by some other cause not yet considered. You then compare this probability with that of recovering from a bout of back pain when you do take the drug, $P(E|C)$. If you were correct in suspecting that the drug causes recovery, you should observe that your recovery is more likely to occur when you follow the treatment than when you do not (i.e., ΔP would be positive).

Different values of ΔP indicate different degrees of association between the potential cause and the effect. When ΔP is positive, the contingency suggests generative causation, which means that the cause produces or increases the likelihood of the effect (e.g., the drug causes relief from back pain; smoking causes lung cancer). When it is negative, however, it suggests preventive causation; that is, the cause reduces the likelihood of the effect (e.g., exercise prevents back pain; a suntan lotion prevents sunburn). The absence of association, given by a ΔP value of zero, corresponds to those situations in which the effect is equally likely to occur in the presence and in the absence of the potential cause (as when a miraculous bracelet neither increases nor reduces the likelihood of back pain).

To conclude, we have described the three principles governing all causal relationships. These rules have no exception in the physical world, which means that, in principle, they should be good cues with which to infer causality. However, as we have mentioned, this implication is not realistic. Not every relationship that meets the priority, contiguity, and contingency principles is in fact causal.

Biased Estimation of Causality

The psychological research on causality estimation is based primarily on experimental procedures in which human participants are asked to judge a causal relationship between two events, such as using a fictional drug and patients' recovery (Blanco, Matute, and Vadillo, 2013). Participants are usually presented with a series of trials corresponding to the cells in the contingency matrix (figure 3.3). Thus, they may receive, for instance, fifty or a hundred trials, one at a time. Each trial may represent a different day in which a fictitious patient either took or did not take a certain drug (potential cause, C) and then reported on whether or not he felt better (potential effect, E). The experimenters may manipulate the frequencies of each trial type in the contingency matrix to produce different levels of contingency, as computed with the ΔP rule. At the end of the experiment, participants provide a subjective judgment about the degree of causality they perceive between the potential cause and the effect.

Years of research indicate that, most of the time, human and nonhuman animals are sensitive to manipulations of the contingency, thus correctly detecting and judging causal relationships (Shanks and Dickinson, 1987; Wasserman, 1990). However, the results of many experiments support the idea that the mechanisms used for causal estimation—like those used in visual perception—contain biases that favor efficiency, economy, and fast judgment, but that can also lead to errors.

Most of these biases are related to the occurrence of fortuitous coincidences between causes and effects. A very salient feature of positive (or generative) causal relationships is that they typically imply frequent co-occurrences of the cause and the effect (i.e., cell a in figure 3.3). Therefore, focusing on these coincidences could be a sensible heuristic that animals may use to detect and estimate causality with very little effort. In other words, instead of computing ΔP or some other contingency index, animals would use the number of coincidences (cell a trials). This heuristic would approximate a valid or "good enough" conclusion. However, contingency and number of coincidences do not always go hand in hand. Because of this, this cue to causality is a very misleading

	Effect (feeling better)	¬Effect (not feeling better)
Cause (taking the drug)	64	16
¬Cause (not taking the drug)	16	4

Figure 3.4
Contingency matrix containing trial frequencies that suggest a positive causal relationship, but yield a null contingency value (i.e., $\Delta P = 0$).

one, and can result in completely incorrect conclusions under some circumstances. For instance, consider the contingency table in figure 3.4.

Figure 3.4 shows an example of a contingency matrix in which the frequencies of each type of trial are depicted. The trial that is observed most frequently in the example is the co-occurrence of the potential cause and the effect, that is, trial a (e.g., taking the drug and feeling better). In this situation, a heuristic based on coincidences would favor the conclusion that the two events are, indeed, causally related. However, applying equation 1 to compute the actual contingency for the data in this figure reveals that such an impression is only apparent, since ΔP in this case is zero: $\Delta P = \frac{64}{64+16} - \frac{16}{16+4} = 0.8 - 0.8 = 0$.

Thus, using coincidences as a cue to causality may lead to the impression that the two events are causally related, when in fact they are not. After all, coincidences can occur by mere chance, as in this example.

This situation parallels that of visual illusions of the sort described previously. Animals and people may use heuristics, which are cognitive shortcuts to rapidly and efficiently arrive at conclusions. When trying to infer causality from the information available to the senses, using heuristics such as priority, contiguity, or the number of coincidences would be successful most of the time. However, as in the Müller-Lyer illusion, under special circumstances these heuristics produce wrong conclusions, with a typical outcome being that one may believe in a causal relationship that actually does not exist. This false-positive error is known as *causal illusion* (Matute et al., 2015), and it has been proposed to underlie many irrational behaviors such as belief in pseudoscience

(Blanco, Barberia, and Matute, 2015; Matute, Yarritu, and Vadillo, 2011) or the use of pseudo-medicine (Blanco, Barberia, and Matute, 2014).

Factors That Produce the Causal Illusion

Under what circumstances, then, would we expect the causal illusion to appear? In principle, we assume that the heuristic that most people use to assess causality is based on the frequency of cause-effect coincidences, as described above. Therefore, any experimental manipulation that results in frequent spurious coincidences (or type a events) would increase the causal illusion. In particular, two of these manipulations deserve comment:

(a) *Frequency of the effect*. Given a null-contingency situation (i.e., $\Delta P=0$), in which one should conclude that there is no causal connection between the potential cause and the effect, presenting the effect with high probability (i.e., increasing the frequency of type a and c trials) produces strong overestimations of contingency and false-positive errors in the perception of causality. This is called the *outcome-density bias* and is a robust result that has been repeatedly reported in the literature (Alloy and Abramson, 1979; Buehner, Cheng, and Clifford, 2003). A real-life example would be a bogus treatment that appears to work because it is used as a remedy for diseases that have a high rate of spontaneous recovery. These cases are particularly prone to induce illusory perceptions of causality.

(b) *Frequency of the cause*. This manipulation consists of increasing the frequency with which the potential cause appears—that is, the number of type a and b trials. This produces cause-density bias, which is analogous to outcome-density bias and has also been reported in many laboratories worldwide (Blanco, Matute, & Vadillo, 2011, 2012; Hannah and Beneteau, 2009).

Moreover, the causal illusion will be even stronger when the two manipulations (frequency of the cause and frequency of the effect) take place at the same time, because this notably increases the number of coincidences. In fact, the frequencies in the example matrix shown in figure 3.4 correspond to this situation with high probability of the cause and high probability of the effect (thus, many accidental cause-effect coincidences occur despite a null contingency), which will typically produce a strong illusion of causality (Blanco et al., 2013).

When the Potential Cause Is the Animal's Behavior

One interesting aspect of the cause-density bias is that, under certain circumstances, it becomes dependent on the animal's behavior (Blanco et al., 2011; Hannah and Beneteau, 2009). To understand this, imagine how a person from the Neolithic ages would feel when trying to understand the occurrence of rain. For early farmers and gatherers, rain was a valuable, highly significant outcome that they would like to predict or even produce. Some people would pay attention to the sky, as several elements (e.g., clouds, lighting) could be treated as potential causes of rain whose actual contingency could then be examined. However, in addition to this, many cultures developed rituals and other actions (e.g., dance, sacrifices) aimed at producing rain. Importantly, these are potential causes of rain that, unlike clouds or lighting, are not external events that can only be observed. Rather, they are controllable actions that are executed by people.

As a direct consequence, the cause-density bias, and hence the causal illusion, can appear just because some actions are repeated very often. Imagine that these farmers from the past suspected that dancing in a particular way could bring rain. Given that this ritual was an inexpensive action requiring little effort, it could be repeated with high frequency, perhaps every day for long seasons. Eventually, when it rained, the chances that the dance had been performed that very same day, or the day before, were high. It is merely a coincidence, but a very likely one. Therefore, when assessing the potential causal link between the dance ritual and the rain, the cause-density bias would produce systematic overestimations of contingency, leading to a causal illusion; even if dancing does not actually produce rain, those people performing the ritual often enough would end up believing that it works. In turn, this belief may fuel the habit of performing the ritual, increasing the chances of further (accidental) coincidences. And this might just be the origin of a superstition.

Compare this situation with another human group in which the supposedly rain-bringing ritual entails offerings of large amounts of food—a very expensive sacrifice that prevents the ritual from being performed with high frequency. This will necessarily decrease the number of fortuitous coincidences between the potential cause (the ritual offering) and the effect (observing the rain), thus diminishing the strength of the causal illusion. This is how behavior—or at least the frequency with which an action is performed to obtain a desired effect—can affect the likelihood of falling prey to causal illusions.

An Example from Pseudo-medicine

Based on this rationale, Blanco et al. (2014) reported an experiment on the perception of the effectiveness of pseudo-medicines. A pseudo-medicine is a treatment that produces no benefit for health beyond the placebo effect. Surprisingly, the popularity of pseudo-medicines is on the rise, and they are frequently perceived as very effective (Matute et al., 2015; Nyhan and Reifler, 2015). Homeopathy, for instance, is a pseudo-medicine whose mechanism lacks any plausibility, and has never been found to be effective beyond placebo for any disease (House of Commons Science and Technology Committee, 2010; Hermes, this volume; National Health and Medical Research Council, 2015), and yet it is still considered for inclusion in some official health systems. Matute et al. (2011) proposed that a causal illusion could underlie the anomalous belief in the effectiveness of pseudo-medicines. They reported that the perceived effectiveness of a completely useless pill increases when people observe proportionally more patients who have taken the pill (and happen to feel better) than when they observe fewer patients who have taken the pill. In other words, in spite of the fact that the pill is equally ineffective in all cases, it appears to be more effective when more patients have taken it. This result can be explained as an instance of a cause-density bias that we have described above.

Interestingly, Blanco et al. (2014) observed that homeopathy and other similar remedies are typically advertised as "free from side effects," as opposed to most conventional treatments. This would encourage massive usage. In particular, given that they have no side effects, the patient can consume these pseudo-treatments without restriction and whenever they so desire—after all, there is no harm in taking them. Additionally, pseudo-medicines are most likely to be used to treat mild diseases or conditions from which the patient would usually rapidly recover without needing treatment (e.g., headache, back pain).

Blanco et al.'s (2014) experiment represented this scenario in a computer program. They asked seventy-four undergraduate psychology students to participate in a game-like experiment, in which the participants had to imagine that they were medical doctors working in an emergency care facility. They were then presented with a series of fictitious patients suffering from a disease. For each patient, participants were given the option of using a fictitious medicine or not. Unbeknownst to the participants, the medicine was completely ineffective in improving the conditions of the patients. It was even unable to produce a placebo effect, because the contingency between using the treatment and observing the patient's remission was null (i.e., $\Delta P = 0$). Nonetheless, patients very frequently reported experiencing spontaneous respite from the symptoms; that is, the frequency of occurrence of the desired effect was high. This condition would presumably produce a strong causal illusion, as evidenced by the results of many previous studies using a similar procedure with computerized tasks (Barberia, Blanco, Cubillas, and Matute, 2013; Blanco et al., 2011).

The crucial manipulation was that, for half of the participants, the medicine was described as producing severe side effects (high-cost group), while for the other half it was presented as being free from side effects (no-cost group). The results of the experiment showed that, as expected, those participants who had the opportunity to use a medicine that would not produce bad consequences tended to use it more often than those having to be more careful due to potential side effects (see figure 3.5, left panel). Similarly, the medicine without side effects was perceived as being moderately effective, whereas the medicine with side effects was judged as less effective (see figure 3.5, right panel). This happened despite the fact that none of the medicines was effective at all (i.e., the probability of the patient feeling better was identical regardless of whether

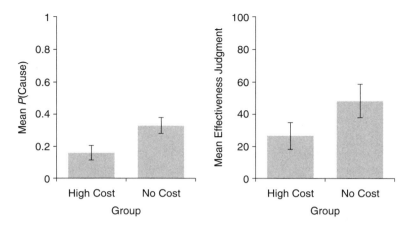

Figure 3.5
Main results of the experiment reported in Blanco et al. (2014). The left panel shows the mean proportion of patients to which the participants gave the fictitious drug—or, in other words, the probability with which they administered the cause, P(Cause), in each of the two groups. The right panel shows the mean effectiveness judgments given by participants in the two groups. Error bars depict 95 percent confidence intervals for the means. Blanco, F., Barberia, I., and Matute, H. (2014). The Lack of Side Effects of an Ineffective Treatment Facilitates the Development of a Belief in Its Effectiveness. *Public Library of Science ONE, 9*(1): e84084.

she took it or not). Thus, both groups showed a false-positive error, although the error was stronger in the group with no side effects. Finally, a mediational analysis showed that the reason why the medicine with no side effects appeared to be more effective was the high frequency with which it was used. The more often people decided to use the drug, the higher their judgment of effectiveness for that drug. Real patients who resort to homeopathy (free from any effect, be it beneficial or otherwise) in their daily lives do not normally use it with caution, but rather tend to take it on a daily basis, and therefore they should be highly sensitive to this bias.

From the results of the experiment, we can deduce that one reason why people believe in the effectiveness of homeopathy (and related pseudo-medicines) is because these practices are presented and marketed as being harmless. Homeopathic pills and other alternative treatments share the common claim that they produce no side effects. Consequently, because people believe that there is no harm in taking these treatments, they use them frequently. After all, even if these treatments do not work, there is supposedly no risk in trying. This highly frequent use of the alternative treatments results in the previously described cause-density bias, and the mistaken belief that the treatments are working effectively. By presenting homeopathy and other alternative treatments as harmless practices, people are indirectly increasing their subjective perception that these treatments are effective, even if they are completely useless, as in the described study.

Therefore, the experiment by Blanco et al. (2014) illustrates how certain causal beliefs that people hold in their daily lives can be the result of a causal illusion—a false-positive error in which people conclude that two events are causally related when they actually are not. Moreover, in these cases, the illusion is often modulated by the participants' behaviors. As shown in the experiment, the illusion was stronger when they used the fictitious medicine (i.e., the potential cause) with high frequency.

Simple Conditioning Models and Causal Illusions

The principles we have described so far that are used to infer causality are so clear and ubiquitous that even relatively simple animals, such as invertebrates, could use them. In fact, the view on causal learning we have provided in the previous sections can be understood as a special case of *associative learning*, or conditioning. Therefore, the vast knowledge that is now available regarding the mechanisms, mathematical models, and neural basis of conditioning can also be applied to causal learning, in both humans and other animals.

Conditioning was first described by Pavlov, who discovered that repeated pairings of a neutral stimulus (e.g., a tone) and a biologically significant stimulus (e.g., food) led to the occurrence of a conditioned response when the neutral stimulus was subsequently presented alone. Almost a century of psychological research has provided us with detailed knowledge about conditioning, or associative learning, in a broad variety of phenotypically distant species, from humans to snails (Loy, Fernández, and Acebes, 2006) and honeybees (Blaser, Couvillon, and Bitterman, 2006).

According to the associative framework that is typically used in the field of conditioning (Shanks and Dickinson, 1987), the presentation of any stimulus activates a mental representation of such a stimulus. In addition, the presentation of an unconditioned stimulus, US (e.g., food), is able to produce, additionally, an unconditioned response (UR) or physiological reflex (e.g., salivation by hungry dogs presented with food). In the conditioning procedure, one must present a neutral stimulus immediately before the US (Gunther, Miller, and Matute, 1997). The repeated pairings of these two stimuli have the consequence of creating/strengthening an association between the mental representations of both of these cues. This principle is well known in the fields of neuroscience, biology, and psychology, as it is the core of Hebb's law (Hebb, 1949). By means of the conditioning procedure and the strengthening of the association, the representation of the neutral stimulus becomes more and more efficient to spread its activation to the US representation. Eventually, the neutral stimulus becomes a conditioned stimulus, CS, and it is able to produce a conditioned response even in the absence of the US (e.g., salivating when the tone is presented, before any food is given).

From a functional viewpoint, conditioning can be seen as a capacity aimed at predicting, or anticipating, significant outcomes. Consequently, it is not surprising that many researchers proposed that causal learning is just another type of associative learning, or at least that it can be understood by appealing to similar mechanisms (Miller and Matute, 1996; Shanks and Dickinson, 1987). After all, both of them seem to serve the common goal of adapting behavior to better predict and produce relevant events. Therefore, theories of associative learning and conditioning have been used to model causal learning. In the common associative learning framework, inferring causality is akin to forming and strengthening an association between two stimuli, a potential cause (CS) and its effect (US).

Arguably the most prominent model of associative learning is the Rescorla-Wagner model (Rescorla and Wagner, 1972). This is a conditioning model that has been successfully applied to causal-learning phenomena (Tangen and Allan, 2004), including biased estimation of causality (Blanco et al., 2013; Blanco and Matute, 2015; Matute, Vadillo, Blanco, and Musca, 2007). The Rescorla-Wagner model proposes that the learning of contingencies between stimuli is driven by a prediction error-correction mechanism. When the CS, or potential cause, is presented, the animal makes a prediction of how likely the US, or the effect, is to appear. This expectation is then adjusted in the light of the actual occurrence (or nonoccurrence) of the US. This updating algorithm closely corresponds to the delta rule developed to train artificial neural networks, and takes the following form:

$$V_{CS,t} = V_{CS,t-1} + \Delta V_{CS,t} \qquad \text{(Equation 2)}$$
$$\Delta V_{CS,t} = \alpha\beta(\lambda_t - V_{CS,t-1}) \qquad \text{(Equation 3)}$$

The parameter $V_{CS,t}$ represents the strength of the association between the CS and the US at the end of trial t. In particular, the greater the magnitude of this parameter, the stronger the estimated contingency. V can also be understood as a prediction: when V is large, the animal will predict that the US will be very likely to occur if the CS is present, and this means that a conditioned response, such as salivating, will be observed in anticipation of the US. Typically, the model assumes that the CS is initially neutral, that is, the associative strength on trial 1 is zero or close to zero. The model describes how this parameter changes in light of the animal's experience. On each trial, V will change in ΔV units (equation 2). The amount of change is computed as a correction of the prediction error. If the animal predicts that the US will occur (i.e., V is large), but the US does not occur, then this error is used to adjust the expectation (V) for the next trial. Likewise, consider if the animal does not predict that the US will occur (or it predicts it weakly, i.e., V is small), but the US does occur, then V will be increased for the next trial. Gradually, the algorithm will adjust the associative strength V so that it makes accurate predictions of the US. This adjustment is made by means of the updating rule in equation 3. In this equation, α and β are learning rate parameters representing the salience of the CS and the salience of the US, respectively, λ is the actual value of the US on a given trial (it is usually assumed to be 1 in trials in which the US is present, and 0 when it is absent), and $V_{CS,t-1}$ is the strength of the CS-US association at the end of the previous trial. Thus, as in the widely used Delta rule, the Rescorla-Wagner model is based on the prediction error, which is in equation 2 presented between brackets: if the value of V is accurate, it would produce a small difference with respect to the actual value of the US, λ, and it will thus require only minimal correction. If V strongly differs from the actual value of the US, then this large prediction error will cause a strong adjustment in the prediction for the next trial.

Despite its apparent simplicity, the Rescorla-Wagner model has been successfully used to account for many phenomena in the field of associative learning, by extension becoming a benchmark for understanding causal learning. How can the Rescorla-Wagner model represent this causal-learning process? The short answer is: because it is heavily based on at least one of the principles of causality that we have described above—contingency.

The principle of priority is only indirectly encoded in the model. When one assigns the roles of potential cause and effect to the two stimuli, the assumption is that the former (CS) is used to make a prediction about the occurrence of the latter (US). Thus, the Rescorla-Wagner model is unable to account for certain associative phenomena, such as backward conditioning (Chang, Blaisdell, and Miller, 2003), that imply reversing the order of the stimuli. Temporal/spatial contiguity is, on the other hand, not present in the Rescorla-Wagner model, as it does not encode any temporal or spatial property (but see Arcediano and Miller, 2002, for alternative associative models that incorporate temporal features into the learning algorithm). In any case, priority and contiguity are properties of a relationship that can be readily perceived by the observation of events.

In contrast, the principle of contingency is present in the model. Essentially, the Rescorla-Wagner model captures predictive relationships based on their contingency; in fact, provided with enough training, the algorithm converges with the value of ΔP (Chapman and Robbins, 1990). Thus, if animals actually use a learning mechanism that resembles the Rescorla-Wagner rule, they would eventually reach causal estimations that are in line with the correct predictions most of the time. As the reader is probably anticipating now, the rule works in such a way that, under certain circumstances, it produces a systematic overestimation of the contingency, and therefore a potential illusion of causality.

Machines Can Also Suffer from the Illusion of Causality
In order to illustrate this point, we performed four computer simulations of the Rescorla-Wagner learning model. They are shown in figure 3.6. The simulations include one hundred trials in which a potential cause and an effect were either present or absent. Typically, the model successfully captures the actual contingency between these two

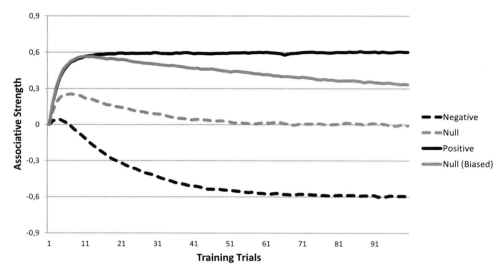

Figure 3.6
Four simulations of the Rescorla-Wagner learning model. Using this model we simulated learning under four different contingency tables: one positive, one negative, and two involving null-contingency matrices. One of the null-contingency matrices included both cause- and outcome-density manipulations, which resulted in an overestimation of the null contingency. This would be equivalent to a causal illusion. The parameters used for these simulations are the same as those used in the simulations of the Rescorla-Wagner model reported by Vadillo, Blanco, Yarritu, and Matute (2016): $\alpha_{CS}=.30$, $\alpha_{Context}=.10$, $\beta_{Outcome}=\beta_{\neg Outcome}=.80$.

stimuli, regardless of whether it is positive, null, or negative (e.g., $\Delta P=0.6$, 0, or -0.6, respectively). The model only needs a certain amount of training before it arrives at correct solution. However, when the manipulations of cause and outcome-density described in the section above are conducted, the judgments of causality produced by the model become biased, so that even a null contingency is overestimated as if it were positive. In this sense, any learning system that incorporates the Rescorla-Wagner or a similar adaptive learning rule behaves in a way that is not so different from animal and human participants in psychological experiments.

The way the Rescorla-Wagner model produces biased estimations depends, again, on the number of coincidences between the potential cause and the effect. It is reasonable to assume that the occurrence of a stimulus attracts the attention of the animal to a greater extent than the absence of such a stimulus. For the same reason, coincidences between the two stimuli would be even more significant for the animal than the presence of only one of them. This means that, in fact, not all the cells in the contingency matrix of figure 3.4 are equally salient. Coincidences (cell *a*) could be given more weight or receive more attention than cells *b* and *c*, whereas cell *d* would be the least interesting for animals, as they convey relatively little information (Mckenzie and Mikkelsen, 2007). Many experiments have shown that this is indeed the case in human judgments of causality (Kao and Wasserman, 1993; White, 2003). The Rescorla-Wagner model is sensitive to this asymmetry between the cells, because α and β in equation 2 represent the salience (or the extent to which a stimulus grabs attention and produces learning) of the stimuli and their absence. When learning a null contingency, if the two stimuli are present, the product $\alpha \times \beta$ is typically higher than when only one, or neither, of them is present. If the amount of coincidences is disproportionately high (i.e., when the cause- and outcome-density manipulations are conducted), these coincidences will exert a greater impact than the other cells, and the model will make an overestimation. In other words, the Rescorla-Wagner model is capable of mirroring the causal illusions that humans and other animals show in psychological experiments, because it contains the very same heuristic that has been said to underlie causal inference: a disproportionate weight of coincidences.

The Bright Side of Causal Illusions

So far, we have described the causal illusion as a false-positive error that appears as a consequence of how the cognitive system infers causal relationships from ambiguous or uncertain data. We have even described artificial systems that are able to learn in a flexible way similar to that of humans and other animals, because they incorporate the same heuristics and, as a consequence, suffer from a similar bias. Humans like to think that they are almost perfect reasoners, but they fall prey to the same errors (causal illusions) that affect almost any other learning system, be it natural or artificial. Why then has this heuristic survived our long evolutionary history? In this section, we will outline some of the benefits of holding a biased causal inference machine, which might help understand why it has survived and endured over the passage of time (Blanco, 2017).

Fast and Frugal Decision Making
Perhaps one of the most exciting psychological research programs to emerge in the late twentieth century revolves around the concept of heuristics. The term was originally coined by Herbert Simon (1957), and was then extensively researched by renowned scientists such as Amos Tversky and Daniel Kahneman (Kahneman, 2013; Tversky and Kahneman, 1981). An important assumption of this framework is that reasoning usually favors efficiency over accuracy. Instead of using costly, computationally demanding rules to perceive geometry or to assess causality with mathematical precision (e.g., computing contingency), our cognitive systems prefer to rely on simpler rules that embody simpler principles (e.g., counting the number of cause-effect coincidences). In doing so,

we sacrifice accuracy (at least under certain circumstances) to obtain two advantages, economy and speed. It is easy to understand why natural selection would favor mechanisms that make such an efficient use of resources, even if they are sometimes liable to produce mistakes. The most adaptive cognitive system is that one which is able to rapidly adapt and respond to changes in the environment, instead of relying on knowledge that was correct in the past, but becomes outdated every time the environmental variables change. Fast learning and responding is the key to success. Accuracy means little in a fast-changing world.

Least Costly Mistake
Although the tendency to jump to a conclusion (eventually leading to the perception of a causal link where there are only random patterns) would seem a problematic trait for any organism, it can in fact be adaptive in many circumstances. Some researchers noticed that the false-positive error that appears in most cognitive illusions frequently represents a "least-costly mistake" (Haselton and Nettle, 2006; Matute, Steegen, and Vadillo, 2014).

Imagine a bunch of primitive people resting close to an open field, with tall grass. One of them sees something strange hidden in the grass. It could be a dangerous predator or it could be a harmless animal. It could even be the wind. Our ancestors were relatively disadvantaged in comparison with their potential predators, both in terms of physical strength and speed. Therefore, they would have always been alert, always watching their surroundings, and always ready to react upon detection of any unusual movement. Thus, if someone suddenly yells, then the whole group gathers, upon which they immediately flee to a safer zone, far from the grass. On the basis of incomplete, uncertain information, these people guessed that a predator was hidden and watching, just as people in today's society may see a warning face in the clouds, or may believe that an impotent pseudo-medicine works just because they saw a few coincidences between taking it and feeling relief. In the former case relying on quick impressions makes a lot of sense, but it may be quite harmful in the latter. Mistaking an inoffensive movement in the grass for a dangerous predator would lead to unnecessarily fleeing and wasting energy, whereas making the opposite mistake (ignoring the signs when there was an actual predator watching) might result in much more serious consequences, even death. In ancestral environments, a false-positive error, which leads to action, is the least costly mistake. It therefore comes as little surprise that evolution has shaped us to instantly react and jump to conclusions, even when armed with only partial information.

In today's knowledge-rich society, this advantage of biased causal reasoning is substantially reduced or even reversed. Consider, for instance, scenarios in which people have to decide whether or not to take a pseudo-medicine to heal a serious disease. We have better remedies today, so favoring those that intuitively seem useful over the evidence-based treatments is not without serious cost. In these cases, thinking carefully and avoiding false positives is clearly preferable. But again, sometimes this process fails because this requires special awareness and efforts given the shaping of our mind (and behavior) over the course of evolution.

Behavioral Persistence

If our cognitive systems are built in a way that favors quick learning and adaptation, they should also be oriented toward action, as adaptive behavior is, ultimately, the goal of the entire learning process. Crucially, action is at the core of the process of learning by reinforcement. If an action is followed by a positive consequence, it will become more probable in the future (and conversely, it will become less probable if followed by an undesired, or aversive, consequence). Even though learning is still possible just by observing the behavior of others, it progresses faster when people can learn from their actions. This fact favors a bias toward action in the behavior of an animal.

Consider the case of the early farmers, described previously. They were unable to produce rain, which was a key element to succeed in their goal of obtaining their food. If they had become aware of this limitation, they would have probably felt helpless and depressed, and thus decided to abandon any attempt to plant their crops, as predicted by learned helplessness theory (Abramson, Seligman, and Teasdale, 1978). This, in turn, would have halted the development of agriculture and civilization. Fortunately, instead of sitting on their hands, these early farmers persevered with their actions, continuing with their attempts to plant crops despite their inability to fully control the situation. Causal illusions quite possibly helped them to be resistant in situations that might otherwise have led them to give up hope. The mere occurrence of coincidences between rituals and rain might have resulted in superstitions and beliefs that eventually sustained their behavioral persistence.

By virtue of reinforcement, once a causal illusion has been formed, the actions that are thought to be useful to produce an outcome become more and more likely, which in turn strengthens the illusion (Matute et al., 2007). This is how irrational rituals—although useless to produce the intended effect—may be beneficial because they foster active behavioral styles (Beck and Forstmeier, 2007).

Illusion of Control, Justice, and Emotional Consequences

A classic line of research in experimental psychology is related to Langer's (1975) studies on the illusion of control. The illusion of control occurs when people believe that they are controlling environmental events that are, in reality, outside of their control. This can certainly be regarded as a special case of the illusion of causality, one in which the potential target cause is one's behavior. This illusion of control is the exact opposite to the learned helplessness effect (Abramson et al., 1978) and can even play a role in protecting against it. If people do not detect that important and desired events in their environment are outside of their control, then they cannot feel helpless. Developing the belief that they control their environment, they will not develop the harmful expectation that the future is uncontrollable. Therefore, they will avoid developing the deficits (emotional, cognitive, motivational, and self-esteem) that characterize the learned helplessness effect (Matute, 1994).

There is a large body of evidence supporting the existence of an association between this need for control and people's emotional states. Indeed, it has even been suggested that the illusion of control could have a prophylactic effect, not only against learned helplessness, but also in the general maintenance of self-esteem and emotional well-being (Taylor and Brown, 1988). For instance, people who feel anxious and lacking in control tend to see patterns and meaningful images when they look at visual stimuli consisting of totally random dot clouds (Whitson and Galinsky, 2008). Further, in a classic series of experiments, Alloy and Abramson (1979) used a procedure that was very similar to the one we described above in relation to causal illusions, comparing the judgments of control of people with mild depression versus people without depression. The group without depression showed a typical illusion-of-control effect. Interestingly, however, the group with mild depression appeared to be protected against the illusion of control. This phenomenon, now known as the *depressive realism* effect, has been replicated in many laboratories worldwide (Moore and Fresco, 2012; Msetfi, Wade, and Murphy, 2013) and is another well-known example of the relationship between mood and the development of illusions that give causal meaning to the many coincidences that we encounter by mere chance in our daily lives.

Research investigating the belief in a fair world is also worth discussing in relation to this point. Many experiments have shown how people tend to believe that we live in a fair and well-ordered world, in which effort and moral actions become reinforced, with little room for randomness. People particularly dislike the idea that reward and punishment are subject to the unpredictable and uncontrollable laws of chance, or that efforts directed toward obtaining a desired event might produce no effect while other people might obtain that same reward without even trying (Lerner, 1980). Interestingly, these beliefs about a fair world have been shown to give rise to a biased interpretation of the information available to the senses. For instance, many experiments have shown that people tend to believe that when they experience positive and desired events, this is exactly what they deserve (a reward for their good actions), whereas negative events constitute a punishment for bad actions (Lerner, 1970). The impact of these beliefs is so intense that it has even been observed that people predict a better and more favorable future for someone who buys an object (e.g., a jacket) that belonged to someone who was highly moral and not immoral, even when the buyer was unaware of the person's moral standards (Stavrova and Meckel, 2016). Thus, causal illusions could also serve the purpose of enhancing this belief in a fair and well-ordered world, which should in turn protect people from the uncomfortable feeling that they are mere puppets in the hands of chance (Whitson and Galinsky, 2008).

Conclusions

Throughout this text we have been discussing what we believe to be one of the central problems underlying pseudoscience, namely, causal illusions. The detection of causal relationships has been largely investigated in experimental psychology, both with humans and other animals, and we now have enough data to conclude that causal illusions are part of a healthy and well-adapted cognitive system. As we have discussed so far, causal illusions can be regarded as part of the associative process that animals and humans use to infer causal relationships. This way of inferring causal relationships may not be perfect, but it is highly efficient and fast. Thus, it confers important survival advantages and has therefore been selected by evolution. But the price for this efficiency and speed is accuracy, and the mechanism sometimes leads to errors. These occasional errors can be categorized into at least two critical subgroups. On the one hand, some errors may consist of missing a hazard when there is one, such as missing an association between a shadow and a predator. This could certainly be a lethal error. Those individuals who, in the past, were not able to detect those links and jump to the rapid conclusion that there was a predator from ambiguous and uncertain evidence were not likely to survive. On the other hand, there might be false-positive errors, which would consist of detecting a relationship where one does not exist. Of these two types of errors, the second one is certainly less costly. When the evidence is ambiguous, detecting a pattern and acting upon the available, yet scarce, information is certainly more adaptive than doing nothing while waiting for conclusive data. Those individuals who had a lower threshold for detecting patterns and relationships were more likely to survive, and therefore they are more likely to be our ancestors.

Thus, in the light of evolution, it is not surprising that today's society embraces so many superstitions, illusions of causality, pseudosciences, and ungrounded beliefs. We are made to hold false beliefs and to jump to conclusions on the basis of very little evidence. In addition, beliefs that provide order to our world help us maintain our self-esteem and psychological well-being. Last but not least, they help us keep active in our search for reinforcers and desired events; that is, they are absolutely necessary for us to avoid passivity and depression, and the lack of reinforcers that both imply.

However, we now live in the twenty-first century, and humanity as a whole has developed a science that should help us overcome many of these problems with much greater accuracy and safety, and with fewer errors. We now understand that in times gone by it would have made perfect sense to accept any herb that a shaman could offer us when suffering from a disease, or to accept the word of the people we trusted when they said that a certain remedy had worked for them. But today we know how fallible human experience is, particularly when judging causal relationships. In particular, we know how little these experiences and anecdotes of family and friends really mean (because they can be completely incorrect), and we know that there is a scientific method that can help us humans, as a group, decide whether or not a certain remedy is effective. Thus, we live in a time when trusting our old intuitions and those of our friends against those based on scientific evidence is no longer adaptive. Indeed, as discussed in the introduction, there are people today suffering from diseases that could have been avoided had they relied on science rather than on their causal illusions.

In conclusion, is there anything that could be done to reduce causal illusions and allow us to make more rational decisions in today's world? To our knowledge, there have so far been relatively few attempts to reduce causal illusions. As noted by Lilienfeld, Ammirati, and Landfield (2009), the study of cognitive biases has been more focused on demonstrating the existence of these biases than on finding ways to combat them. However, there have been a few studies in recent years that should shed some light on how these biases could be reduced. Some interesting methods have made use of homeopathy, paranormal phenomena, and heavy metal music in class to highlight examples where biased reasoning often occurs, as an intervention that might be effective in debiasing the judgments of students (Schmaltz, 2016; Schmaltz & Lilienfeld, 2014; see also Herried this volume).

We ourselves have focused our research on understanding the bias of causality and the illusion of control that we have described throughout this chapter, and on trying to develop effective strategies to reduce them. A practical educational application that has been shown to be successful in debiasing adolescents against the illusion of causality is described in detail in Barberia et al. (2013). This intervention was aimed at high school students and included several steps to ensure they were successfully debiased. Initially, we set up a staging phase in which a miracle product was presented to the students. The product was a small piece of metal without any special property, but it was described as a new material capable of making wonderful improvements in people's performance in a variety of scenarios (e.g., intellectual puzzles, physical activity). Then, once the students were convinced that the piece of metal could lead to important improvements in their abilities, we showed them how we had fooled them and, importantly, taught them how they could have learned about the actual powers of the metal. In brief, we taught them the rudiments of rigorous experimental testing and the importance of using control conditions, and showed them the type of questions that they should have asked when presented with the product. After this brief seminar (approximately one hour long), we conducted a standard causal illusion experiment, similar to the ones described through this chapter. The result was that the adolescents that had received the seminar showed a reduced illusion of causality, as compared to a control group who had not (Barberia et al., 2013).

So, yes, there are ways of reducing the illusion of causality, the illusion that is so often at the heart of pseudoscience. This can be achieved through educational strategies, some of which have been validated through evidence, are already published, and can thus be implemented. If we were to speculate about the factor or factors underlying all successful debiasing strategies, we will say that the idea is, first, to show people that they cannot trust their intuitions about causality, so that they will be willing to learn about potential tools they can use. Second, they must be shown the basics of scientific thinking and sound methodology. In other words, we should teach them to be skeptical and doubt both their intuitions and those of others, and show them how to distinguish their hypotheses from evidence-based knowledge, and also how to test their hypotheses. Learning scientific thinking early in life and frequently practicing these principles during school years should also help people acquire the habit of thinking critically. This requires practice and determination, but once acquired it should become extremely useful in daily life.

Acknowledgments

Support for this research was provided by Spain's Dirección General de Investigación (Grant No. PSI2016-78818-R), and by the Basque government's Departamento de Educación, Universidades e Investigación (Grant No. IT955-16).

References

Abramson, L. Y., Seligman, M. E. P., and Teasdale, J. D. (1978). Learned Helplessness in Humans: Critique and Reformulation. *Journal of Abnormal Psychology*, *87*(1): 49–74.

Allan, L. G. (1980). A Note on Measurement of Contingency between Two Binary Variables in Judgment Tasks. *Bulletin of the Psychonomic Society*, *15*(3): 147–149. DOI: 10.3758/BF03334492

Allan, L. G., and Jenkins, H. M. (1980). The Judgment of Contingency and the Nature of the Response Alternatives. *Canadian Journal of Experimental Psychology*, *34*(1): 1–11. DOI: 10.1037/h0081013

Alloy, L. B., and Abramson, L. Y. (1979). Judgment of Contingency in Depressed and Nondepressed Students: Sadder but Wiser? *Journal of Experimental Psychology: General*, *108*(4): 441–485. DOI: 10.1037/0096-3445.108.4.441

Arcediano, F., and Miller, R. R. (2002). Some Constraints for Models of Timing: A Temporal Coding Hypothesis Perspective. *Learning and Motivation*, *33*(1): 105–123. DOI: 10.1006/lmot.2001.1102

Barberia, I., Blanco, F., Cubillas, C. P., and Matute, H. (2013). Implementation and Assessment of an Intervention to Debias Adolescents against Causal Illusions. *PLOS ONE*, *8*(8): e71303. DOI: 10.1371/journal.pone.0071303

Beck, J., and Forstmeier, W. (2007). Superstition and Belief as Inevitable By-products of an Adaptive Learning Strategy. *Human Nature*, *18*(1): 35–46. DOI: 10.1007/BF02820845

Blanco, F. (2017). Positive and Negative Implications of the Causal Illusion. *Consciousness and Cognition, 50*(1): 56–68. DOI: 10.1016/j.concog.2016.08.012

Blanco, F., Barberia, I., and Matute, H. (2014). The Lack of Side Effects of an Ineffective Treatment Facilitates the Development of a Belief in Its Effectiveness. *PLOS ONE, 9*(1): e84084. DOI: 10.1371/journal.pone.0084084

Blanco, F., Barberia, I., and Matute, H. (2015). Individuals Who Believe in the Paranormal Expose Themselves to Biased Information and Develop More Causal Illusions Than Nonbelievers in the Laboratory. *PLOS ONE, 10*(7): e0131378. DOI: 10.1371/journal.pone.0131378

Blanco, F., and Matute, H. (2015). Exploring the Factors That Encourage the Illusions of Control: The Case of Preventive Illusions. *Experimental Psychology, 62*(2): 131–142. DOI: 10.1027/1618-3169/a000280

Blanco, F., Matute, H., and Vadillo, M. A. (2011). Making the Uncontrollable Seem Controllable: The Role of Action in the Illusion of Control. *Quarterly Journal of Experimental Psychology, 64*(7): 1290–1304. DOI: 10.1080/17470218.2011.552727

Blanco, F., Matute, H., and Vadillo, M. A. (2012). Mediating Role of Activity Level in the Depressive Realism Effect. *PLOS ONE, 7*(9): e46203. DOI: 10.1371/journal.pone.0046203

Blanco, F., Matute, H., and Vadillo, M. A. (2013). Interactive Effects of the Probability of the Cue and the Probability of the Outcome on the Overestimation of Null Contingency. *Learning & Behavior, 41*(4): 333–340. DOI: 10.3758/s13420-013-0108-8

Blaser, R. E., Couvillon, P. A, and Bitterman, M. E. (2004). Backward Blocking in Honeybees. *Quarterly Journal of Experimental Psychology, 57*(4): 349–360. DOI: 10.1080/02724990344000187

Blaser, R. E., Couvillon, P. A, and Bitterman, M. E. (2006). Blocking and Pseudoblocking: New Control Experiments with Honeybees. *Quarterly Journal of Experimental Psychology, 59*(1): 68–76. DOI: 10.1080/17470210500242938

Buehner, M. J. (2005). Contiguity and Covariation in Human Causal Inference. *Learning & Behavior, 33*(2): 230–238.

Buehner, M. J., Cheng, P. W., and Clifford, D. (2003). From Covariation to Causation: A Test of the Assumption of Causal Power. *Journal of Experimental Psychology: Learning, Memory, and Cognition, 29*(6): 1119–1140. DOI: 10.1037/0278-7393.29.6.1119

Bullock, M., and Gelman, R. (1979). Preschool Children's Assumptions about Cause and Effect: Temporal Ordering. *Child Development, 50*(1): 89–96.

Carroll, R. (2015). Too Rich to Get Sick? Disneyland Measles Outbreak Reflects Anti-vaccination Trend. *Guardian*. Retrieved from http://www.theguardian.com/us-news/2015/jan/17/too-rich-sick-disneyland-measles-outbreak-reflects-anti-vaccination-trend

Chang, R. C., Blaisdell, A. P., and Miller, R. R. (2003). Backward Conditioning: Mediation by the Context. *Journal of Experimental Psychology: Animal Behavior Processes, 29*(3): 171–183. DOI: 10.1037/0097-7403.29.3.171

Chapman, G. B., and Robbins, S. J. (1990). Cue Interaction in Human Contingency Judgment. *Memory & Cognition*, *18*: 537–545.

Danks, D. (2003). Equilibria of the Rescorla-Wagner Model. *Journal of Mathematical Psychology*, *47*(2): 109–121. DOI: 10.1016/S0022-2496(02)00016-0

Domjan, M. (2014). *The Principles of Learning and Behavior* (7th ed.). Stanford, CT: Cengage Learning.

Ernst, E. (2002). A Systematic Review of Systematic Reviews of Homeopathy. *British Journal of Clinical Pharmacology*, *54*(6): 577–582.

European Commission. (2005). *Special Eurobarometer 224: Europeans, Science and Technology*. Retrieved from http://ec.europa.eu/commfrontoffice/publicopinion/archives/ebs/ebs_224_report_en.pdf

European Commision. (2010). *Special Eurobarometer 340: Science and Technology*. Retrieved from http://ec.europa.eu/public_opinion/archives/eb_special_en.htm#340

Freckelton, I. (2012). Death by Homeopathy: Issues for Civil, Criminal and Coronial Law and for Health Service Policy. *Journal of Law and Medicine*, *19*: 454–478.

Gilovich, T., Griffin, D., and Kahneman, D. (2002). *Heuristics and Biases: The Psychology of Intuitive Judgment*. New York: Cambridge University Press.

Gunther, L. M., Miller, R. R., and Matute, H. (1997). CSs and USs: What's the Difference? *Journal of Experimental Psychology: Animal Behavior Processes*, *23*(1): 15–30.

Hannah, S. D., and Beneteau, J. L. (2009). Just Tell Me What to Do: Bringing Back Experimenter Control in Active Contingency Tasks with the Command-Performance Procedure and Finding Cue Density Effects Along the Way. *Canadian Journal of Experimental Psychology*, *63*(1): 59–73. DOI: 10.1037/a0013403

Haselton, M. G., and Buss, D. M. (2000). Error Management Theory: A New Perspective on Biases in Cross-Sex Mind Reading. *Journal of Personality and Social Psychology*, *78*(1): 81–91.

Haselton, M. G., and Nettle, D. (2006). The Paranoid Optimist: An Integrative Evolutionary Model of Cognitive Biases. *Personality and Social Psychology Review*, *10*(1): 47–66. DOI: 10.1207/s15327957pspr1001_3

Hebb, D. O. (1949). *The Organization of Behavior*. New York: Wiley & Sons.

House of Commons Science and Technology Committee. (2010). *Evidence Check 2: Homeopathy. HCSTC Report No. 4 of Session 2009–2010*. London: The Stationary Office.

Howe, C. Q., and Purves, D. (2005). The Müller-Lyer Illusion Explained by the Statistics of Image-Source Relationships. *Proceedings of the National Academy of Sciences of the United States of America*, *102*(4): 1234–1239. DOI: 10.1073/pnas.0409314102

Hume, D. (1748). *An Enquiry Concerning Human Understanding*. (Reprinted 2008, London: Oxford Press.)

Hutson, M. (2015). The Science of Superstition. *Atlantic*. Retrieved from http://www.theatlantic.com/magazine/archive/2015/03/the-science-of-superstition/384962/

Johnson, D. D. P., Blumstein, D. T., Fowler, J. H., and Haselton, M. G. (2013). The Evolution of Error: Error Management, Cognitive Constraints, and Adaptive Decision-Making Biases. *Trends in Ecology and Evolution, 28*(8): 474–481. DOI: 10.1016/j.tree.2013.05.014

Kahneman, D. (2013). *Thinking, Fast and Slow*. New York: Penguin Books.

Kao, S., and Wasserman, E. A. (1993). Assessment of an Information Integration Account of Contingency Judgment with Examination of Subjective Cell Importance and Method of Information Presentation. *Journal of Experimental Psychology: Learning, Memory, and Cognition, 19*(6): 1363–1386.

Katagiri, M., Kao, S., Simon, A. M., Castro, L., and Wasserman, E. A. (2007). Judgments of Causal Efficacy Under Constant and Changing Interevent Contingencies. *Behavioural Processes, 74*(2): 251–264. DOI: 10.1016/j.beproc.2006.09.001

Lagnado, D. A., and Sloman, S. A. (2006). Time as a Guide to Cause. *Cognition, 32*(3): 451–460. DOI: 10.1037/0278-7393.32.3.451

Lagnado, D. A., and Speekenbrink, M. (2010). The Influence of Delays in Real-Time Causal Learning. *Open Psychology Journal, 3*: 184–195.

Langer, E. J. (1975). The Illusion of Control. *Journal of Personality and Social Psychology, 32*(2): 311–328.

Lerner, M. (1970). "The Desire for Justice and Reactions to Victims." In J. Macaulay and L. Berkowitz (Eds.), *Altruism and Helping Behavior*, pp. 205–229. New York: Academic Press.

Lerner, M. (1980). *The Belief in a Just World: A Fundamental Delusion*. New York: Plenum.

Lewandowsky, S., Ecker, U. K. H., Seifert, C. M., Schwarz, N., and Cook, J. (2012). Misinformation and Its Correction: Continued Influence and Successful Debiasing. *Psychological Science in the Public Interest, 13*(3): 106–131. DOI: 10.1177/1529100612451018

Lilienfeld, S. O., Ammirati, R., and Landfield, K. (2009). Giving Debiasing Away: Can Psychological Research on Correcting Cognitive Errors Promote Human Welfare? *Perspectives on Psychological Science, 4*(4): 390–398.

Lilienfeld, S. O., Ritschel, L. A., Lynn, S. J., Cautin, R. L., and Latzman, R. D. (2014). Why Ineffective Psychotherapies Appear to Work: A Taxonomy of Causes of Spurious Therapeutic Effectiveness. *Perspectives on Psychological Science, 9*(4): 355–387. DOI: 10.1177/1745691614535216

Loy, I., Fernández, V., and Acebes, F. (2006). Conditioning of Tentacle Lowering in the Snail (Helix aspersa): Acquisition, Latent Inhibition, Overshadowing, Second-Order Conditioning, and Sensory Preconditioning. *Learning & Behavior, 34*(3): 305–314.

Malmendier, U., and Tate, G. A. (2005). CEO Overconfidence and Corporate Investment. *Journal of Finance, 60*: 2661–2700. DOI: 10.1111/j.1540-6261.2005.00813.x

Matute, H. (1994). Learned Helplessness and Superstitious Behavior as Opposite Effects of Uncontrollable Reinforcement in Humans. *Learning and Motivation, 25*(2): 216–232.

Matute, H., Blanco, F., Yarritu, I., Diaz-Lago, M., Vadillo, M. A., and Barberia, I. (2015). Illusions of Causality: How They Bias Our Everyday Thinking and How They Could Be Reduced. *Frontiers in Psychology*, 6(888). DOI: 10.3389/fpsyg.2015.00888

Matute, H., Steegen, S., and Vadillo, M. A. (2014). Outcome Probability Modulates Anticipatory Behavior to Signals That Are Equally Reliable. *Adaptive Behavior*, 22: 207–216.

Matute, H., Vadillo, M. A., Blanco, F., and Musca, S. C. (2007). "Either Greedy or Well Informed: The Reward Maximization—Unbiased Evaluation Trade-Off." In S. Vosniadou, D. Kayser, and A. Protopapas (Eds.), *Proceedings of the European Cognitive Science Conference*, pp. 341–346. Hove, UK: Erlbaum.

Matute, H., Yarritu, I., and Vadillo, M. A. (2011). Illusions of Causality at the Heart of Pseudoscience. *British Journal of Psychology*, 102(3): 392–405. DOI: 10.1348/000712610X532210

McKenzie, C. R. M., and Mikkelsen, L. A. (2007). A Bayesian View of Covariation Assessment. *Cognitive Psychology*, 54: 33–61. DOI: 10.1016/j.cogpsych.2006.04.004

Mercier, P. (1996). Computer Simulations of the Rescorla-Wagner and Pearce-Hall Models in Conditioning and Contingency Judgment. *Behavior Research Methods, Instruments, & Computers*, 28(1): 55–60.

Miller, R. R., and Matute, H. (1996). "Animal Analogues of Causal Judgment." In D. R. Shanks, K. J. Holyoak, and D. L. Medin (Eds.), *The Psychology of Learning and Motivation*, vol. 34, pp. 133–166. San Diego, CA: Academic Press.

Moore, D. W. (2005). Three in Four Americans Believe in Paranormal. *Gallup News Service*. Princeton. Retrieved from http://www.gallup.com/poll/16915/three-four-americans-believe-paranormal.aspx

Moore, M. T., and Fresco, D. M. (2012). Depressive Realism: A Meta-analytic Review. *Clinical Psychology Review*, 32(6): 496–509. DOI: 10.1016/j.cpr.2012.05.004

Msetfi, R. M., Wade, C., and Murphy, R. A. (2013). Context and Time in Causal Learning: Contingency and Mood Dependent Effects. *PLOS ONE*, 8(5): e64063. DOI: 10.1371/journal.pone.0064063

Müller-Lyer, F. (1889). Optische Urteilstäuschungen. *Archiv Für Psychologie*, 2: 263–270.

National Health and Medical Research Council. (2015). *NHMRC Statement on Homeopathy and NHMRC Information Paper: Evidence on the Effectiveness of Homeopathy for Treating Health Conditions* (NHMRC Publication No. CAM02). Retrieved from https://www.nhmrc.gov.au/guidelines-publications/cam02

Newport, F., and Strausberg, M. (2001). Americans' Belief in Psychic and Paranormal Phenomena Is Up Over Last Decade. *Gallup News Service*. Princeton. http://news.gallup.com/poll/4483/americans-belief-psychic-paranormal-phenomena-over-last-decade.aspx

Nyhan, B., and Reifler, J. (2015). Does Correcting Myths about the Flu Vaccine Work? An Experimental Evaluation of the Effects of Corrective Information. *Vaccine*, 33: 459–464. DOI: 10.1016/j.vaccine.2014.11.017

Pineño, O., and Miller, R. R. (2007). Comparing Associative, Statistical, and Inferential Reasoning Accounts of Human Contingency Learning. *Quarterly Journal of Experimental Psychology*, *60*(3): 310–329. DOI: 10.1080/17470210601000680

Rescorla, R. A., and Wagner, A. R. (1972). "A Theory of Pavlovian Conditioning: Variations in the Effectiveness of Reinforcement and Nonreinforcement." In A. H. Black and W. F. Prokasy (Eds.), *Classical Conditioning II: Current Research and Theory*, pp. 64–99. New York: Appleton-Century-Crofts.

Schmaltz, R. (2016). Bang Your Head: Using Heavy Metal Music to Promote Scientific Thinking in the Classroom. *Frontiers in Psychology*, *7*(146). DOI: 10.3389/fpsyg.2016.00146

Schmaltz, R., and Lilienfeld, S. O. (2014, April). Hauntings, Homeopathy, and the Hopkinsville Goblins: Using Pseudoscience to Teach Scientific Thinking. *Frontiers in Psychology*, *5*: 336. DOI: 10.3389/fpsyg.2014.00336

Seligman, M. E. P., and Maier, S. F. (1967). Failure to Escape Traumatic Shock. *Journal of Experimental Psychology*, *74*(1): 1–9.

Shang, A., Huwiler-Müntener, K., Nartey, L., Jüni, P., and Dörig, S. (2005). Are the Clinical Effects of Homeopathy Placebo Effects? Comparative Study of Placebo-Controlled Trials of Homeopathy and Allopathy. *Lancet*, *366*, 726–732. DOI: 10.1016/S0140-6736(05)67177-2

Shanks, D. R., and Dickinson, A. (1987). "Associative Accounts of Causality Judgment." In G. Bower (Ed.), *The Psychology of Learning and Motivation: Advances in Research and Theory*, vol. 21, pp. 229–261. San Diego, CA: Academic Press.

Simon, H. A. (1957). "A Behavioral Model of Rational Choice." In *Models of Man, Social and Rational: Mathematical Essays on Rational Human Behavior in a Social Setting*. New York: Wiley & Sons.

Solokoff, L. (1960). "The Metabolism of the Central Nervous System In Vivo." In J. Field, H. Magoun, and V. Hall (Eds.), *Handbook of Physiology*, pp. 1843–1864. Washington, DC: American Physiological Society.

Stavrova, O., and Meckel, A. (2016). The Role of Magical Thinking in Forecasting the Future. *British Journal of Psychology*. DOI: 10.1111/bjop.12187

Swaine, J. (2011). Steve Jobs "Regretted Trying to Beat Cancer with Alternative Medicine for So Long." *Telegraph*. Retrieved from http://www.telegraph.co.uk/technology/apple/8841347/Steve-Jobs-regretted-trying-to-beat-cancer-with-alternative-medicine-for-so-long.html

Tangen, J. M., and Allan, L. G. (2004). Cue Interaction and Judgments of Causality: Contributions of Causal and Associative Processes. *Memory & Cognition*, *32*(1): 107–124.

Taylor, S. E., and Brown, J. D. (1988). Illusion and Well-Being: A Social Psychological Perspective on Mental Health. *Psychological Bulletin*, *103*(2): 193–210.

Tversky, A., and Kahneman, D. (1981). The Framing of Decisions and the Psychology of Choice. *Science*, *211*(4481): 453–458.

Vadillo, M. A., Blanco, F., Yarritu, I., and Matute, H. (2016). Single- and Dual-Process Models of Biased Contingency Detection. *Experimental Psychology*, *63*(1): 3–19. DOI: 10.1027/1618-3169/a000309

Waldmann, M. R., and Holyoak, K. J. (1992). Predictive and Diagnostic Learning within Causal Models: Asymmetries in Cue Competition. *Journal of Experimental Psychology: General*, *121*(2): 222–236.

Walton, A. G. (2011). Steve Jobs' Cancer Treatment Regrets. *Forbes*. Retrieved from http://www.forbes.com/sites/alicegwalton/2011/10/24/steve-jobs-cancer-treatment-regrets/#169d97c53594

Wasserman, E. A. (1990). "Detecting Response-Outcome Relations: Toward an Understanding of the Causal Texture of the Environment." In G. H. Bower (Ed.), *The Psychology of Learning and Motivation*, vol. 26, pp. 27–82. San Diego, CA: Academic Press.

Wasserman, E. A., Elek, S. M., Chatlosh, D. L., and Baker, A. G. (1993). Rating Causal Relations: Role of Probability in Judgments of Response-Outcome Contingency. *Journal of Experimental Psychology: Learning, Memory, and Cognition*, *19*(1): 174–188. DOI: 10.1037/0278-7393.19.1.174

White, P. A. (2003). Making Causal Judgments from the Proportion of Confirming Instances: The pCI rule. *Journal of Experimental Psychology: Learning, Memory, and Cognition*, *29*(4): 710–727. DOI: 10.1037/0278-7393.29.4.710

Whitson, J. A., and Galinsky, A. D. (2008). Lacking Control Increases Illusory Pattern Perception. *Science*, *322*(5898): 115–117. DOI: 10.1126/science.1159845

Wiseman, R., and Watt, C. (2006). Belief in Psychic Ability and the Misattribution Hypothesis: A Qualitative Review. *British Journal of Psychology*, *97*(3): 323–338. DOI: 10.1348/000712605X72523

Yarritu, I., Matute, H., and Luque, D. (2015). The Dark Side of Cognitive Illusions: When an Illusory Belief Interferes with the Acquisition of Evidence-Based Knowledge. *British Journal of Psychology*: 1–12. DOI: 10.1111/bjop.12119

4 Hard Science, Soft Science, and Pseudoscience: Implications of Research on the Hierarchy of the Sciences

Dean Keith Simonton

When I was teaching the history of psychology—the capstone course for the psychology major at the university where I worked—a whole lecture was devoted to discussing two notorious pseudosciences in the discipline's history, mesmerism and phrenology. I used this topic as a transition lecture between psychology as philosophy and psychology as science. These pseudosciences somehow manage to be neither philosophical nor scientific, often lacking the logical rigor and systematic coherence of the former while missing the latter's empirical foundations and consilience with genuine sciences. I would also try to use these two historic cases to infer some of the general reasons for the emergence of pseudosciences, with the hope that we can avoid such embarrassments in present and future research or practice. After all, a major theme of the course is that history has lessons that should enable psychology to become more scientific.

I had been conveying these historical lessons for more than twenty years without having any direct experience with contemporary pseudosciences. To be sure, as a Darwinist by inclination, I was certainly aware of such pseudoscientific developments as "creation science" and "intelligent design." Yet those anti-evolutionary movements always took place in somebody else's discipline rather than in mine. Only very recently was I forced to deal with living pseudoscientists who directly (and often aggressively) challenged me to consider their bizarre claims. The impetus for most of these encounters was a brief speculative essay I had written about whether geniuses like Albert Einstein had become extinct in the natural sciences (Simonton, 2013). Published in *Nature*, a high-profile science journal, the piece attracted lots of media attention. Much of this attention entailed a serious, responsible, and informed debate about the status of contemporary astronomy, physics, chemistry, and biology, especially whether any of these sciences will ever again be in the position to experience a revolution such as those launched by Copernicus, Newton, Lavoisier, or Darwin.

But there also came an inundation of emails from "neglected geniuses." These individuals advised me that the more than three decades I had devoted to the topic, especially as consolidated in my 1988 *Scientific Genius* and 2004 *Creativity in Science*

(Simonton, 1988, 2004a), has been decisively and derisively discredited. After all, genius could not have gone extinct in science if the email message had been sent by an even greater genius who had already proven that Einstein was plain wrong. The famous formula $E=mc^2$ had to be replaced by the stranger but putatively more comprehensive and powerful $E=mQ^2$ (believe me, you don't want to know what Q stands for). Einstein may have shown that time and space are relative rather than absolute, and that they are integrated as four-dimensional space-time, but these modern revolutionaries have conclusively established that neither time nor space actually exists. Instead, events in space and time represent illusions generated by a hologram concentrated into an infinitesimally small point. Something vaguely like that.

Admittedly, I did not completely understand these theories—but it's not that I didn't try. I would ask for pdf's of the peer-reviewed publications announcing these revolutionary ideas, only to be told that this new work was too original to earn acceptance in top-notch scientific journals. So I found myself referred to the alternative publication venues used by these ignored geniuses: (a) non-peer-reviewed open access journals (aka vanity presses, such as the *General Science Journal*); (b) personal websites most often of very unprofessional quality; (c) pdf versions of unpublished manuscripts just uploaded on some nondescript server; and even (d) posted YouTube presentations hoping to go viral and earn their makers a Nobel Prize. Sometimes these "publications" were riddled with equations, diagrams, and concepts that would only seem impressive to anyone who hasn't taken high school courses in physics, chemistry, or biology. For example, one webpage contained a mumbo jumbo of formulas integrating all of the fundamental constants of nature and much else besides (the kitchen sink excepted). Remarkably, where Einstein was obliged to use highly advanced mathematics to work out the implications of his ideas, especially in the general relativity and the unified field theories, Einstein's lifework is claimed to have been utterly overthrown by equations that are not even as advanced as those taught in high school algebra!

The above illustrations have emphasized overlooked "revolutions" in theoretical physics, yet the onslaught involved alleged breakthroughs in other sciences as well. By a different route, I even had a heated confrontation with a person who argued that he had produced a general theory that integrated all of psychology, solely using the rudimentary ganzfeld effect to deduce predictions and explanations for every critical cognitive and behavioral phenomenon. Nevertheless, Einstein seems the prime target for pseudoscientists. Because Einstein outdid Newton, and was even declared *Time* magazine's "man of the century," undoing Einstein makes you the next VIP in the succession of scientific prophets. Amazingly, no science appears immune from pseudoscience. Even mathematics has instances. A notorious example is the claim of the philosopher Thomas Hobbes to have solved the ancient problem of "squaring the circle," a claim that was torn to shreds by John Wallis, the great mathematician. There's a reason why *squaring the circle* is often used as a metaphor for attempting the impossible, such as designing a perpetual motion machine. It has been mathematically proven to be absolutely impossible (hint: π is a transcendental number).

Although I have said that pseudoscience may be found in any science, in this chapter I want to discuss the possibility that some disciplines are more vulnerable to pseudoscientific ideas. Furthermore, even if the sciences might not differ in the number of historic examples, they might radically differ in what happens to those examples when they see the light of day. This discussion is predicated on recent research on the hierarchy of the sciences, a conjecture originally put forth by August Comte, the French philosopher.

Comte's Hierarchy of the Sciences

Comte (1855) proposed that the empirical sciences could be arranged into a "hierarchy" (Cole, 1983). At one end of the hierarchy are found the "inorganic" sciences, namely, astronomy, physics, and chemistry, whereas at the other end are located the "organic" sciences, namely, physiology (or biology) and "social physics" (or sociology). Sciences within each of these major groups could also be ordered according to their increasing complexity and decreasing generality or abstractness, as well as their enhanced dependence on the other sciences. For instance, Comte viewed astronomy as less complex, more independent, and more general than physics. As a consequence, a mature astronomical science could develop long before a full-fledged physics. In a similar manner, because sociology grappled with more complex phenomena, it could not emerge until after biology. Where biology handled individual organisms, sociology scrutinized interactions among individuals in social systems. Although the concept of hierarchy would seem judgmental, this hierarchical configuration should be considered more descriptive than evaluative. Just because chemistry deals with phenomena that are more complex and must partly depend on physics does not oblige it to be an inferior field. In fact, nothing prevents us from saying that sociology sits at the top of the hierarchy while astronomy stands at the bottom. Comte himself saw sociology as the last and greatest of the sciences to evolve out of the historical development of European civilization.

Ironically, although Comte advocated a "positive" (meaning "empirical") philosophy, he based his hierarchy on logic rather than data. Nevertheless, starting in the 1980s and continuing up to the present day, some researchers have determined whether scientific data support the theory (e.g., Cole, 1983; Fanelli, 2010; Fanelli and Glänzel, 2013; Smith, Best, Stubbs, Johnston, and Archibald, 2000). Some of this recent research has even focused on where psychology fits within the hierarchy (Simonton, 2002, 2004b, 2015a). Because psychology did not even free itself from philosophy until decades after Comte passed away, he never addressed this issue (cf. Allport, 1954). In any case, empirical research has demonstrated that the sciences of physics, chemistry, biology, psychology, and sociology can be objectively arrayed into the hierarchical arrangement depicted in figure 4.1.

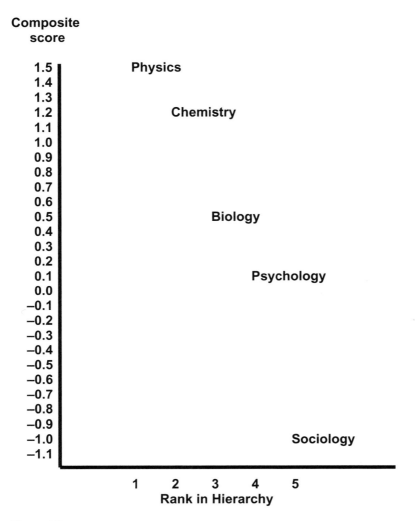

Figure 4.1
The disciplines of physics, chemistry, biology, psychology, and sociology placed in a Comtean hierarchy of the sciences. The horizontal axis indicates the rank and the vertical axis indicates the composite score on the definitional factor.

Hard Science, Soft Science, and Pseudoscience

The physical sciences dwell at the top, with physics just above chemistry. After a noticeable gap, biology appears, followed by psychology. Then we see an even bigger gap before sociology pops up. Interestingly, this manner of describing the hierarchies is more precise than simply saying that the natural sciences (physics, chemistry, and biology) score higher than the social sciences (psychology and sociology). That greater precision ensues from the fact that biology's placement is closer to psychology than to chemistry and that psychology's position is closer to biology than to sociology.

But what are the empirical factors that predict a science's location in the Comtean hierarchy?

Empirical Indicators of Hierarchical Placement

A large number of discipline characteristics were used to define and validate the results reported in figure 4.1 (Simonton, 2004b, 2015a). Nevertheless, these can be reasonably grouped into the following seven categories: (a) rated hardness and paradigm development; (b) consultation rate and graph prominence; (c) theories-to-laws ratio and lecture fluency; (d) peer-evaluation consensus, early impact rate, citation concentration, and h index; (e) multiples probability and anticipation frequency; (f) obsolescence rate and citation immediacy; and (g) confirmation bias favoring positive results. Although all seven categories are germane to the hierarchy of the sciences, some categories will prove most relevant to understanding the occurrence of pseudoscientific events.

Rated Hardness and Paradigm Development

Let us start with the only subjective assessment in the set of all indicators, namely the rated hardness of the five disciplines according to a survey of psychologists that was reported in Smith et al. (2000). What this correlation implies is that perceptions of the relative status of the five disciplines are not mere stereotypes out of line with objective reality. Rather, those judgments closely reflect that reality. Physics really is harder than biology, just as sociology is truly softer than psychology (see also Fanelli, 2010).

Rated hardness is nicely complemented by the paradigm development indicator. This indicator is a composite based on three objective measures (Ashar and Shapiro, 1990). Two of these measures gauge the "wordiness" of the titles and abstracts of doctoral dissertations published in the sciences, the more verbose theses more likely hailing for disciplines toward the bottom of the hierarchy (see also Fanelli and Glänzel, 2013). Significantly, Albert Einstein's own 1905 dissertation, with a German title merely five words long (six words in English), produced a journal article less than twenty pages if we omit the blank and duplicate pages (but include the dedication page; see http://e-collection.library.ethz.ch/eserv/eth:30378/eth-30378-01.pdf). In contrast, when Robert K. Merton, the distinguished sociologist, published his doctoral dissertation, it weighed in at 273 pages of print, with a thesis title 60 percent longer than Einstein's (Merton, 1938).

The third objective measure entering the assessment of a discipline's paradigm development concerned the logical structure of its undergraduate major (Ashar and Shapiro, 1990). On the one hand, we have majors in the physical sciences where virtually all courses have explicit prerequisite courses that themselves demand prerequisite courses, forming sequences that may last a year or more to complete (often designated with catalog numbers like 101A and 101B). On the other hand, we see undergraduate majors such as psychology and sociology that more frequently permit students to take a smorgasbord of upper-division courses with the sole proviso that they first complete an introductory course and possibly one on methods. This contrast suggests that the former sciences are defined by a unified inventory of cumulative knowledge that stipulates that the student master the material in steps, moving from elementary to advanced topics, just like learning a new language. In the latter sciences, in contrast, students who have acquired the basics are adequately prepared for any other topic.

Before turning to the next group of indicators, it is worth noting that rated hardness and paradigm development themselves correlate very highly, indicating a firm concordance between subjective and objective measures (Simonton, 2015a). The harder sciences are more paradigmatic.

Consultation Rate and Graph Prominence
The next two objective indicators, namely the consultation rate (Suls and Fletcher, 1983) and graph prominence (Cleveland, 1984), concentrate on research and publication practices in scientific disciplines. In the first case, it seems probable that scientists who conduct research further down the hierarchy will be less confident about the merits of their ideas and thus might feel obligated to find outside opinions before submitting their results for publication. Because the measure of the consultation rate controls for the number of coauthors, this difference cannot be an artifact of coauthors serving the same function in sciences higher in the hierarchy (cf. the "N. of authors" measure used in Fanelli and Glänzel, 2013). The consultation rate really appears a repercussion of relative uncertainty. This uncertainty is partly founded in the weak disciplinary consensus, a problem that will be returned to shortly.

In the instance of graph prominence, this uncertainty might manifest itself in the actual empirical results rather than in the disciplinary consensus about the significance of those results. In the physical sciences, and to a somewhat lesser extent the biological sciences, findings tend to be so clean, and the effect sizes so large, that the results can be readily depicted in visual form. The error bars around a fitted curve are small, often even trivial. In comparison, psychology and especially sociology deal with phenomena so complex that the findings cannot be so simply and cleanly portrayed. Hence, the results will more likely show up in statistical tables, with the number of asterisks lined up in a row rather deceptively indicating importance (Meehl, 1978). Printing a scatterplot for a statistically significant Pearson product-moment correlation of .30 would show much more noise than signal, and would thus serve to visually undermine the strength of the abstract conclusion. Even so, $r = .30$ qualifies as a reasonably *large* association in psychology (Rosenthal, 1990). Almost 10 percent of the variance would be shared between the two variables.

Complicating this picture even more, another reason why graphs are far more prominent at the top of the hierarchy than at the bottom could be that scientists at the upper end need fewer words to say what needs to be said to narrate their results. If both titles and abstracts become increasingly wordy as we descend the hierarchy, then the same trend would likely hold for text. Consequently, scientists in psychology and sociology might take more text even to explain whatever graphs that may appear in their journal articles. For soft disciplines, a single picture is *not* worth a thousand words but rather requires a thousand words of explanation.

Whatever the factors involved, it may be instructive to note the sheer magnitude of the contrasts in graphic prominence (Cleveland, 1984). The proportion of space devoted to graphs in the physical sciences is more than three times higher than seen in psychology, and seventeen times higher than in sociology. Biology's visual presentation practices are much closer to the physical sciences than to the social sciences, even just psychology.

Theories-to-Laws Ratio and Lecture Fluency

Most sciences can be separated into two divergent parts. At one extreme, the discipline's *core* entails "fully evaluated and universally accepted ideas which serve as the starting points for graduate education" (Cole, 1983, p. 111). At the other extreme, the discipline's research *frontier* encompasses "all research currently being conducted" (Cole, 1983, p. 111) at the leading edge of that science. For instance, with one minor exception—the length of the prerequisite chains that formed part of the paradigm development measure—all of the measures treated thus far have focused on the research frontier. In contrast, the next two measures concentrate not just on the core, but also on the core's very core, the lower-division introductory courses most frequently numbered 1 or 101 (with or without letters to indicate consecutive semesters or quarters of the same course). The first indicates the textbooks the students purchase for these courses, the second the lectures that their instructors deliver in the classrooms.

The introductory textbook measure is the theories-to-laws ratio, which provides an inverse measure of a discipline's place in the hierarchy (Roeckelein, 1997). This particular measure features an extremely high correlation with a science's placement in the hierarchy depicted in figure 4.1 (Simonton, 2004b, 2015a). Accordingly, sciences at the top of the hierarchy, such as physics, have a large proportion of laws relative to theories; those at the bottom, such as sociology, have a large proportion of theories relative to laws. Observe that this correlation may also be considered compatible with graph prominence. Theories tend to be largely verbal—as illustrated by Darwin's theory of natural selection—whereas laws are more likely expressed in terms of mathematical functions that can be easily represented visually—such as Fechner's law in psychology.

The introductory lecture indicator is a fluency measure, the inverse of the disfluency measure that assesses the rate of filled pauses—such as "uh," "er," and "um"—during the course of the lecture (Schachter et al., 1991). At the upper end of the hierarchy, terms and concepts are so precisely defined that the lecturer rarely struggles with recalling the right word to identify or describe a phenomenon, a linguistic accessibility not available at the lower end of the hierarchy. Even words that might be used by two or more disciplines will introduce variable constraints on their meaning. The word "force" has many more synonyms in sociology than it does in physics. It is conceivable that this semantic contrast is partly responsible for the consultation rate. Frequently colleagues are consulted for advice on whether specific ideas were effectively communicated, including whether the best word was selected for a given idea. This same contrast is certainly connected with the greater wordiness of doctoral theses as we proceed down the hierarchy. If words do not have precise definitions, then more adjectives and adverbs are necessary to clarify what the writer is trying to say.

The foregoing results are significant because frequently researchers argue that psychology really does not differ from the "hard" sciences respecting the research frontier (e.g., Hedges, 1987). Maybe so, but it definitely holds that the cores substantially differ. Furthermore, as seen already, the frontiers differ as well. That contrast will be witnessed further on.

Peer-Evaluation Consensus, Early Impact Rate, Citation Concentration, and *h* Index
The first three measures in this group (namely peer-evaluation consensus, early impact rate, and citation concentration) are homogeneous in more than one way (Simonton, 2004a). More specifically, I should make these four points (cf. Simonton, 2015a):

First, all three measures involve the degree of consensus exhibited by scientists working in the same discipline. This disciplinary consensus is merely assessed three different ways: (a) the degree to which peer evaluations concur when applied to the same colleagues, (b) the propensity for citations to favor specific publications rather than being distributed in a more egalitarian fashion, and (c) the probability that a young scientist receives early recognition for their work as judged by citations.

Second, all three consensus indicators were taken from Cole's (1983) seminal examination of the hierarchy question. That reliance reflects the fact that Cole placed considerable emphasis on the place of consensus in defining the hierarchy. His emphasis should not be surprising, given that Cole was a sociologist by training.

Third, all three feature very high correlations with the composite measure that defined figure 4.1, albeit all three also contributed directly to defining that composite, making the correlations essentially the same as factor loadings (Simonton, 2015a). Because the three measures also highly correlate with each other, we can say that the consensus component of the hierarchical placement must be highly reliable.

Fourth, all three correlate very highly with paradigm development. Hence, the disciplinary consensus is intimately associated with undergraduate and graduate education.

These four points notwithstanding, the second point might raise some concern—the three measures all coming from the same original source. Therefore, I conducted a follow-up study using a totally different definition of disciplinary consensus (Simonton, 2015a). The analysis took advantage of data on interdisciplinary contrasts in the mean h index attained by researchers (see Grosul and Feist, 2014, for the study also using these same raw data). Here h is the highest number of publications that have been cited h or more times. As an example, when $h = 60$, the investigator has sixty publications that have received at least sixty citations (Hirsch, 2005). The average h value not only varies across disciplines (see also Iglesias and Pecharromán, 2007), but also does so in line with the hierarchy depicted in figure 4.1. Scientists who contribute to disciplines higher up in the hierarchy will, on average, publish much more highly cited work, implying a superior consensus on what represents the high-impact research in the discipline.

The significance of consensus will be reiterated in the next section, but in a strikingly different manner.

Multiples Probability and Anticipation Frequency

The present pair of measures go back even earlier into the process of making contributions to the research frontier of a scientific discipline—the point when a scientist initiates inquiry into a given question or makes the discovery that will be offered for peer evaluation.

The first of these concerns a distinctive event in the history of science and technology, namely, multiple discovery and invention (Merton, 1961). This event takes place when two or more scientists (or inventors) independently make the same discovery (or invention). Traditionally, the occurrence of multiples is taken to prove that scientific discoveries are inevitable, attaining a probability of unity at a specific place and time in history (Lamb and Easton, 1984). Nevertheless, this sociocultural deterministic interpretation has been undermined by recent research that demonstrates that the central features of such events can be predicted according to a stochastic combinatorial model (Simonton, 2010). Of particular importance here is the demonstration that the distribution of multiple grades is precisely predicted by a Poisson model (Simonton, 1979; see also Price, 1986). The grade of a multiple is simply the number of independent scientists or inventors who can be credited with the discovery or invention. Characteristic of this distribution is its similarity to an inverse power function, so that nonmultiple "singletons" are the most common, followed by doublets, triplets, quadruplets, and so on, in a decelerating fashion, so that octets and nonets are extremely rare. Importantly, the Poisson distribution is defined by a single parameter, μ, which essentially defines the expected grade of multiple in a given sample of discoveries and inventions. The critical observation is not only that μ varies across scientific disciplines (Simonton, 1978), but also that the magnitude of μ decreases as we progress down the hierarchy shown in figure 4.1 (Simonton, 2015a). In concrete terms, this correlation means that in the hard sciences multiples are relatively common, whereas in the soft sciences multiples are extremely rare if not nonexistent. This relationship makes sense in terms of what was noted in the previous section: scientists in the upper section of the hierarchy will display a firm consensus on the key questions and the methods to answer those questions, a circumstance that more likely than not results in more than one scientist working on the same problem. Often the consequence is a multiple where two or more independent contributors are obliged to share credit for the same discovery or invention.

Although such an outcome might seem unpleasant, especially when it provokes priority disputes—the accusations exchanged between Newton and Leibnitz over who actually conceived the calculus were exceptionally acrimonious—a frequent alternative is no less unpleasant. This alternative outcome is captured by the second indicator, anticipation frequency (Hagstrom, 1974). Anticipation occurs when one scientist or inventor beats the other to the goal line, publishing first and thereby preventing the other from getting any credit for making the same contribution. Occurrences like these led Merton (1961) to assert that many "singletons" are actually multiples because we can never know how many other scientists would have made the same discovery had they not been preempted. For example, in all likelihood Linus Pauling would have deciphered the structure of DNA had Watson and Crick not gotten there first.

Although the anticipation indicator was based on survey data, and the multiple probability measure on the historical record, the two variables correlate highly enough to suggest that they tap into the same underlying factor: a strong disciplinary consensus concerning theoretical questions and the empirical methods needed to address those questions. In short, multiples and anticipation are indicators of a discipline's strong paradigmatic, or "hard," status as depicted in figure 4.1.

Obsolescence Rate and Citation Immediacy
This pair of indicators taps into what presumably occurs at the research frontier of the science (Cole, 1983). For disciplines at the top of the hierarchy, this leading edge advances quickly, accepting or rejecting current hypotheses, incorporating what needs to be added to the core, and then advancing to the next set of big questions (often provoked by the answers to the earlier set of questions). One result is that citations tend to focus on the most current contributions to the discipline. This concentration departs strikingly from what takes place in disciplines lower in the hierarchy, where researchers will still cite work now decades old—such as the "citation classics" of Sigmund Freud or Jean Piaget in psychology (Endler, Rushton, and Roediger, 1978).

Another repercussion, naturally, is that disciplinary knowledge becomes obsolete much faster in the sciences situated higher in the hierarchy (McDowell, 1982). Taking leave from original research to take on administrative responsibilities has far greater career costs when the leading edge has moved on in the meantime. Although these two indicators have somewhat smaller correlations with the composite measure used to create figure 4.1, the correlations are still substantial, especially when citation immediacy is corrected for range restriction (Simonton, 2015a). Moreover, these two indicators correlate quite highly with each other (even when uncorrected for range restriction). Hence, the indicators really are assessing the same construct two different ways.

Confirmation Bias Favoring Positive Results
The final indicator is perhaps the most fascinating from the standpoint of understanding the pseudosciences. In an ideal world, scientists first formulate hypotheses and then subject them to rigorous empirical tests. The probability of affirming or disproving the hypotheses should be contingent on nothing more than whether the hypotheses are "true" or "false" when weighed against the relevant empirical data. Hypothesis testing is supposed to be an objective act, heavily constrained by both theory and method, and thus immune from the intrusion of the scientist's own personal or subjective inclinations. To be sure, a scientist might deliberately commit fraud, forging data rather than collecting real data. Yet, ironically, such deceit itself constitutes an objective act, an act where the perpetrator knows better than anybody else that the hypothesis has not been tested according to proper scientific conduct. Indeed, it requires a high degree of scientific expertise to carry out such a deception without immediate detection. The failure is due more to lapses in personal morality than to deficiencies in professional competence.

That said, the supposed theoretical and methodological constraints imposed on the working scientist may vary according to the discipline's placement in the hierarchy graphed in figure 4.1. For hard sciences, the constraints would be precise and strong, whereas for soft sciences the constraints would be vague and weak. If so, then the way is open for subtler human cognitive and motivational inclinations to intrude between the hypothesis and the "scientific" test of that hypothesis.

This outcome was demonstrated in a study conducted by Fanelli (2010). He specifically analyzed more than 2000 journal articles published in every major contemporary science, from the physical sciences to the social sciences, with the biological sciences defining the middle of the hierarchy. The sole restriction on the sample was that each paper had to explicitly state that it was testing a specific hypothesis (just as Fanelli's paper was doing). Then the article was scrutinized to determine whether or not the hypothesis was supported (fully or partially). Fanelli's own hypothesis was confirmed. The higher a science stands in the hierarchy, the lower the odds of confirming the paper's hypothesis. In contrast, hypotheses are more likely to be confirmed when the scientist is operating in a "soft" rather than "hard" science. Indeed, the odds of reporting a confirmation in the social sciences was 2.3 times higher than in the physical sciences!

Fanelli (2010) hastened to point out that the confirmation rates in the social sciences are not so high as to invalidate the scientific status of these disciplines. The social sciences differ from the physical sciences only in degree, not in kind. This additional finding is critical given that Fanelli's own study represents social science (namely social science of science studies). If it were excessively easy to confirm social scientific hypotheses, Fanelli's own study would be cast into doubt. It might then even count as pseudoscience!

Genuine versus Pseudoscience

Before I can treat the implications of the hierarchy for understanding the pseudosciences, I first must define what I mean by pseudoscience. Probably the best place to begin is to specify what pseudoscience is not. As already implied earlier, scientific misconduct is not taken as pseudoscientific because the agent deliberately passes off falsity as truth: the hypothesis was said to have been confirmed even though objective and replicable data were not collected (or were collected but then ignored or manipulated). A notorious example is that of Piltdown man, in which the fragment of the cranium of an adult modern human was buried with the lower jawbone of an orangutan. This combination purported to confirm the then popular hypothesis that *Homo sapiens* evolved a large brain prior to becoming an omnivore. The fraud was so skillful that it took paleoanthropologists about forty years to debunk this false evidence of the famous "missing link."

In addition, although I previously allowed pseudoscientific episodes to occur in mathematics, such events will not be treated here. Because Comte's hierarchy deals with the *empirical* sciences, and mathematics is nonempirical—the truth or falsity of hypotheses (aka theorems) established by pure logic—any potential relevance reduces to zero. One does not "test" the Pythagorean theorem by collecting data on a random sample of right triangles (albeit if we did so, the theorem would be confirmed within a given amount of measurement error). Recall that part of the foundation for the hierarchy in the first place is the increased complexity of the phenomena that define the domain. Yet because mathematics contains no empirical phenomena, that critical characteristic has no meaning. The only exception concerns *applied* mathematics, where pure mathematics is applied to solve problems in the empirical sciences. Those empirical sciences that involve the simplest phenomena will lend themselves most readily to mathematical treatment. Hence, the hard sciences tend to be the most mathematical as well—indeed physics without math is unimaginable. As we descend down the hierarchy, the likelihood of a successful mathematical treatment declines, albeit macroeconomics provides an interesting exception to this general pattern.

Even then, when people speak of the "mathematical sciences," they seldom include economics.

Having just distinguished between pure and applied mathematics, it is also necessary to distinguish between the pure and applied disciplines of any science in Comte's hierarchy. The original hierarchy conceived the sciences in their "pure" form, with the goal being to understand the phenomena regardless of any applications. This separation is critical because the status of the pure science may not necessarily correspond to its applied version. The most dramatic example is the current status of astronomy relative to astrology. Although Comte placed astronomy at the apex of the hierarchy (a placement having some empirical support; Ashar and Shapiro, 1990; Fanelli, 2010), astrology is now considered a prototypical example of a pseudoscience. Yet Johannes Kepler, one of the greatest astronomers during the Scientific Revolution that opened the seventeenth century, earned his living as the official court astrologer to the Holy Roman emperor. Furthermore, empirical research has already shown that the applied sciences must be treated separately from their corresponding pure sciences (Fanelli, 2010; Hemlin, 1993; Klavans and Boyack, 2009). Nor is that surprising, given that the applied sciences have to overcome a constraint unknown to their pure counterparts: the hypothesis must not only be empirically confirmed, but it must yield practical results. Nuclear fusion has been a demonstrated fact for a very long time, but except for blowing whole metropolitan areas into smithereens, its practical applications are minimal. Despite the billions of dollars invested in controlled fusion reactors, not one single region on this planet has ever been powered by nuclear fusion. For that matter, nuclear fusion has never been used to power the light bulbs in a nuclear fusion laboratory!

Those stipulations in place, let me now enumerate what I consider the defining attributes of a genuine science. To the extent that a purported "science" or a "scientific" discovery departs from these attributes, then it can be considered a pseudoscience or a pseudoscientific discovery. The characteristics mentioned earlier have the highest importance in revealing the intrusion of pseudoscience.

1. *Naturalism*—Genuine sciences attempt to provide naturalistic explanations for observed phenomena, explanations that are free from the intrusion of gods, spirits, miracles, and other mysterious forces. This requirement dates back to the ancient Greek philosopher Thales, who attempted to provide naturalistic reasons for phenomena that would previously have been explicated in terms of the gods and spirits of Greek mythology—such as the lightning bolts thrown by Zeus. Although these naturalistic explanations can take many forms, most often they involve the application of such concepts as natural laws and cause-effect relationships. An obvious illustration is when Charles Darwin replaced the Genesis account of the origins of life and humans, where God plays the key role, with an explanation that required only the application of natural selection to variation within species. Because Creationism and Intelligent

Design attempt to reintroduce the Divine Being as the agent of organic evolution, these must be considered pseudosciences by this criterion alone (albeit the argument from Intelligent Design is more philosophically subtle and less obviously contingent on religious scripture). Because naturalism is fundamentally secular rather than religious, science becomes a universal that transcends the dictates of any particular faith. A nuclear physicist can be a practicing Muslim, Christian, Hindu, Buddhist, or Jew—just to name a few world religions—without compromising his or her beliefs.

2. *Parsimony*—This second characteristic requires the application of Occam's razor: anything superfluous in a theory or explanation should be cut off, like trimming the fat off of meat. Given a complex and a simple interpretation of the same data, the latter is preferred. Parsimony is connected to naturalism because naturalistic explanations tend to be more parsimonious. Ancient Greek natural philosophers attempted to explain all observed phenomena by means of a small number of basic elements (water, earth, air, and fire). The physicians of those times tried to explain human ailments in terms of the four humors (blood, yellow bile, black bile, and phlegm). In modern science, parsimony adopts many forms. One common version is a naming system that allows the phenomena that define a science to be identified in a consistent and unambiguous manner, such as the binomial nomenclature introduced by Carl Linnaeus for biological species. Another guise of parsimony is some organizing principle that coordinates a complex set of facts in a relatively simple yet informative way, as seen in the periodic table of the elements that decorates every chemistry lecture hall. Of course, often the parsimony is found in an integrative law or theory that reduces the apparent complexity of the observed phenomena. Einstein's famous $E=mc^2$ affirms a relation between energy and mass, which hitherto were thought to exist as separate phenomena. Lastly, I should mention a form of parsimony that is especially germane to Comte's hierarchy of the sciences: reductionism. When a phenomenon at one level in the hierarchy (say, biology) can be explicated in terms of principles that have been established at a higher level (to wit, chemistry), then science gets a two-for-one deal. Molecular biology provides many examples, such as DNA molecules zipped together by hydrogen bonds.

3. *Falsifiability*—This characteristic was famously introduced by Popper (1959) to solve the demarcation problem of how to separate science from nonscience. To be scientific, a hypothesis or theory must prove susceptible to an empirical test; it must yield a prediction that can be disconfirmed by observed data. Just as parsimony bears some connection to naturalism, so does the falsification principle connect with parsimony. By requiring falsifiability, scientists immediately reduce the number of hypotheses or theories they have to consider. Besides ignoring all conjectures that are inconsistent with empirical findings, scientists can also ignore all conjectures that cannot possibly be inconsistent with any empirical findings. For example, if someone claims that extrasensory perception exists but that it cannot be studied under controlled laboratory conditions because those very rigorous circumstances cause psychic powers to dissipate, then

Hard Science, Soft Science, and Pseudoscience

how can the claim possibly be disproven? Of course, the falsifiability criterion assumes that a science be sufficiently rigorous that it is in fact possible to make an up-or-down decision based on empirical results—an assumption that we will also return to later.

4. *Objectivity*—It is perhaps ironic that the fourth and last defining attribute is not easy to define objectively. That difficulty arises because the concept's definition is contingent on what ontology and epistemology one favors. If someone believes that there is a "real world" out there (the ontology of philosophical realism) that is immediately and reliably available to everyone's five senses (the epistemology of empiricism), then "objective truth" is equivalent to a belief being empirically veridical. Presumably if the traffic light turns green, then all drivers see green and press their feet on the gas pedal. Yet that view of objectivity becomes increasingly problematic for many scientific facts and theories. Who among scientists has ever actually seen an atom or viewed a black hole in the same sense that one sees a green traffic light? Hence, objectivity in science requires something more—most often a consensus on what is considered objective and what is no more than subjective. In the hard sciences this consensus involves an agreement on the instruments that most accurately measure key variables, such as temperature, pressure, and volume with respect to the gas laws. In the soft sciences a consensus may also be reached, albeit not as strong. For example, most personality psychologists today would consider a standardized self-report measure like the Minnesota Multiphasic Personality Inventory to be far more objective than the projective Rorschach test using ink blots.

That concludes my conception that naturalism, parsimony, falsifiability, and objectivity are essential in constituting a bona fide science. To the extent that ideas depart from one or more of these four stipulations, we move into the pseudosciences. Actually, we should be more precise by using the term "psychoscientific" ideas or beliefs rather than pseudosciences. The latter sweepingly implies that a whole domain, rather than particular parts of the domain, fails to meet these standards.

Pseudoscientific Beliefs within the Hierarchy

Given the hierarchy of the sciences summarized in figure 4.1 and my four criteria for what constitutes genuine science, we should be able to predict the frequency and nature of the pseudoscientific ideas across scientific disciplines. The only departure from the empirically established hierarchy depicted in the figure is that I will not dip any lower than psychology. Many of my choicest illustrations will come from the psychological sciences simply because that happens to be the discipline I know best as a psychologist. But that should not be taken as an assertion that sociology and other social sciences are immune from this problem. No science is completely off the hook.

Most obviously, I will claim that pseudoscientific beliefs will increase in prominence as we descend from the hard sciences at the top of the hierarchy to the soft sciences at the bottom. This expectation ensues from all four assumed attributes of bona fide scientific ideas, but let us start with the falsifiability criterion. That treatment will lead us into the remaining three standards as the discussion dictates.

Falsifiability

Clearly, if hypotheses are more likely to be confirmed in sciences lower in the hierarchy, then it follows that hypotheses are accordingly less falsifiable. As a consequence, the softer sciences will be riddled with false but still accepted results. We end up with the unfortunate situation where "most published research findings are false" (Ioannidis, 2005, p. 40). Many illustrations come from the recent history of psychological research, and most particularly from social psychology. According to the recent findings, a very large proportion of published findings cannot be replicated even though those results appeared in top-tier peer-reviewed journals and thus were considered empirically established (Open Science Collaboration, 2015). We are not talking about outright fraud here, albeit the permissive methodological standards render that problem all too common as well (e.g., the notorious Diederik Stapel case in social psychology). Instead, well-meaning psychologists find it all too easy to let confirmatory bias get the best them.

To provide the most dramatic recent example, the distinguished social psychologist Daryl Bem (2011) was purportedly able to establish the empirical reality of psi phenomena in a series of nine experiments published in the high-prestige *Journal of Personality and Social Psychology* (*JPSP*), where "the term *psi* denotes anomalous processes of information or energy transfer that are currently unexplained in terms of known physical or biological mechanisms" (p. 407). Despite the prima facie implausibility of such phenomena, statistically significant results were obtained for eight out of the nine experiments, with a mean effect size of 0.22 (Cohen's *d*), which is small but not trivial. To be sure, *JPSP* was eventually obliged to publish replication failures that showed, using Bem's own procedures, that the actual effect size could not be significantly greater than zero (Galak, LeBoeuf, Nelson, and Simmons, 2012). Yet if psychology were more rigorous in falsifying hypotheses, Bem's original study would not have been published in the first place, not even in a lower-tier journal (LeBel and Peters, 2011). Too many methodological constraints would automatically prevent confirmatory bias to override empirical truth.

Naturalism

From the standpoint of the pseudosciences, Bem's (2011) hypotheses violate a second characteristic of true science, namely naturalism. Whether called parapsychology, extrasensory perception, or psi phenomena, such supposed effects lack a naturalistic explanation. This explanatory failure is especially problematic in the case of precognition and premonition, the two psi phenomena specifically tested in Bem's series of experiments—and disconfirmed in (Galak et al., 2012) replication attempts. The antinaturalism of these two phenomena is apparent in the title of Bem's article: "Feeling the Future: Experimental Evidence for Anomalous Retroactive Influences on Cognition and Affect." The researcher is testing the hypothesis that people can mysteriously experience events that haven't happened at the time of the experience, whether through cognition or affect. The title even admits that such occurrences would be considered anomalous. After all, if these psychological phenomena were truly real, they would violate concepts of cause-effect sequences dating back to the beginning of natural philosophy. Effects could occur before the causes! Although believers in precognition and premonition might speculate on various naturalistic explanations, such as incorporating some of the more mysterious features of quantum theory, the bottom line remains that the capacity to predict future events must rely on the extrapolation from past cause-effect sequences, an extrapolation that is explicitly excluded in the very definition of these two psi phenomena (Bem, 2011).

Returning to the hierarchy of the sciences, it should be evident that naturalism is highest in the *natural* sciences, lowest in the behavioral and social sciences—I almost said the "unnatural" sciences. Indeed, the physical sciences such as physics and chemistry pretty much epitomize what can be considered naturalistic disciplines. That is why they score higher in rated hardness and paradigm development, as mentioned earlier, and also why these disciplines have a much higher ratio of established laws to speculative theories. Laws specify exactly how nature behaves, whereas theories merely conjecture about the causes of that behavior. Not surprisingly, the biological sciences are a little less naturalistic than the physical sciences. Certainly biologists have been trying, given their extirpation of creationism and vitalism from the mainstream biological disciplines in the nineteenth century. But naturalism in biology has to fight a constant battle against the forces of supernaturalism. This fight is especially conspicuous in applied domains such as medicine, where popular culture promulgates herbal remedies, new age therapies, and miracle drugs with complete disregard for natural laws. Yet even in pure scientific discussion we can see examples of antinaturalism attempting to come out of the shadows. A case in point is the controversy over whether mutations can be directed or not (Lenski and Mittler, 1993). If mutations could truly be guided toward greater fitness, then we would have the biological analog of precognition in parapsychology. Lacking a naturalistic explanation, directed mutations would seem to evoke a return to vitalism—if not to Intelligent Design. A supreme being is somehow looking out for the welfare of its creations each time there's a DNA replication error!

Parsimony

In the absence of strong supportive empirical evidence, any argument that mutations can be directed would also seem to violate Occam's razor. Biological evolution would then have two sources of variation, one requiring selection to separate the fit from the unfit, and the other obviating selection because the variation is already "known" to be adaptive. The most brilliant feature of the Darwinian theory of evolution by natural selection is that it did not suppose that variations had to be generated with any foresight regarding whether they would survive the struggle of existence. Of course, some species can bypass the variation by engaging in certain forms of asexual reproduction, such as budding, but that can only happen if the species has already become well adapted to a stable niche via past generations of variation selection. If otherwise, the whole species could vanish at the slightest environmental change. Similarly, organisms can respond to environmental stressors by increasing variation in potential adaptive responses (such as new enzymes when the usual nutrient sources become scarce), but that enhanced variation only serves to provide more raw material for natural selection. Hence, most evolutionary biologists have opted to stick to undirected variation. The latter not only has a more naturalistic postulate but also entails a more parsimonious one.

The problem with psychology and most social sciences is that they so often lack theoretical parsimony. The problem is not that there are so many theories, particularly relative to laws, for the greater complexity of phenomena studied by these disciplines might require more theories in proportion to the greater number of phenomena requiring explanation. The trouble is that many fields have proliferated a large number of theories chasing the *same* phenomenon. A favorite illustration comes from my own specialty area: creativity. After conducting research on this topic for more than forty years, I have become frustrated with the huge inventory of processes and procedures that have been put forward as essential for creative thought (Simonton, 2015b). Even if we overlook those frequent instances where two or more terms have been assigned to the same mechanism—another betrayal of the law of parsimony—the inventory still probably contains at least a dozen or more distinct processes and procedures. Furthermore, my attempts to integrate these diverse mechanisms into a single coherent predictive and explanatory theory have often been unfavorably received (Simonton, 2010, in press). In part, colleagues in this area do not take kindly to any attempts to subordinate their favorite process or procedure to special cases of a single unifying theory of creativity. So I imagine that new mechanisms will continue to be discovered willy-nilly.

Objectivity

One might argue that the reason why creativity researchers have been so fertile in coming up with novel mechanisms is that each researcher tends to favor those processes or procedures that each has found most valuable for his or her own personal creativity. In short, the preferences are highly subjective rather than objective. Any unified theory would only serve to show that their favorite route to creativity enjoys no special status, others displaying equal creativity by totally different paths. In addition, the same subjectivity permeates other specialty areas in psychology. For example, it is often said that the number of personality theories is equal to the number of theorists with different personalities. Whether true or not, it is apparent that a psychologist may more likely lapse into subjectivity than any biologist, chemist, or physicist. After all, psychologists have spent an entire lifetime becoming highly familiar with their own psychology. The resulting egocentric perspective may even help explain the confirmatory bias mentioned earlier. It is much easier to overlook methodological or theoretical issues that might interfere with an implicit self-confirmation.

As already noted, objectivity in the hard sciences is largely ensured by a strong consensus about the concepts, nomenclature, measures, methods, theories, and phenomena that define the science. This superior agreement can be seen in how sciences at the top of the hierarchy score higher on peer-evaluation consensus, early impact rate, citation concentration, h index, multiples probability, anticipation frequency, obsolescence rate, citation immediacy, and lecture fluency, yet score lower on consultation rate. For instance, when scientists agree on the important research questions, they are more likely to make the same discoveries or anticipate the discoveries of others. Likewise, when investigators all concur on the most important findings, those results will become recognized far more quickly and more rapidly incorporated into the discipline's core knowledge—thus removing the obligation to cite the now established work. Indeed, the leading edge of this consensus moves so fast that pseudoscientific thinking barely pokes its head above ground before it being pulled out by the roots. When the supposed demonstration of cold nuclear fusion first made headlines in 1989, the scientific community needed less than a year to reach a consensus that the supposed phenomenon was specious. It probably would have been discredited even earlier had the two supposed discoverers not been reputable scientists to start with.

In fact, for high consensus sciences, the cruel work of objectivity prevents many pseudoscientific ideas from even getting past the gatekeepers—such as the examples presented at the beginning of this chapter. Anybody who uploads a YouTube video to prove that he or she is a greater scientist than Albert Einstein can be objectively certified as someone who doesn't know what they're talking about, whether through ignorance or mental illness. No chance whatsoever that the clip will go viral among theoretical physicists—except perhaps as a joke.

Discussion

This chapter has been devoted to the thesis that the Comtean hierarchy of the sciences can help us appreciate the likelihood that any given scientific discipline will be plagued by pseudoscientific ideas. Those sciences at the top of the hierarchy depicted in figure 4.1 will be less susceptible to such events than those at the bottom. This expectation should not be taken as a dogmatic claim that pseudoscientific thinking cannot occur at all in the upper levels. On the contrary, some might argue that any superstring theory already has that suspicious status in theoretical physics. Although highly naturalistic and objective, its parsimony can be questioned insofar as the only way the theory can integrate all the fundamental forces of nature is to posit many more dimensions than the standard four (three of space and one of time). Worse still, no superstring theory has yet generated an empirically falsifiable prediction. A "theory of everything" then becomes a theory that claims nothing.

Would a string theorist then insist, "Judge not lest ye be judged!" Good point. How well does my theorizing do when matched against its own criteria? To begin with, the original basis for figure 4.1 certainly passed the fourfold test of naturalism, parsimony, falsifiability, and objectivity (Simonton, 2015a). More importantly, the application of the hierarchy to pseudoscientific ideas is not just empirically testable but falsifiable. Just assess a representative sample of potentially pseudoscientific beliefs on the four criteria (presumably on a quantitative scale, however approximate), and then tabulate the scored cases across the sciences in the hierarchy (using the four criteria scores as weights). The final step would be to calculate the correlation between the status in the hierarchy and the resulting tabulation. If the resulting correlation indicates no relation whatsoever or, even worse, a tendency for the harder sciences to exhibit more pseudoscientific ideas than the softer sciences, then my conjectures have been falsified—period.

Drawing upon Comte's positive philosophy, I might venture a second prediction, but one which I am unwilling to put forward as a critical test. As the philosopher would have it, positivism would gradually work its way through the entire hierarchy until all sciences would have the same status. Naturally, the sciences could still be arranged according to the complexity of the phenomena and the degree of dependence on sciences that had attained positive status earlier, but eventually psychology and sociology would be every bit as scientific as physics and chemistry. By implication, the various sciences in the hierarchy would gradually become more equal with respect to the appearance of pseudoscientific beliefs. Consequently, we could examine the changes to determine whether such positivistic progress takes place.

But I won't make that prediction. In my more skeptical moments I wonder whether psychology and the social sciences will ever attain the same degree of naturalism, parsimony, falsifiability, and objectivity that has become so conspicuous in the natural sciences. I am especially inclined to believe that psychology forces a break in the required progression. For reasons hinted at earlier, the immediate experiences that psychologists have with their own psychology will continue to obstruct progress. Seeing the discipline's phenomena from both the inside (subjective) and the outside (objective) ends up introducing unresolvable problems, such as the mind-body relation and free will. Perhaps the optimism of contemporary cognitive neuroscientists is justified so that both consciousness and volition (the subjective inside) will be empirically and theoretically equated with brain scans and overt behavior (the objective outside). Perhaps not. Instead psychology might end up with another pseudoscience, a modern phrenology with more expensive equipment. Hence, rather than make an up-down prediction, it is far wiser to wait and see. Will positivism eventually banish all pseudoscientific ideas from the psychological sciences?

References

Allport, G. W. (1954). The Historical Background of Modern Social Psychology. In G. Lindzey, (Ed.), *Handbook of Social Psychology*, vol. 1, pp. 3–56) Reading, MA: Addison-Wesley.

Ashar, H., and Shapiro, J. Z. (1990). Are Retrenchment Decisions Rational? The Role of Information in Times of Budgetary Stress. *Journal of Higher Education, 61*: 123–141.

Bem, D. J. (2011). Feeling the Future: Experimental Evidence for Anomalous Retroactive Influences on Cognition and Affect. *Journal of Personality and Social Psychology, 100*: 407–425.

Cleveland, W. S. (1984). Graphs in Scientific Publications. *American Statistician, 38*: 261–269.

Cole, S. (1983). The Hierarchy of the Sciences? *American Journal of Sociology, 89*: 111–139.

Comte, A. (1855). *The Positive Philosophy of Auguste Comte* (H. Martineau, trans.).

Endler, N. S., Rushton, J. P., and Roediger, H. L., 3rd. (1978). Productivity and Scholarly Impact (Citations) of British, Canadian, and U.S. Departments of Psychology (1975). *American Psychologist, 33*: 1064–1082.

Fanelli, D. (2010). "Positive" Results Increase Down the Hierarchy of the Sciences. Public Library of Science One *5*(4): e10068. DOI: 10.1371/journal.pone.0010068

Fanelli, D., and Glänzel, W. (2013). Bibliometric Evidence for a Hierarchy of the Sciences. Public Library of Science One, *8*(6): e66938. DOI: 10.1371/journal.pone.0066938

Galak, J., LeBoeuf, R. A., Nelson, L. D., and Simmons, J. P. (2012). Correcting the Past: Failures to Replicate Psi. *Journal of Personality and Social Psychology, 103*: 933–948.

Grosul, M., and Feist, G. J. (2014). The Creative Person in Science. *Psychology of Aesthetics, Creativity, and the Arts, 8*: 30–43.

Hagstrom, W. O. (1974). Competition in Science. *American Sociological Review, 39*: 1–18.

Hedges, L. V. (1987). How Hard Is Hard Science, How Soft Is Soft Science? *American Psychologist, 42*: 443–455.

Hemlin, S. (1993). Scientific Quality in the Eyes of the Scientist: A Questionnaire Study. *Scientometrics, 27*: 3–18.

Hirsch, J. E. (2005). An Index to Quantify an Individual's Scientific Research Output. *Proceedings of the National Academy of Sciences, 102*: 16569–16572.

Iglesias, J. E., and Pecharromán, C. (2007). Scaling the H-index for Different Scientific ISI fields. *Scientometrics, 73*: 303–320.

Ioannidis, J. P. (2005). Why Most Published Research Findings Are False. *Chance, 18*(4): 40–47.

Klavans, R., and Boyack, K. W. (2009). Toward a Consensus Map of Science. *Journal of the American Society for Information Science and Technology, 60*, 455–476.

Lamb, D., and Easton, S. M. (1984). *Multiple Discovery*. Avebury, UK: Avebury.

LeBel, E. P., and Peters, K. R. (2011). Fearing the Future of Empirical Psychology: Bem's (2011) Evidence of Psi as a Case Study of Deficiencies in Modal Research Practice. *Review of General Psychology, 15*: 371–379.

Lenski, R. E., and Mittler, J. E. (1993). The Directed Mutation Controversy and Neo-Darwinism. *Science, 259*: 188–194.

McDowell, J. M. (1982). Obsolescence of Knowledge and Career Publication Profiles: Some Evidence of Differences among Fields in Costs of Interrupted Careers. *American Economic Review, 72*: 752–768.

Meehl, P. E. (1978). Theoretical Risks and Tabular Asterisks: Sir Karl, Sir Ronald, and the Slow Progress of Soft Psychology. *Journal of Consulting and Clinical Psychology, 46*: 806–834.

Merton, R. K. (1938). Science, Technology and Society in Seventeenth-Century England. *Osiris, 4*: 360–632.

Merton, R. K. (1961). Singletons and Multiples in Scientific Discovery: A Chapter in the Sociology of Science. *Proceedings of the American Philosophical Society, 105*: 470–486.

Open Science Collaboration. (2015). Estimating the Reproducibility of Psychological Science. *Science*, 349 (6251) DOI: 10.1126/science.aac4716

Popper, K. (1959). *The Logic of Discovery*. New York: Basic Books.

Price, D. (1986). *Little Science, Big Science . . . and Beyond*. New York: Columbia University Press.

Roeckelein, J. E. (1997). Psychology among the Sciences: Comparisons of Numbers of Theories and Laws Cited in Textbooks. *Psychological Reports, 80*: 131–141.

Rosenthal, R. (1990). How Are We Doing in Soft Psychology? *American Psychologist, 45*: 775–777.

Schachter, S., Christenfeld, N., Ravina, B., and Bilous, R. (1991). Speech Disfluency and the Structure of Knowledge. *Journal of Personality and Social Psychology, 60*: 362–367.

Simonton, D. K. (1978). Independent Discovery in Science and Technology: A Closer Look at the Poisson Distribution. *Social Studies of Science, 8:* 521–532.

Simonton, D. K. (1979). Multiple Discovery and Invention: Zeitgeist, Genius, or Chance? *Journal of Personality and Social Psychology, 37*: 1603–1616.

Simonton, D. K. (1988). *Scientific Genius: A Psychology of Science*. Cambridge: Cambridge University Press.

Simonton, D. K. (2002). *Great Psychologists and Their Times: Scientific Insights into Psychology's History*. Washington, DC: APA Books.

Simonton, D. K. (2004a). *Creativity in Science: Chance, Logic, Genius, and Zeitgeist*. Cambridge: Cambridge University Press.

Simonton, D. K. (2004b). Psychology's Status as a Scientific Discipline: Its Empirical Placement within an Implicit Hierarchy of the Sciences. *Review of General Psychology*, *8*: 59–67.

Simonton, D. K. (2010). Creativity as Blind-Variation and Selective-Retention: Combinatorial Models of Exceptional Creativity. *Physics of Life Reviews*, *7*: 156–179.

Simonton, D. K. (2013). After Einstein: Scientific Genius Is Extinct. *Nature*, *493*: 602.

Simonton, D. K. (2015a). Psychology as a Science within Comte's Hypothesized Hierarchy: Empirical Investigations and Conceptual Implications. *Review of General Psychology*, *19*: 334–344.

Simonton, D. K. (2015b). "So We Meet Again!"—Replies to Gabora and Weisberg. *Psychology of Aesthetics, Creativity, and the Arts*, *9*: 25–34.

Simonton, D. K. (in press). Domain-General Creativity: On Producing Original, Useful, and Surprising Combinations. In J. C. Kaufman, J. Baer, and V. P. Glăveanu (Eds.), *Cambridge Handbook of Creativity across Different Domains*. New York: Cambridge University Press.

Smith, L. D., Best, L. A., Stubbs, D. A., Johnston, J., and Archibald, A. B. (2000). Scientific Graphs and the Hierarchy of the Sciences. *Social Studies of Science*, *30*: 73–94.

Suls, J., and Fletcher, B. (1983). Social Comparison in the Social and Physical Sciences: An Archival Study. *Journal of Personality and Social Psychology*, *44*: 575–580.

II What Pseudoscience Costs Society

5 Food-o-science Pseudoscience: The Weapons and Tactics in the War on Crop Biotechnology

Kevin M. Folta

Dinner, an Easy Target for Charlatans

The topic of food cooks up frequent morsels of pseudoscience—from fad diets to unfounded fears to age-old traditions that social archaeologists trace back to their ancient foundations in 1972. In recent times it is hard to find a more flaming insult to the scientific method than the loose claims centered on the topic of food. Within that space the tools of genetic engineering (GE; the scientifically precise term for what is familiarly referred to as "GMO") are among the most maligned of food technologies. The ability to cut and paste DNA sequences from one organism to another has been with us for almost half a century and is universally considered a revolutionary technology—and you ain't seen nothin' yet. We have been using such technologies to make drugs like insulin, enzymes for cheesemaking, and all kinds of compounds ranging from vitamins to nutritional supplements to chemotherapeutic agents. Even the most ardent activists do not express too much concern about such applications, as they expand healthy options that enrich our lives. In many cases they make life itself possible; not only that, they give us cheese. The world does not have enough slaughterhouse floor pancreases to meet the modern demand for insulin. Genetic engineering makes the availability of therapeutic drugs like insulin safer and more reliable. The decreased medical demand for pancreases probably keeps hotdog prices down too. -But apply GE technology to crops that produce food ingredients, and the conversation drastically shifts.

Most commercial applications of GE have been in the area of agronomic crops, where over eighteen million acres of GE plants are grown worldwide (Fernandez-Cornejo, Wechsler, and Milkove, 2016). As of today the major crops that end up in the human diet are field corn (not sweet corn, the other stuff fed to cows and car engines), canola (for canola oil), sugar beets (for sugar), and soybeans (National Academies of Sciences, 2016). The products derived from these GE crop plants end up in about 80 percent of grocery store products in North America. Essentially these ingredients represent the dusting of corn starch, sugar, or oil used in mass-produced foods. If you really think about it, although the plants are engineered to do something atypical (like resist insects or herbicides), the engineering process does not directly influence the end products that go into our mouths. While the plants are engineered, the products we purify from them are exactly identical to those from non-GE counterparts, except possibly for a nanogram speck of stowaway DNA or maybe the protein it encodes. The sugars, oils, and starches are chemically identical to those that were derived from non-GE plants. That is the majority of ingredients from GE plants in human food. Plenty of GE cotton is grown for fiber, and a rapidly expanding acreage of GE alfalfa is grown for cattle feed. In the United States there are a few GE plants that produce food for humans, such as sugar from sugar beets, and those products are identical to those from conventional or organic counterparts (except they cost less). There also is a tiny acreage of virus-resistant papaya in Hawaii and a few acres of virus-resistant squash around that nobody cares about, because people do not really like squash much. New products like safer, nonbrowning potatoes and nonbrowning apples have been approved and their parent companies are simply ramping up plant numbers before the products are widely available. Just in these first paragraphs a few myths have been busted.

Among scientists, there is remarkable consensus that these products are safe (Fahlgren et al., 2016; Nicolia, Manzo, Veronesi, and Rosellini, 2014), and they allow farmers to produce more food with fewer inputs (like insecticides), which translates into larger profits on the farm (Klümper and Qaim, 2014). Since their first implementation in 1996, there has not been a single case of illness or death related to these products in humans or other animals (Panchin and Tuzhikov, 2016; Van Eenennaam and Young, 2014), a brilliant safety record that underscores the success of careful design and stringent regulatory hurdles (Strauss and Sax, 2016).

Certainly science recognizes some limitations and even negative environmental impacts of overdependence on some GE strategies (Lombardo, Coppola, and Zelasco, 2016). The emerging problems of herbicide-resistant weeds and insect resistance to plant-produced insect controls are very real indeed. Those of us working in science when the products were released did not need crystal balls to make such bold predictions. Anytime in life that we rely on a single strategy to solve a problem we end up creating more problems, and herbicide and insect resistance to GE solutions is certainly a problem. With costly regulation's hands around the neck of progress, the three to four chemistries that should have been developed were not, leaving farmers with no easy way to rotate herbicides in a manner that would kill weeds yet leave crops unmolested.

The emergence of resistant weeds and insects is a residue of a well-financed pseudoscientific tsunami that has paralyzed new product development. When we create a paranoid development environment rife with manufactured risk, we oftentimes spawn new issues by deploying half-solutions. However, even with these shortcomings, the benefits of GE crops far outweigh the risks (Kathage and Qaim, 2012; Klümper and Qaim, 2014). Many environmental benefits have been made possible in light of the limitations (Brookes and Barfoot, 2015), and the promise of these technologies holds great potential to solve some of the world's most critical food security and environmental impact issues of growing food.

Despite the widespread adoption and paucity of harm, there is a burgeoning public controversy afoot (Blancke, Van Breusegem, De Jaeger, Braeckman, and Van Montagu, 2015). What would possibly lead people to ignore good data, double down on kludge evidence, rally around the words of Internet food celebs, and condemn the words of credentialed scientists? Manufactured risk. It is fomented from a vocal and well-financed movement sworn to ending the sustained use of truly sustainable technology. Extremists want to end any form of tinkering with plant genetics; they wish a return to the pastoral charm of isolating insulin from pancreases pulled from gut piles and human growth hormone extruded from buckets of cadaver pituitaries—the stuff granddad waxed fondly about. Others hold anticorporate or antiquated farming agendas, and want the use of GE expunged from the farmer's field even though they enjoy its benefits with a Coke at Chipotle or a wedge of cheddar made with enzymes from GE microbes. Condemning GE's use and enjoying its utility are incompatible ideas but consistent themes in the anti-GE movement.

Objection to new technology is not unnatural. Humans have shown time and time again that the emergence of a novel and powerful breakthrough frequently travels with the gremlin of ill-placed skepticism. Major paradigm shifts, from the Earth traveling around the sun to human conception in a petri dish, have always broken scientific harmony with a sour note of dissent. In the modern day this pushback against progress manifests as pseudoscientific claims that oftentimes are difficult to decipher from actual science. Examination of the literature over the last twenty years uncovers some startling trends. Politically motivated "sortascience" and the emerging trend of predatory publishing abuses (see Beall, this volume) allow soft claims to creep into scientifically weak publication venues that appear legitimate. Consumers and casual readers are challenged to separate actual science from something sloppy that looks science-shiny. The Internet is no friend to that foggy clarity. A well-orchestrated and typically well-financed manipulation of Internet space amplifies pessimistic messages of gloom, translating illegitimate science into a format for wide distribution and credulous consumption. This is why consumers are in the confusing place they are today, and why so many are concerned about their food.

The purpose of this chapter is to highlight the flimsy science that pollutes the scientific discussion of genetic engineering, a technology that unfortunately shared a birth and toddlership with the Internet. These were two good technologies, poised to change the world. But today one is used to malign the other, and their relationship is only becoming more contentious as they transition into adolescence. The chapter will demonstrate the role of poorly designed science, lofty misinterpretations, or even deliberate misrepresentation of legitimate science to bolster an ideological or political food, or anticorporate agenda. Discussion of blatantly bogus claims and the effects of the pseudoscience around agricultural biotechnology will be presented. Ultimately the thesis of this work is that GE is a powerful technology that, like any technology, has the ability to do many good things. It certainly could be used for malevolent acts if one was so inclined. Like any technology it has things it does well and things it could do better. It is not magic, it is not doom, and it is not a panacea. However, we are over twenty years into its widespread deployment. Where science should be dominating the discussion of future application, crank pseudoscience in the popular press clouds the public discussion of risk and benefit and slows deployment of potentially useful technologies, with profound effects on those in desperate need. We miss an opportunity to soothe the needs of farmers, the environment, and those frozen in poverty, as well as the affluent industrialized world consumer. Time will show that the pseudoscience used to conjure risk around this sound technology has left a considerable body count and unfortunate environmental footprint.

The Danger of Soft Sortascience

Opponents of biotechnology will claim that the technology is simply a vehicle for a handful of multinational corporations to obtain great wealth and ultimately control the food supply. However, these technologies have actually been shown to principally benefit farmers (Areal, Riesgo, and Rodriguez-Cerezo, 2013; Brookes and Barfoot, 2014), and have been extremely valuable in the developing world (Azadi et al., 2015; Azadi et al., 2016). Genetically engineered papaya is a splendid example of how these technologies can help farmers and the environment, with the most conspicuous benefit being saving the productivity of a small local industry (Gonsalves, Lee, and Gonsalves, 2004). GE eggplant seeds are freely distributed among farmers in Bangladesh and change hands royalty-free every single day (Giri and Tyagi, 2016). Promises from the laboratory show potential solutions for the developing world, where supplying a single nutrient may deter blindness and death (Beyer et al., 2002). A small genetic tweak can (well, in theory; the technology has not been deployed yet) supply micronutrients needed for healthy babies, normal neural development, and general health (Kiekens et al., 2015; Naqvi et al., 2009; Saini, Nile, and Keum, 2016). These applications of biotechnology innovation are not pie-in-the-sky dreams. They are proven technologies that exist and have been tested—and shelved. Shelved on a high shelf of overblown precaution with no step stool in sight.

Their hindered deployment drags an anchor on public sentiment, an anchor attached by the "good guys" in the well-fed industrialized world. Some of these folks would rather construct an imaginary demon before allowing technology to serve one desperate child. This pseudoscience has a body count, and therein lies the supreme irony. Those claiming harm where none exists are perpetuating harm where it does exist. Those standing up and proclaiming themselves to be champions of people are actually the ones killing them. The soft or false claims issued by the anti-biotech movement impede the dissemination of sound technologies, which prompts the necessity for this chapter.

Human Misery Spawns Agriculture

Humans in the most industrialized world have enjoyed ample access to food truly only in the last century. In fact, 99.9 percent of *Homo sapiens* history teetered at the edge of starvation, spending the wealth of work and leisure time foraging for roots, tender young leaves, grubs, and other critters for basic sustenance. Reality probably looked like small bands of folks wandering from spot to spot, scrounging for calories, and not so much loinclothed warriors spearing mammoths and enjoying saber-toothed tiger jerky. If you were lucky you would survive to reproduce, as some of the primitive human's daily diet discoveries unquestionably contained a good dose of natural toxic compounds to complement the meager nutritional gains. Food trends were defined by the products that did not kill us or make us ill, and produced some edible bits throughout the year. At some point about twenty thousand years ago, humans realized that they could override nature's sad plan and took plants into their control. Plants were moved from where nature put them to where humans needed them. Traits that nature found valuable were forgotten as humans selected and propagated the plants bearing traits they found most useful. Human intervention into manipulating plant genetics was born.

Since then, humans have redistributed plants from where evolution put them and repatriated them throughout the world. It is all relatively recent history. Science minds like Charles Darwin, Nikolai Vavilov, and others examined the evidence of plant domestication, and made the observation that food crops tended to originate around population centers (Gepts, 2010). The implication is that humans made plants better. Whether they realized it or not, just by clearing land, inadvertently dropping seeds, or scattering garbage, early human civilizations began impinging on the natural order of plant evolution. They were the first to modify plant genetics.

Another piece of socially accepted sortascience comes from the notion of what is natural. The term "natural" is meaningless, yet is touted as a term of piety, defining a product as somehow superior to peer consumer products. Mother Nature gave humans some rather ratty raw materials. Our antecedents did not find beefsteak tomatoes, Cavendish bananas, and big ears of corn growing in organic urban local plots. Instead, nature's offerings were pretty rough. Corn's distant forefodders were little hard beads of grain on a plant called *teosinte*, occurring only in a small region of modern-day Mexico (Benz, 2001). Tomatoes are nightshades, and the fruits were usually green and growing on toxic vines in the Andes Mountains and Galapagos Islands (Pease, Haak, Hahn, and Moyle, 2016). Bananas were similar to today's dessert banana, only small and full of rock-hard seeds and found mostly in Southeast Asia (Heslop-Harrison and

Schwarzacher, 2007). In short, nothing you eat today is identical to its *natural* form. Everything you eat has been changed genetically (dare we say genetically modified?) by humans. Everything.

Even if you swore only to eat today's major food crops in their natural state, you would have to leave North America and most locations outside of equatorial and subtropical regions. Virtually all major fruits and vegetables in the grocery store produce section have origins scattered over a narrow range of the globe. Broccoli, trendy kales, and the rest of the brassicas are from the Mediterranean. Tomatoes, potatoes, and peanuts are from South America. Squash, beans, and corn are from modern-day Mexico. Citrus species radiated from Southeast Asia. Apples are from Kazakhstan, strawberries from China. Only a few crops, like sunflowers, some brassicas, blueberry, and a few others, can call North America a genetic home. Not only have humans changed plant genetics, we have also affected ecology, moving plants from the small regions where they originated to farms where they arguably do not belong.

The pseudoscience and abundant misunderstandings around food origins are based on the false notion that food and nutrition were once magically better. This misconception defines the backdrop by which modern technologies are judged. Basically, the critics of modern genetics and production techniques have a screwy denominator. Longing for a concept that never existed drives today's disdain for modern food technology.

Directed Human Intervention in Genetic Improvement

Only in the last century, and really only in the last fifty years, did plant genetic improvement put the petals to the metal. Deliberate derivation of commercial agronomic hybrid plants made its debut in the 1920s with the introduction of hybrid corn. Throughout the 1800's corn production held at fifteen to twenty bushels per acre. Little was known about the genetics of the varieties used, and keeping the same seed from self-fertilizing plants led to a process called *inbreeding depression*. Inbreeding depression is a phenomenon where constant cycles of self-fertilization result in a loss of genetic variability within a genetic lineage. The canine mutt has its resilience because of a rangy genetic composition. Breeding against direct relatives makes the genetic foundation thinner, and mutt-ism is lost to genetic homogeneity, resulting in more uniform offspring with less vigor. Compared with mixed-breed dogs, the purebreds exhibit a higher incidence of several disorders related to heart disease and epilepsy (Bellumori et al., 2013). The same thing happens to plants. Not heart disease and epilepsy, but other more planty health issues.

The corn lines of the day came from open-pollinated fields. While there were hundreds of popular varieties adapted to various regions, they were not productive; they exhibited poor resistance to disease and low product quality. However, hybrid corn was developed by depositing the pollen of one inbred on the silks of another, and something remarkable happened—the next generation grew with tremendous vigor, reflecting the high degree of genetic variability in its genome. These strong hybrids, resulting from crosses of weenie inbreds, were remarkably uniform and outproduced any of the best open-pollinated varieties of the day.

But when farmers tried to save the seeds from hybrid plants, genetics kicked in and wrecked the party. In the next generation some corn looked like one inbred parent, other corn resembled the other parent, and a crazy range of other corn plants fell in between. This crop was useless because the strong characters of the hybrid were lost to another generation of genetic shuffling, and the plants were not uniform.

An industry was born. Hybrid corn quickly doubled yields, so it made economic sense to purchase those seeds year after year (Griliches, 1958). Companies sprang up, using their best inbred lines to create hybrid seed that could be annually sold to farmers. This is the genesis of the position that modern seed companies are somehow immoral because they require farmers to buy seeds every year. However, companies only profit as much as they satisfy the high demands of farmers. Continual, expensive improvement of inbreds and resulting hybrids is an endless pursuit, as the foundational plant's performance is essential, long before genetic engineering ever enters the picture.

The Atomic Age brought new ways to wreck DNA. Radiation sources were implemented in plant genetic improvement, as sublethal doses of radiation could cause breaks in DNA, wild chromosome rearrangements, and other genetic alterations that occasionally led to useful traits. Plants were grown in "atomic gardens" adjacent to a radiation source, or perhaps seeds were bombarded with radiation. Many of these radiation-induced changes from long ago reside in the nuclei of today's elite crop plants, from barley to wheat to citrus products. What changes did the radiation make? Who knows? Nobody really cared—it was the magical age of atomic energy and genetic improvement. Every once in a long time a seed would survive the bombardment and produce a plant with a highly favored trait. There was no way to identify where the genetic changes took place or how many additional collateral alterations came along as baggage. The bottom line was that the breeder, the farmer, and the consumer got the trait they wanted. Mission accomplished.

Chemicals were also used to induce mutations in DNA, similar to the random damage radiation does. But the neatest trick in genetic improvement came from polyploidization, or the process of doubling, quadrupling, or x-tupling chromosomes (Stebbins, 1955, 1956; Veilleux, 1985). Cells divide through the process called mitosis, with chromosomes doubling in one cell before splitting into a pair of equal daughter cells. The process requires tiny molecular ropes called "spindle fibers" to attach to chromosomes and then pull them into polar opposite ends of cellular real estate, so when cells divide they each get the full complement (Picket-Heaps, Tippit, and Porter, 1982). Scientists then discovered that the chemical colchicine disrupts spindle fiber formation. When plant tissues were treated with these compounds, genetic material would fail to split evenly between two new compartments, leaving all of it in one cell and none of it in the other (Elgsti and Dustin, 1955). This new chromosome-heavy cell was pretty happy and could be regenerated into a whole new plant with double the chromosomes. Polyploids were a huge score for plant genetic improvement, as polyploid plants, flowers, and fruits were typically bigger and more vigorous than their simpler genetic forms (Stebbins, 1940). They also would be the basis of creating new varieties like seedless watermelons (Oh et al., 2015). Again, a chemically induced DNA freakout—too many chromosomes. Yet the world greeted this with open arms and open mouths.

These kinds of crude techniques were the only early methods that plant breeders could use to control genetic variation. If they wanted a new trait, they could make it with random damage, or they had to go find it in a wild plant and hope that it could be introduced via plant sex, followed by selection, and then years of more plant sex. The advent of molecular biology and gene transfer techniques opened a world where plant genetic improvement could take place with great precision and speed. Finally a needed and understood trait could be integrated into just about any plant background in months rather than years.

The first traits were engineered into tobacco, with an eye on traits that could hook kids earlier and make people smoke more. No, not really. Tobacco is remarkably simple to transform (to add genes) and fast to grow, so it was a great surrogate for the first engineered genes. But soon the technologies ended up in other crops, mostly just large-scale agricultural crops that were engineered to be resistant to insects or herbicides. A minor acreage of papaya and squash was planted that was engineered to resist viruses. Farmers found these technologies useful, since the herbicides were safer and cheaper and they could use fewer insecticide treatments (Klümper and Qaim, 2014), translating to savings in fuel, time, and labor.

The details of what was genetically engineered and how it was engineered are discussed in great detail elsewhere. The purpose of this prelude is to dispel one of the central misunderstandings of crop genetic improvement. As figure 5.1 shows us, genetic engineering is by far the most predictable, precise, and least-invasive method of genetic change. Compared with homogenization of disparate genomes from plant breeding, the use of x-rays and chemicals to whack chromosomes, or other techniques that cause whole-scale doubling of genetic materials, the modern tools of molecular breeding and transgenic crop development are the most precise, most understood, and also the most feared.

The Food-o-scientist Toolbox

How do enemies of modern farming and crop technology achieve their ends? One of the central tools is a well-contrived series of arguments that are heavy in scientific misgivings and logical fallacy. They are promoted by a familiar merry-go-round of personalities and organizations. Some of these organizations sport grand web-based facades

	Traditional Breeding	Hybrids	Polyploids	Mutation Breeding	Transgenic (GMO)	Gene Editing
Examples	Many	Field corn, some tomatoes	Strawberries, sugar cane, bananas	Pears, apples, barley, rice, mint	Corn, canola, soy, sugar beets, papaya	Many coming!
Number of genes affected	30,000 - >50,000	30,000 - >50,000	30,000 - >50,000 (times number of genomes)	No way to assess	1-3	1
Occurs in Nature?	Yes	Yes	Yes	Yes	Some examples	No
Knowledge of genes affected	No	No	No	No	Yes	Yes
Environmental assessment?	No	No	No	No	Yes	Undetermined
Tested before marketed?	No	No	No	No	Extensively	Undetermined
Organic acceptable?	Yes	Yes	Yes	Yes	No	Undetermined, probably no.
Label wanted?	No	No	No	No	Yes	Undetermined
Adverse effects?	Yes	Yes	?	?	No	Not marketed
Time to new variety?	5-50 years	5-30 years	> 5 years	> 5 years	< 5 years	< 5 years

Figure 5.1

The Frankenfood Paradox. Techniques of plant genetic improvement that are less invasive and more predictable, and that affect the fewest genes, are the ones that bother people most. Ranging methods with unpredictable outcomes are well accepted.

screaming for public concern, when they really are a tiny office in rural Iowa. Some are run by someone selling books and documentaries, using the sham organization as a front for a lucrative pseudoscience empire. Others exploit legitimate scientific training or academic affiliations to provide ivory tower credibility to flimsy scientific claims. The following section documents just a few of the main ways that these dubious people and organizations exploit myth and fallacy, tailored to tweak the noncritical thinker, or even people with a zero-tolerance approach to food issues.

Confusing Correlation and Causation

Just because one thing happens before the other, does that mean it causes the other? One of the most common vehicles used to deliver genetic engineering pseudoscience is confusing correlation with causality. Numerous authors and activist claims have keyed off the fact that the last twenty years have seen tremendous advances in technology, including the widespread cultivation of genetically engineered crops. The rapid adoption of genetically engineered soy, corn, canola, and sugar beets by farmers, and their emergence in 80 percent of grocery store foods, has led many to fall for the fallacy that these products must be linked to modern maladies. It is a classic case of logical leaps that con the malleable mind into a position of fear, with the hope of further propagating the misinterpreted information.

The perception is that diseases like cancer, Parkinson's, and Alzheimer's are increasing, and that the perceived increases match the clear rise in allergies and diagnosis of autism. Of course, people are not dying of heart attacks and pneumonia like they used to, so that opens the gate for higher incidence of the luxury of long-term degenerative or developmental diseases (Hoyert, 2012). These differences in disease incidence, real or perceived, are the basis for conflating GE crop adoption as their cause.

The confusion of correlation and causation is the root of insane cherry-picking expeditions that selectively harvest nuggets from the expert and crank literature. Tenuous associations are drawn, and then propagated with the sole intention of creating doubt and generating fear. Such actions, performed by people that are formally trained in science, and under the name of a prestigious institution, really test the edge of ethics to achieve a science-free political outcome.

Twisting Real Science to Fit a Corrupt Agenda
If you do not like the data, are there ways to massage them into something more compatible with your deeply held beliefs? Such egregious offenses are common in pseudoscientific circles as they warp legitimate research to fit a predetermined agenda. Certainly real scientists are not allowed to do this, as interpretations and conclusions should be built on properly interpreted data from good experiments with proper design. But what if the good data inconveniently do not match the activist-desired outcome? There are many examples where good research was performed and came to solid conclusions that reflected well-designed experiments, good controls, and rigorous execution. These works were performed by reputable scientists and were published in decent journals.

But what if real data fail to reinforce a belief? There are many flaming examples where a legitimate paper with solid conclusions was *reinterpreted* in the activist media with a crooked slant, arriving at completely different results. It is like looking at the data that show the Earth goes around the sun, but claiming that the Earth still is the center of the universe. In short, it is falsely reinterpreting science to fit an agenda, in defiance of what the data and the authors say.

While some studies performed in petri dishes appear to be set up to produce the results the authors are primed to overinterpret, other in vitro work is done with beautifully sound methods, great rigor, and interpretations within the bounds of the results. However, these reports are prone to fresh evaluation through a tainted lens, as even good science goes bad when insidious efforts bend mundane data into untold tomes of danger.

A great example is work by Dr. Fiona Young, an endocrine toxicologist in Australia. Dr. Young and colleagues (Young, Ho, Glynn, and Edwards, 2015) tested immortalized placental cells for the production of estrogen in response to glyphosate, the active ingredient in herbicides used with some GE crops. Over the past few years, as GE crops have proved safe and reliable, those opposed to the crops started to fabricate claims about the chemicals used in association with them. Chemicals are scary.

In this set of experiments, Dr. Young's group tested both the solo chemical and one of its many commercial preparations, in this case one known as Roundup. The experiments were rather simple: grow placental cells in culture and provide increasing amounts of herbicide. In such an experiment researchers expect to find what is called a dose-response relationship; that is, as more of the compound is applied, the more toxic effect should be seen. In the confines of a glass dish that is true with herbicides, drugs, sugar or distilled water. In this case, the effects measured were cell viability and detectable levels of estrogen in response to the herbicide. The experiment showed that you can add glyphosate to the culture medium to a hundred times field-applied levels, and over 720,000 times higher than levels found on raw produce (if it is there at all) and there is no effect.

However, if you use Roundup—which is glyphosate plus a surfactant (a chemical with detergent-like properties that helps the active ingredient penetrate the tough leathery leaves)—at its full weed-killing concentration, the cells eventually stop producing hormone. The smoking gun? Not really. The cells that did not produce hormone were already dead, so yes, death is the endocrine disruptor. Glyphosate? Not so much. In this scenario the authors interpreted the lack of estrogen as being due to dead cells, and cells were killed by high levels of soapy surfactants in the herbicide mixture. These detergent-like compounds make short work of the biological scum layer in a petri dish. Membranes do not hold up well in their presence, and that leads to dead cells that do not do squat. The authors correctly concluded that dead cells do not produce estrogen.

However, the title of the work was *Endocrine Disruption and Cytotoxicity of Glyphosate and Roundup in Human JAr Cells In Vitro*. Within days the Internet exploded with claims that glyphosate was an endocrine disruptor and was causing all kinds of health havoc (Global Research News, 2015; Rowlands, 2015). The anti-GE websites exploded with news that glyphosate was shown to be an endocrine disruptor, even when the conclusions clearly showed otherwise. They did not read past the title, and interpreted it in a manner that confirmed their biases.

This is a common theme. Real scientists do good experiments and get solid results that are interpreted correctly in their own work, yet they are later intentionally misinterpreted by those that use the Internet's fuzzy science pipeline to promote the outcomes critics want to find.

One-Off Reports and DOA Science
Breakthroughs are rare, and when they finally happen two research races begin. First, the lab with the breakthrough mines every conceivable fast extension of the new findings. Once published, there will be a mad race to be number two, birthing intense new

competition to extend the findings and apply new knowledge to other questions. It is a common theme in science. The lab that publishes the work starts the next tangents and lays the foundations for the following papers. Grant proposals are written; preliminary data are gathered. Upon publication other laboratories with related interests spring to life, hoping to contribute their talents and expertise to extending the new discovery. Within a year's time the primary findings expand in many directions, opening new avenues at the speed of research. This is the normal course of legitimate scientific discovery.

What about the highly celebrated "breakthroughs" showing evidence of harm from GE crops? These reports make a splash and then quickly circle the drain, only to disappear with no follow-through. These are what are referred to as one-off studies. They are typically statistically underpowered, and they feature conclusions that overstep the data that they do actually generate. Most of all, they are the end of the line. There are no further examinations, no follow-up, and while activists parade the heroic findings around the Internet, defending them against any scientific interrogation, the scientific community reads critically with an open mind, rolls its eyes, and moves along with real science.

One thing these one-off studies all share is that they disappear in science's rearview mirror, surviving only from self-citation and eternal content of activist websites. There are some super examples, most of which are held up as high evidence and outstanding research by anti-GE folks. However, the activist community claims that the work was never reproduced or continued due to a scientific conspiratorial cabal that crushed any chance of repetition.

Perhaps the supreme example of a one-off report was the 2012 work by Séralini and colleagues (Séralini et al., 2012), the work known among scientists as the Lumpy Rat Paper. While some of the experiments in his papers over the years are done well and have merit, they are frequently overinterpreted to lofty levels that somehow evade peer review. In the 2012 work, the experimental design was to feed rats GE-based rat food, non-GE rat food, Roundup in the drinking water, and a combination of both. After two years rats subjected to these experimental treatments were covered in massive tumors, and they were paraded around the Internet as grotesquely disfigured warnings of GE technology.

Scientists reacted harshly. First, the rats were clearly suffering and looked more like socks stuffed with billiard balls than living mammals. Next, these were Sprague-Dawley rats, a rat lineage that spontaneously develops tumors; high numbers of the rats will have tumors after two years regardless of diet (Prejean et al., 1973). But the real kicker was that the control rats grew tumors too (as shown in the study's table 2), with the lumpy control animals curiously being omitted from figure 3 of the work. It is hard to fathom how that ever passed peer review.

Many scientists have criticized the interpretations and experimental design, so we do not need to do that here. It is more appropriate to put this into the context of the race to be number two as an indicator of quality science. When Séralini's paper broke, the best cancer labs did not spring to work hoping to expand the new link between 70 percent of the United States' food and tumors. Scientists did not take it seriously. Researchers and farmers raising animals fed almost 100 percent GE feed must have been confused, as their animals were doing just fine (Van Eenennaam and Young, 2014). Even after five years Séralini's lab had never taken the work to the next level, never attained funding for their breakthrough discovery and reproducing the work with tremendous rigor, massive numbers, and statistical relevance. That is what would have happened on the heels of a real breakthrough.

And, of course, the logical follow-up experiment connects a phenomenon observed to presumed causes mechanistically. In other words, if GE feed causes tumors, how does it do it? If Roundup causes tumors, how does it do it? What are the *molecular mechanisms*, well established in rat models, that are directly affected by these treatments? Given that the rat is at the center of understanding cancer development, how do these new claims mesh with known models? The authors show no evidence of deeper investigation, and that is inconsistent with how legitimate research proceeds.

And is it not kind of weird that both the food and the herbicide were alleged to cause the exact same problem? That is what scientists asked, and the follow-up should have been published hot on the skinny tail of the Lumpy Rat Paper. The scientific community yawned, and to this day the lab that did the work never expanded the study. All in all, damning evidence that this one-off work likely holds little scholarly merit, despite the credence it receives from activists.

Séralini's work is perhaps the most famous of the one-off papers that pollute the genetic engineering conversation. The works represent unsinkable rubber duckies of pseudoscience, held in deep esteem by the activist communities, but eschewed by legitimate scientists due to their inherent flaws and lack of reproducibility. What are some other widely celebrated findings that never go away? How about pesticides in umbilical cords (Aris and Leblanc, 2011)? Irritation of pig stomachs (Carman et al., 2013)? Formaldehyde in corn and soybeans (Ayyadurai and Deonikar, 2015)? All of these topics are based on the flimsiest fantasy evidence, and all share a common plot in the graveyard of one-off research that accumulates more headstones with time.

The race to be number two is a brilliant predictor of scientific reality. High-quality science with profound impacts grows. Great work inspires new investigations, some well under way long before the primary reports are published. The reports discussed here slowly pass into irrelevance, touted only by those so sworn to believe the conclusions that they fail to even consider the important shortcomings.

Overstepping the Results

Another major hallmark of flawed scientific work is when researchers publish interpretations that exceed what really may be learned from the data. Every study has experimental limitations based on the design, and statistics may only be interpreted so far.

Researchers are usually extremely careful to not venture beyond those limits when discussing the work or its implications. Good editors and reviewers are typically quick to point out when authors have made such errors.

However, overstepping the data is an all-too-common theme in pseudoscientific claims, and occasionally sneaks past editorial oversight in good journals. In other cases, predatory publishers allow overstepping and perhaps even encourage it. It happens on at least two levels. One is when the authors fail to abide by the technical restrictions of the experimental design or limitations of the data. Despite peer review that typically filters such occurrences, such work does occasionally squeak into publication, even in legitimate journals. Statistics is hard but p-hacking is easy. The other common abuse is that when scientific work is published, it is frequently misrepresented in hyperbolic ways. In these cases a pseudoscientific network, either deliberately or out of ignorance, extends findings to new extremes, thereby popularizing bad information that supports a particular bias. In the following examples there are cases where good science is stretched to support a particular view, and cases where limited data are tortured into new dimensions that fit and reinforce an ideological point.

We commonly see this happening with experiments performed in vitro. What happens in the petri dish stays in the petri dish, at least in terms of how far we can legitimately interpret results. Cells growing in a nutrient soup are fragile critters in most cases; that is why they are an excellent system to start to explore mechanisms associated with chemically induced dystrophy. When you subject them to perturbations or insults, they react. It is a great starting point to begin exploring the fudge between chemical insults and biology. However, to the anti-GE world, findings based on the reaction of cells bathed in relatively benign chemistry are oftentimes held up as high evidence of danger in the whole organism. In the genetic engineering world, in vitro reports abound, and oftentimes the authors use these simple tests to tie chemical products to dreaded human diseases or disorders. It is a pretty big leap. Killing brain cell lines in a dish with high doses of a drug does not mean that the drug causes Parkinson's disease. Killing testicular cells in a dish does not mean the product is harming world fertility, yet that is the claim that was implied (Clair, Mesnage, Travert, and Séralini, 2012). To take this to the extreme, I can kill the most aggressive cancer cells in a dish with a hammer. It does not mean hammers cure cancer. When we examine the anti-GE literature we can find many examples of such tenuous conclusions made comfortably and with great authority.

There are several important caveats in interpreting in vitro data. First, as anyone that has attempted cell culture can tell you, mammalian cells are not simple to work with. Minor deviations in media, changes in environment, and contamination by fungal fuzzies are constant battles that can significantly affect outcomes. Therefore, as cells are incubated with increasing amounts of any compound, there likely is going to be some effect to observe. It does not matter if cells are incubated with herbicides, food additives, window cleaner, distilled water, or pesticides used in organic crop production—eventually effect on the culture will be observed. How that relates to a biological system lies firmly on the researcher, and how they wish to extrapolate cell goo behaviors to whole organisms is more a barometer of ethics than hard evidence of toxicity.

Furthermore, in cases where biologically relevant levels induce reproducible and clear results, there is another important question. Many studies in this area examine effects on hormone production or tumorogenic potential, so they study cell lines derived from the placenta, testes, or breast tissue. In the case of the anti-GE literature much of the work examines the effects of glyphosate and its holoformulation Roundup. These studies on cell lines operate in a vacuum that ignores the biological system as a whole. In other words, a thin film of goo in a dish is not people. The pharmacological fate of glyphosate in the body is well understood, with the vast majority moving through the body and eliminated in the urine, with some in the stool. The remainder is metabolized in the liver by a specific set of enzymes called cytochromes. With this reality, how much of the active ingredient would possibly be available to cells in the testes, placenta, or breast tissues? The amounts would be extremely low, and in these precautionary reports the levels are never measured. However, that does not preclude wild speculation about the dangers of the herbicide and its effects on humans and human populations.

Should we automatically discount all in vitro data? Absolutely not. Such studies are easily performed and provide important primary data that allow researchers to begin honing the scope of the work. For instance, if you want to know how toxic a compound might be, or if it has adverse effects on cell viability or gene expression, an in vitro test is a great place to start. However, it is only a start. Extending the findings to the whole organism requires much more evidence, evidence that never seems to surface.

Dismissing the Experts

The ad hominem argument is the last and most desperate weapon in defending soft-science conclusions. When technology opponents fail to possess real evidence, and when they are confronted with legitimate questions about techniques, statistical rigor, or interpretations, the response is not to satisfy the inquiry with sound defense of the science. Instead, promoters lash out at critics as paid agents of big agricultural companies. Like clockwork. I will not name any companies specifically, as industry consolidation is rampant; next we will have one big company, MonDuDowBayGento-SF. You need not venture further than the comments section of any agricultural news article or any article penned by a science scholar for popular press. The topic is covered here only to point out its prevalence, as it will be the subject of much forthcoming work.

Ironically, learned scholars with lifelong devotion to research and education are recast as malleable industry pawns. At the same time the lucrative haunts of websites have been a nursery for nonexpert "experts" that gain massive followings and celebrity gravitas simply by confirming the false claims throngs of victims wish to hear. From ballroom dancers to self-appointed nutrition experts to TV doctors to popular chefs to Internet quacks to yogic flying instructors, the Internet hosts many personalities that are prepared to rescue the credulous from actual scientists that know stuff. These dirty conduits serve to propagate pseudoscience through its many channels, with twisted views on everything from farming to food to fitness to medicine. Whereas a few activists posing as scientists create the manure, the Internet's pseudoscience dealers are quick to spread it, trading their bandwidth and fan bases for cash to build bogus empires that misguide many and harm the poorest among us.

Specific Examples of Food-o-science

Huber's Molecular Bigfoot

In January 2011, a letter slid across the desk of Agriculture Secretary Tom Vilsack. It bore the title, "Researcher: Roundup or Roundup-Ready Crops May Be Causing Animal Miscarriages and Infertility." The letter then magically appeared on countless anti-genetic-engineering websites, allegedly leaked from Vilsack's office. The letter can be read on the Farm and Ranch Freedom website (www.farmandranchfreedom.org) where it resides in aging glory. The letter comes from the desk of a former academic scientist with credentials as long as your arm. This is serious cause for concern. Maybe.

Here is an excerpt from that letter, as it appears on many websites. Remember, this letter came from a well-established scientist and was sent to the agriculture secretary of the United States. This is pretty serious stuff:

Dear Secretary Vilsack,

A team of senior plant and animal scientists have recently brought to my attention the discovery of an electron microscopic pathogen that appears to significantly impact the health of plants, animals, and probably human beings. Based on a review of the data, it is widespread, very serious, and is in much higher concentrations in Roundup Ready (RR) soybeans and corn—suggesting a link with the RR gene or more likely the presence of Roundup. This organism appears NEW to science!

This is highly sensitive information that could result in a collapse of U.S. soy and corn export markets and significant disruption of domestic food and feed supplies. On the other hand, this new organism may already be responsible for significant harm (see below) . . .

This previously unknown organism is only visible under an electron microscope (36,000X), with an approximate size range equal to a medium size virus. It is able to reproduce and appears to be a micro-fungal-like organism. If so, it would be the first such micro-fungus ever identified. There is strong evidence that this infectious agent promotes diseases of both plants and mammals, which is very rare.

It is found in high concentrations in Roundup Ready soybean meal and corn, distillers meal, fermentation feed products, pig stomach contents, and pig and cattle placentas.

Laboratory tests have confirmed the presence of this organism in a wide variety of livestock that have experienced spontaneous abortions and infertility. Preliminary results from ongoing research have also been able to reproduce abortions in a clinical setting.

The pathogen may explain the escalating frequency of infertility and spontaneous abortions over the past few years in US cattle, dairy, swine, and horse operations.

I have studied plant pathogens for more than 50 years. We are now seeing an unprecedented trend of increasing plant and animal diseases and disorders. This pathogen may be instrumental to understanding and solving this problem. It deserves immediate attention with significant resources to avoid a general collapse of our critical agricultural infrastructure.

Yikes! If it sounds like an emergency to you, just jump on YouTube and search for his videos. As a credentialed scientist, Huber provides a compelling, emotional tale of the dangers of transgenic crops, along with the herbicides used to grow them. He has credibility, he has grandfatherly charm, and he certainly knows his stuff.

But the letter is strangely devoid of some key details. Who are the collaborators? Why does the letter show no data? Why are there no citations? Why are the remedies simply to stop using genetically engineered crops, as opposed to working with the Centers for Disease Control and Prevention (CDC) to immediately quarantine infected plants, cattle, and people? The sky is allegedly falling and Dr. Huber wants to clean the gutters.

The only evidence presented was fuzzy images from electron microscopy. Like Bigfoot, this critter that Huber claims to be ubiquitous has somehow eluded any other scientific detection. Despite all of the genomes sequenced, all of the proteomes (the entire suite of proteins in the cell) studied, nobody has found evidence of this prevalent creature that Huber claims causes almost every human disease, kills plants, and causes abortions in cattle.

To most of us this seemed like an odd ruse. Huber has been around. Any scientist working in plant pathology can play six-degrees-of-Don-Huber and connect every time. I know people that have authored books with him, and some that had him as a teacher. They liked him fine. But others are not as impressed. Former colleagues at Purdue University responded with public skepticism (Camberato et al., 2011). Most scientists, as scientists do, simply shrugged the whole thing off as a wild claim in the absence of any evidence and offered much more elegant and scientifically plausible ways to explain the observations.

But the anti-GMO warriors continue to be sold that Huber's magical new class of "virus-like micro-fungus" is killing them, killing plants, and causing spontaneous abortions in livestock (Murphy, 2011; Warner, 2011).

On November 12, 2013, I attended a talk by Dr. Huber, eager to hear about his claims. I sat in a room with about thirty people, most of them interested in Huber's call for shunning the products of Big Ag, especially agricultural chemicals. Dr. Huber was an excellent speaker, clearly seasoned from years of work in the area and thousands of talks delivered. He did a skillful presentation and knew how to work an audience.

I watched his slides and took my notes, pages of them. I found data being misrepresented, claims made without evidence. And with each claim, the anti-GMO-heavy audience gasped in disgust. The images of the aborted calves drew palpable reactions. This "micro-fungus" was one badass pathogen, and Monsanto was 100 percent to blame.

At the end of the talk the emcee of the evening, Marty Mesh, the leader of Florida Organic Growers, took the microphone. Marty is a good guy, works hard, and is deeply committed to sustainable food production. I love what he does for organic agriculture in my state and support his efforts in many ways. Marty certainly was aware of my background as a scientist and my criticism of Huber's pseudoscientific, cryptozoological claims.

Marty started the question and answer period. He recognized me in the audience and said, "I know you probably disagree with everything he presented, but can you please maybe start with just one question?"

I replied, "I don't have a question, but I do have an offer. I was the contributing author on the strawberry genome sequence, and my lab does that kind of bioinformatics work all the time. I would like to offer Dr. Huber the opportunity to determine the exact identity of his organism. I just need a bit of the culture. All credit goes to him. I'm glad to do the work at my expense, we'll make all data public, and most of all, we solve this crisis."

Like a volley over the net in a tennis match, all heads turned in unison away from me, and over to Huber. He was clearly rattled. He stammered and stumbled, verbally meandering and not answering the question. Eventually the rationale to reject my kind offer boiled down to "The organism has no DNA, so there's nothing to sequence," which ultimately does sort of poop the genome sequencing party.

When I asked him about his collaborators he said that they all want to remain secret, and that they were in Austria and China. Great, a guy who has held positions integrating with national security is now shipping a deadly pathogen to China. But he will not share it with a fellow scientist here in the States.

After the session I went to speak with him privately, but his local handler shuffled him rapidly toward the door. I reached out my hand to shake his, and asked him to collaborate on solving the crisis.

"Go isolate it and do it yourself!" he said using his *get off my grass* voice.

Of course, with no published protocols it would be difficult to isolate. And, with no hard biological description, how would I know if I did isolate it? I really did want to give him the benefit of the doubt. As a scientist, participating in the isolation and characterization of a new life-form is certainly intriguing, and while my rational brain thought the idea was silly, the cosmic wonderer in me always hopes to find a rule breaker.

However, since I could not round up a culture of the organism, had no idea of what it looks like or how to culture it, and was provided no instructions on how to isolate it, I was feeling quite up the glyphosate creek without a microfungus.

A few weeks later I was speaking to one of the deans at my university and he handed me a letter. It was to the university administration, my bosses, from Dr. Huber. The letter was to "alert you (university administration) to an apparent ethics concern and unprofessional conduct" as (Folta) "made every effort to be disruptive and disparaging" during (Huber's) "fair and informative discussion." The letter goes on to mention "considerable question of [Folta's] scientific competence" and Folta's presumed role as the "belligerent, self-appointed ultimate arbiter of truth and censorship." Huber recommends "counseling in anger management, ethical behavior, professional conduct, courtesy and respect." Luckily, I recorded his entire seminar. One can hear that I never made a peep and was certainly courteous with my request to help solve the crisis of the mystery pathogen. I was really disappointed because I was deliberately kind and cooperative. I am glad I recorded the lecture and could demonstrate his demonizing of me. If I had not recorded it, or if I were a young faculty member, I might have been on thin career ice.

Since I was not likely to get cooperation from Dr. Huber, I thought I would contact Purdue University. If they were housing this agent, they must have found it had some very special needs for isolation, safe handling, and storage. They never returned an email or a call. I contacted the CDC. They never heard of such a thing, and when I pressed they eventually sent me to a person at the USDA. When I called her on the phone, I could hear her eyes rolling back in her head as we discussed the secret creature. Her opinion was that it was a pathetic hoax and one that has cost the department considerable time, money, and hassle.

The infamous "leaked" letter to Vilsack was mailed six years before this chapter was written. Six years after that letter and Huber was still on the speaking circuit, giving talks about his Molecular Bigfoot, which had become not a microfungus but "a prion or biomatrix." He still claimed it obeys Koch's postulates and can be cultured, so how it self-assembles and becomes infectious would be a new trick of biology for sure. Three years after Huber's letter about me, there was still zero-point-zero evidence of this deadly organism, a threat he continued to use constantly in his lectures, speeches, and presentations in agriculture areas and to public audiences.

Years after I asked him for proof of his wild claims, he still does not like me much. I was attending the Plant/Animal Genome meeting in San Diego, California, in January 2016. PAG is a strange conference, yet everyone goes, and part of its charm is that scientists can propose sessions that are almost always accepted for the program. In late 2015 I saw that there was a session called "OneHealth Epigenomics, from Soil to People," and one of the speakers was Dr. Don Huber!

I went to the session, which included several scientists speaking. Some presented perfectly legit work, but the agenda was driven by the moderator, who tied all of the speakers' talks together with threads of environmental and human poisoning. Huber did his usual shtick, focusing more on glyphosate and claims that do not fit well with the rest of science. But the best part was that I had made his slide deck. He had a slide with me wearing a court jester hat, with some derogatory text I neglected to capture.

For scientists in genomics, PAG is an annual event. Good ol' Dr. Huber certainly knew that and it was a good chance to try to make me look bad in front of the scientists in my field. Such things do not make me look bad; they make him look worse. From his apparent fabrication of a deadly organism to his frequent talks using science to scare, to his petty poke at a scientist who asked him to kindly support his claims. That is pathetic.

How does the story end? Only time will tell. There is one genuine side of me that hopes Dr. Huber publishes a breakthrough paper on the cover of *Science* magazine and gets his picture on a box of Wheaties. We can use a new kingdom of biological microschmootz to study. Another side of me would like to see him just come clean if the claims are indeed false. In either case, he would be a hero. Science affords us the opportunity to reshape our thinking, to interrogate ourselves, and to reevaluate our claims. The scientific field is generally a forgiving one. I think people would consider his history of good work and feel compelled to give him a pass. Look at Mark Lynas, a long-time activist who switched from tearing up research plots to following evidence; he is deeply respected by the scientific community he once abhorred.

The only way I hope it does not end is for him to continue a cloaked charade, making claims of a fantastical beast that only he sees, yet millions fear. Telling a story that messes with people's heads and scares them about their food has some serious ethical overlays if it is not true.

Still Dr. Huber jet-sets around the nation, flaunting photos of aborted calves and dying crops plants, along with tales of a deadly pathogen that threatens the planet and spawns our darkest human diseases. The CDC does not know of it, and he does not release it to the broader scientific and medical community. I do hope we see evidence of existence or nonexistence soon, as carrying on a suspect story designed to erode public trust in their food, farming, and scientists is unbecoming for a credentialed scientist. In fact, it borders on a soft, nonviolent form of terrorism—at least some people are afraid of food, and fear for their health because of the claims that he makes.

Formaldehyde and Glyphosate—Sham Data and Flimsy Computational Claims
Genetically engineered crops have been used safely for over two decades, and genetically engineered medicines (insulin) and food products (cheese enzymes) even longer. Predictions of catastrophe were rife in the 1980's and 1990's, but time would show that the engineered plants would generally cut insecticide use and switch herbicide from traditional herbicides to glyphosate, a relatively benign molecule (National Academies of Science, 2016). As opponents of technology found themselves on the losing side of every argument against the technology itself, they switched to attacking the chemical partners that permit its use. They also just make up crazy talk.

Such is the case of formaldehyde. Formaldehyde is the punching bag of organic molecules. A natural product of metabolism, this small one-carbon molecule is unfortunately familiar from its historical role in pickling laboratory specimens and other embalmed sundries. This role has given formaldehyde a histrionic yuck factor that places it at the ready in the holster of the pseudoscientist. People who want to malign genetic engineering find ways to mention formaldehyde, just like the vaccine foes who talk about "toxic formaldehyde," so the public will think their food is being used to pre-embalm them. The sad part is that these claims based on manufactured data tables and lousy experiments are being used to scare people away from food choices, as activists foist the formaldehyde flag to scream that food is somehow toxic.

The "Stunning Corn Comparison"   In the spring of 2013 the Internet exploded with a warning of impending maize disaster. The "Stunning Corn Comparison" was presented on the Moms Across America website, and told the story of the composition of GE corn relative to a non-GE counterpart. The data told a disturbing and shocking story—that "non-GMO corn" is remarkably different from "GMO corn." However, when examined with an eye of scientific scrutiny the claimed composition of "GMO corn" is not even biologically feasible. But that did not stop this activist group from assembling a plague of false claims that echoed loudly and rapidly through the Internet.

The table offered by Moms Across America features some odd parameters. "Cation Exchange Capacity" is a term we would not consider when testing corn. In fact, if one looks at all of the parameters that were claimed to be measured, one quickly realizes that this is a *panel from a soil test*, and not one of corn or any other biological material.

The website blows the alarm on the formaldehyde levels observed, a nice, round number of 200 ppm, which is not biologically sustainable. The authors do not mention how they measure formaldehyde, which is not an easy task. Formaldehyde occurs only briefly, as cells use to "formyl-ate" (or add formyl- groups to) other biological molecules. What did they actually measure, free formaldehyde? How did they do it? Glyphosate is reported present at 13 ppm, which could be possible, but is hundreds of times below biological relevance. There are 60 ppm of "Anerobic biology" and it is unclear what that might be. There is no "chemical content" in the "Non-GMO Corn," which leads the reader to question what the hell it is made of. They report soluble solids (sugars) as "% Brix," and provide the value of 1 percent for GMO corn, a number that is so low that it might be from a plastic corn cob. They report Non-GMO Corn with a % Brix of 20, which could be possible in well-raised sweet corn. However, the genetic identity of the materials used was not revealed, and genetically engineered corn is typically field corn with modest Brix, not sweet corn.

While the numbers in the chart are far from fathomable, they were presented with the authority of a legitimate experiment. Literally hundreds of websites still display the material and tout its veracity. The sites stand by these data as legitimate, despite no independent replication since 2012 and no publication in a legitimate scientific journal. Many of these sites are familiar haunts of credulous claims, such as Natural News and the Mercola.com website.

Of course, I approached the authors and dipped a toe in the water of replication. To their credit, my willingness to replicate the data (at my personal expense) was embraced by the authors and their collaborators at first. However, as the experimental plan grew real and questions started to emerge about what corn varieties were tested, the promoters of the table withdrew from the challenge, leaving us only with a soil-test table with manufactured values not known to biology.

But the Computer Says It Is There . . .   In the summer of 2015 a paper was published that made a strong claim—that genetically engineered crops contained high amounts of (wait for it) formaldehyde. We know that plants produce formaldehyde in trace amounts as a product of metabolism—every organism does. It is a toxic intermediate that is rarely found in its free form. Upon production, it is rapidly changed to other intermediates or attached to other biomolecules during "formylation," a process that creates new compounds and detoxifies free formaldehyde. But that is the science, and it is not so freaky.

Enter Dr. Shiva Ayyadurai, a PhD-level computer scientist who at the time was married to actress Fran Drescher. Drescher has long been recognized for her pseudoscientific leanings, leanings that are more like falling-overs. Ayyadurai was the corresponding author on a work called, "Do GMOs Accumulate Formaldehyde and Disrupt Molecular Systems Equilibria? Systems Biology May Provide Answers" (Ayyadurai and Deonikar, 2015). The title itself is *Just Asking Questions*, a way to create an association without actually showing evidence of formaldehyde, which they never did. Still, in the anti-GMO community and social media, the paper's title and conclusion spread like measles in children whose parents listened to Drescher's anti-vax rants.

The first major tipoff is the term "GMO" in the title. Scientists do not use this term. It is imprecise, meaningless, and used simply to serve as a beacon luring the credulous as they surf the web. According to Ayyadurai's little computer program, any plant that is genetically engineered is going to produce more formaldehyde.

The major issue with the paper is an obvious one. If the computer prediction reveals an increase in formaldehyde, *shouldn't the researcher actually test soybeans for formaldehyde levels to actually verify that the computer prediction is correct?* This validation step is an expected addition to any decent computational biology paper, yet was not included in Ayyadurai's work. It is the same issue as using a computational algorithm to predict the location of Munich and learning that this German city is underwater somewhere off the coast of Sarasota, Florida. Would you take the drive into the Gulf of Mexico to hit Oktoberfest? This claim from a computer output, like the Stunning Corn Comparison table, had nothing to do with biology and everything to do with satisfying the preconceived bias of an author on a mission to scare people about their food. After all, who cares about actual biology when the computer spits out exactly what you believe?

The activist media was quick to seize on these announcements. The untested computational claim contrived from a biased computational input should not hold much weight, and scientists in systems biology turned the page and went on to actual science.

Typically, papers making claims of metabolic dystrophy require the authors to demonstrate the validity of their computational test by actually measuring something. However, here the prediction alone confirmed the suspicions of the throngs that are pleased to be told what to think, and the claim of food containing a dangerous compound exploded into the anti-GE literature and associated websites.

Formaldehyde should be a red chemical flag when thinking about pseudoscientific claims about genetically engineered crops. From a biological perspective, it is too easy of a target. It provides the shock value activists crave, and the biological implausibility scientists understand.

Roundup Is Everywhere! Today's analytical chemistry instrumentation allows scientists to detect the presence of some molecules at levels just this side of not there. We are talking parts per trillion. That is equivalent to being able to tell you exactly what happened during a particular few seconds that elapsed at some point after the birth of Christ. This stuff is sensitive.

The glory of sensitive detection is that scientists can tell you exactly what compounds are present in and on the food we eat. We can be damn sure that agricultural chemicals are being used safely, and that detected levels are thousands, if not millions, of times below levels that actually would have physiological relevance.

But to the anti-GE activist, identifying a microdusting of chemistry is a way to convince the public that the needle in the haystack is really a hatchet in their kid's head. The fact that molecules are *detected*, even if barely there, can be sold as scary with activist misamplification. From cholesterol meds, to doughnuts, to water, the dose makes the poison, and agricultural chemicals are no exception. Long before they are released for use, we understand how they are metabolized and what their mechanisms of action are. Thousands of rats and rabbits suffered rather miserable experiences to help us understand their relative toxicity. From these data, tight thresholds are established for occupational or residual exposure, and then exposure tolerances are set orders of magnitude below those levels.

But activists do not treat detection as the presence of a wisp of stuff at the edge of nothing. They play it out like cancer—if it is detected, at any level, it is bad. In such zero-tolerance cases "detection" is warped into a potentially life-threatening problem. They blur the definition of risk, which requires two things—hazard and exposure. For instance, we know that agricultural chemicals can be a hazard; that is why we have handling guidelines, and users are extremely careful to limit exposure. A hazard with no exposure poses no risk. It is why we do not worry much about shark attacks when we are walking down the street in Milwaukee in the winter. The sharks do represent a hazard, but your chances of meeting one are quite slim in that setting.

There are so many cases of misappropriating risk, but two stick out because of how they were used to manipulate public perception and trust. One of these exploded into the popular soft press, and soon claims were everywhere that kids were dancing in a rain of herbicides. But is that what the data really said?

A website claimed, "Roundup Weedkiller in 75% of Air and Rain Samples, Gov't Study Finds." The story, published on the scientifically limp website GreenMedInfo, claimed that "the GM farming system has made exposure to Roundup herbicide a daily fact of our existence, and according to the latest US Geological Survey study its [*sic*] probably in the air you are breathing." Come to think of it, my lungs are weed-free . . .

The information presented cited a peer-reviewed paper by Majewski et al. (2014) published in *Environmental Toxicology and Chemistry*. After reading the GreenMedInfo article I immediately tried to access the cited work, only to find that it was not yet available. How did GreenMedInfo's crack scientific staff get access? Turns out they did not. In typical fashion, they read a title, parsed an abstract, ignored the materials, methods, and results, and simply drew an inflammatory conclusion consistent with their mission to misinform the curious.

Taking the route frequently evoked by scholars, I contacted the paper's authors and they provided me with the first electronic copy distributed by request. What did the story really say? The author team is composed of experts with the duty to monitor environmental persistence of farm chemicals. These are experts that possess the world's most sensitive equipment to survey presence or absence of insecticides and herbicides in the environment. The study compared samples between 1995 and 2007, before and after genetically engineered crops were widespread.

The results of the paper showed that glyphosate was in fact detected in air—a few nanograms per cubic meter. That means a few billionths of a gram pulled out of one million cubic centimeters of air. Using the time analogy again, that is a few seconds in thirty-two years. Is that what was found in downtown Manhattan, at the St. Louis Arch, or deep in The Food Babe's veranda? Nope. It was the level measured a few steps from the side of an agricultural field where the product was recently applied.

Remember the sensational headline and story—weed killer found in 75 percent of air and water. The GreenMedInfo (and countless other pages that continued to promote the story) failed to tell you that the herbicide was (almost not) detected a few meters (in three-quarters of samples) from where it was applied. The authors detected farm chemicals on the farm, almost none in air, and next to almost none in water runoff. The sensationalist anti-GMO media in their typical dishonesty did not bother to disclose that. In fairness, they did not know enough to know that. They did not bother to read and understand the paper before drawing the conclusion they found acceptable.

If anything, the scientific paper had a very pro-GE thesis, as clearly ag chemical profiles in the environment have changed from 1995 to 2007. The data actually show a very positive result in terms of environmental sustainability. Before genetically engineered plants, the same survey team routinely detected insecticides and herbicides like methyl parathion, malathion, and chlorpyrifos. After implementation of GE plants that produce their own insect protection, the use of these chemicals decreased to a wee fraction of 1995 levels. Glyphosate was certainly detected in 2007 (it was not measured in 1995), but it was present in vanishingly small amounts, while herbicides found in 1995 were found at much higher levels. What this report really said is that the authors are good at finding chemicals in air and water, and that the profiles have changed from old-school insect and weed controls to more sustainable alternatives.

However, this is not what was reported. The news on hundreds of activist websites showed kids in raincoats playing in the rain, with the words "It's Raining Roundup." They showed artistically shot puddles with the phrase "Government study finds Roundup in 75 percent of air and water samples." The flimsy pseudoscientific press accepted and expanded the conclusions presented on GreenMedInfo, bending the story even further because the reporters did not read the original paper.

If you read the news articles on ag chemistry and scroll through the comments sections, you can usually find people complaining that their lives have been destroyed by glyphosate. They never used it, they eat organic food to avoid it. But when it is in every breath, every glass of water, you can not avoid it, right? This is the glyphophobia that a pseudoscientific article based on misinterpreting the title and abstract of a scientific paper can generate.

If It Is Everywhere, It Might Not Be Anywhere The Stunning Corn Comparison was the first volley in a number of reports claiming detection of glyphosate where it should not be, at levels that are not biologically feasible. In subsequent years we saw continual reports of glyphosate being detected in everything from bodily fluids to beverages. The reports were not gracing the pages of the finest scholarly journals or even gutter predatory ones. Instead they were found on websites, Twitter feeds, and in every possible tentacle of the anti-GMO disinformation network. The point? To scare the hell out of people by telling them that their families are eating pesticides. The goal is to stop them from trusting their food and farmers, and to drive policy change to curtail use of helpful technologies.

The next section describes this fear campaign. The punchline to follow: they failed to use the correct detection method, and are only reading noise of the assay. All of these claims are based on results obtained from a kit used to detect glyphosate in water. Why this is insufficient is explained in detail at the end of the section.

Moms Across America, the same folks that presented the fabricated information in the Stunning Corn Comparison, published a "study" that purports identification of glyphosate in breast milk. They report it on their website on April 7, 2014, with no peer review or professional vetting. Why test breast milk? To scare the public, especially parents that take on the mantle of protecting their children at any cost. Moms

Across America specifically targets parents. The group knows that if its plan to end genetic engineering is going to get traction, it will need to hit people where it hurts—protecting their families from harm.

The report is worth a look as it features commentary by scientists. Of course, they roll out the usual cadre of science obfuscators, leading off with Dr. Don M. Huber (the guy with the Molecular Bigfoot). The results show no detection of glyphosate in most samples, with the term the authors use for no detection (<75 µg/L) managing to suggest some low amount is actually present. Of course, <75 µg/L is the detection limit of the kit, so a zero signal is interpreted by these folks as "clearly something there, we just can't see it, obviously." In the three samples of the ten they assessed, one was at the detection limit (at 76 µg/L) and another at 99 µg/L, which is essentially at the edge of detection. Another was at 166 µg/L and comes from Florida (figures). This highest level they claim to detect is 166 parts per billion. That is like saying you won $10 million and you noticed $1.66 was missing. Yes, it is that small, and even if the claim were true, the figure would be far below any biologically meaningful threshold.

Yet this flimsy, awful evidence on a website, with no replication or statistics, is the basis for a scare campaign to frighten new mothers from feeding their children the best possible neonatal nourishment—breast milk.

Good scientists do not take such claims lightly, and some are particularly affected by campaigns to frighten mothers. Enter Dr. Shelley McGuire, lactation specialist. Dr. McGuire is an expert in examining the chemicals in breast milk, a complicated solution as far as biological exudates go. She and her team skillfully developed a method to extract glyphosate from breast milk and demonstrate its faithful detection with the gold standard of sensitive equipment (Jensen, Wujcik, McGuire, and McGuire, 2016). She published the results in peer-reviewed journals. The results? No glyphosate detected (McGuire et al., 2016). Science stood in conflict with the Moms Across America website and its team of credulous supporting scientists, who then claimed that McGuire was simply (wait for it) a shill for Monsanto. Shortly thereafter, her emails were collected using Freedom of Information Act (FOIA) laws. She was publicly defamed and trashed both as a scientist and as a mother in the venomous social media space.

The same group of glypho-sleuths also claimed to find glyphosate in urine. Again, they used the same kit to claim to find infinitesimally small amounts. Now, if there was a place you would actually find glyphosate, it would be in the urine. It is a small, soluble molecule that is rapidly eliminated from the body. Still, these assays show amounts that straddle the limit of detection. There is no replication, no statistical treatment, just a single data point of noise suggesting something might be present. Again, the assay is not made for use with urine. Not all urine is equal. Ask anyone that ate asparagus, took B vitamins, or ate fava beans last night. Yet the authors claim to faithfully detect amounts akin to 8.1 cents in $10 million and interpret that as the poisoning of humankind.

Another report claims to detect the herbicide in wine. Nine samples were tested from a variety of wines from different wineries in California. All tested positive, even organic grapes that by definition must be grown without this herbicide. To the scientist, this would represent a great negative control, as no glyphosate would be present. A positive detection means something is seriously wrong with the assay. That is what a scientist thinks. To an activist mom that has no clue how science works, a positive signal in the organic wines means that the contamination is more widespread than could be imagined, and even organic wines are literally brimming with herbicides. But how much are they detecting? According to their claim, it is between 0.659 and 18.74 parts per billion. There are no replicates, no statistics, and no clear discussion of the control. Because there is none, it is probably water, and that matters.

They say that the 18.74 parts per billion of glyphosate is an existential threat, even though it is far below any threshold of biological activity. At the same time, wine is 13 percent alcohol, which is a known carcinogen. So why the generally benign herbicide is claimed to be there at about 19 ppb, the known threat of ethanol is there at 130,000,000 ppb. I'll bet Moms Across America members pop a cork now and then without worry.

Why do they detect a herbicide everywhere they look? Simple. *They want to find it everywhere*, so they do not question the methods or the results from an improperly used assay that gives them the outcome they desperately seek. But is the method of measurement valid? The method is a kit made by a company called Abraxis. It is what they call a *competitive ELISA* (enzyme-linked, immunosorbent assay). It is like a mass game of molecular musical chairs, with the chairs being antibodies directed against glyphosate. They are bound to the surface of a tiny well on a petri dish with ninety-six wells. Each well is about half a centimeter across and is a little cylinder. The sample to be tested (urine, wine, whatever) is added to the cylinder along with a known quantity of an enzyme-linked version. In other words, the two compounds, the one from your sample and the spiked in, enzyme-linked one, will compete for the same targets. The more of the agent that is there to be detected, the more it will bind instead of the enzyme-linked control. It is competition between a known amount and an unknown amount.

Once bound to the plate you can imagine a collection of two groups of molecules—those with the enzyme and those without—that are now decorating the surface of the plate. A simple addition of some reagents forms a blue precipitate from the linked enzyme. In other words, if your sample is loaded with the compound to be detected, then the reaction will appear to have almost no color. If there is nothing present, it will be all color, because each of the jillions of spots on the plate will be occupied by a linked enzyme and not the target itself.

Now here's the rub. What if there is 100 percent of the control glyphosate, the enzyme-bound one, bound to antibodies on the plate? There is zero glyphosate in the sample. But what if there is a molecule that cross-reacts with the antibody, competing for that space? It will take up space on the plate and will be inferred as a positive. What if something in a complicated matrix like breast milk, wine, or eggs binds the enzyme and inhibits the reaction? These are mixtures of many proteins and metabolites, and methods of extraction of the target molecule need to be developed, as well as faithful measurement techniques. So if something in breast milk, urine, beer, or wine interferes with any aspect of that enzymatic assay (for instance, it might bind the substrate, interfere with the reaction, or the product), then the reaction would produce less color, leading to a false-positive signal. There is no adequate control for breast milk, wine, or urine, because these are not standard solutions that can be tested with or without glyphosate added.

So why would a company make such a kit? It is designed for environmental assessment. It is produced to examine the amount of glyphosate in water, and the company that makes it tells you flat out that it is a cursory first step. It is only designed to work with water, not urine, not grape jelly, not breast milk, and not eggs.

But does that matter to Zen Honeycutt and the Moms Across America? Of course not! They have a result that is compatible with what they believe, so it must be correct.

The bottom line: these people are misusing an analytical method to obtain the results they wish to find. The kit shows results that are near the limit of detection, far below any dangerous levels by thousands to millions of times. But the finding is the outcome they want. Even if it is wrong, to them it is exactly right. It fits their religion. They take to the Internet and parade the results around like they are real, manufacturing fear, conjuring mistrust, and scaring people about their food.

Glyphophobia and Formalde-HIDE! GM plants, their health implications, and their environmental impacts have taken up little space in this chapter. The anti-GM movement knows that the arguments about these things were lost long ago. Scary herbicides and insecticides are the movement's new focus. The chemical legacy of pesticides is scary and complex, and here exploited for political gain. With this villainization they can appeal to popular chemophobia, the notion that all chemicals are evil and deadly, especially those made by companies that the activists do not like. But for those of us thinking critically, two points must be remembered. *When something is claimed to be everywhere, that probably means it is nowhere. When something is claimed to cause everything, it probably causes nothing.* People who propagate such claims are dangerous fearmongers, pushing loopy messages for political gain, no matter who may be harmed by the words.

Conclusion

Another interesting parallel can be made between food-o-science pseudoscience and the other areas of plain-wrapper pseudoscience. Just as UFOs, Nessie, Sasquatch, and other rare phenomena disappeared as the technology poised to capture their evidence became widespread, we see the same thing happening in anti-GMO claims. Over the twenty years that these products have been deployed, the science around them has grown in complexity and sensitivity. Epidemiological trends and huge data sets have been complied. If there really was a problem with the products on the market, it would have been seen in livestock, humans, or both. Actually, it would have been found long before the plant was commercialized. The resolution of tomorrow's tools will further reinforce efforts to create safe food products, and allow us to continually monitor those that are already with us.

New genome-editing technologies (e.g., CRISPR/Cas9, TALEN) stand to revolutionize plant and animal genetic improvement, as well as produce profound benefits in human health. The challenge of the next decade will be for scientists, physicians, farmers, and dietitians to grab control of this topic, to understand it, and to communicate its awesome power to a skeptical public. The anti-GMO movement is poised to ambush and denounce this good technology. In their mind, if a corporation uses it to improve crops, there is nothing good that can be said about it, even if science says good things about it. Those that say good things about it should be ignored, as they are obviously paid to say good things about it.

There is a deliberate, well-funded, and organized movement sworn to tarnish the perception of a useful technology, the farmers that use it, and the scientists that teach it. Unlike some other forms of pseudoscience, this active fight against science and reason has a daily body count. While today it may be fun to marvel at the silly claims and twisted memes made by activists, Internet celebrities, and discredited scientists, we have to remember that the suspicion and doubt they have seeded have arrested a generation of new crops from ever reaching the field. Solutions have been blocked from serving those they were meant to serve, blocked from protecting the planet they were designed to protect. The technologies hypothesized by a scientist lying awake at night, furiously staring at a ceiling for an answer, never ever materialized. This brand of pseudoscientific blind science slander hinders progress, and hurts the poor, the planet, and the farmer. One day we will look back on this time as a tragedy, when governments and companies, populations and people, elected to diminish progress based on the thin thoughts of charlatans over the reproducible truths of scholarly science. We abandoned solutions for many in need, in exchange for comforting precaution of a few. It is a sad testament to the power of pseudoscientific claims and how they shape our decisions about food.

References

Areal, F. J., Riesgo, L., and Rodriguez-Cerezo, E. (2013). Economic and Agronomic Impact of Commercialized GM Crops: A Meta-Analysis. *Journal of Agricultural Science, 151*(01): 7–33.

Aris, A., and Leblanc, S. (2011). Maternal and Fetal Exposure to Pesticides Associated to Genetically Modified Foods in Eastern Townships of Quebec, Canada. *Reproductive Toxicology, 31*(4): 528–533. DOI: 10.1016/j.reprotox.2011.02.004

Ayyadurai, V. A. S., and Deonikar, P. (2015). Do GMOs Accumulate Formaldehyde and Disrupt Molecular Systems Equilibria? Systems Biology May Provide Answers. *Agricultural Sciences, 6*(7): 630.

Azadi, H., Ghanian, M., Ghoochani, O. M., Rafiaani, P., Taning, C. N. T., Hajivand, R. Y., and Dogot, T. (2015). Genetically Modified Crops: Towards Agricultural Growth, Agricultural Development, or Agricultural Sustainability? *Food Reviews International, 31*(3): 195–221.

Azadi, H., Samiee, A., Mahmoudi, H., Jouzi, Z., Rafiaani Khachak, P., De Maeyer, P., and Witlox, F. (2016). Genetically Modified Crops and Small-Scale Farmers: Main Opportunities and Challenges. *Critical Reviews in Biotechnology, 36*(3): 434–446.

Bellumori, T. P., Famula, T. R., Bannasch, D. L., Belanger, J. M., and Oberbauer, A. M. (2013). Prevalence of Inherited Disorders among Mixed-Breed and Purebred Dogs: 27,254 cases (1995–2010). *Journal of the American Veterinary Medical Association, 242*(11): 1549–1555.

Benz, B. F. (2001). Archaeological Evidence of Teosinte Domestication from Guilá Naquitz, Oaxaca. *Proceedings of the National Academy of Sciences, 98*(4): 2104–2106.

Beyer, P., Al-Babili, S., Ye, X., Lucca, P., Schaub, P., Welsch, R., and Potrykus, I. (2002). Golden Rice: Introducing the β-carotene Biosynthesis Pathway into Rice Endosperm by Genetic Engineering to Defeat Vitamin A Deficiency. *Journal of Nutrition, 132*(3): 506S–510S.

Blancke, S., Van Breusegem, F., De Jaeger, G., Braeckman, J., and Van Montagu, M. (2015). Fatal Attraction: The Intuitive Appeal of GMO Opposition. *Trends in Plant Science, 20*(7): 414–418.

Brookes, G., and Barfoot, P. (2014). Economic Impact of GM Crops: The Global Income and Production Effects, 1996–2012. *GM Crops & Food, 5*(1): 65–75.

Brookes, G., and Barfoot, P. (2015). Environmental Impacts of Genetically Modified (GM) Crop Use, 1996–2013: Impacts on Pesticide Use and Carbon Emissions. *GM Crops & Food, 6*(2): 103–133.

Camberato, J., Casteel, S., Goldsbrough, P., Johnson, B., Wise, K., and Woloshuk, C. (2011). Glyphosate's Impact on Field Crop Production and Disease Development. *Purdue Extension Weed Science, 2*: 4.

Carman, J. A., Vlieger, H. R., Ver Steeg, L. J., Sneller, V. E., Robinson, G. W., Clinch-Jones, C. A., . . . Edwards, J. W. (2013). A Long-Term Toxicology Study on Pigs Fed a Combined Genetically Modified (GM) Soy and GM Maize Diet. *Journal of Organic Systems, 8*(1): 38–54.

Clair, É., Mesnage, R., Travert, C., and Séralini, G.-É. (2012). A Glyphosate-Based Herbicide Induces Necrosis and Apoptosis in Mature Rat Testicular Cells In Vitro, and Testosterone Decrease at Lower Levels. *Toxicology in Vitro, 26*(2): 269–279. DOI: http://dx.doi.org/10.1016/j.tiv.2011.12.009

Elgsti, O. J., and Dustin, P. (1955). *Colchicine—in Agriculture, Medicine, Biology and Chemistry*. The Iowa State College Press.

Fahlgren, N., Bart, R., Herrera-Estrella, L., Rellán-Álvarez, R., Chitwood, D. H., and Dinneny, J. R. (2016). Plant Scientists: GM Technology Is Safe. *Science, 351*(6275): 824–824.

Fernandez-Cornejo, J., Wechsler, S., and Milkove, D. (2016). The Adoption of Genetically Engineered Alfalfa, Canola, and Sugarbeets in the United States.

Gepts, P. (2010). Crop Domestication as a Long-Term Selection Experiment. *Plant Breeding Reviews, 24*(Part 2): 1–44.

Giri, J., and Tyagi, A. K. (2016). Genetically Engineered Crops: India's Path Ahead. *Nature India*, March 4.

Global Research News. (2015). Monsanto Roundup Is an Endocrine Disruptor in Human Cells at Levels Allowed in Drinking Water. Unpublished manuscript.

Gonsalves, C., Lee, D., and Gonsalves, D. (2004). Transgenic Virus-Resistant Papaya: The Hawaiian "Rainbow" Was Rapidly Adopted by Farmers and Is of Major Importance in Hawaii Today. APSnet Feature, American Phytopathological Society, August–September: http://www.apsnet.org/online/feature/rainbow.

Griliches, Z. (1958). Research Costs and Social Returns: Hybrid Corn and Related Innovations. *Journal of Political Economy*, 419–431.

Heslop-Harrison, J. S., and Schwarzacher, T. (2007). Domestication, Genomics and the Future for Banana. *Annals of Botany, 100*(5): 1073–1084.

Hoyert, D. L. (2012). 75 Years of Mortality in the United States, 1935–2010. *Age, 1200*: 1400.

Jensen, P. K., Wujcik, C. E., McGuire, M. K., and McGuire, M. A. (2016). Validation of Reliable and Selective Methods for Direct Determination of Glyphosate and Aminomethylphosphonic Acid in Milk and Urine Using LC-MS/MS. *Journal of Environmental Science and Health, Part B, 51*(4), 254–259.

Kathage, J., and Qaim, M. (2012). Economic Impacts and Impact Dynamics of Bt (*Bacillus thuringiensis*) Cotton in India. *Proceedings of the National Academy of Sciences, 109*(29): 11652–11656.

Kiekens, F., Blancquaert, D., Devisscher, L., Daele, J., Stove, V. V., Delanghe, J. R., . . . Stove, C. P. (2015). Folates from Metabolically Engineered Rice: A Long-Term Study in Rats. *Molecular Nutrition & Food Research, 59*(3): 490–500.

Klümper, W., and Qaim, M. (2014). A Meta-Analysis of the Impacts of Genetically Modified Crops. *PLOS ONE, 9*(11): e111629.

Lombardo, L., Coppola, G., and Zelasco, S. (2016). New Technologies for Insect-Resistant and Herbicide-Tolerant Plants. *Trends in Biotechnology, 34*(1): 49–57.

Majewski, M. S., Coupe, R. H., Foreman, W. T., and Capel, P. D. (2014). Pesticides in Mississippi Air and Rain: A Comparison between 1995 and 2007. *Environmental Toxicology and Chemistry, 33*(6): 1283–1293.

McGuire, M. K., McGuire, M. A., Price, W. J., Shafii, B., Carrothers, J. M., Lackey, K. A., . . . Vicini, J. L. (2016). Glyphosate and Aminomethylphosphonic Acid Are Not Detectable in Human Milk. *American Journal of Clinical Nutrition, 103*(5): 1285–1290.

Murphy, D. (2011). Dr. Huber's Warning. Food Democracy Now. http://action.fooddemocracynow.org/sign/dr_hubers_warning/ (accessed 3/1/2016)

Naqvi, S., Zhu, C., Farre, G., Ramessar, K., Bassie, L., Breitenbach, J., . . . Capell, T. (2009). Transgenic Multivitamin Corn through Biofortification of Endosperm with Three Vitamins Representing Three Distinct Metabolic Pathways. *Proceedings of the National Academy of Sciences, 106*(19): 7762–7767.

National Academies of Sciences. (2016). *Genetically Engineered Crops: Experiences and Prospects.* Washington, DC: National Academies Press.

Nicolia, A., Manzo, A., Veronesi, F., and Rosellini, D. (2014). An Overview of the Last 10 Years of Genetically Engineered Crop Safety Research. *Critical Reviews in Biotechnology, 34*(1): 77–88.

Oh, S. A., Min, K. H., Choi, Y. S., Park, S. B., Kim, Y. C., and Cho, S. M. (2015). Development of Tetraploid Watermelon Using Chromosome Doubling Reagent Treatments. *Korean Journal of Plant Resources, 28*(5): 656–664.

Panchin, A. Y., and Tuzhikov, A. I. (2016). Published GMO Studies Find No Evidence of Harm When Corrected for Multiple Comparisons. *Critical Reviews in Biotechnology*, 1–5. DOI: 10.3109/07388551.2015.1130684

Pease, J. B., Haak, D. C., Hahn, M. W., and Moyle, L. C. (2016). Phylogenomics Reveals Three Sources of Adaptive Variation during a Rapid Radiation. *PLOS Biology, 14*(2): e1002379.

Pickett, Heaps, J. D., Tippit, D. H., and Porter, K. R. (1982). Rethinking Mitosis. *Cell 29*(3): 729–744.

Prejean, J. D., Peckham, J. C., Casey, A. E., Griswold, D. P., Weisburger, E. K., and Weisburger, J. H. (1973). Spontaneous Tumors in Sprague-Dawley Rats and Swiss Mice. *Cancer Research, 33*(11): 2768–2773.

Rowlands, H. (2015). Dr Young: Roundup Herbicide Is Endocrine Disruptor in Human Cells at Drinking Water Levels. *GMO Evidence.* http://www.gmoevidence.com/dr-young-roundup-herbicide-is-endocrine-disruptor-in-human-cells-at-drinking-water-levels/ (accessed 2/12/2016)

Saini, R. K., Nile, S. H., and Keum, Y.-S. (2016). Food Science and Technology for Management of Iron Deficiency in Humans: A Review. *Trends in Food Science & Technology, 53*: 13–22.

Séralini, G.-E., Clair, E., Mesnage, R., Gress, S., Defarge, N., Malatesta, M., . . . De Vendômois, J. S. (2012). RETRACTED: Long-Term Toxicity of a Roundup Herbicide and a Roundup-Tolerant Genetically Modified Maize. *Food and Chemical Toxicology, 50*(11): 4221–4231.

Stebbins, G. L. (1940). The Significance ofPpolyploidy in Plant Evolution. *American Naturalist, 74*(750): 54–66.

Stebbins, G. L. (1956). *Artificial Polyploidy as a Tool in Plant Breeding.* In Genetics in plant breeding. Brookhaven Symposia in Biology. pp. 37–52.

Strauss, S. H., and Sax, J. K. (2016). Ending Event-Based Regulation of GMO Crops. *Nature Biotechnology, 34*(5): 474–477.

Van Eenennaam, A. L., and Young, A. E. (2014). Prevalence and Impacts of Genetically Engineered Feedstuffs on Livestock Populations. *Journal of Animal Science*, *92*(10): 4255–4278.

Veilleux, R. (1985). Diploid and Polyploid Gametes in Crop Plants: Mechanisms of Formation and Utilization in Plant Breeding. *Plant Breeding Reviews*, *3*: 253–288.

Warner, M. (2011). Mystery Science: More Details on the Strange Organism That Could Destroy Monsanto. *CBS Moneywatch*. Retrieved from http://www.cbsnews.com/news/mystery-science-more-details-on-the-strange-organism-that-could-destroy-monsanto/ (accessed 3/1/216).

Young, F., Ho, D., Glynn, D., and Edwards, V. (2015). Endocrine Disruption and Cytotoxicity of Glyphosate and Roundup in Human JAr Cells In Vitro. *Integr Pharmacol Toxicol Genotoxicol*, *1*(1): 12–19.

6 An Inside Look at Naturopathic Medicine: A Whistleblower's Deconstruction of Its Core Principles

Britt Marie Hermes

Introduction

I stopped calling myself a doctor after realizing I was never one in the first place.

For three years, I practiced as a licensed naturopath in two US states.[1] I earned my naturopathic medicine doctorate degree (ND) at Bastyr University and then completed an additional year of supervised clinical training in a private clinic. I stayed at this clinic until I moved to Tucson, Arizona, where I practiced alongside other naturopaths, acupuncturists, and a chiropractor.

According to the naturopathic profession and state law, I was a primary care doctor and well trained to work in an outpatient setting. I was allowed to use the title "naturopathic medical doctor" (NMD) and call myself a physician in the state of Arizona. In addition to a medical license, I held a national provider identification number and a Drug Enforcement Administration (DEA) number for midlevel practitioners. Some of my services were even reimbursable by health insurance. By all appearances, I was a doctor.

Looks can be deceiving.

A Brief History of Naturopathic Theory

Naturopathy was founded during the late nineteenth and early twentieth centuries by naturalistic healers who confused human physiology with vitalism (Atwood, 2003; Pizzorno and Murray, 2013). Vitalism is the belief that all living beings embody a magical energy force. Naturopaths believe this force can cure disease and restore health if activated using remedies found in nature. The rise of naturopathy was contemporaneous with advances in science and medicine, including germ theory, hygiene, vaccines, analgesics, x-rays, blood transfusions, antibiotics, and many other developments that greatly improved health care. While scientists and physicians were making great gains that moved the social consciousness away from vitalism and toward a system of medicine based on science, naturopaths clung tightly to their antiquated philosophy. They insisted that patients didn't require medical intervention in order to recover from illness; in its place, naturopaths employed water, herbs, homeopathy, and other eclectic "energy" therapies (Pizzorno and Murray, 2013). As medical care evolved, proponents of naturopathy continued to believe medicine had not progressed beyond the dangerous and barbaric era of centuries past. It was as if naturopaths focused their criticism on the age of bloodletting with a blind eye to logic, reason, and empirical inquiry.

Early naturopaths named the magical healing life force inside one's body the *vis*; this name is still in use today. Other alternatives to medicine have concepts similar to this unproven internal force, but they employ different names. Traditional Chinese practitioners, for example, call it *chi*. Ayurvedic texts call these forces *prana*. Modern medicine has no such term. The body's innate ability to heal itself is simply known to be a result of a properly functioning immune system. No magic is required.

Even in the twenty-first century, naturopaths embrace and teach vitalism in their curricula, which are required to become licensed. According to naturopaths, the *vis* can help direct the energetic essence of a treatment, allowing the treatment to work on levels deeper than the physical body. Let me provide an example.

The botanical medicine hawthorn is widely believed by naturopaths to be a heart tonic and generally supportive of the cardiovascular system (Alternative Medicine Review, 2010). Naturopaths commonly recommend hawthorn for patients with cardiovascular diseases or those who have also experienced grief, loss, or a broken heart. The idea is that the *vis* will coordinate the medicinal properties of the plant, helping the patient heal on both physical and emotional levels. In circumstances where the patient is too frail to be helped by naturopathic treatments, hawthorn is used to strengthen the body for healing. This is like a naturopathic pretreatment, if you will. To some, allowing patients to believe a berry can heal a broken heart may seem innocent and sweet. I remember feeling comforted by believing that nature could heal. But let's be honest: it is utter nonsense to rely on a supernatural force in the practice of medicine. Such occultism and disregard for scientific knowledge exemplify naturopathy.

Naturopathic Medicine Today and Its Six Principles

Naturopathy is best known today as naturopathic medicine. The word "medicine" has been coupled with naturopathy in order to make the profession appear as a credible medical specialty. The naturopathic profession claims naturopathy is a distinct field of primary care that employs the best of both traditional healing and modern science (American Association of Naturopathic Physicians, 2011a). Having worked with and observed many naturopaths, I can say that the overwhelming majority of naturopathic care relies extensively on dubious alternative therapies, rather than established protocols based on an ever-increasing body of medical and scientific research.

Naturopathic practitioners seem stuck between two worlds. They fully embrace antiquated methods, such as homeopathy, herbalism, and acupuncture, which have been repeatedly demonstrated to be ineffective or unsafe (Atwood, 2003). On the other hand, naturopaths desperately want to be considered a scientific medical profession but lack the training required to understand research and assess the plausibility and effectiveness of their own practices (Hermes, 2015). This is why a patient of a naturopath can easily be found taking a variety of homeopathic remedies while receiving an intravenous drip of high-dose vitamin C, in addition to having a long-term antibiotic prescription to treat "chronic Lyme disease."[2] The practice of blending old-timey remedies and debunked treatments with real medicine does not make naturopathy a specialized field of medicine. This pseudoscientific amalgam makes naturopathy a particularly dangerous form of quackery.

Naturopaths have defined their brand using a set of "healing" principles that sound pleasant to the ear. The naturopathic principles, readers will learn, are concepts that describe an archaic and sometimes twisted understanding of medicine. As a naturopathic student, I was required to extensively study these principles through a series of courses called "naturopathic philosophy," in which my fellow students and I were taught to adhere to the profession's tenets in the following order, with few exceptions (e.g., Association of Accredited Naturopathic Medical Colleges, n.d.-b; Bastyr Center for Natural Health, 2016a; Pizzorno and Murray, 2013; American Association of Naturopathic Physicians, 2011a):

- Healing power of nature
- Identify and treat the cause of disease
- First, do no harm
- Doctor as teacher
- Treat the whole person
- Prevention

First of all, why is "First, do no harm" not first? This is a glaring red flag. How can the "healing power of nature" be prioritized over the most fundamental aim of medicine? I wish I had caught this blunder much earlier in my career.

It is terribly frustrating to recall my naturopathic training, during which we were told that as clinicians, if we were not addressing the principles of naturopathy in treating a patient, we were not practicing naturopathy. It seems to me that any clinician's priority should be the safety and well-being of patients, not the honoring of lore and mythology.

There are two other problems to point out about the naturopathic principles as a collective set. Despite numerous statements from the naturopathic profession that these six principles are unique to naturopathy, these ideas are already a part of established medical philosophy; except, of course, for the one about a natural healing force.

In real medicine, ethical and professional tenets have not been transformed into sound bites to market a commercial brand. The other problem is that naturopaths constantly seem to violate their own value system, as if the words of their principles actually convey alternative meanings or work in an alternative reality. How else can one identify a patient with a fake disease called adrenal fatigue (e.g., Clinton, 2016a), recommend to parents that childhood vaccines should be avoided (Gavura, 2014), or run bogus diagnostic tests to look for mysterious vitamin and neurotransmitter imbalances in your body (Atwood, 2010)?

Rather than operationalizing these principles into a standardized code of ethics, naturopaths seem to be using them as a moral sleight of hand that deceives not only the naturopaths but their patients. I'd argue that the naturopathic principles appeal to the emotional desires of many for health and medicine to be elegant and comforting. I cannot help but draw connections to the strong proclivity many of us have to find ways to resonate with "pseudo-profound bullshit" (Pennycook, Cheyne, Barr, Koehler, and Fugelsang, 2015). Naturopaths are experts at stringing together buzzwords, e.g., "Naturopathic doctors (NDs) blend centuries-old knowledge and a philosophy that nature is the most effective healer with current research on health and human systems (Bastyr University, 2016)." Superficially, these statements seem very sage, but in reality they are hollow.

Although the six naturopathic principles seem to convey profound insights into health and wellness, they are really just describing what medical doctors already do. There is nothing novel about naturopathy's principles, other than naturopaths flaunting them to draw in business and invoke magic. The bottom line is that the guiding principles of naturopathic medicine disguise the systematic deception of students, patients, lawmakers, and too many medical professionals. The general population seems to be buying it, and this is alarming and depressing.

I'm not going to lie. Writing this chapter hurts my heart. I was one of these students duped into believing this pseudoprofound philosophy. I became a practitioner who followed these principles when diagnosing and treating patients. I helped parents delay and avoid immunizations for their children by signing vaccination exemption forms and designing alternative vaccine schedules. I counseled patients on following strict gluten-free diets, often in the absence of medical indications to restrict wheat intake. I prescribed homeopathic remedies. I used herbs instead of antibiotics. By all professional measures, I was a good naturopath.

I now know that I was simply dangerous. I was just another quack. I knew enough medicine to appear qualified as a doctor, maybe even to some MDs in the right settings. It wasn't until I began witnessing egregious acts of medical malpractice that I saw for the first time the sharp limits of naturopathic medicine. Once I began questioning my own beliefs, it became clear that I had made a terrible mistake. And no, hawthorn would not help.

Walking away from the naturopathic profession and speaking out cost me many things in addition to my career. I lost friends. I lost years of my life. I lost an enormous amount

An Inside Look at Naturopathic Medicine

of money—the same sum that it costs to attend a real medical school. I want to prevent others from being fooled into thinking that naturopathy is something that it is not.

I have been accused of lying about my experiences and misrepresenting naturopathic education and training (Ernst, 2016; Senapathy, 2016). There is no doubt these accusations will continue. Naturopaths are very good at ignoring criticism, scientific realities, and their own shortcomings. They also hate me vehemently (Gorski, 2016; Thielking, 2016). I am sure this chapter will not mend my broken relationships with naturopaths.

I'd like to take the reader through a deconstruction of each principle of naturopathic medicine based on my inside perspective. Perhaps I can prevent others from making the same mistakes I made. As far as I can tell, I am the only whistleblower to emerge out of this secretive community.

First, Do No Harm

First things first. Do no harm. I find it foolish for naturopaths to demote the most fundamental medical principle to the third position in their list. Let's fix that and address the principles in the most logical order. Above all else, medical practitioners have this ethical obligation: "either help or do not harm the patient" (Lloyd, 1983). I cannot see how naturopaths could ever honor this core value if they can't even put it in the right place.

Naturopaths are not trained to be medically competent. Indeed, they can cause great harm by adhering to the crooked order of their principles instead of prioritizing the welfare of a patient. Faced with such criticism, naturopaths would probably insist that the sequence of their principles is fluid, not fixed. But the order of their principles is consistently reproduced across the websites and printed material of naturopathic schools, organizations, and leaders (e.g., American Association of Naturopathic Physicians, 2011a; Association of Accredited Naturopathic Medical Colleges, n.d.-b; Bastyr Center for Natural Health, 2016a; Bastyr University, 2008a; Council on Naturopathic Medical Education, 2016; National University of Natural Medicine, 2016; Pizzorno and Murray, 2013). This listing sends a strong message: the body's ability to cure itself of disease and the priority of uncovering the root cause of illness are more central to the practice of naturopathy than ensuring that harm is minimized. This hardly matters, though, for when harm does occur, naturopaths have a convenient cop-out: blame the patient. I've seen this blame take many forms, from accusing patients of not pursuing naturopathic care early enough to lacking faith and persistence in the treatment. One horrific experience has stuck with me.

While practicing in Tucson, Arizona, I had a very sweet and quirky woman as a patient. She was anxious about her health and did not like to visit any kind of doctor. When she admitted to me during a checkup that she had been experiencing troublesome symptoms for several months, my ears perked up.

During my exam, I immediately became concerned. There were abundant, hard-to-miss red flags of cancer. She had a palpable liver mass. Her spleen was enlarged. She complained of fatigue. She was thin and pale. I ordered a few key laboratory tests but already knew this patient needed proper medical care, not a naturopath. I began organizing a referral for the patient to see a gastroenterologist as soon as possible. The lab results arrived the next day. They indicated hepatitis at best, cancer at worst. My clinical assessment remained the same: the care of a medical doctor was needed urgently.

This did not go over well with the patient. She was hesitant to transfer care to a medical doctor. I insisted this was the only appropriate course of action. We were at a stalemate. She eventually accused me of trying to bully her. (She even accused me of being "too allopathic."[3]) Unfortunately, as a long-time patient of naturopathy, she believed the drivel that a medical doctor would treat her as a disease instead of a person and not consider other contributing factors to illness, such as her self-reported high stress level.

I countered this argument with my concern for her life. Her medical condition was serious and beyond my training. I suspected this was the case before ordering her blood work, which I had done only to prove she needed a medical doctor. Now, with the grim results in hand, I was surprised the patient disagreed with me. I was even more taken aback when she instead chose to continue her care with another naturopath at the clinic where I worked. At least, I thought, I would easily get this naturopath's support in referring this patient out, as I thought it was obvious that needed to happen. This naturopath had been in practice for over thirty years.

I specifically and unambiguously told the naturopath who replaced me that I was worried that the patient had liver cancer. The naturopath had access to my chart notes, the lab results, my assessment, and gastroenterology referral. This other naturopath outright disagreed with me. To my horror, the patient left with a diagnosis of anorexia, a whey protein supplement, and instructions on how to manage stress.

Less than one month later, the patient was hospitalized for complications from cancer that had metastasized to the brain; it was reported to me that the patient had slipped into a coma. I read in the obituary section of a local newspaper that the patient died months later.

It is the ethical duty of a treating practitioner, naturopath or otherwise, to make every effort possible to ensure treatments provided to patients are clinically relevant, proven to be safe, and effective. Even when the cancer symptoms are jumping off of a patient's chart, naturopaths are not trained well enough to recognize such obvious illness. Nor do decades of experience, it seems, improve clinical skills, for at least this one naturopath. Instead of considering the impact of her own actions on this outcome, the naturopath went on to blame the patient. I will never forget this naturopath commenting to me about how it was this patient's anxiety that killed her, as she sauntered off to make sure her chart notes "were in order" after learning the patient was in a coma. The naturopath confided in me that there could be a malpractice lawsuit.

The harm in this story is obvious. A patient missed an important opportunity for proper diagnosis and treatment. Earlier medical care may not have extended her life, given the advanced stage of her disease. But a timely diagnosis may have given her the opportunity to prepare for death and spend more quality time with her loved ones. A naturopath's inadequate diagnosis and treatment possibly expedited her death.

What is clear from this story is that more years spent practicing naturopathy does not equate to more medical competency—witness my older colleague. Perhaps the two are negatively correlated, as age fortifies the brazen attitude that naturopathy is capable of curing anything and "allopathy" is to be avoided at all costs.

The Healing Power of Nature

Naturopathy claims the body possesses an innate ability to heal. This means, without medical intervention, the body often heals itself. The problems here are best illustrated in how naturopaths attempt to treat cancer (that is, when they know about it).

Cancer is complicated by anyone's assessment. It is well-established that cancerous tumors are very heterogeneous: they can harbor many different types of cancer markers, contain cells in various stages of growth and differentiation, and display aberrant functions that have been acquired through a large number of genetic mutations (Hanahan and Weinberg, 2000). As a result, appropriate cancer treatment provided by medical oncologists is personalized as much as possible for each patient. Medical cancer care involves different types of anticancer therapies, such as chemotherapies, hormone-blocking medications, radiation, surgery, and so forth. It does not include special foods, high-dose vitamins, supplements, saunas, enemas, or other common naturopathic treatments. This is because the evidence of using naturopathy to treat cancer is very clear. No natural product, diet, or alternative therapy has ever been shown to effectively and safely cure cancer (Cassileth and Yarett, 2012; Vickers, Kuo, and Cassileth, 2006). In fact, patients who use alternative therapies in cancer treatment have reported a worse quality of life (Yun et al., 2013) and have higher rates of mortality compared to cancer patients who do not use alternative or naturopathic treatments (Ernst, 2013; Johnson, Park, Gross, and Yu, 2017; Risberg et al., 2003; Yun et al., 2013). Despite the clear evidence that medical oncology offers the best chances of survival (National Institutes for Health, 2013), naturopaths believe otherwise and fervently market their ineffective treatments to cancer patients (e.g., American Board of Naturopathic Oncology, n.d.-a).

The naturopathic profession offers a specialty in naturopathic oncology with its own board certification. Practitioners who pass a naturopathic oncology exam are given the title "Fellow of American Board of Naturopathic Oncology." These naturopaths can then display a "FABNO" acronym after their naturopathic title, a model copied from established medical specialty boards.

Naturopathic oncologists believe that "modern medicine has made little advance in its War on Cancer," and so it seems "many people choose to also include complementary and alternative medicine in their fight against cancer" (American Board of Naturopathic Oncology, n.d.-a). It is blatantly wrong that modern medicine has had little success in treating cancer. If one only considers the most basic statistic, the five-year relative survival rate, it is immediately obvious that one's chances of surviving cancer after diagnosis have improved greatly over the past ninety years (National Cancer Institute, 2016; National Institutes for Health, 2013; Schattner and Waldman, 2010). The five-year survival rate for leukemia, as an example, has skyrocketed since the 1960's, rising from about 14 percent to 61.7 percent due to medical research and modern treatment (Leukemia and Lymphoma Society, 2015). This is a beautiful example of the success of modern medicine. But, of course, cancer is scary, and reporting false facts is an easy way to unnerve patients who, by the nature of their disease, are extraordinarily vulnerable to extraordinary claims. This scare tactic seems to be par for the course in naturopathy.

Supposedly, earning the FABNO certification demonstrates competency in both naturopathic and conventional oncology (American Board of Naturopathic Oncology, n.d.-a). I have never seen the FABNO exam, but competency in conventional oncology seems impossible for a naturopath to achieve, given what I know about the training and practices of naturopathic oncologists.

In order to sit for the FABNO board certification exam, a naturopath needs to have completed a two-year "residency" in a cancer clinic that offers naturopathic cancer treatments. In comparison, medical oncologists spend three years in an oncology training program *after* completing a prerequisite medical residency. This equates to more than twenty thousand hours in clinical training with tens of thousands of patients. Alternatively, a naturopath can demonstrate competency in naturopathic oncology by spending a minimum of five years doing any of the following: 1) achieving five thousand clinical hours with 2,250 cancer patients, 2) conducting naturopathic cancer research, 3) teaching naturopathic oncology, or 4) overseeing a naturopathic oncology residency (American Board of Naturopathic Oncology, n.d.-b). A real medical resident clocks at least five thousand hours in patient care in just one year.

Additionally, naturopaths seeking FABNO certification must also publish case reports and complete continuing education hours in oncology (American Board of Naturopathic Oncology, n.d.-b). A favorite venue for naturopathic oncologists to publish is *Integrative Cancer Therapies*, which is a peer-reviewed journal that focuses on complementary and alternative cancer treatments. Notably, a naturopath sits on the editorial board for the journal (Editorial Board of Integrative Cancer Therapies, 2016). The journal is not high-ranking (Scimago Journal and Country Rank, 2014).

Medical doctors understand that it is overwhelming, and maybe impossible, to know everything about cancer, especially as researchers continually add to the scientific body of knowledge of cancer. For this reason, the field of oncology is broken up into several subspecialties, such as hematologic oncology, radiation oncology, and pediatric oncology, to name a few. Oncologists even specialize in specific types of cancer, such as breast, colon, or lung cancer, each being too complex for cancer surgeons or oncologists to treat multiple types. Again, this is why an extraordinary amount of postgraduate medical training is required to specialize in oncology.

Knowing this, think for a moment about the absurdity of a naturopath claiming to know enough conventional oncology to treat any type of cancer in patients of any age. Most medical oncologists would never make such a bold claim; they'd be asking for a malpractice lawsuit or appear like the infamous Stanislaw Burzynski, the medical doctor known for his one-size-fits-all experimentally dubious cancer treatments using a compound found in human urine.[4] Yet this is exactly what naturopathic oncologists claim to do with their own customized treatments.

Naturopathic oncology is rife with quackery based on the idea that medicine is not always necessary to cure cancer. The special healing powers of nature are required. Naturopathic treatment alternatives include dietary restrictions, such as cutting out sugar, and injections of vitamins or even baking soda. Naturopaths want to fill the system with things that are claimed to support detoxification,[5] like coffee enemas, special teas, and supplements, lots of supplements. These therapies are often touted as being safe and clinically effective alternatives to chemotherapy and radiation (e.g., Huber, 2016), despite the fact that many complementary and alternative treatments are understudied or have evidence to suggest they are harmful (Tascilar, de Jong, Verweij, and Mathijssen, 2006). And, of course, it wouldn't be naturopathy unless herbs were involved.

A common naturopathic cancer treatment is the intravenous injection of mistletoe extract. My former boss used an extract of mistletoe with his cancer patients. The plant was even painted on the wall of the waiting room in his clinic. Mistletoe remains unapproved for the treatment of cancer (National Cancer Institute, 2015b), although it has shown some anticancer effects in petri dish experiments and animal studies (Mengs, Göthel, and Leng-Peschlow, 2002). A lack of rigorous and well-conducted clinical trials for mistletoe in cancer patients has led experts to consider mistletoe as a potentially promising but understudied therapy that requires additional research to assess its safety and effectiveness (Wuesthof and CAM-Cancer Consortium, 2005). It is important to remember that many substances kill cancer cells in a petri dish but don't work well in the body. Naturopaths do not seem to mind selling or administering these substances to their patients (Hermes, 2016a). When I was in practice in Arizona, I learned that some naturopaths seem to have little respect for the US Food and Drug Administration (FDA), which regulates medicines in order to protect patients from ineffective and unsafe substances—such as those recommended by naturopaths.

I was standing in the lab area with my former boss one afternoon when he casually mentioned that Ukrain, the cancer medication he was administering to patients, had not arrived in the mail as it usually does. Curious, I asked, "Why not?"

He replied, "It's probably been confiscated."

"Why would it be confiscated?"

His answer was vague. Something about the FDA doing so from time to time. But I shouldn't worry, he said, the next shipment would arrive soon.

It never did.

Michael Uzick, my ex-boss, had been importing Ukrain from Austria through a system of money wiring that took place outside of the clinic. I do not remember the exact cost of the treatments, but it was several thousand dollars per round; patients were getting regular injections, some multiple times per week. Given what I soon learned about Ukrain, it was fortunate for both Uzick and his patients that this last order never came, although these patients had paid the clinic for their Ukrain injections in advance.

Ukrain is not an FDA approved medication for any disease (Memorial Sloan Kettering Cancer Center, 2014). This status made the importation and administration of Ukrain unlawful, and because Uzick was using it on cancer patients, it was exceptionally unethical. Under Uzick's medical orders, I had administered Ukrain to cancer patients before I knew his operation was illegal.

It is difficult to describe how it feels to learn one accidentally played a role in criminal activity. It is painful, nauseating, and scary. Rightly so, I was worried I would find myself in deep trouble. And I was angry; it turned out Uzick knew his actions were (to use his words) "legally questionable." Most of the patients receiving Ukrain in our office were very ill. Some even died in the midst of treatment. With sadness, I often think of our patients and the tens of thousands of dollars they paid for the false hope of a cancer cure.

About forty-eight hours after learning the gravity of the situation, I hired a lawyer and reported my boss to the Arizona attorney general and the state's naturopathic board. Soon after, I quit the practice. This event was the impetus to my retirement from naturopathy. The decision to speak about this experience came after the former president of the American Association of Naturopathic Physicians, Michael Traub, urged me not to report what I had learned to the authorities, as I was a "*naturo*path after all." His statement implied that naturopathy is made up of practitioners knowingly using a plethora of unethical treatments. He was right.

Despite statements on the Ukrain website that it is designated as an orphan drug for the treatment of pancreatic cancer, a search for the drug's designation in the FDA orphan drug database returns no results. It is marketed by its manufacturer as a miracle drug, capable of treating *all types* of cancer (Nowicky Pharma, n.d.). Through some molecular mystery, the manufacturer reports that Ukrain only targets and kills cancer cells, leaving healthy cells unharmed (Nowicky Pharma, n.d.). Had I been paying attention to basic consumer warnings, I would have quickly recognized this type of marketing as a common ploy used by cancer quacks (Federal Trade Commission, 2008; US Food and Drug Administration, 2008).

During my final meeting with my former boss, he said all naturopaths walk the line between legal and illegal practices. He felt that his ability to assess evidence and safety was better than that of the FDA, and it bore no weight that the FDA, as well as most European countries and Australia, had denied approval for the use of Ukrain due to lack of effectiveness, coupled with safety and quality assurance problems (Boehm, Ernst, and CAM-Cancer Consortium, 2015). I understand that all professions contain individuals who display what I consider to be delusions of grandeur regarding their abilities and intelligence. I can accept that some people may think that they know better than major medical institutions around the world. What I do not understand is the naturopathic medical board failing to recognize the gravity of Uzick's misconduct. A misleading report by the naturopathic board states Uzick had voluntarily stopped treating patients with Ukrain prior to my complaint (Arizona Naturopathic Physicians Medical Board, 2014). This statement misses the point that Uzick was waiting on an order of Ukrain to arrive in order to treat patients at the time I filed my complaint. And obviously, it misses the fact that he used Ukrain on patients for about seven months knowing he was doing something illegal.

Had the board properly investigated Michael Uzick and his use of Ukrain on cancer patients, this information would have been revealed. By not taking my complaint seriously, the board effectively kept Uzick from being adequately punished. He received a mere letter of reprimand (Arizona Naturopathic Physicians Medical Board, 2014). As far as I know, nothing has come from the investigation by the attorney general's office. Uzick continues to be a respected naturopathic cancer doctor within the community.

Naturopaths operate in their own reality when it comes to science and medicine. Ukrain is thought to be a natural product, derived from the herb chelidonium. It does contain alkaloids extracted from the herb, but these are mixed with a known chemotherapeutic agent called thiotepa in a manner that raises serious concerns about purity and toxicity (Boehm et al., 2015). There have been a handful of randomized, controlled trials investigating Ukrain's anticancer effectiveness, but the low quality of the studies, most having been performed in Belarus and funded by the manufacturer, do not allow independent researchers to make any solid conclusions (Ernst and Schmidt, 2005).

Ukrain was made in a do-it-yourself pharmaceutical lab in Vienna that was shut down after its inventor was found guilty on charges of fraud related to relabeling and selling expired product (Local, 2015). Based on the time line of events, it seems likely that our patients received injections of this expired medication. I assume this information would not have made a difference in disciplinary decisions, sadly, as it seems the naturopathic board acted to protect the image of the profession, rather than patients.

Identify and Treat the Root Cause of Disease

Outside of treating cancer, naturopaths thrive on patients with diseases and symptoms that are not well explained or treated by conventional medicine. Naturopaths believe their unique training gives them skills that are superior to that of medical doctors. Make no mistake, naturopaths are not omnipotent healers. Since leaving the profession, I have carefully reviewed my education in comparison to official claims about the quality of my naturopathic training. Naturopaths are not adequately educated or equipped to be primary care physicians.

I attended Bastyr University from 2007 to 2011. During the 2007 summer term, I finished the remainder of the prerequisite naturopathic courses, which included abbreviated versions of organic chemistry and physics. Physics class included many circle discussions to review homework that could be completed in groups. We took a class field trip to an ice skating rink but delved simplistically into marvels of motion and matter. Overall, it was the easiest college-level coursework I ever had.

Gaining entrance into Bastyr University's naturopathic program was not difficult. Compared with medical schools, naturopathic schools do not have selective admissions. Acceptance rates at US naturopathic schools can range from 75 percent at the Southwest College of Naturopathic Medicine and Health Sciences in Tempe, Arizona (Princeton Review, 2017b) to 82 percent at National University of Natural Medicine in Portland, Oregon (Princeton Review, 2017a). I was accepted and offered a nominal scholarship to every school to which I applied. I graduated from San Diego State University cum laude, but did not take a graduate school exam or the medical school entrance exam. It seems that the only real determining factor for matriculating into an accredited naturopathic program is the individual's ability to pay tuition.

Despite some of the obvious clues that naturopaths are not attending real medical school, naturopaths genuinely believe their education is similar to a legitimate medical education. Figures 6.1 and 6.2 show the most commonly shared pieces of naturopathic propaganda supporting this fantasy—charts comparing the coursework hours between an MD student and an ND student in an accredited program (Hermes, 2015).

Anyone reviewing such numbers would conclude that, compared with MD students, ND students

- take more hours of coursework in anatomy, embryology, biochemistry, physiology, and pathology;
- take more than twice as many hours in "clinical and modality training";
- have about the same number of overall hours of instruction.

It seems, on paper, that naturopathic students are receiving not just the same training as medical doctors, but actually a more "well-rounded" education. Indeed, the

An Inside Look at Naturopathic Medicine   149

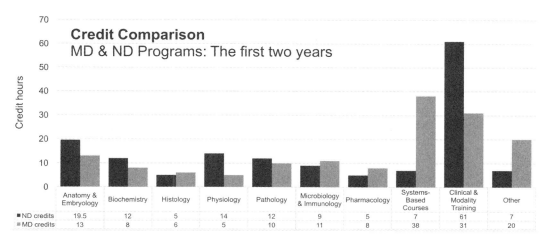

Figure 6.1
A comparison chart of MD and ND credits in the first two years of each program, published and widely used by the naturopathic profession (Association of Accredited Naturopathic Medical Colleges, n.d.-a). The numbers are based on funny accounting, which seems to inflate naturopathic training. The graphic is adapted from a graph that originally appeared on the website of the Association of Accredited Naturopathic Medical Colleges. It is no longer available. The authors of the original graph claimed that it accurately compared the credit load for the medical degree program at the University of Washington and the naturopathic degree program at Bastyr University for the 2010 academic year.

differences between medical and naturopathic school could be explained by the different focus of the programs: natural versus conventional medicine. I remember reviewing these charts prior to starting school at Bastyr and being impressed by the seemingly extensive curriculum. I also liked the idea of not having to study for the Medical College Admission Test.

It is easy to say naturopaths have more training by simply comparing numbers on a chart. The truth is that this extra training is not even a little bit medical or scientific. Understandably, education comparison charts such as these always fail to mention what the "other courses" include, what "modality training" really means, and that clinical sciences are taught by naturopaths, not experts with real medical training. Here is a quick look at what some of this extra training entails (Hermes, 2015):

- 88 hours in homeopathy classes
- 198 hours in physical medicine classes[6]
- 33 hours in traditional Chinese medicine class
- 146 hours in botanical medicine classes[7]
- 165 hours in naturopathic philosophy classes

The American Association of
Naturopathic Physicians

Naturopathic Medical Education Comparative Curricula
Comparing Curricula of Naturopathic Medical Schools and
Conventional Medical Schools

National College of Naturopathic Medicine	Bastyr University- Naturopathic Medicine	Yale University	Johns Hopkins	Medical College of Wisconsin
Federally and Regionally Accredited Naturopathic Medical School	Federally and Regionally Accredited Naturopathic Medical School	Federally and Regionally Accredited Conventional Medical School	Federally and Regionally Accredited Conventional Medical School	Federally and Regionally Accredited Conventional Medical School

Basic and Clinical Sciences:
Anatomy, Cell biology, Physiology, Histology, Pathology, Biochemistry, Pharmacology, Lab diagnosis, Neurosciences, Clinical physical diagnosis, Genetics, Pharmacognosy, Bio-statistics, Epidemiology, Public Health, History and philosophy, Ethics, and other coursework.

1548	1639	1420	1771	1363

Clerkships and Allopathic Therapeutics:
including lecture and clinical instruction in Dermatology, Family Medicine, Psychiatry, Medicine, Radiology, Pediatrics, Obstetrics, Gynecology, Neurology, Surgery, Ophthalmology, and clinical electives.

2244	1925	2891 (+thesis)	3391	2311

Naturopathic Therapeutics:
Including Botanical medicine, Homeopathy, Oriental medicine, Hydrotherapy, Naturopathic manipulative therapy, Ayurvedic medicine, Naturopathic Case Analysis/Management, Naturopathic Philosophy, Advanced Naturopathic Therapeutics.

588	633	0	0	0

Therapeutic Nutrition

144	132	0	0	0

Counseling

144	143	Included in psychiatry (see above)	Included in psychiatry (see above)	Included in psychiatry (see above)

TOTAL HOURS OF TRAINING

4668	4472	4311 (+thesis)	5162	3674

Sources:
Curriculum Directory of the Association of American Medical Colleges

818 18th St, NW, Suite 250   •   Washington, DC 20006   •   202-237-8150   •   www.naturopathic.org

Figure 6.2
A chart comparing the naturopathic curricula to the medical curricula at top universities in the United States (California Naturopathic Doctors Association, 2013). The total number of training hours for naturopathic students comprises time spent learning "naturopathic therapeutics," which include debunked and unproven therapies like homeopathy and chiropractic techniques. Naturopaths use this misleading figure alongside the amount of real training completed in medical schools to make it seem as though naturopaths receive equivalent training as medical doctors.

Basically, the naturopathic curriculum is presented in a way that lumps pseudoscience into an educational category, called "clinical and modality training" or "naturopathic therapeutics," so that it seems legitimate. It's a glaring red flag that I received fifty-five hours of pharmacology education but also eighty-eight hours of homeopathy coursework! All of these hours spent memorizing homeopathic remedies or learning superstitious practices will never be medically applicable. Regardless, mischievous charts like these are widely disseminated and used politically by the profession to convince lawmakers that naturopathy is a distinct form of primary care medicine. In fact, most of the false information I once believed about my ND education can be traced to my experiences lobbying for the naturopathic profession to the federal government.

In 2015, I publicly condemned the content of the naturopathic training I received at Bastyr University. In doing so, I critically analyzed several of the education comparison charts produced by the naturopathic profession, including the two above. I pointed out that the charts provided misleading and inaccurate information to the public by falsely comparing credit hours while ignoring the content of the education. The naturopathic profession seems to have responded to my criticism by removing these comparison charts from many, but not all, of their websites. I can only hope naturopaths have stopped disseminating them to lawmakers as well.

Much of what is presented to lawmakers by those lobbying for naturopathic medicine is information about the accreditation of the naturopathic schools. Bastyr University is one of five schools that offer an accredited graduate program in naturopathic medicine in the United States. There are also two universities in Canada that are accredited by the profession. In my opinion, the naturopathic profession fools others into believing that their accreditation means an internationally renowned and rigorous curriculum that is on par with that of medical school.

The North American naturopathic institutions are accredited by the Council on Naturopathic Medical Education (CNME), which has been granted programmatic accrediting status by the US Department of Education for meeting specific administrative criteria, such as keeping their finances in order and publishing a mission statement, among other markers of an organized institution (US Department of Education, 2016). The CNME is mostly made up of naturopaths and alternative medicine advocates (Council on Naturopathic Medical Education, 2015). Despite being flaunted as such by naturopaths, the fact that the U.S. Department of Education recognizes the CNME is not a testament to the quality of the naturopathic curriculum or any other educational program (US Department of Education, 2016).

In addition to naturopaths being confused about what accreditation means, they have also circulated misconceptions about how the degree compares to professional doctoral degrees. In a 2011 lobbying document, the naturopathic professional organization (American Association of Naturopathic Physicians) describes the ND degree as a "US Department of Education and Carnegie Institute First-Professional Degree under Doctorate-Profession (Clinical), on par with MD and DO" (American Association of Naturopathic Physicians, 2011b). This document was used by naturopaths, including myself, to lobby for access to the same loans, scholarships, and residency funding that MDs and DOs receive, since our ND education, after all, was apparently just like that of a medical doctor.

First-professional degrees are considered comprehensive doctoral and professional programs that offer the highest degrees available in the fields of humanities, social sciences, and science, technology, engineering, and mathematics (STEM) fields, plus graduate degrees in business, engineering, law, and medicine (Carnegie Classification of Higher Institutions, 2015c). I looked up how Bastyr's ND degree was classified by Carnegie in 2015, and it is not in one of the first-professional degree categories (Carnegie Classification of Higher Institutions, 2015a). Rather, it is classified as coming from a "special focus institution," which includes acupuncture schools, theology programs, and ITT Technical Institute (Carnegie Classification of Higher Institutions, 2015b). As far as I can tell, the American Association of Naturopathic Physicians has been lying to lawmakers and students about this esteemed classification.

It seems that as long as naturopaths can convince lawmakers that they are similar enough to medical doctors, they will continue to make political gains. The naturopathic profession relies on the benefits of a closed-loop accreditation system that distances itself from external review. If lawmakers and policymakers really knew what was being taught in naturopathic schools, the profession might face serious opposition and regulatory sanctions. I think one place to start is to demand that the Naturopathic Physicians Licensing Examination (NPLEX) undergo a critical review by experts of medical education. Naturopaths claim their "board exam" is on par with the United States Medical Licensing Examination (USMLE), which is used to license medical doctors (e.g., Krumbeck, 2011; Mountain, 2011). This comparison is perverted. The NPLEX includes numerous questions on homeopathy and other naturopathic modalities when faced with cases about life-threatening medical conditions that should be treated with immediate emergency care (Hermes, 2016c; North American Board of Naturopathic Examiners, 2016). Although naturopathic students often use USMLE study materials to prepare for the NPLEX, this study strategy does not mean the exams are similar; it only speaks to the fact that the USMLE study materials and exam are superior to their naturopathic counterparts.

As a result of aggressive lobbying by the profession, as of 2017, naturopaths are licensable in nineteen US states, the District of Columbia, Puerto Rico, and the US Virgin Islands with varying scopes of practice (American Association of Naturopathic Physicians, 2017). Since I started exposing the misrepresentations by the profession, naturopathic political advances seem to have slowed. Naturopathic organizations have begun to take information, like the education comparison charts shown above, off of their websites and changing how they publicly describe the naturopathic doctorate degree. This may be in part due to lawmakers demanding to see proof of the claims made by the naturopathic profession.

Doctor as Teacher

Naturopaths believe an integral part of the doctor-patient relationship includes educating patients about medicine and illness in order to provide them with resources and knowledge to maintain their own health. Naturally, naturopaths teach patients information they learned in their schooling. Naturopathic education teaches a lot of pseudoscience alongside some real science. For example, I took one class in pharmacology, but three in homoeopathy; I took one class in orthopedic medicine, but six in chiropractic techniques. This poses a problem. It is incredibly difficult, perhaps impossible, to differentiate between what is science-based and what is not. How, then, can naturopaths teach their patients useful medical information if most of their own medical education is dubious? I find that much of the naturopathic advice taught to patients is nothing more than beautifully packaged nonsense. As my mother-in-law says, good packaging can sell anything. Bogus medicine is no exception.

Naturopaths depend heavily upon marketing to sell their brand. Examples of successful naturopathic marketing include the designing and selling of naturopathic wellness packages. The strategy of bundling naturopathic services with promises of improved health is so ubiquitous in naturopathy that even the naturopathic teaching clinics do it.

Southwest College of Naturopathic Medicine and Health Sciences (SCNM) in Tempe, Arizona, for example, promotes a springtime detoxification program that patients should complete *at least* once per year (Southwest College of Naturopathic Medicine, 2016a). According to SCNM, our bodies are machines that get tired of performing their daily detoxification and other normal tasks and need a break a few times a year. For $199, one of the clinic's naturopaths will take you through an intense food elimination protocol:

Some of the top foods to avoid during the 21-day period include dairy, eggs, sugar, wheat and gluten, soy, peanuts, corn, specified forms of processed meats and even certain vegetables. Many people don't recognize how vegetables in the nightshade family—tomatoes, chilies, eggplants and peppers, for example—can sometimes negatively affect the body, especially in people with autoimmune disorders, according to Dr. [Christina] Youngren. Patients who commit to the cleanse will also avoid caffeine, which can sometimes pose a challenge.

Youngren then goes on to make wild claims. She says a healthy digestive system will improve your outlook on life; she also says that the elimination of offending foods may alleviate autoimmune and neurological disorders (Southwest College of Naturopathic Medicine, 2016a). If that's true, then this detox is a miracle cure.

The National University for Natural Medicine (NUNM) in Portland, Oregon, has offered wellness packages through a GroupOn for naturopathic care (National University of Natural Medicine, 2012). For $250, patients received three naturopathic skin-rejuvenation treatments. Apparently, "the Clinic's certified physicians harness the exfoliating forces of photons, delivering Light and Heat Energy (LHE) phototherapy to activate the body's natural healing abilities." Yeah, right. While the flavors of these different health packages are different, one thing is the same: each is a way to receive payment in advance.

Yes, naturopathic clinics are cleverly marketing wellness packages that consist of unnecessary medical treatments. Naturopaths are taught such schemes through a series of practice development courses and business workshops. In addition to the mandatory practice management courses at Bastyr, I took one of these naturopathic business development workshops offered through a seminar series called the Health of Business Business of Health (HBBH) program.

Dickson Thom, a dentist and naturopath, is HBBH's founder (Thom, n.d.). He was my instructor for the HBBH seminar I took. Thom is also a former dean of naturopathic medicine at NUNM. He was a "professor" there for a wide range of subjects such as radiology and gastroenterology. He is a natural medicine author. He is also an avid believer in energy medicine. Not only does Thom seem to do it all, he can also teach you how to sell it all.

One of the key concepts I learned at HBBH is that the naturopathic patient visit needs to be carefully planned in order to maximize return business. We were taught, for example, how to physically set up an office to make patients feel comfortable. I also learned how to recognize emotional cues from my patients that helped me sell naturopathic treatment programs. I learned how to design these wellness programs in a way that ensured repeat business all year long.

The HBBH wellness program template specifies health checkups every season of the year, thereby guaranteeing four patient visits per year per customer. This cycle continues indefinitely until the naturopath deems the patient healthy enough to reduce their visits. Nothing in this ploy is medically necessary. Yet the language used in these programs suggests that in order to achieve optimum health, the patient must diligently follow naturopathic advice. Based on the knowledge gained in my HBBH marketing training, I helped develop a similar wellness program.

The Right Detox was a food-based "detoxification" program developed and sold at my former practice in Arizona. The marketing advanced the idea that no matter your age or health status, this detox program was right for you. There were several tiers of The Right Detox. The higher the tier, the "deeper" the detoxification. Supplements and services were bundled after we had chosen the price points of each tier, which guaranteed a profit. It was recommended that patients follow this detox program four times per year, once per season.

If you are thinking this experience is just one cherry-picked example and not necessarily indicative of how naturopaths create detoxification programs, think again. I have been part of other detoxification programs at other clinics, and the services and products always come down to one factor: profit. Detoxification scams are not the only business ploys naturopaths use to bait and hook patients.

Naturopaths are often acknowledged as having decent training in nutrition, even by their critics. I will concede that at Bastyr, students do receive a great deal of nutritional education. But they, like all naturopaths, are also taught to use invalidated tests, which can be used to increase supplement sales and to interpret results using fabricated medical theories in order to increase patient visits.

When I was at Bastyr, naturopathic students received 143 hours in nutritional coursework and even more in clinical science classes. For instance, a former colleague of mine was once recommended by a naturopath to take over fifty nutritional supplements per day, based on deficiencies noted through the bogus diagnostic method of muscle testing.

Muscle testing, also known as applied kinesiology, involves the comparison of muscle strength under two different conditions. The first condition is the patient's ability to resist pressure applied to an outstretched arm. This is considered a control. The second condition is the patient's ability to resist pressure applied to an outstretched arm while holding a substance in his or her hand. If the tester perceives a weakness in the second condition, this means this substance affects the patient. It is thought that the patient might be deficient in the tested substance, or that this substance needs to be detoxified out the patient's body. It's up to the naturopath to decide.

Another common invalidated test used to make dietary recommendations is the IgG food allergy panel (Gavura, 2012). This is an unproven food allergy test that uses the presence of memory antibodies (IgG) instead of true allergy antibodies (IgE) in order to diagnose what naturopaths call food intolerances. Memory antibodies are not associated with allergies or any kind of adverse reaction to antigens. IgG antibodies increase after exposure to foods, which is normal (Gavura, 2012). This is why patients who eat the same foods frequently, like a banana every morning, will have high levels of IgG antibodies for antigens found in bananas. But a naturopath will see high IgG levels on a results panel and diagnose the patient with a bogus banana allergy. You can probably see how this test is a vicious cycle.

IgG food allergy results are usually confusing for patients; they often don't know with certainty whether they have any unpleasant symptoms after eating certain foods. Naturopaths use this doubt in order to convince patients of a lurking problem. For example, the naturopath will suggest that the "food intolerance" may be causing migraines, fatigue, joint pain, or any quirky symptom. I frequently, and successfully, used this tactic in my practice when reviewing IgG food panel results with patients.

Naturopaths are taught to recommend a food allergy panel for their patients because it is a convenient business model that gets patients to commit to multiple office visits into the future. The IgG food allergy scheme went like this: the appointment and blood draw for the panel would require a specific office visit and a hefty out-of-pocket expense. Then the patient would need to schedule a follow-up visit two weeks later to review the results. Finally, I would recommend a variety of naturopathic treatments, which often included expensive nutritional supplements and strict dietary interventions. Sometimes another visit was required in order to assess if the patient had improved. I was unknowingly abusing my platform as a teacher. I was practicing according to what I learned while in school.

The naturopathic principle of doctor as teacher portrays naturopaths as medical experts selflessly dedicated to sharing naturopathic wisdom about healthy living and natural medicine. In reality, they are not natural health experts. They are selling modern-day snake oils disguised by a thin veil of scientific trappings. Naturopaths have adopted just enough scientific knowledge to leverage it against the best interests of patients.

Treat the Whole Person

Naturopaths say they treat the whole person, not diseases. This is absurd. They diagnose their patients with fake diseases. What are they treating there? A person with the fake disease?

Medical doctors and other health practitioners also treat people as a whole. Why do you think they ask for detailed histories? This sounds like a frivolous point to make against naturopaths, but it is not. Naturopaths say their training is extraordinary, when in fact it appears to be oriented toward profit maximization relying on heavy doses of pseudoscience. The problem is that just enough hollowed-out science is taught to make the whole curriculum seem valid to those without a critical eye.

My clinical science courses at Bastyr University, such as dermatology, gastroenterology, or cardiology, were predominately taught by naturopaths who had declared themselves "specialists" in these fields. Aside from the exceptions of oncology, pediatrics, and homeopathy, naturopathic education and training does not offer specialties. The naturopathic schools claim that all students are trained as primary care practitioners (Thielking, 2016). In order to become a naturopathic "specialist," the naturopath more or less selects a patient base and then sometime afterward transforms into an expert in that field. Naturopaths like to say their specialties chose them.

Although some classes taught at Bastyr have the same titles as courses in medical programs, the content is not the same. My neurology teacher (a naturopath who calls herself a neurological disease specialist) frequently expressed her opinion that high doses of antioxidants and vitamins delivered intravenously could treat diseases such as Parkinson's and Alzheimer's. I also learned how to use plants, some of which are poisonous, instead of pharmaceutical drugs to manage many diseases. I learned how to dose the flower lobelia as a substitute for the life-saving medication albuterol in asthma attacks. I learned that the plant digitalis may be an effective and safe alternative for managing serious heart conditions. I learned how the tasty berry from the elderberry plant might be a smart replacement for the influenza vaccine.

If only using plants and natural substances to replace medicine was that simple. But the use of plant extracts for the treatment of disease has unpredictable results. Herbal products are not regulated as drugs in the United States by the Food and Drug Administration (US Food and Drug Administration, 2016). Supplement, homeopathic, and herbal medicine manufacturers do not have to comply with the same strict rules as drug companies. As a result, herbal products and dietary supplements may contain adulterated ingredients, hidden ingredients, or nothing at all. In a life or death situation, such as a severe asthma attack, the use of an herbal medication is very dangerous. But even in seemingly benign circumstances, alternative medicines can be risky.

In 2016, homeopathic teething tablets came under investigation after reports that the tablets caused serious adverse events including vomiting, fever, and lethargy and may have been related to the deaths of ten children (Christensen and Gumbrecht, 2016). There was concern over the amount of poisonous compounds from belladonna, the plant used to make the tablets. The FDA issued a statement warning parents to not use the homeopathic teething tablets. Yet, the Southwest College of Naturopathic Medicine and Health Sciences responded to the FDA's warnings with a statement that homeopathic remedies are inherently much safer than conventional drugs (Southwest College of Naturopathic Medicine, 2016b).

In January 2017, the FDA confirmed inconsistent levels of belladonna compounds in the teething tablets, sometimes in amounts substantially larger than what was reported on the label (US Food and Drug Administration, 2017). The product was recalled from the market in April 2017 (Howard, 2017).

As any student, I implemented what I learned in school into my naturopathic practice. I remember the look of trepidation on a mother's face as I tried to explain how holding half of a heated onion against her child's ear would somehow alleviate his ear infection (e.g., Neff, 2015). I got lost in the explanation, realized with embarrassment the ridiculousness of it all, and never recommended this folk remedy again. I am amazed that other naturopaths can convince parents to try this trick, instead of opting for a real treatment or just adopting a watch-and-wait strategy. Fortunately, most childhood ear infections are viral and self-limiting.

Accordingly, the accredited naturopathic schools publicly appear to have set standards appropriate for producing qualified primary care naturopaths. Students must complete a minimum of 850 hours in patient care (Council on Naturopathic Medical Education, 2016), but they really receive far fewer hours in direct patient contact than this already small number. In my own training, for example, hours counted when students reviewed a case with peers or when we observed advanced students performing physical exams. One hour out of every four-hour clinical shift (meaning a block of time spent at the teaching clinic under the supervision of a naturopath) was spent discussing patient cases. In a roundtable-type format, students would present a patient case and rely upon naturopathic philosophy to devise a diagnosis and treatment. A man suffering from headaches might be treated with a homeopathic remedy associated with desires to travel because he likes to take vacations. Yes, non sequiturs are common. We had lots of these discussions. According to the naturopathic accrediting agency, these counted as hours spent in patient care and were "primary care medicine."

Students could even count their hours sitting in clinic when there were no patients. This was common. On average, I saw about two patients per clinical shift. I sometimes had no patients at all. I remember students were not supposed to have more than four patients scheduled per four-hour clinical shift. In comparison to a regular medical student shift in a hospital, this is diluted to the point of being homeopathic.

In addition to counting hours for training standards, my class cohort was required to have contact with a minimum of 350 (now, 450) patients with diseases across various categories (Bastyr University, 2008b; Council on Naturopathic Medical Education, 2016). This created a bit of a problem because the patients using the Bastyr teaching clinic were generally healthy and presented only a limited range of minor issues. In other words, I was trained to treat the worried well. There are some specialty shifts, where the supervisor only sees patients with a specific medical condition, such as diabetes, cancer, or HIV/AIDS. But these shifts were rare, and the patients were typically there to take peat baths or discuss healthy diets.

In order to reach the graduation requirement of treating patients with certain conditions, it was necessary for students to earn credit by giving oral presentations on the diseases we had not seen. I don't remember whether I saw a patient suffering from acute chest pain. If not, I could have gotten cardiovascular competency by presenting information about the condition to my classmates. I do not know for certain, but it seems to me that any real medical student completing her clinical training in a real teaching hospital would treat at least one patient suffering from a heart attack. This is why accreditation by a naturopathic organization approved by the US Department of Education does not guarantee a high-quality medical education.

It would be unjust to call the education provided in Bastyr's naturopathy program anything close to medical training. Students at Bastyr could perform physical exams on other students in order to show competency. Without this leeway, students would have never earned the tallies needed to qualify them for graduation. I do not remember writing a single prescription for a medication as a student clinician. I do, however,

An Inside Look at Naturopathic Medicine

remember writing dozens of scripts for homeopathic remedies and herbal preparations. Naturopaths do not complete a medical school education; they attend naturopathy school, where they are trained to provide alternatives to medicine and smother critical thinking.

Prevention

A good naturopath, it is said, will help patients avoid the need for medicine altogether. There is a quote by Thomas Edison that naturopaths love to cite:

> The doctor of the future will give no medicine, but will instruct his patient in the care of the human frame, in diet and in the cause and prevention of disease.

I find this reference incredibly comical, because naturopaths, by and large, do not fully support the most reputable and proven preventative medical intervention: immunizations (Hermes, 2016b).

Naturopaths commonly claim to support vaccines but frequently subscribe to anti-vaccine conspiracy theories. In practice, naturopaths deviate from the established immunization schedule or discourage vaccines altogether. In some states where naturopaths are licensed, naturopaths have the authority to write immunization exemptions for personal beliefs. This practice has been associated with higher rates of measles and whooping cough infections (Atwell et al., 2013). These naturopaths call themselves vaccine-neutral or say there are "truths and lies" on both sides (e.g., Vancouver Island Naturopathic Clinic, n.d.; Zimmermann, n.d.). They give far more weight to vaccine misinformation and misguided parental choice than the overwhelmingly clear scientific evidence that vaccines are safe, effective, and necessary.

When discussing vaccines, these "vaccine neutral" naturopaths use buzzwords like "individualized" or "compassionate" to describe their perspective (e.g., Krumbeck, 2014; Specialized Natural Health Care of Vermont, n.d.). They pander to unsubstantiated fears that vaccines contain toxins, are linked to developmental disorders, or can overload a child's immune system. They sell supplements with misleading claims that using the product will reduce vaccine reactions (e.g., Clinton, 2016b; Krumbeck, 2014). Naturopaths have created a profitable business by capitalizing on bad science and parental fears that have often been uncritically covered by the mainstream media. These unnecessary customized immunization schedules and supplements provide no benefit for the patient, but they mean more income for naturopaths.

Perhaps worst of all, vaccine-neutral naturopaths suggest that vaccine-preventable diseases can be avoided or treated with natural medicines and a healthy lifestyle (e.g., Baral, 2015). I frequently reiterated these notions when I was in practice. It is as if naturopaths are ignoring the entire human history of disease, which brought early and horrible death until the development of modern medicine. Health and life expectancies would backslide if we were all to follow naturopathic advice of spacing out, delaying, or avoiding vaccines.

As with all health care issues, naturopaths have seemingly made up their own nonstandard standard on immunizations. Not surprisingly, naturopaths are taught how to present a false equivalence between the science supporting vaccinations and the unsubstantiated fears about vaccinations. The very first sentence from my naturopathic treatment notebook entry about vaccines is a quote from my naturopathic pediatrics teacher. It reads, "When discussing vaccinations with parents, these questions need to be discussed: Which ones to give? When to give chosen immunizations?" This is not surprising, considering that we read Dr. Bob Sears's *Vaccine Book,* a discredited book that encourages parents to avoid or delay vaccines (Offit and Moser, 2009), and that we were lectured on flawed reasons why vaccines are dangerous.

According to naturopathy, not all children should receive vaccinations following the immunization schedule set by the Centers for Disease Control and Prevention (CDC). Of course, there are medically valid reasons for avoiding or delaying vaccinations. But when I learned this information, my former teacher was not referring to any of these reasons. She was referring to parent preference and conspiracy theories regarding the safety and effectiveness of immunizations. Naturopaths are failing in their roles as medical professionals who prevent disease. You are not "pro-vaccine" simply because you agree with administering some vaccines, to some children, under some circumstances. Providing parents with wrong information about vaccinations is not empowering, it is deceiving.

A small percentage of naturopaths do believe the evidence for the CDC vaccination schedule. I personally know only three or four. When I was in practice, I believed the CDC vaccination schedule was a decent option for children, but I thought an alternative, delayed childhood vaccination schedule was safer and therefore better. I believed that the CDC created the schedule of childhood vaccines to convenience physicians in scheduling fewer office visits. I did not believe, nor do most naturopaths, that the immunization schedule is based on a proven track record of safety and effectiveness with the goal of providing immunity to dangerous diseases as quickly as possible.

Alternative vaccination schedules are typically designed through a collaborative effort between naturopaths and parents. I always noticed that parents appreciated the time and detail given to their concerns and questions. In my former practice in Seattle, parents and I usually agreed on the delayed vaccination schedule designed by Dr. Bob Sears. Even though I was not following the CDC guidelines, I was following the CDC's schedule closer than my colleagues were. Thus, I self-identified as an evidence-based practitioner.

In reality, I was nothing of the sort. I was pandering to anti-vaccine ways of thinking. Without realizing it, I was also simultaneously increasing my income by asking parents to schedule more office visits to accommodate their child's custom alternative vaccination schedule. As much as I would want to believe that I was providing personalized care for my patients based on the naturopathic training, I was offering quackery through a clever business scheme. This is painful to admit. I suffered from a very dangerous kind of cognitive dissonance, the very same that continues to characterize my former profession.

Conclusion

A friend, who is a German medical doctor, and I were talking about my experiences in naturopathy. I was explaining some of its most unethical practices: homeopathy to treat infectious diseases, fanciful cancer treatments, high-dose vitamins to treat mental illness, and so forth. She seemed a bit confused. "But you can only use natural medicine for illnesses that are going to get better on their own," she said. Undoubtedly, naturopaths are overstating the usefulness of their medicine.

Most of the conventional medical profession knows nothing, or little, about the political machinations of naturopaths. Medical associations get worn down trying to repeatedly stave off these efforts. Physicians often do not have time to show up and argue against the claims purported by naturopaths, who arrive in droves to committee hearings, along with other alternative health care providers, and often times, dozens of die-hard patients. Laws favoring naturopaths seem to slip through without anyone outside of the naturopathic profession noticing. It's as if lawmakers rationalize this support for approaches assumed to be safe and natural by falling back on the question "What's the harm?"

Patients of naturopaths are missing opportunities for timely medical care or are wasting thousands of dollars on fake treatments, like the patients who received Ukrain injections for their cancer. Patients are also being harmed by naturopaths.

Here are three widely publicized examples from 2016 and 2017:

1) A young boy with autism in the United Kingdom was hospitalized after being treated with silver, herbs, vitamins, and minerals by a naturopath. The boy had dangerously elevated levels of calcium and medical doctors needed to preform lifesaving interventions (Boyd and Moodambail, 2016).

2) A Canadian toddler with bacterial meningitis died in 2012 after his parents treated him with echinacea that they purchased from a naturopath's clinic. They called emergency services once the boy stopped breathing (Dormer, 2016). The parents believed natural remedies were effective for treating what they suspected to be meningitis (Gibson, 2016).

3) A woman died after a naturopath in San Diego gave her an intravenous injection of a solution containing curcumin, a constituent in the herb turmeric. The woman was being treated for eczema, a benign, but bothersome, skin condition (Hermes, 2017).

Sadly, there are many more stories like these. Children seem to be an especially vulnerable patient population.

In North America, naturopaths are posing as primary care doctors and getting away with it. Naturopaths do not attend medical school, do not take medical licensing exams, or complete medical residencies, yet naturopaths are asking the government for the privilege of practicing medicine. Despite how ridiculous this sounds, naturopaths have enjoyed political success. Every year naturopaths lobby the US government for broader scopes of practices, inclusion in Medicaid and Medicare, and insurance reimbursement parity with medical doctors. Once naturopaths gain licensing in a state, it is just a matter of time until their scope of practice expands to allow them to perform many of the same functions as medical doctors. State licensing almost never gets repealed.

Yet, in order to become as mainstream as naturopaths want, they would have to do away with too much of their philosophy. At the very least, they would have to publicly reject homeopathy, a field the profession has defended, endorsed, and taught as a valid treatment for decades. Naturopathy cannot be a science-based profession and an ally to homeopathy. It cannot claim to follow scientific evidence and then subscribe to the healing powers of nothing.

There are naturopaths who agree with my critical analysis of the naturopathic profession. Unfortunately, they are not in a position to jeopardize their careers or relationships by speaking out against their colleagues. But for the most part, I think it is fair to say that naturopaths do not know what they do not know. They likely never will.

It has taken two years of investigation into the naturopathic profession, as well as starting over in a master of science program in biomedicine, to understand how deeply and effectively I was brainwashed by my naturopathic schooling. The naturopathic belief system is strong. Many naturopaths I used to know would prefer to reject me outright, rather than consider a perspective that challenges their worldview. I have accepted this. It is better to be ostracized from this cult-like profession than to practice quackery using guiding principles that lay the foundation for patient exploitation.

Notes

1. In the United States, there are licensed naturopaths, like the kind I was, and unlicensed lay naturopaths. These providers are separate groups with different political agendas and perspectives on public health. In this chapter, when I speak of naturopaths, I am referring to licensed naturopaths, unless stated otherwise.

2. Chronic Lyme disease is a controversial diagnosis. To date, there is no reputable scientific evidence demonstrating a relationship between the commonly reported symptoms of chronic Lyme and the acute Lyme infectious agent *B. burgdorferi*. Medical authorities recommend against long-term antibiotic treatment for *B. burgdorferi* infection (Baker, 2010; Feder et al., 2007). Additional

articles written about chronic Lyme disease from a science-based medicine perspective can be found online at https://www.sciencebasedmedicine.org/tag/chronic-lyme-disease.

3. Allopathic medicine is a term commonly used by alternative practitioners and proponents of alternative medicine. It refers to the suppression of symptoms with prescription drugs rather than treatment. It has a negative connotation.

References

Alternative Medicine Review. (2010). Crataegus oxycantha Monograph (Hawthorn). *Alternative Medicine Review*, *15*(2): 164–167.

American Association of Naturopathic Physicians. (2011a). Definition of Naturopathic Medicine. Retrieved from http://www.naturopathic.org/content.asp?contentid=59

American Association of Naturopathic Physicians. (2011b). Equity in Education for Naturopathic Medicine. Retrieved from http://www.naturopathic.org/Files/Events/DC_FLI/equity.education.pdf

American Association of Naturopathic Physicians. (2017). Licensed States and Licensing Authorities. Retrieved from http://www.naturopathic.org/content.asp?contentid=57

American Board of Naturopathic Oncology. (n.d.-a). About Naturopathic Oncology. Retrieved from http://fabno.org/about.html

American Board of Naturopathic Oncology. (n.d.-b). Requirements for ABNO Certification. Retrieved from http://www.fabno.org/requirements.html

Arizona Naturopathic Physicians Medical Board. (2014). Michael Uzick Disciplinary Actions. Retrieved from http://directorynd.az.gov/agency/pages/directorySearchDetail.asp?holderID=102

Association of Accredited Naturopathic Medical Colleges. (n.d.-a). Comparing the ND & MD Curricula. Retrieved from https://aanmc.org/resources/comparing-nd-md-curricula/

Association of Accredited Naturopathic Medical Colleges. (n.d.-b). Naturopathic Medicine's Six Principles. Retrieved from https://aanmc.org/6-principles/

Atwell, J. E., Van Otterloo, J., Zipprich, J., Winter, K., Harriman, K., Salmon, D. A., . . . Omer, S. B. (2013). Nonmedical Vaccine Exemptions and Pertussis in California, 2010. *Pediatrics*, *132*(4): 624–630. DOI: 10.1542/peds.2013-0878

Atwood, K. (2003). Naturopathy: A Critical Appraisal. *Medscape General Medicine*, *5*(4). Retrieved from http://www.medscape.com/viewarticle/465994

Atwood, K. (2010 April 29). Bogus Diagnostic Tests. Retrieved from https://www.sciencebasedmedicine.org/bogus-diagnostic-tests/

Baker, P. J. (2010 November 1). Chronic Lyme Disease: In Defense of the Scientific Enterprise. *The FASEB Journal*, *24*(11): 4175–4177. DOI: 10.1096/fj.10-167247

Baral, M. (2015). Measles: A Different Perspective from Another Phoenix Physician. Retrieved from http://drmatthewbaral.com/archives/1699

Bastyr Center for Natural Health. (2016a). About Naturopathy. Retrieved from http://bastyrcenter.org/services/naturopathic-medicine/about-naturopathy

Bastyr Center for Natural Health. (2016b). Physical Medicine. Retrieved from http://www.bastyrcenter.org/services/naturopathic-medicine/physical-medicine

Bastyr University. (2008a). Bastyr University Course Catalog 2008–2009. Retrieved from http://www.bastyr.edu/sites/default/files/images/pdfs/course-catalog/catalog-archive/Catalog-08-09.pdf

Bastyr University. (2008b). Student Clinician Handbook—Global Module 2008–2009.

Bastyr University. (2016). What is Naturopathic Medicine? Retrieved from http://bastyr.edu/academics/naturopathic-medicine/about-naturopathic-medicine

Boehm, K., Ernst, E., and CAM-Cancer Consortium. (2017 February 17). Ukrain. [online document]. Retrieved from http://www.cam-cancer.org/The-Summaries/Herbal-products/Ukrain

Boyd, C., and Moodambail, A. (2016). Severe Hypercalcaemia in a Child Secondary to Use of Alternative Therapies. *BMJ Case Reports*, *2016*, bcr2016215849. hDOI: 10.1136/bcr-2016-215849

California Naturopathic Doctors Association. (2013). MD vs ND Comparative Curricula. Retrieved from http://www.calnd.org/files/CNDA%20Comparative%20Curricula%2012-11-13.pdf

Carnegie Classification of Higher Institutions. (2015a). Carnegie Classification Institution Lookup. Retrieved from http://carnegieclassifications.iu.edu/lookup/lookup.php

Carnegie Classification of Higher Institutions. (2015b). Carnegie Classifications Standard Listings. Retrieved from http://carnegieclassifications.iu.edu/lookup/srp.php?clq=&limit=50&unit_id=235547&start_page=lookup.php&submit=FIND+SIMILAR

Carnegie Classification of Higher Institutions. (2015c). Graduate Instructional Program Methodology. Retrieved from http://carnegieclassifications.iu.edu/methodology/grad_program.php

Cassileth, B. R., and Yarett, I. R. (2012). Cancer Quackery: The Persistent Popularity of Useless, Irrational "Alternative" Treatments. *Oncology*, *26*(8): 754.

Christensen, J., and Gumbrecht, J. (2016, October 13). Teething Tablets May Be Linked to 10 Children's Deaths. Retrieved from http://www.cnn.com/2016/10/12/health/hylands-teething-tablets-discontinued-fda-warning/index.html

Clinton, C. (2016a). Adrenal Fatigue Explained. Retrieved from http://www.naturopathic.org/content.asp?pl=14&contentid=314

Clinton, C. (2016b). VacciShield Nutritional Support for Infants and Kids during Vaccination. Retrieved from https://wellfuture.myshopify.com/pages/vaccishield-faqs

Council on Naturopathic Medical Education. (2015). Council on Naturopathic Medical Education Board of Directors. Retrieved from http://www.cnme.org/directors.html

Council on Naturopathic Medical Education. (2016). 2016 CNME Handbook of Accreditation for Naturopathic Medicine Programs. Retrieved from http://www.cnme.org/resources/2016_cnme_handbook_of_accreditation.pdf

Department of Education (US). (2016). College Accreditation in the United States [Educational Guides]. Retrieved from http://www2.ed.gov/admins/finaid/accred/accreditation_pg2.html

Dormer, D. (2016). Alberta Dad Confused by Guilty Verdict in Son's Meningitis Death, But Has No Plans to Appeal. Retrieved from http://www.cbc.ca/news/canada/calgary/david-collet-stephan-ezekiel-meningitis-lethbridge-1.3618485

Editorial Board of Integrative Cancer Therapies. (2016). Retrieved from https://uk.sagepub.com/en-gb/eur/integrative-cancer-therapies/journal201510#editorial-board

Ernst, E. (2013). Cancer Patients Who Use Alternative Medicine Die Sooner. Retrieved March 9, 2016, from http://edzardernst.com/2013/04/cancer-patients-who-use-alternative-medicine-die-sooner/

Ernst, E. (2016). Naturopaths: Rubbish at Healthcare, Excellent at Character-Assassination. Retrieved from http://edzardernst.com/2016/06/naturopaths-rubbish-at-healthcare-excellent-at-character-assassination/

Ernst, E., and Schmidt, K. (2005). Ukrain—A New Cancer Cure? A Systematic Review of Randomised Clinical Trials. *BMC Cancer*, 5: 69. DOI: 10.1186/1471-2407-5-69

Feder, H. M., Johnson, B. J. B., O'Connell, S., Shapiro, E. D., Steere, A. C., Wormser, G. P., and the Ad Hoc International Lyme Disease Group. (2007 October 4). A Critical Appraisal of "Chronic Lyme Disease." *The New England Journal of Medicine*, 357(14): 1422–1430. DOI: 10.1056/NEJMra072023

Federal Trade Commission (US). (2008). Anatomy of a Cancer Treatment Scam Consumer Information. Retrieved from https://www.consumer.ftc.gov/media/video-0032-anatomy-cancer-treatment-scam

Food and Drug Administration (US). (2017). FDA Consumer Updates: Products Claiming to "Cure" Cancer Are A Cruel Deception. Retrieved from https://www.fda.gov/ForConsumers/ConsumerUpdates/ucm048383.htm

Food and Drug Administration (US). (2016). Dietary Supplements [WebContent]. Retrieved from http://www.fda.gov/Food/DietarySupplements/

Food and Drug Administration (US). (2017, January 27). FDA Confirms Elevated Levels of Belladonna in Certain Homeopathic Teething Products [WebContent]. Retrieved from https://www.fda.gov/newsevents/newsroom/pressannouncements/ucm538684.htm

Gavura, S. (2012, February 2). IgG Food Intolerance Tests: What Does the Science Say? Retrieved from https://www.sciencebasedmedicine.org/igg-food-intolerance-tests-what-does-the-science-say/

Gavura, S. (2014, April 24). Naturopathy vs. Science: Vaccination Edition. Retrieved from https://www.sciencebasedmedicine.org/naturopathy-vs-science-vaccination-edition/

Gavura, S. (2015 January 1). Detox: What "They" Don't Want You To Know. Retrieved from https://sciencebasedmedicine.org/detox-what-they-dont-want-you-to-know/

Gibson, J. (2016, April 25). Parents on Trial in Meningitis Death of Toddler Defended Use of Natural Remedies in Police Interview. Retrieved from http://www.cbc.ca/news/canada/calgary/stephan-david-collet-lethbridge-meningitis-police-interview-toddler-death-1.3537887

Gorski, D. (2016, February 22). What Naturopaths Say to Each Other When They Think No One's Listening, Part 2. Retrieved from https://www.sciencebasedmedicine.org/what-naturopaths-say-to-each-other-when-they-think-no-ones-listening-part-2/

Hanahan, D., and Weinberg, R. A. (2000). The Hallmarks of Cancer. *Cell, 100*(1): 57–70.

Hermes, B. (2007). Class Notes: Hawthorne Use in Naturopathic Practice. Unpublished.

Hermes, B. (2015, March 13). ND Confession, Part 1: Clinical Training Inside and Out. Retrieved from https://www.sciencebasedmedicine.org/nd-confession-part-1-clinical-training-inside-and-out/

Hermes, B. (2016a, February 23). Is Your Naturopathic "Doctor" Talking about You on the Internet? Retrieved from http://www.naturopathicdiaries.com/is-your-naturopathic-doctor-talking-about-you-on-the-internet/

Hermes, B. (2016b, April 14). Naturopathic Pediatrics Is Not Safe. Retrieved from https://www.naturopathicdiaries.com/naturopathic-pediatrics-is-not-safe/

Hermes, B. (2016c, May 25). A Question Off the Naturopathic Licensing Exam (NPLEX). Retrieved from http://www.naturopathicdiaries.com/one-question-nplex-exam/

Hermes, B. (2017, April 10). Confirmed: Licensed Naturopathic Doctor Gave Lethal 'Turmeric' Injection. Retrieved from https://www.forbes.com/sites/brittmariehermes/2017/04/10/confirmed-licensed-naturopathic-doctor-gave-lethal-turmeric-injection/#37009acf6326

Howard, J. (2017, April 15). Hyland's Homeopathic Teething Tablets Recalled Nationwide. Retrieved from http://edition.cnn.com/2017/04/14/health/hylands-teething-tablet-fda-recall-bn/

Huber, C. (2016). Natural Cancer Treatments. Retrieved from http://natureworksbest.com/

Johnson, S. B., Park, H. S., Gross, C. P., and Yu, J. B. (2018, January 1). Use of Alternative Medicine for Cancer and Its Impact on Survival. *JNCI: Journal of the National Cancer Institute, 110*(1). https://doi.org/10.1093/jnci/djx145

Krumbeck, E. (2011, May 19). So You're Going to Be, Like, Almost a Doctor, Aren't You? Retrieved from http://www.mtwholehealth.com/2011/05/so-youre-going-to-be-like-almost-a-doctor-arent-you

Krumbeck, E. (2014). Alternate and Delayed Schedule Vaccines, Missoula Montana. Retrieved from http://www.mtwholehealth.com/vaccines

Leukemia and Lymphoma Society. (2015). Childhood Facts and Statistics from the Leukemia and Lymphoma Society. Retrieved from http://www.lls.org/http:/llsorg.prod.acquia-sites.com/facts-and-statistics/facts-and-statistics-overview/facts-and-statistics

Local. (2015, January 1). Ukrainian Chemist Lands in Court for Cancer "Cure." Retrieved from http://www.thelocal.at/20150128/ukrainian-chemist-lands-in-court-for-cancer-cure

Lloyd, G. (1983). *Hippocratic Writings* (2nd ed.). London: Penguin Books.

Memorial Sloan Kettering Cancer Center. (2014). Ukrain. Retrieved from https://www.mskcc.org/cancer-care/integrative-medicine/herbs/ukrain

Mengs, U., Göthel, D., and Leng-Peschlow, E. (2002). Mistletoe Extracts Standardized to Mistletoe Lectins in Oncology: Review on Current Status of Preclinical Research. *Anticancer Research, 22*(3): 1399–1407.

Mountain, R. (2011, July 9). Discovering Naturopathy. Retrieved from http://discoveringnaturopathy.blogspot.com/2011/07/nplex-usmle.html

National Cancer Institute (US). (2015a). Antineoplastons [Cancer Information Summary]. Retrieved from https://www.cancer.gov/about-cancer/treatment/cam/patient/antineoplastons-pdq#section/all

National Cancer Institute (US). (2015b). Mistletoe Extracts [Cancer Information Summary]. Retrieved from http://www.cancer.gov/about-cancer/treatment/cam/patient/mistletoe-pdq

National Cancer Institute (US). (2016). SEER Cancer Statistics Factsheets: Cancer of Any Site. Retrieved from http://seer.cancer.gov/statfacts/html/all.html#survival

National Institutes for Health (US). (2013). NIH Fact Sheets—Cancer. Retrieved from https://report.nih.gov/nihfactsheets/viewfactsheet.aspx?csid=75

National University of Natural Medicine. (2012). Skin-Rejuvenation Treatments at the National University of Natural Medicine Clinic. Retrieved from https://www.groupon.com/deals/ncnm-integrative-skincare-clinic

National University of Natural Medicine. (2016). National University of Natural Medicine. Retrieved from http://nunm.edu/naturopathic-principles-of-healing/

Neff, T. (2015, January 11). Ear Infections and Naturopathic Medicine. Retrieved from http://naturopathicpediatrics.com/2015/01/11/ear-infections-naturopathic-medicine/

North American Board of Naturopathic Examiners. (2016). NPLEX Examination Overview. Retrieved from https://www.nabne.org/home/exam-overview/

Nowicky Pharma. (n.d.). What Is Ukrain? Retrieved from http://www.ukrain.ua/mainnew.html#regstat

Offit, P. A., and Moser, C. A. (2009). The Problem With Dr. Bob's Alternative Vaccine Schedule. *Pediatrics, 123*(1): e164–e169. DOI: 10.1542/peds.2008–2189

Pennycook, G., Cheyne, J. A., Barr, N., Koehler, D. J., and Fugelsang, J. A. (2015). On the Reception and Detection of Pseudo-Profound Bullshit. *Judgment and Decision Making, 10*(6): 549.

Pizzorno, J. E., and Murray, M. T. (Eds.). (2013). *Textbook of Natural Medicine*. St. Louis: Elsevier.

Princeton Review. (2017a). National College of Natural Medicine Admissions, Average Test Scores & Tuition. Retrieved from http://www.princetonreview.com/schools/1041674/med/national-college-natural-medicine

Princeton Review. (2017b). Southwest College of Naturopathic Medicine Admissions, Average Test Scores & Tuition. Retrieved from http://www.princetonreview.com/schools/1041675/med/southwest-college-naturopathic-medicine

Risberg, T., Vickers, A., Bremnes, R. M., Wist, E. A., Kaasa, S., and Cassileth, B. R. (2003). Does Use of Alternative Medicine Predict Survival from Cancer? *European Journal of Cancer 39*(3): 372–377.

Schattner, E., and Waldman, K. (2010). Who's a Survivor? Slate. Retrieved from http://www.slate.com/articles/health_and_science/medical_examiner/2010/10/whos_a_survivor.html

Scimago Journal and Country Rank. (2014). Journal Rankings on Oncology. Retrieved from http://www.scimagojr.com/journalrank.php?category=2730&area=0&year=2014&country=&order=sjr&min=0&min_type=cd&page=3

Senapathy, K. (2016). Why Is Big Naturopathy Afraid Of This Lone Whistleblower? Retrieved from http://www.forbes.com/sites/kavinsenapathy/2016/05/31/why-is-big-naturopathy-afraid-of-this-lone-whistleblower/

Southwest College of Naturopathic Medicine. (2016a). "Spring Cleaning" for Optimal Health. Originally retrieved from http://www.scnm.edu/news-stories-and-upcoming-events/news/spring-cleaning-for-optimal-health/. Archived version at https://web.archive.org/web/20160531025043/http://www.scnm.edu/news-stories-and-upcoming-events/news/spring-cleaning-for-optimal-health/#.WTQbPsklGl4

Southwest College of Naturopathic Medicine. (2016b). Hyland's Baby Teething Tablets: Are They Safe? Originally retrieved from http://www.scnm.edu/news-stories-and-upcoming-events/news/hyland-s-baby-teething-tablets-are-they-safe/. Archived version at https://web.archive.org/web/20170313083928/https://www.scnm.edu/news-stories-and-upcoming-events/news/hyland-s-baby-teething-tablets-are-they-safe/#.WTQc-8klGl4

Specialized Natural Health Care of Vermont. (n.d.). Specialized Natural Health Care of Vermont. Retrieved from http://specializednaturalhealthcarevt.com/individual-pediatric-vaccination-plans/

Tascilar, M., de Jong, F. A., Verweij, J., and Mathijssen, R. H. (2006). Complementary and Alternative Medicine during Cancer Treatment: Beyond Innocence. *Oncologist, 11*(7): 732–741.

Thielking, M. (2016, October 20). "Essentially Witchcraft": A Former Naturopath Takes On the Field. Retrieved from https://www.statnews.com/2016/10/20/naturopath-critic-britt-hermes/

Thom, D. (n.d.). Dickson Thom DDS ND. Retrieved from http://www.hbbhealth.net/dickson-thom-dds-nd

Vancouver Island Naturopathic Clinic. (n.d.). Children's Health. Retrieved from http://islandnaturopathic.com/childrens-health/

Vickers, A. J., Kuo, J., and Cassileth, B. R. (2006). Unconventional Anticancer Agents: A Systematic Review of Clinical Trials. *Journal of Clinical Oncology*, *24*(1): 136–140.

Wuesthof, M., and CAM-Cancer Consortium. (2005). Mistletoe (Viscum album). Retrieved from http://www.cam-cancer.org/The-Summaries/Herbal-products/Mistletoe-Viscum-album/Does-it-work

Yun, Y. H., Lee, M. K., Park, S. M., Kim, Y. A., Lee, W. J., Lee, K. S., . . . Park, S. R. (2013). Effect of Complementary and Alternative Medicine on the Survival and Health-Related Quality of Life among Terminally Ill Cancer Patients: A Prospective Cohort Study. *Annals of Oncology: Official Journal of the European Society for Medical Oncology / ESMO*, *24*(2): 489–494. DOI: 10.1093/annonc/mds469

Zimmermann, A. (n.d.). Vaccination Information by Anke Zimmermann, ND. Retrieved from http://www.drzimmermann.org/vaccine-controversy.html

7 Risky Play and Growing Up: How to Understand the Overprotection of the Next Generation

Leif Edward Ottesen Kennair, Ellen Beate Hansen Sandseter, and David Ball

Introduction

The term "play" is often correlated in adult minds with "having fun." While this is often a fair presumption, it can be taken too far. Play also has its dark side, about which relatively little is known, in which children and young people push the boundaries by carrying out physical, emotional, and social experiments. The gathering evidence is that this risk-taking behavior is an essential contributor to development. However, if you subscribe solely to "the play is fun" doctrine, any kind of risk is seen as undesirable and a target for elimination. Thus, there are many websites that warn against the risks of play and emphasize strategies of control. For example, the US National Safety Council maintains a page entitled "Playgrounds Don't Have to Hurt," which is concerned with the type of surfacing present in conventional playgrounds (National Safety Council, 2016). A consequence of this way of thinking has been a continuing drive by some professional interests, including safety engineers and pediatricians (e.g., Canadian Paediatric Society, 2012), to mitigate or eliminate playground risks, which has in turn generated its own research agenda focusing on risk minimization. In addition, this has further raised questions on the balance between safety legislation, safety standards, and litigation versus the benefits such play offers to children's development (Ball, 1995, 2002, 2004; Chalmers, 2003; Freeman, 1995; Furedi, 2001; Little, 2006; Satomi and Morris, 1996; S. J. Smith, 1998; Stephenson, 2003). An exaggerated safety focus in children's play is problematic. Children should avoid injuries, but children also need challenges and varied stimulation to develop normally, both physically and mentally.

Play's Complexity—More Than Having Fun

Pellegrini and Smith (2005) argue that play is primarily something that children and young people do, whereas Sutton-Smith (1997) views play as a lifelong activity that occurs in different forms at all ages. Research has also revealed explicit individual and cultural differences in play (Fromberg and Bergen, 2006).

A common characterization of play is that it is internally directed, with the activity being more important than its ends (Bekoff and Byers, 1981; Martin and Caro, 1985; Pellegrini and Smith, 2005). Children themselves describe play as voluntary, self-controlled, self-initiated, fun, spontaneous, free, and unrestricted activity where adults have not decided what they should do (Wiltz and Fein, 2006). Sutton-Smith (1997) also states that play activity provides children with an optimal experience of arousal, excitement, fun, merriment, joy, and light-heartedness, often in a way that makes them repeat the activity over and over again.

Within the landscape of different play types, risky play would partly share characteristics with several different play types included in prior categorizations. For example, it could involve elements from locomotor (Sawyers, 1994) and physical activity play (P. K. Smith, 2005), rough-and-tumble play (Blurton Jones, 1976; Humphreys and Smith, 1987; P. K. Smith, 2005), as well as play with objects (P. K. Smith, 2005).

Several studies have found that climbing to substantial heights, sliding, jumping down, balancing, and swinging with high speed are perceived by children as fun and thrilling (Coster and Gleave, 2008; Readdick and Park, 1998; S.J. Smith, 1998). Letting the children venture out on their own away from the surveillance of caretakers is also seen as risky (S. J. Smith, 1998). Research also shows that rough-and-tumble play, as described by several researchers, includes a potential of harm to the participants (Blurton Jones, 1976; Humphreys and Smith, 1984; P. K. Smith, 2005). It involves the chance of children unintentionally hurting each other while wrestling, fighting, fencing, and so on, and there is a fine balance between the activity maintaining play and a real fight. Drawing on these studies, as well as her own research, Sandseter has defined risky play as thrilling and exciting forms of physical play that involve uncertainty and a risk of physical injury, and identified six categories of risky play (Sandseter, 2007): 1) play with great HEIGHTS—danger of injury from falling, 2) play with high SPEED—uncontrolled speed and pace that can lead to a collision with something (or someone), 3) play with dangerous TOOLS—that can lead to injuries, 4) play near dangerous ELEMENTS—where you can fall into or from something, 5) ROUGH-AND-TUMBLE play—where the children can unintentionally harm each other, 6) play where the children can DISAPPEAR/GET LOST, for instance, when the children are without supervision.

Benefits of Risky Play

On a conscious level, the benefits and rewards of risky play are the positive experiences one can gain. Intense exhilaration is one of the potential rewards of engaging in risky situations (Cook, 1993; Cook, Peterson, and DiLillo, 1999). The joy of mastering new and challenging tasks, often on the borderline of control, is found to be a driving force and rewarding experience when children engage in risky play (Stephenson, 2003).

Similarly, Coster and Gleave's (2008) study on children's views of risk taking in play revealed that feelings such as fun, enjoyment, excitement, thrill, pride, achievement, and good self-esteem were reasons for engaging in risky play. The children in both Coster and Gleave's (2008) and Sandseter's (2010a) studies clearly stated that this kind of play was both fun and scary at the same time, and that experiencing these contrasting feelings was exciting. In risk-taking behavior, the excitement, exhilaration, and intense pleasure one can experience from mastering risks and gaining a high level of arousal will be experienced as a reward (Adams, 2001; Apter, 2007; Zuckerman, 1994).

Other benefits of children's engagement in risky play are the lessons for life that they unconsciously learn while practicing handling risks. Risky play, as several researchers argue, is a way for children to enhance their risk-mastery skills. Children approach the world around them through play, they are driven by curiosity and a need for excitement, they rehearse handling real-life risky situations through risky play, and they discover what is safe and not safe (Adams, 2001; Apter, 2007).

Boyesen (1997) states that in order for a child to "learn" how to master a risk situation, he or she will necessarily need to somehow approach the situation, thereby increasing the risk. Also, Ball (2002) and Stutz (1995) emphasize the importance of letting children develop a sound sense of risk through taking risks in play, and a study investigating play providers' views of children's risky play in the United Kingdom reported that enabling children to test their abilities, develop skills for use in the wider world, and learn about the real consequences of risk taking are the most important benefits of risks and challenge in play (Greatorex, 2008). Aldis (1975) shows how children progressively encounter risky play and seek out thrills in a gradual manner, which allows them to master the challenges involved. Research has also indicated that through physical activity and risk taking in play, children show improved motor skills and spatial skills and learn risk assessment and how to master risk situations; their subjective perception of the risk becomes more realistic (Ball, 2002; Boyesen, 1997; Fjørtoft, 2000; S. J. Smith, 1998; Stutz, 1995).

Evolutionary Function of Play and Risky Play

In an evolutionary selective model, Sutton-Smith (1997) argues, play creates uncertainties and risks that children rehearse when managing both fictive and real play situations. Similarly, according to Bruner (1976), play provides a less risky situation than "real life," thus minimizing the consequences of one's actions.

Bjorklund and Pellegrini (2000) discuss children's play as an ontogenetic adaptation. In their opinion, the function of play is an interesting issue since the earlier literature defined play as an activity that serves no apparent purpose, with the means of the behavior being more important than the ends (Martin and Caro, 1985; Pellegrini and Bjorklund, 2004; P. K. Smith and Vollstedt, 1985).

Bjorklund and Pellegrini (2000) ask how a behavior can be developmentally important, yet serve no apparent purpose. This is in particular an interesting question when considering risky play, where the possible outcome may potentially be injury and sometimes even death. Bjorklund and Pellegrini state that the benefit of behavior is its function, and that the cost is the risk it imposes. From an evolutionary perspective, the behavior will be naturally selected if the benefit of the behavior is greater than the cost. Bekoff and Byers (1981) state that play in general would have been eliminated, or never would have evolved, unless it had beneficial results that outweighed its disadvantages.

There is now a consensus that play can have both deferred and immediate benefits (Bekoff and Byers, 1981; Pellegrini and Bjorklund, 2004; Pellegrini and Smith, 1998). It is not an imperfect version of adult behavior (Bjorklund and Pellegrini, 2000). Through play, children learn skills that are important for adulthood, but play also adapts individuals to their current environments, play being a specific adjustment to childhood (Pellegrini and Bjorklund, 2004; Pellegrini and Smith, 1998).

The risky play observed by researchers such as Stephenson (2003), Greatorex (2008), and Sandseter (2007) involves play at heights, with speed, seeking the unknown, and rough-and-tumble play. The benefits of these kinds of play may be learning about one's ecology, exploring the environment (Bjorklund and Pellegrini, 2002), and practicing and enhancing different motor/physical skills for developing muscle strength, endurance, skeletal quality, and so on (Bekoff and Byers, 1981; Bjorklund and Pellegrini, 2000; Byers and Walker, 1995; Humphreys and Smith, 1987; Pellegrini and Smith, 1998). All physical practice and training might be relevant for the developing child. These kinds of play also involve training on perceptual competencies, such as depth, form, shape, size, and movement perception (Rakison, 2005), and general spatial-orientation abilities (Bjorklund and Pellegrini, 2002). These are important skills both for survival in childhood (immediate benefits) and for handling important adaptive tasks in adulthood (deferred benefits).

Children's venturing out on their own away from the surveillance of caretakers is also mentioned as a risky kind of play (Sandseter, 2007; Smith, 1998). Bjorklund and Pellegrini (2002) argue that children come to know their environment through continuously exploring new areas and objects. According to Bjorklund and Pellegrini, the fact that boys engage more than girls in exploration and also explore larger areas than girls is related to what Bowlby called the environment of evolutionary adaptedness (EEA), where males were hunters and had to be able to safely move around in diverse and large areas away from home. This is in accordance with the research of Silove, Manicavasagar, O'Connell, and Morris-Yates (1995) arguing that a lower level of separation anxiety among boys than girls is due to the adaptive pressure for boys to learn hunting skills and the courage to venture far from the home, while girls were adapted to learn skills for nurturing and creating safe environments for child-rearing. It seems that children attain enhanced familiarity and competence about their environment, its potentials and its dangers through exploring its features (Bjorklund and Pellegrini, 2002).

Rough-and-tumble play also involves great physical and motor stimulation (Bekoff and Byers, 1981; Bjorklund and Pellegrini, 2000; Byers and Walker, 1995; Humphreys and Smith, 1987; Pellegrini and Smith, 1998). Another possible function of rough-and-tumble play is to enhance social competence. Flinn and Ward (2005) argue that childhood play is favored by natural selection, that it's part of a lifetime process of learning how to use manipulation and the projection of personal superiority to gain control over the people and resources in one's ecology. This will in the long run enhance survival and reproduction. Social physical play, like rough-and-tumble play, enhances children's social competencies, such as affiliation with peers, social signaling, as well as good managing and dominance skills within the peer group (Humphreys and Smith, 1987; Pellegrini and Smith, 1998). It also provides for practice of complex social skills, such as bargaining, manipulating, and redefining situations (P. K. Smith, 1982).

Another evolutionary function of children's risky play is the antiphobic effect such play may have (Sandseter and Kennair, 2011). This suggested function of children's risky play is based on research suggesting that several of humans' fears and phobias—such as separation anxiety, fear of heights, and fear of water—appear naturally at a developmentally relevant age as a part of the child's maturation due to interplay between genes and the environment, and that these fears vanish again due to a natural interaction with the relevant environment and the anxious stimulus as part of normal development (Poulton and Menzies, 2002). Poulton and Menzies suggest that the susceptibility to these fears and phobias is nonassociative and innate, and that the fears originated as adaptive features necessary to keep the child safe, alert, and careful when dealing with potentially dangerous situations.

Research on fear of heights has shown that sustaining injury due to falls both before age five and between ages five and nine is associated with the absence of fear of heights at age eighteen (Poulton, Davies, Menzies, Langley, and Silva, 1998). Thus, risky play with great heights will provide a desensitizing or habituating experience, resulting in less fear of heights later in life (Sandseter and Kennair, 2011). Similarly, research on separation anxiety shows that the number of separation experiences before age nine correlates negatively with separation anxiety symptoms at age eighteen (Poulton, Milne, Craske, and Menzies, 2001), and research on fear of water has concluded that there is no relationship between experiencing water trauma before age nine and the symptoms of water fear at age eighteen (Poulton, Menzies, Craske, Langley, and Silva, 1999). These findings suggest that risky play where children separate from their caretakers, and where they explore new and unknown areas and play near and in water, also has habituating effects on the innate fears of separation and water (Sandseter and Kennair, 2011). As such, Sandseter and Kennair suggest that one of the most important aspects of risky play is the antiphobic effect of exposure to typical anxiety-eliciting stimuli and contexts, in combination with positive emotions (thrills, excitement, and fearful joy) and relatively safe situations. The children learn to cope with potentially dangerous situations, and therefore not to fear them..

Attempts to Regulate Play's Riskiness

The other side of the coin is the study of harmful outcomes of play, the most tangible being the toll of injuries. Thus, many academic papers on playground safety commence with a review of injury statistics; gathered on a national basis, these often provide impressively large numbers. The US Centers for Disease Control and Prevention reported over 200,000 emergency department visits resulting from play in one year (2012), and the equivalent figure for the United Kingdom, with its smaller population, is 42,000 (Ball, 2002). Prima facie, these numbers suggest the need for urgent intervention, and have no doubt triggered many professional interests, such as standards writers, injury researchers, manufacturers, and lawyers.

Unraveling the science, however, is not nearly so straightforward, and the situation becomes yet more complex when the science is fed into policy, where other considerations should come to bear. Herrington and Nicholls (2007) have described how, in the 1970s and 1980s, lobbyists for Canadian playground standards apparently misapplied injury data from the American Consumer Product Safety Commission and failed to take account of studies that showed that injury levels on playgrounds were not significantly high. They speculated that this may have been done to fulfil primary objectives of the Canadian Standards Association (2003), which were to "foster and promote voluntary standardization as a means of advancing the national economy" and to "facilitate domestic and international trade."

The fact is that the number of injury cases is insufficient to justify an intervention. For one thing, even if the number of cases is large, it is necessary to consider the size of the population exposed and the duration of the exposure; if they too are large, then the risk per person per outing may actually be small. Thus, in the United Kingdom it emerged that although there were 42,000 accidents per year on UK playgrounds, the risk per child was very small, especially when compared with accident rates for other popular pastimes; these, incidentally, had not achieved the notoriety of playgrounds as sources of injury. Figure 7.1 shows a compilation of such data based on UK statistics in which it can be seen that playgrounds come low down on the spectrum.

As noted, the role of science in policy is yet more complex. As an illustration, we refer to one of the best papers ever written on the epidemiology of playground accidents (Chalmers et al., 1996). This case-control study of injury from playground falls in New Zealand found a statistically significant increase in risk of injury from playground falls as a function of height of fall. The authors concluded by recommending that

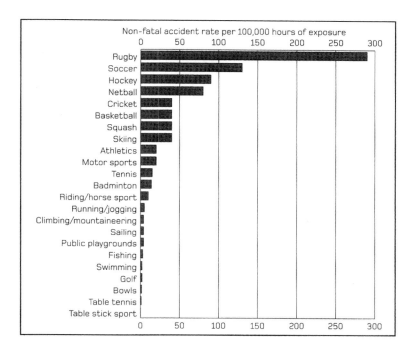

Figure 7.1
Non-fatal injury rates for selected activities based on A&E attendances in the United Kingdom (Play Safety Forum, 2012, p. 11).

consideration be given to lowering the maximum permissible fall height on public playgrounds to 1.5 meters. Such a proposition, if written into standards, would have a massive impact on play provision, and its consequences can only be speculated upon. What might children, especially teenagers, make of that, and how would they respond? The likelihood is that they would seek their thrills elsewhere.

Also with significant consequences has been the requirement in many countries to carry out risk assessments of play spaces. A difficulty has been that risk assessment protocols originate primarily from workplaces where the task has often been seen as one of risk minimization or elimination. Furthermore, those who carry out risk assessments often have been trained in industrial settings, so the transfer of their experiential thinking and workplace protocols into a children's play environment can lead to outcomes at loggerheads with the understanding of play and child development specialists. The presence, for example, of a fireman's pole or something similar (like a wobbly bridge) in a children's play space might seem quite shocking, given it has been debated whether they are safe for use in fire stations (Health and Safety Executive, 2008; Priceonomics, 2014).

The Science of Risk Perception

There is a surprising number of "actors" involved in the provision of children's play opportunities. These range over the children themselves and their parents, but also include neighbors, playground providers, equipment manufacturers, authors of play equipment standards, insurance companies, and even the courts (in some cases when accidents have been reported). The media may also play a part, reporting on accidents and accident frequency. All of these "actors" have, however, a somewhat different agenda and also different perceptions of the level of risk and how it should be judged (Ball, 2000). The perception of risk by children and young people is perhaps most important, but tends to be overlooked or little understood by adults.

Thus, when children engage in risky play, they are in fact continuously making risk-taking decisions. Adams (2001, p. 26) broadly defines the term "risk," as used in everyday life, as "unquantifiable danger, hazard, exposure to mischance or peril." With this definition, Adams emphasizes the variability and uncertainty in the issue of risk, and even though he acknowledges that there is an objective risk, he states that this risk cannot be quantified or predicted: "The problem for those who seek to devise objective measures of risk is that people to varying degrees modify their level of vigilance and their exposure to danger in response to their *subjective* perceptions of risk" (Adams, 2001, p. 13).

Still, Adams (2001) further points out that many of the risk decisions involving children are taken by adults because children are most of the time under surveillance of adults. Children's risk-taking decisions are therefore influenced also by supervising adults' evaluation of the risk situation and how the adults choose to act upon children's risk taking in play. This is a problem also highlighted by other researchers (Allin, West and Curry, 2014; Smith, 2014) because it can restrict children from gaining valuable experiences with freedom to move around, to take responsibility of their own actions, and to face risk situations in play and learn to handle risk and uncertainty. Allin et al. (2014) found that mothers' risk decisions concerning their children reflected uncertainty and somewhat contradictory feelings, and that the mothers' decisions were influenced by media (e.g., reporting of child abductions that heightened the "stranger danger" concern) and their community's dominant views of good or bad mothering.

Another problem about individuals' risk perception and the introduction of safety regulations and safety gear is the aforementioned function of compensating behavior: risk compensation. Risk compensation is a term used to describe the tendency of individuals to unconsciously increase their risk taking when wearing safety gear (Hedlund, 2000; Lasenby-Lessard and Morrongiello, 2011; Phillips, Fyhri, and Sagberg, 2011). Even in parents' evaluation of children's risk raking in play we find risk compensation, with the parents allowing significantly greater risk taking by their children under safety gear than non-safety gear conditions (Morrongiello and Major, 2002).

Beyond the immediate players (the children) and their guardians, one encounters the perceptions of all those other "actors" with fingers in the pie. Modern-day playgrounds tend to be dominated by manufactured equipment; when accidents occur, equipment tends to be involved. This can lead to claims against providers and manufacturers via the courts, and the claims in turn involve insurers. Thus, all of these bodies have an interest in risk, namely the risk of liability. They try to protect themselves by means of international playground equipment standards, and by having formal and documented processes of risk assessment in place. Play becomes part of a safety-management system driven by forces that have little connection with play's fundamental needs and purposes (Spiegal et al., 2014).

Normal Fear and Pathological Anxiety

To understand the emotional aspects that influence our perceptions we need to consider how emotionally dysregulated fear may cause unhelpful behavior. We believe it is important to consider how features of anxiety and worry disorders may inform us about how parents and institutions perceive danger and monitor threat. Anxiety motivates safety behaviors, including threat monitoring, avoidance, and worry. On the other hand, treatment involves reducing avoidance and safety behaviors, increasing exposure, and learning that one may cope with previously feared situations.

The Function of Fear: Avoidance and Safety Behaviors

Fear motivates us to avoid specific dangers. This avoidance is adaptive, as it prevents us from actually being harmed. That is the function of fear: to keep us out of danger or, if necessary, drive us to fight or defend against the source of danger (Kennair, 2007, 2011, 2014; Marks and Nesse, 1994; Öhman and Mineka, 2001; Poulton and Menzies, 2002).

The fear mechanisms are comparable to an alarm system that is regulated by consequences and avoidance. One of the features of this feedback system is that it is overreactive (Nesse, 2001). The cost of reacting and wasting a small amount of calories is always smaller than the cost of not reacting and ending up dead. This emotional system has therefore evolved to err on the side of being safe rather than sorry (Nesse, 2001).

The problem is that fear itself is aversive. Fear of fear creates emotional avoidance. We may start acting in ways that stop making sense from an entirely functional perspective, as we start avoiding or defending against nondangerous situations. The combination of an overactive alarm system, with an interest in avoiding aversive emotional states, may cause us to fear perceived but not real danger; that is, start displaying behavior typical of anxiety disorders, such as phobias, generalized anxiety, or panic disorder.

Phobias

Fear keeps us safe from evolutionary-relevant predators and dangers, but these menaces belong to prehistory. Modern threats and dangers are another matter. Most people are afraid of strangers, but most rapes and assaults are perpetrated by someone the victim knows. People fear the dark and the monsters that magically appear once the light goes out. They do this despite the dark not being dangerous in any specific way. (Our species has the poorest eyesight for darkness, so roaming around in the dark is dangerous. But going to bed in a dark room is not.) People fear animals, heights, open and closed spaces, and (especially) snakes and spiders.

Governments and health and insurance agencies want us to fear real dangers. They want us to fear smoking, drugs, alcohol, sugar, fat, unprotected sex, sunbathing without suntan lotion, and driving without a seat belt. Most people are not phobic about such phenomena. Actually, for many the list constitutes the ingredients of a successful spring break. That is why these things are dangerous: because we like them and then seek them in excess. But given that excess of speed, sex, sugar, and fat was not likely in our species' evolutionary past, we have evolved to find that excess is desirable, not aversive. It also shows how difficult it is to teach a population to fear a thing simply for the danger it presents. We are not rational. We have evolved predispositions to learn to fear some stimuli more than others (Öhman and Mineka, 2001).

The treatment of phobias has been known and available for many years: exposure therapy (Wolitzky-Taylor, Horowitz, Powers, and Telch, 2008; see also Foa and McLean, 2016). Systematic desensitization is recommended, where one gradually exposes oneself to stimuli that provide continuously more anxious discomfort. The mechanism at work is considered to be a downregulation of anxiety responses as a result of finding stimuli less harmful than anticipated, with the organism learning that there is no danger through exposure.

It is important to include response prevention: the discontinuation of rituals or compulsion or safety behaviors. Establishing safety behaviors is typically part of the etiology of anxiety disorders. When anxious, we attempt to protect ourselves. Sometimes the safety behavior itself exacerbates symptoms and is the primary maintaining factor in the disorder. When panic patients lie down on the floor, fearing they will faint, this safety behavior is not necessary, as they will not faint. But in lying down they believe they have saved themselves from falling. For people to reduce their anxiety and improve their functioning, those who treat them must understand the key roles played by exposure and the removal of unnecessary and symptom-aggravating safety behaviors.

The Problem with Believing Worry Is Helpful: Parents Worry

Worry is normal. In one study, all participants (79 percent of the total) who handed in their worry diaries reported having worried during the previous week (Wells and Morrison, 1994). Maybe those who worried less were less compliant; maybe the less compliant also worried. Worry is fearful or anxious anticipation of future events. One of the problems of predicting future events through worry is that one exaggerates the negative future and life turns into a horror movie. Danger monitoring ensues. Worry is often considered to be verbal problem-solving behavior (Borkovec and Inz, 1990; Borkovec, Ray, and Stober, 1998). But this cognitive focus of attention also functions as a cognitive avoidance strategy that prevents emotional processing (Borkovec et al., 1998; Borkovec, Robinson, Pruzinsky, and DePree, 1983). Some may worry because it actually makes them feel a little better—initially. On the other hand, worry rarely solves real problems. Rather, fearful apprehension is more likely to lead to procrastination and inaction.

Parents worry. One wants the best for one's children, and there are enough sources of guilt for not being the world's best mother or father. One way many parents believe they can show how much they care is by worrying.

There is worry in institutions and in government agencies, too. Sometimes because worried groupthink processes make us believe that any accident that anyone can foresee may also be prevented. With liability and the threat of bad public opinion hanging over them, the people answerable for children's safety—people in governments and at institutions, daycare centers, and schools—all believe they have reason to worry. But worry does not necessarily aid realistic predictions about the future. Some bumps and bruises will be avoided by trying to cut out risk, but society as a whole may suffer much more because children's freedom and natural development have been curtailed (Sandseter and Kennair, 2011). Of course, this last statement will probably cause many to start worrying.

Generalized Anxiety Disorder: Exaggerated Worry

There is a specific anxiety disorder that consists of exaggerated worry. It is called generalized anxiety disorder (GAD). This worry disorder is diagnostically defined by exaggerated worry, worry that the patients experience as being out of control.

One of the most promising treatments for GAD is metacognitive therapy (van der Heiden, Muris, and van der Molen, 2012). This approach is interesting because the theory behind the therapy suggests that worry is regulated by both positive and negative metabeliefs about the benefits or harm of worry (Wells, 2009).

Negative metabeliefs concern the perceived uncontrollability and harmfulness of worry in itself. Beliefs that excessive worry may cause stress and different sicknesses or disorders cause worry about worry—metaworry. Beliefs that worry may not be controlled similarly cause metaworry. Ineffective strategies and inflexible attentional processes maintain the latter. Negative metabeliefs are probably the most relevant features of GAD as a mental disorder. For an understanding of an excessive but diagnostic subthreshold, normal worry, the positive metabeliefs are probably most relevant.

Positive metabeliefs about worry cause worry. Worry is actually a coping strategy, an attempt at dealing with negative catastrophic intrusive thoughts (Wells, 2009).

Patients start worrying due to several different metabeliefs (e.g., "Worry will make me prepared," "Worry keeps me safe," "Worry will prevent bad things from happening"). This also causes increased threat monitoring, which in itself increases the perception of danger and more triggers to worry about.

One thing that is surprising to worried patients is that many therapists believe they worry as safety behavior. This means they worry so they can avoid the uncomfortable emotional state that catastrophic trigger thoughts elicit (Borkovec et al., 1998; Borkovec et al., 1983; Wells, 2009). The problem is that this distraction or false comfort is fleeting.

The very odd thing here is that worry actually makes people anxious, not safe, over time. And preparing by worrying does not increase clairvoyance; also, there is a clear bias toward increased negative expectations and predictions. When bad things happen, one will rarely feel prepared due to having been anxious for months ahead about everything that could go wrong. Moreover, one cannot even imagine everything that possibly can go wrong, resulting in the paradox of the things one worried about not coming true while some unthought-of catastrophe actually might strike. The magical thinking behind worry being able to actually change the future may sound unusual, but consider how many people believe they can change the results of academic or medical examinations by hoping for positive results (or negative, as the case may be). If you worry now, somehow fate owes you a break—for the misery you have brought upon yourself.

People believe worry is helpful. Maybe some instances of worry may be; maybe it keeps us from being overconfident some times. In general, planning and caring without worry probably will be more helpful, less negatively biased, and less emotionally disturbing. That is where science and knowledge enters. Rather than using worry to predict the future and manage emotions one should discontinue worry (Wells, 2009).

The Knock-On Implications for Older Members of Society: The Infantilization of Our Young Adults and the Peculiar Case of Trigger Warnings

Once children are past the playground, there is the matter of letting them grow up. We suggest that the current movement for trigger warnings and safe places is an illustration of academic emotional overprotection. As such it is an infantilization of our young adults. It promotes a culture of vulnerability in a world that has never been safer.

There seems to be some misconceptions about how trauma, anxiety, and worry should be dealt with. Rather than increasing safety behavior and its mindset (avoidance, the belief that one is vulnerable, and the metabeliefs that go with this belief), one should decrease safety behavior, expose oneself to feared situations, learn to master aversive feelings—in short, discover that one is not as vulnerable as this latest fad would have young adults believe they are. The last important point is that the combination of overprotection from aversive feelings, coupled with vetting of academic contents and discussion, causes a lack of learning, coping, and mastery. It hampers a truly academic university education, and stunts emotional and intellectual growth.

It is not so much pseudoscience as lack of holistic science and intellectual, academic approaches to learning and growing up that makes "trigger warnings" and "safe places" a relevant illustration. The same poor application of available knowledge seems to be motivated by a desire to have the students not come in touch with undesirable thoughts, regardless of the important opportunities for intellectual growth that these bring. This is therefore a problem that attacks the foundation of intellectual freedom. At the same time, the infantilization of the young adults causes greater dependency on parents, teachers, and counselors.

Students Are Not More Vulnerable Than the General Population

Research would suggest that a student population is not more vulnerable to emotional distress and disorder than the general population. If anything, students as a group are a little less vulnerable (see Blanco et al., 2008—although one needs to consider the tables and not just the conclusions). Socioeconomic status and high academic functioning probably explain why there are fewer problems in this group in general. This has at least been the general picture for many disorders. For example, students have half the risk of committing suicide compared to a matched subsample of the general population (Suicide Prevention Resource Center, 2014). These facts have not stopped talk of a crisis of mental disorder among college students.

In a piece on counseling in colleges and universities, Gray (2015) writes that there has been a steady increase in students seeking counseling at US colleges. He refers to reports that suggest that there is an increase in mental problems, but also a general fragility among students, one that causes them to rely too much on having college institutions and staff continue to act like overprotective parents. This mirrors the perceptions of Lukianoff and Haidt (2015) and Marano (2015).

There might be evidence of increased mental disorder among students seeking counseling and also a greater perceived need for counseling. c But one might suspect that this upsurge is just as much a result of a culture of psychotherapy, of a belief in the need to talk about problems, and thereby also of faulty metacognitions about one's own vulnerability and psychological damage. The latter would be the specific targets for change in a metacognitive approach to treatment (Wells, 2009). This suggests that, while students might be reporting an increase in mental disorder, these emotional distress responses may be treatable. They may be the result of perceptions of fragility and mental safety behaviors such as rumination and worry and faulty negative metacognitions. If such were the case, providing even more safety behaviors and strengthening the negative metacognitions and promoting avoidance would not be recommended, especially if they get in the way of what the students are doing at university. We suggest this mirrors what younger children ought to be experiencing while developing through risky activities in playgrounds.

Universities Need Academic and Intellectual Freedom

Are you at university to be safe? No one is there to be unsafe, but the point of play (or learning) is not safety. It is to have fun, to challenge yourself, and to mature physically, mentally, and behaviorally. To have proper fun, and to learn how your body moves and how your mind works and what you can cope with or cannot cope with—for these things, challenge is a must.

Lecturers and researchers are employed to challenge current wisdom. If a government body suggests undersurfacing protects children, and policy makers and manufacturers design playgrounds with new and more "caring" undersurfacing, we need researchers to study whether that claim is true, and to challenge the intuitive conclusions as necessary. We need researchers who dare and are supported in challenging all aspects of our reality and society, and how we understand human nature. We do not burn academics we disagree with; we argue against them and provide better evidence.

The same goes for students. They must play with new and dangerous ideas. They need to be exposed to things they have not learned before, and their lecturers need to be allowed to teach these things. Yes, warn about obvious and intersubjective discomfort or specific discomfort, but do not coddle students or limit their exposure to ideas because of preset beliefs about the students' fragility. There is no reason to believe they are fragile. Even if they have strong emotional reactions, they are not harmed and have not been harmed by them; we should not be guided by such negative metacognitions (Wells, 2009). Sensible, polite, and humane communication is not a problem. A culture of trigger warnings and no-platforming is, and it presents a threat to freedom of speech, intellectual and academic freedom, and mental health (Lukianoff and Haidt, 2015).

How to Treat Exaggerated Negative Emotional Reactions

The most efficient approaches to treating anxiety disorders including phobias, social phobia, GAD, and trauma (including full-fledged PTSD) are the many evidence-based cognitive behavioral manuals (Kaczkurkin and Foa, 2015; Norton and Price, 2007). In general, these manuals will have patients expose themselves to the feared stimuli, either in vivo (in real life) or in vitro (in thought or imagination). These therapies also discuss patients' unhelpful and often faulty thinking about how their body and mind interact (as in panic disorder), how they appear to others (as in social phobia) or how helpful worry is (as in GAD). This is discussed by social psychologist Jonathan Haidt and science writer Hara Estroff Marano in a piece that considers how the way we are treating children and students may be detrimental to their mental health and development as adults and academics (Marano, 2015). They ask, as we do in this chapter, what mental health care has shown us about handling stressful situations and emotions. We fully agree with their conclusion: that exposure, not avoidance, is best. As Lukianoff and Haidt (2015) put it: "According to the most-basic tenets of psychology, helping people with anxiety disorders avoid the things they fear is misguided." As with playgrounds, universities are not improved by becoming coddlers.

Not Learning to Cope Is to Not Mature Adequately

One needs to learn how to climb, swing, jump, hang from monkey bars, navigate labyrinths, and crawl through tunnels. The thrilling experience of two-year-olds is boring to the average three-year-old. Therefore, they invent new ways of using old toys, or move on to larger, faster, and higher apparatus, seeking new thrills. Indeed, this is central to the creative process.

Haidt points out in a discussion of how the antifragility of children demands exposure to stressors, and how being gravity-free may actually cause more fragility. He says about children: "They spend years and years without gravity, so when they get to adulthood, they are fragile, rather than anti-fragile" (Marano, 2015).

At universities one needs to be challenged in the same way. Living away from parents, one needs to learn how to handle one's own economy, to start earning money, to become adult and autonomous. One needs to learn how to meet diversity with interest and respect, including diversity of ideas. One allows one's preconceptions to be challenged. Academics are supposed to change their minds in the face of new evidence and new ideas that are better argued and supported than the previous concepts and beliefs one held about the world. The high school student may have been asked to learn the contents of the curriculum. At university one should be able to criticize the contents (though this challenge should not be based on ideology or safe simple truths).

Conceptual challenges are to be sought out. They are ever more challenging, and thus intellectually thrilling. Yes, you may disagree, but that is only a challenge to argue more coherently why the opposition is wrong, based on evidence. And in this feast of knowledge and ideas one cannot shy away from or ban specific positions or concepts. One may not burn books. One may not censor. Such overprotection stunts growth. Such safety behaviors stunt the learning environment of others. Such fear and gut reaction hamper intellectual discourse. Eventually they get in the way of the intellectual and academic freedom needed for a university to truly be a place of higher learning.

Creating a university environment that smacks of the most hyperprotective and boring kindergarten playground, even one with the most caregiving surfacing, may still fail to keep metaphorical sprained wrists and bruised knees away. But just as this kind of play environment may not do much for children's physical safety, while perhaps harming them psychologically, similar overprotective universities may be harmful for young adults. As Gray (2015) notes, there is probably an overconsumption of counseling among college students. Even if there should be an increased need, we should look for causes, and approach increased anxiety responses with helpful rather than maintaining responses.

There is a new need to highlight that universities are places of learning and academic and intellectual freedom. We need to increase exposure and reduce avoidance of risky ideas and triggering perspectives. In the culture of emotional safety dangerous ideas need protection. For the next generations' sake, as they mature academically in places of higher learning through intellectual risky play.

How Too Much Safety and Protection Can Be Harmful

During the last decade there has been an increasing interest in how children growing up with intensive parenting, sometimes called "helicopter parenting" or "overparenting," cope when they emerge adulthood.

Overparenting is associated with internalizing difficulties among seven year-olds (Bayer et al., 2010), and a higher level of child anxiety and less child perceived competence at ages six through thirteen (Affrunti and Ginsburg, 2012). Studies suggest that overparenting and too much control from parents restrict children's access to their environment and reduces the opportunity for the child to develop competence or mastery over things in their environment, particularly novel and threatening situations, and the ability to develop appropriate problem-solving skills (Affrunti and Ginsburg, 2012).

LeMoyne and Buchanan (2011) found helicopter parenting to be negatively associated with psychological well-being and positively associated with recreational pain pill use and taking prescription medications for anxiety and depression among university students. Segrin and colleagues (2012) found a negative effect from overparenting, causing young adults' sense of entitlement. The children growing up with helicopter parenting had a stronger belief that others should solve their problems, and as such showed more helplessness at handling things themselves. Similarly, Schiffrin and colleagues (2013) found that students with overcontrolling parents reported significantly higher levels of depression and less satisfaction with life. Helicopter parenting was negatively correlated with students' autonomy, competence and relatedness, and positively correlated with depression. Further, overparenting is associated with higher student levels of narcissism and more ineffective coping skills (e.g., internalizing, distancing), anxiety, and stress (Segrin et al., 2013), and with lower student self-efficacy, maladaptive responses to workplace scenarios (Bradley-Geist and Olson-Buchanan, 2014), and lower levels of internal locus of control (Kwon, Yoo, and Bingham, 2016).

Conclusions

Play is children's way of being together, of having fun together, and learning about themselves and their surroundings. Play is characterized by intrinsic motivation among the players; children themselves describe it as voluntary, self-controlled, self-initiated, fun, spontaneous, free, and unrestricted. Researchers have also pointed out that in play the activity itself is the most important thing, with no focus on results or performances. Children's play often also involves arousal, excitement, thrill, uncertainty, and risk taking, all things that help them develop normally and gain valuable experiences of risk perception and how to handle risks. This development is threatened by the push toward regulation of children's play environments and restriction of the kids' freedom to play.

Injuries are the downside of children's risk taking. But research shows that total injury rates have not decreased after regulation of play was introduced, and research on individuals' risk perception has shown that we, both children and adults, tend to compensate for safety measures by increasing our risk taking. Safety measures may well have made parents and authorities *feel* safer without doing anything for children's safety. Finally, remember that the chance of the children being injured by play is relatively small.

Risky play has a developmental and evolutionary function for children, both immediately and during adulthood. Children are attracted by taking risks in play because it rewards them with joy, excitement, and other positive experiences while they unconsciously practice risk perception and how to handle risks. They also gain motor and physical skills and spatial-orientation abilities, competence about their environment, and social skills, and they benefit from a possible antiphobic effect by learning to cope with anxiety-eliciting situations and to no longer fear stimuli of nonassociative fears and phobias. These are all positive functions of risky play. On the other hand, research on young adults who have grown up with overprotective parents shows that the experience has had a negative effect on their overall psychological health. Among other effects, they have low psychological well-being, autonomy, and independency, and a high prevalence of anxiety, depression, and ineffective coping skills. Overall they seem to have a lower life quality than their peers who have not grown up with overprotective parents. Rather than fleeing to safe places, young people need exposure to those features of their ecology that call for mastering. Anxious parents should take a similar lesson and learn that enforcing avoidance is not the way to feel safe. The same goes for academics who worry about the mental welfare and well-being of their students and let them opt out of hearing about things that challenge them emotionally. To treat anxiety, we expose patients to the things they fear; this reduces unhelpful and disorder-maintaining safety behaviors. Patients learn that they are able to cope with what they believed was too threatening. Children probably learn much the same thing through risky play (Sandseter and Kennair, 2011). We all must learn not to exaggerate the fragility of patients, children, or college students.

References

Adams, J. (2001). *Risk*. London: Routledge.

Affrunti, N. W., and Ginsburg, G. S. (2012). Maternal Overcontrol and Child Anxiety: The Mediating Role of Perceived Competence. *Child Psychiatry and Human Development, 43*(1): 102–112. DOI: 10.1007/s10578-011-0248-z

Aldis, O. (1975). *Play Fighting*. New York: Academic Press.

Allin, L., West, A., and Curry, S. (2014). Mother and Child Constructions of Risk in Outdoor Play. *Leisure Studies, 33*(6): 644–657. DOI: 10.1080/02614367.2013.841746

Apter, M. J. (2007). *Danger: Our Quest for Excitement*. Oxford: Oneworld.

Ball, D. J. (1995). "Applying Risk Management Concepts to Playground Safety." In Christiansen, M. L. (Ed.), *Proceedings of the International Conference of Playground Safety*, pp. 21–26. Pennsylvania: Penn State University, Center for Hospitality, Tourism & Recreation Research.

Ball, D. J. (2000) Ships in the Night and the Quest for Safety. *Journal of Injury Control and Safety Promotion, 7*(2): 83–96.

Ball, D. J. (2002). *Playgrounds—Risks, Benefits and Choices*, vol. 426. London: Health and Safety Executive Contract Research Report, Middlesex University.

Ball, D. J. (2004). Policy Issues and Risk-Benefit Trade-Offs of "Safer Surfacing" for Children's Playgrounds. *Accident Analysis and Prevention, 36*(4): 661–670.

Ball, D. J. (2007). Trends in Fall Injuries Associated with Children's Outdoor Climbing Frames, *International Journal of Injury Control and Safety Promotion, 14*(1): 49–53.

Bayer, J. K., Hastings, P. D., Sanson, A. V., Ukoumunne, O. C., and Rubin, K. H. (2010). Predicting Mid-childhood Internalising Symptoms: A Longitudinal Community Study. *International Journal of Mental Health Promotion, 12*(1): 5–17. DOI: 10.1080/14623730.2010.9721802

Bekoff, M., and Byers, J. A. (1981). "A Critical Reanalysis of the Ontogeny and Phylogeny of Mammalian Social and Locomotor Play: An Ethological Hornet's Nest." In K. Immelmann, G. Barlow, W., L. Petrinovich, and M. Main (Eds.), *Behavioral Development: The Bielefield Interdiciplinary Project*, pp. 296–337. Cambridge: Cambridge University Press.

Bishop, J. C., and Curtis, M. (2001). *Play Today in the Primary School Playground*. Philadelphia: Open University Press.

Bjorklund, D. F., and Pellegrini, A. D. (2000). Child Development and Evolutionary Psychology. *Child Development, 71*(6): 1687–1708.

Bjorklund, D. F., and Pellegrini, A. D. (2002). *The Origins of Human Nature: Evolutionary Developmental Psychology*. Washington, DC: American Psychological Association.

Blanco, C., Okuda, M., Wright, C., Hasin DS, Grant BF, Liu SM, and Olfson M. (2008). Mental Health of College Students and Their Non–college-Attending Peers: Results from the National Epidemiologic Study on Alcohol and Related Conditions. *Archives of General Psychiatry, 65*(12): 1429–1437. DOI: 10.1001/archpsyc.65.12.1429

Blurton Jones, N. (1976). "Rough-and-Tumble Play among Nursery School Children." In J. S. Bruner, A. Jolly, and K. Sylva (Eds.), *Play: Its Role in Development and Evolution*, pp. 352–363. Harmondsworth, UK: Penguin Books.

Borkovec, T. D., and Inz, J. (1990). The Nature of Worry in Generalized Anxiety Disorder: A Predominance of Thought Activity. *Behaviour Research and Therapy, 28*(2): 153–158. DOI: http://dx.doi.org/10.1016/0005-7967(90)90027-G

Borkovec, T. D., Ray, W., and Stober, J. (1998). Worry: A Cognitive Phenomenon Intimately Linked to Affective, Physiological, and Interpersonal Behavioral Processes. *Cognitive Therapy and Research, 22*(6): 561–576. DOI: 10.1023/A:1018790003416

Borkovec, T. D., Robinson, E., Pruzinsky, T., and DePree, J. A. (1983). Preliminary Exploration of Worry: Some Characteristics and Processes. *Behaviour Research and Therapy, 21*(1): 9–16. DOI: 10.1016/0005-7967(83)90121-3

Boyesen, M. (1997). *Den truende tryggheten* (doctoral thesis). Norway, Trondheim: Norwegian University of Science and Technology.

Bradley-Geist, J. C., and Olson-Buchanan, J. B. (2014). Helicopter Parents: An Examination of the Correlates of Over-Parenting of College Students. *Education + Training, 56*(4): 314–328. DOI: 10.1108/ET-10-2012-0096

Bruner, J. S. (1976). "Nature and Uses of Immaturity." In J. S. Bruner, A. Jolly, and K. Sylva (Eds.), *Play: Its Role in Development and Evolution*, pp. 28–63. Harmondsworth, UK: Penguin Books.

Byers, J. A., and Walker, C. (1995). Refining the Motor Training Hypothesis for the Evolution of Play. *American Naturalist, 146*(1): 25–40.

Canadian Paediatric Society (2012). Position Statement: Preventing Playground Injuries. Retrieved from http://www.cps.ca/documents/position/playground-injuries

Canadian Standards Association. (2003). *CAN/CSA-Z614: Children's Playspaces and Equipment Standard*. Toronto.

Centers for Disease Control and Prevention (US). (2012). Playground Injuries: Fact Sheet. Retrieved from http://www.cdc.gov/HomeandRecreationalSafety/Playground-Injuries/playgroundinjuries-factsheet.htm

Chalmers, D. (2003). Playground Equipment Safety Standards. *Safe Kids News* (21): 4.

Chalmers, D. J., Marshall, S. W., Langley, J. D., Evans, M. J., Brunton, C. R., Kelly, A-M. and Pickering, A. F. (1996). Height and Surfacing as Risk Factors for Injury in Falls from Playground Equipment: A Case-Control Study. *Injury Prevention 2*: 98–104.

Cook, S. C. (1993). *The Perception of Physical Risk by Children and the Fear/Exhilaration Response* (MA thesis). Columbia: University of Missouri.

Cook, S. C., Peterson, L., and DiLillo, D. (1999). Fear and Exhilaration in Response to Risk: An Extension of a Model of Injury Risk in a Real-World Context. *Behavior Therapy, 30:*5–15.

Coster, D., and Gleave, J. (2008). *Give Us a Go! Children and Young People's Views on Play and Risk-Taking.* Retrieved from: http://www.playday.org.uk/wpcontent/uploads/2015/11/give_us_a_go___children_and_young_peoples_views_on_play_and_risk_taking.pdf

Fjørtoft, I. (2000). *Landscape and Playscape: Learning Effects from Playing in a Natural Environment on Motor Development in Children* (doctoral thesis). Oslo: Norwegian School of Sport Science.

Flinn, M. V., and Ward, C. V. (2005). "Ontogeny and Evolution of the Social Child." In B. J. Ellis and D. F. Bjorklund (Eds.), *Origins of the Social Mind: Evolutionary Psychology and Child Development*, pp. 19–44. New York: Guilford Press.

Foa, E. B., and McLean, C. P. (2016). The Efficacy of Exposure Therapy for Anxiety-Related Disorders and Its Underlying Mechanisms: The Case of OCD and PTSD. *Annual Review of Clinical Psychology, 12*(1): 1–28. DOI: 10.1146/annurev-clinpsy-021815-093533

Freeman, C. (1995). The Changing Nature of Children's Environmental Experience: The Shrinking Realm of Outdoor Play. *International Journal of Environmental Education and Information, 14*(3): 259–280.

Fromberg, D. P., and Bergen, D. (2006). *Play from Birth to Twelve Contexts, Perspectives, and Meanings* (2nd ed.). London: Routledge.

Furedi, F. (2001). *Paranoid Parenting: Abandon Your Anxieties and Be a Good Parent.* London: Penguin.

Gray, P. (2015). Declining Student Resilience: A Serious Problem for Colleges. *Psychology Today.* Retrieved from https://www.psychologytoday.com/blog/freedom-learn/201509/declining-student-resilience-serious-problem-colleges

Greatorex, P. (2008). *Risk and Play: Play Providers' Experience and Views on Adventurous Play.* Retrieved from: http://www.playday.org.uk/wpcontent/uploads/2015/11/risk_and_play___play_providers%E2%80%99_experience_and_views_on_adventurous_play.pdf

Health and Safety Executive. (2008). Myth: Health and Safety Laws Banned Poles in Fire Stations. Retrieved from http://www.hse.gov.uk/myth/jul08.htm

Hedlund, J. (2000). Risky Business: Safety Regulations, Risk Compensation, and Individual Behavior. *Injury Prevention, 6*(2): 82–90. DOI: 10.1136/ip.6.2.82

Herrington, S., and Nicholls, J. (2007). Outdoor Play Spaces in Canada: The Safety Dance of Standards as Policy. *Critical Social Policy, 27*(1): 128–138.

Humphreys, A. P., and Smith, P. K. (1984). "Rough-and-Tumble in Preschool and Playground." In P. K. Smith (Ed.), *Play: In Animals and Humans*, pp. 241–266. Oxford: Blackwell.

Humphreys, A. P., and Smith, P. K. (1987). Rough and Tumble, Friendship, and Dominance in Schoolchildren: Evidence for Continuity and Change with Age. *Child Development, 58*: 201–212.

Kaczkurkin, A. N., and Foa, E. B. (2015). Cognitive-Behavioral Therapy for Anxiety Disorders: An Update on the Empirical Evidence. *Dialogues in Clinical Neuroscience, 17*(3): 337–346.

Kennair, L. E. O. (2007). Fear and Fitness Revisited. *Journal of Evolutionary Psychology, 5*(1): 105–117. DOI: 10.1556/JEP.2007.1020

Kennair, L. E. O. (2011). "The Problem of Defining Psychopathology and Challenges to Evolutionary Psychology Theory." In D. M. Buss and P. H. Hawley (Eds.), *The Evolution of Personality and Individual Differences*, pp. 451–479. New York: Oxford University Press.

Kennair, L. E. O. (2014). Evolutionary Psychopathology and Life History: A Clinician's Perspective. *Psychological Inquiry, 25*(3–4): 346–351. DOI: 10.1080/1047840X.2014.915707

Kwon, K.-A., Yoo, G., and Bingham, G. E. (2016). Helicopter Parenting in Emerging Adulthood: Support or Barrier for Korean College Students' Psychological Adjustment? *Journal of Child and Family Studies, 25*(1), 136–145. DOI: 10.1007/s10826-015-0195-6

Lasenby-Lessard, J., and Morrongiello, B. A. (2011). Understanding Risk Compensation in Children: Experience with the Activity and Level of Sensation Seeking Play a Role. *Accident Analysis & Prevention, 43*(4): 1341–1347. DOI: dx.doi.org/10.1016/j.aap.2011.02.006

LeMoyne, T., and Buchanan, T. (2011). Does "Hovering" Matter? Helicopter Parenting and Its Effect on Well-Being. *Sociological Spectrum, 31*(4): 399–418. DOI: 10.1080/02732173.2011.574038

Little, H. (2006). Children's Risk-Taking Behaviour: Implications for Early Childhood Policy and Practice. *International Journal of Early Years Education, 14*(2): 141–154.

Lukianoff, G., and Haidt, J. (2015). The Coddling of the American Mind. *Atlantic.* Retrieved from http://www.theatlantic.com/magazine/archive/2015/09/the-coddling-of-the-american-mind/399356/

Marano, H. E. (2015). Where Did Colleges Go Wrong? *Psychology Today.* Retrieved from https://www.psychologytoday.com/blog/nation-wimps/201510/where-did-colleges-go-wrong

Marks, I. M., and Nesse, R. M. (1994). Fear and Fitness: An Evolutionary Analysis of Anxiety Disorders. *Ethology and Sociobiology, 15*(5–6): 247–261. DOI: 10.1016/0162-3095(94)90002-7

Martin, P., and Caro, T. M. (1985). "On the Function of Play and Its Role in Behavioral Development." In J. Rosenblatt, C. Beer, M. Bushnel, and P. Slater (Eds.), *Advances in the Study of Behavior*, vol. 15, pp. 59–103. New York: Academic Press.

Morrongiello, B. A., and Major, K. (2002). Influence of Safety Gear on Parental Perceptions of Injury Risk and Tolerance for Children's Risk Taking. *Injury Prevention, 8*(1): 27–31. DOI: 10.1136/ip.8.1.27

National Safety Council. (2016). Landing Lightly: Playgrounds Don't Have to Hurt. Retrieved from http://www.nsc.org/learn/safety-knowledge/Pages/news-and-resources-playground-safety.aspx

Nesse, R. M. (2001). The Smoke Detector Principle. *Annals of the New York Academy of Sciences, 935*(1): 75–85. DOI: 10.1111/j.1749-6632.2001.tb03472.x

Norton, P. J., and Price, E. C. (2007). A Meta-analytic Review of Adult Cognitive-Behavioral Treatment Outcome across the Anxiety Disorders. *Journal of Nervous and Mental Disease, 195*(6): 521–531. DOI: 10.1097/01.nmd.0000253843.70149.9a

Pellegrini, A. D., and Bjorklund, D., F. (2004). The Ontogeny and Phylogeny of Children's Object and Fantasy Play. *Human Nature, 15*(1): 23–43.

Pellegrini, A. D., and Smith, P. K. (1998). Physical Activity Play: The Nature and Function of a Neglected Aspect of Play. *Child Development, 69*(3): 577–598.

Pellegrini, A. D., and Smith, P. K. (2005). *The Nature of Play: Great Apes and Humans.* New York: Guilford Press.

Phillips, R. O., Fyhri, A., and Sagberg, F. (2011). Risk Compensation and Bicycle Helmets. *Risk Analysis, 31*(8): 1187–1195. DOI: 10.1111/j.1539-6924.2011.01589.x

Play Safety Forum. (2012). Managing Risk in Play Provision: Implementation Guide. Retrieved from http://www.playengland.org.uk/media/172644/managing-risk-in-play-provision.pdf

Play Safety Forum. (2014). Risk Benefit Assessment Forum. Retrieved from http://www.playboard.org/wp-content/uploads/2014/11/PSF-Risk-Benefit-Assessment-Form-Worked-Example.pdf

Policy Studies Institute. (2015). Children's Independent Mobility: An International Comparison and Recommendations for Action. Retrieved from http://www.psi.org.uk/site/publication_detail/1823

Poulton, R., Davies, S., Menzies, R. G., Langley, J. D., and Silva, P. A. (1998). Evidence for a Non-associative Model of the Acquisition of a Fear of Heights. *Behaviour Research and Therapy, 36*: 537–544.

Poulton, R., and Menzies, R. G. (2002). Non-associative Fear Acquisition: A Review of the Evidence from Retrospective and Longitudinal Research. *Behaviour Research and Therapy, 40*(2): 127–149. DOI: 10.1016/S0005-7967(01)00045-6

Poulton, R., Menzies, R. G., Craske, M. G., Langley, J. D., and Silva, P. A. (1999). Water Trauma and Swimming Experiences Up to Age 9 and Fear of Water at Age 18: A Longitudinal Study. *Behaviour Research and Therapy, 37*: 39–48.

Poulton, R., Milne, B. J., Craske, M. G., and Menzies, R. G. (2001). A Longitudinal Study of the Etiology of Separation Anxiety. *Behaviour Research and Therapy, 39*: 1395–1410.

Priceonomics. (2014). The Rise and Fall of the Fireman's Pole. Retrieved from http://priceonomics.com/the-rise-and-fall-of-the-firemans-pole/

Rakison, D. H. (2005). "Infant Perception and Cognition: An Evolutionary Perspective on Early Learning." In B. J. Ellis and D. F. Bjorklund (Eds.), *Origins of the Social Mind: Evolutionary Psychology and Child Development*, pp. 317–353. New York: Guilford Press.

Readdick, C. A., and Park, J. J. (1998). Achieving Great Heights: The Climbing Child. *Young Children, 53*(6): 14–19.

Sandseter, E. B. H. (2007). Categorizing Risky Play—How Can We Identify Risk-Taking in Children's Play? *European Early Childhood Education Research Journal, 15*(2): 237–252.

Sandseter, E. B. H. (2010a). "It Tickles in My Tummy!"—Understanding Children's Risk-Taking in Play through Reversal Theory. *Journal of Early Childhood Research, 8*(1): 67–88.

Sandseter, E. B. H., and Kennair, L. E. O. (2011). Children's Risky Play from an Evolutionary Perspective: The Anti-phobic Effects of Thrilling Experiences. *Evolutionary Psychology, 9*(2): 257–284.

Satomi, T. S., and Morris, V. G. (1996). Outdoor Play in Early Childhood Education Settings: Is It Safe and Healthy for Children? *Early Childhood Education Journal, 23*(3): 153–157.

Sawyers, J. K. (1994). The Preschool Playground: Developing Skills through Outdoor Play. *Journal of Physical Education, Recreation & Dance, 65*(6): 31–33.

Schiffrin, H. H., Liss, M., Miles-McLean, H., Geary, K. A., Erchull, M. J., & Tashner, T. (2013). Helping or Hovering? The Effects of Helicopter Parenting on College Students' Well-Being. *Journal of Child and Family Studies, 23*(3): 548–557. DOI: 10.1007/s10826-013-9716-3

Segrin, C., Woszidlo, A., Givertz, M., Bauer, A., and Murphy, M. (2012). The Association between Overparenting, Parent-Child Communication, and Entitlement and Adaptive Traits in Adult Children. *Family Relations, 61*(2): 237–252. DOI: 10.1111/j.1741-3729.2011.00689.x

Segrin, C., Woszidlo, A., Givertz, M., and Montgomery, N. (2013). Parent and Child Traits Associated with Overparenting. *Journal of Social and Clinical Psychology, 32*(6): 569–595. DOI: 10.1521/jscp.2013.32.6.569

Silove, D., Manicavasagar, V., O'Connell, D., and Morris-Yates, A. (1995). Genetic Factors in Early Separation Anxiety: Implications for the Genesis of Adult Anxiety Disorders. *Acta Psychiatrica Scandinavia, 92*: 17–24.

Smith, K. (2014). Discourses of Childhood Safety: What Do Children Say? *European Early Childhood Education Research Journal, 22*(4): 525–537. DOI: 10.1080/1350293X.2014.947834

Smith, P. K. (1982). Does Play Matter? Functional and Evolutionary Aspects of Animal and Human Play. *Behavioral and Brain Sciences, 5*: 139–184.

Smith, P. K. (2005). "Play: Types and Functions in Human Development." In B. J. Ellis and D. F. Bjorklund (Eds.), *Origins of the Social Mind: Evolutionary Psychology and Child Development*, pp. 271–291. New York: Guilford.

Smith, P. K., and Vollstedt, R. (1985). On Defining Play: An Empirical Study of the Relationship between Play and Various Play Criteria. *Child Development, 56*: 1042–1050.

Smith, S. J. (1998). *Risk and Our Pedagogical Relation to Children: On Playground and Beyond*. New York: State University of New York Press.

Spiegal, B., Gill, T., Harbottle, H., and Ball, D. J. (2014). Children's Play Space and Safety Management: Rethinking the Role of Play and Standards. *Sage Open*, 111. DOI: 10.1177/2158244014522075

Stephenson, A. (2003). Physical Risk-Taking: Dangerous or Endangered? *Early Years, 23*(1): 35–43.

Stutz E. (1995) Rethinking Concepts of Safety and the Playground: The playground as a place in which children may learn skills for life and managing hazards, in Christiansen M.L. (Ed) Proceedings of the International Conference of Playground Safety. Pennsylvania: Penn State University, Center for Hospitality, Tourism & Recreation Research.

Suicide Prevention Resource Center. (2014). *Suicide among College and University Students in the United States*. Retrieved from: http://www.suicideprevention.osu.edu

Sutton-Smith, B. (1997). *The Ambiguity of Play*. Cambridge, MA: Harvard University Press.

van der Heiden, C., Muris, P., and van der Molen, H. T. (2012). Randomized Controlled Trial on the Effectiveness of Metacognitive Therapy and Intolerance-of-Uncertainty Therapy for Generalized Anxiety Disorder. *Behaviour Research and Therapy, 50*(2): 100–109. DOI: 10.1016/j.brat.2011.12.005

Wells, A. (2009). *Metacognitive Therapy for Anxiety and Depression*. New York: Guilford Press.

Wells, A., and Morrison, A. P. (1994). Qualitative Dimensions of Normal Worry and Normal Obsessions: A Comparative Study. *Behaviour Research and Therapy, 32*(8): 867–870. DOI: http://dx.doi.org/10.1016/0005-7967(94)90167-8

Wiltz, N. W., and Fein, G. G. (2006). "Play as Children See It." In D. P. Fromberg and D. Bergen (Eds.), *Play from Birth to Twelve Contexts, Perspectives, and Meanings* (2nd ed.), pp. 127–139. London: Routledge.

Wolitzky-Taylor, K. B., Horowitz, J. D., Powers, M. B., and Telch, M. J. (2008). Psychological Approaches in the Treatment of Specific Phobias: A Meta-analysis. *Clinical Psychology Review, 28*(6): 1021–1037. DOI: 10.1016/j.cpr.2008.02.007

Zuckerman, M. (1994). *Behavioral Expressions and Biosocial Bases of Sensation Seeking*. Cambridge: Cambridge University Press.

Öhman, A., and Mineka, S. (2001). Fears, Phobias, and Preparedness: Toward an Evolved Module of Fear and Fear Learning. *Psychological Review, 108*(3): 483–522. DOI: 10.1037/0033-295X.108.3.483

8 The Anti-Vaccine Movement: A Litany of Fallacy and Errors

Jonathan Howard and Dorit Rubinstein Reiss

Before we begin, let us make three facts clear.

Vaccines are very safe, though not perfectly so.

A systemic review of vaccine safety in 2014 concluded that while there is "evidence that some vaccines are associated with serious adverse events . . . these events are extremely rare" (Maglione et al., 2014, 1).

Vaccines have stopped hundreds of millions of illnesses, preventing needless suffering and countless deaths.

Research done at the University of Pittsburgh and published in 2013 concluded that in the United States alone, vaccines have prevented over a hundred million cases of serious infectious disease since 1924. Another study estimated that in children born in 2009, vaccines will prevent nearly 42,000 early deaths and twenty million cases of disease (van Panhuis et al., 2013; Zhou et al., 2014).

Vaccines save an enormous amount of money.

An economic analysis of vaccines in 2014 found that for children born in 2009, vaccines would lead to a "net savings of $13.5 billion in direct costs and $68.8 billion in total societal costs" (Zhou et al., 2014).

Given this, opposition to vaccines might seem as reasonable as opposition to sunshine and water. Yet, since Edward Jenner vaccinated eight-year-old James Phipps with cowpox in 1796, giving him immunity to smallpox, resistance to vaccines has been a constant. In fact, the first documented instance of vaccination was Benjamin Jesty, an English farmer who vaccinated his family in 1774. Per accounts at the time, he was "hooted at, reviled and pelted whenever he attended markets in the neighbourhood."

The anti-vaccine movement is diverse in some ways, including people with diverse political views and differing motives. Generally speaking, however, vaccine opponents in the United States tend to be white and affluent, though the link to educational achievement is less clear (Yang, Delamater, Leslie, and Mello, 2016). Objections

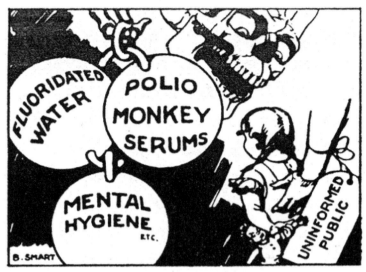

Figure 8.1

A flier first issued in 1955 warning about dangers of the polio vaccine, which had been invented three years prior.

The Anti-Vaccine Movement

Figure 8.2
James Gillray's 1802 caricature of Jenner vaccinating patients who feared it would make them sprout cowlike appendages.

to vaccines vary greatly across time and location, and even within the United States alone, there are different strands of thought among people who oppose vaccines. Vaccine opponents include parents who believe their children were harmed by vaccines, people convinced of the dangers of vaccines for other reasons, and sellers of alternative treatments and products. Many of these sellers benefit financially when fear of vaccines leads people to buy their products and use their services. In spite of these differences, common themes emerge, and allow us to discuss anti-vaccine arguments generally, while acknowledging that specific actors may differ in the emphasis placed in and how strongly they hold the views described here.

In addition to the common themes and common demographic characteristics, the various strands of the anti-vaccine movement have another common feature: they can all cause serious harm to the community. Nonvaccinating increases a child's risk of disease, and widespread nonvaccination undermines herd immunity and leads to outbreaks. The link between nonvaccinating and disease is, therefore, very direct, as multiple studies show ("Personal Belief Exemptions for Vaccination Put People at Risk. Examine the Evidence for Yourself," n.d.).

Because nonvaccinating leads to outbreaks, the anti-vaccine movement has led to needless suffering and death, though we can't compute the toll. For example, after Andrew Wakefield's fraudulent research scared parents in the British Isles from giving MMR to their children, an outbreak of measles swept through the Isles. Children got sick, and four children in England and Ireland died (Offit, 2010).

A more recent—and very clear—example is the measles outbreak in Minnesota. Anti-vaccine activists have visited a community of Somali refugees in Minnesota repeatedly since 2008, fueling fears that the measles-mumps-rubella (MMR) vaccine causes autism. As a result, the rate of vaccination with MMR within this community dropped from over 90% in 2004 to 42% in April 2017. In 2017, an outbreak of measles centered on that community (Offit, 2017).

As of June 2, 2017, seventy-three people in Minnesota were sick with measles—more than the total number of cases for 2016 in the entire United States. Sixty-seven were unvaccinated people, mostly young, American-born children from the Somali community targeted by anti-vaccine activists. Twenty-one people, mostly young unvaccinated children, were hospitalized (Howard, 2017).

Romania, too, has recently seen a measles outbreak attributed, in large part, to anti-vaccine efforts. Thousands have been sickened, and at least 26 unvaccinated people, mostly children, died (Scutti, 2017).

In this chapter, we will give examples of arguments used to justify or generate opposition to vaccines, and explain why each argument is problematic, identifying, where appropriate, the relevant logical fallacy.

Naturalistic Fallacy: A View That What Is Natural Is Inherently Good and What Is Unnatural Inherently Bad

The idea that infectious diseases are benign simply because they are natural is common in the anti-vaccine movement. This common sentiment is expressed by Dr. Jennifer Margulis in the National Public Radio documentary *The Vaccine Wars*.

As a parent, I would rather see my child get a natural illness and contract that the way that illnesses have been contracted for at least 200,000 years that homo sapiens has been around. I'm not afraid of my children getting chickenpox. There are reasons that children get sick. Getting sick is not a bad thing. (Palfreman and McMahon, 2010)

Stating even more strongly that preventable diseases are not a concern, Dr. Kelly Brogan in her essay "Why Vaccines Aren't Paleo" claims: "We coexist with bacteria and viruses to a level of enmeshment that makes the perception of 'vaccine-preventable infections' a laughable notion" (Brogan and Ji, 2014).

Other articles expressly tout alleged benefits of vaccine-preventable diseases. Articles titled "The Unreported Health Benefits of Measles," which appeared on the pseudoscience site GreenMedInfo, are quite common (GreenMedInfo, 2015).

This view understates, dismisses, or ignores the risks associated with preventable diseases. Chickenpox, which Dr. Margulis wants for her children, can kill, though rarely, and can have complications—and the virus stays in the body to reawaken as shingles, which can cause an extremely painful, long-lasting rash and occasionally, serious complications such as blindness and strokes. Measles has killed children in the United States as recently as 1989–1991, and a recent study of the effects of the outbreak in California found that the rates of an always fatal complication, subacute sclerosing panencephalitis, may be as high as 1:660 in infants (Wendorf et al., 2016).

Measles is anything but a benign disease. It has high rates of complications, even in developed countries (Mina, Metcalf, de Swart, Osterhaus, and Grenfell, 2015). It can also erase immune memory and leave a child at risk of other infections to which the child was previously immune (Perry and Halsey, 2004).

Preventable diseases may not look so bad to believers in the naturalistic fallacy, but vaccines are unnatural and therefore (they figure) dangerous. This, in spite of the fact that vaccines are held to an extremely high safety standard, and a risk of, say, a 1:10,000 of a serious complication can lead to a vaccine being taken off the market (CDC, 2011).

Another common tactic of the anti-vaccine movement is to simply list the chemical ingredients in vaccines in order to make them sound as unnatural as possible, a technique pro-vaccine blogger Dr. David Gorski refers to as the "toxins gambit." Just about any everyday product (eggs contains eicosanoic acid and formaldehyde!) can be made to sound terrifying to some people if the chemical names of the ingredients are listed without further context or explanation (Dvorsky, 2014).

Nirvana Fallacy (Perfect Solution Fallacy): The View That If a Solution Is Not Perfect, It Is Worthless

Like everything in this world, vaccines are not perfect, and no serious scientist, doctor, or health official would claim otherwise. Like bulletproof vests and smoke detectors, they do not work 100 percent of the time. Vaccines may not prevent every case of disease. Nor are they completely risk free. Very rarely, vaccines can cause serious harm.

A recurrent theme voiced by anti-vaccine activists is that since vaccines are not perfect, they are useless. As stated by anti-vaccine advocate Sayer Ji, "The only way that the act of refraining from vaccinating could be justifiably characterized as 'insane behavior' is if vaccines were proven effective 100% of the time" (Ji, 2014).

This is doubly wrong. First, 50 percent protection—the rate achieved by the influenza vaccine, one of the least effective—is much better than the zero protection provided by nonvaccination. That's leaving aside the influenza vaccine's ability to reduce the chances of death and hospitalization among those patients it doesn't protect completely. Against other vaccines, many of which are more than 90 or 95 percent effective, the argument is even less true (CDC, 2016).

Second, the perfect effectiveness demand is inconsistent. Other mechanisms to prevent injury are not held to this standard. Seat belts are not perfect. Seat belts can cause serious injuries, and the medical literature has described them in gory detail (Anderson, Rivara, Maier, and Drake, 1991; Reid, Letts, and Black, 1990). But no one in the anti-vaccine movement rejects seat belts. Why then should the substantial benefits vaccines be rejected because they do not meet an unrealistic standard of perfection?

Shifting the Burden of Proof (See *Onus Probandi*): "I Need Not Prove My Claim, You Must Prove It Is False"

Anti-vaccine advocates routinely blame vaccines for a broad range of maladies, ranging from autism and allergies to acne and Down syndrome (Whale.to, n.d.). These advocates then demand that critics prove vaccines do not cause these conditions, rather than providing evidence that vaccines do cause them. To quote the anti-vaccine cardiologist Jack Wolfson, "To all the pediatricians in the world, please show me the study that found 69 doses of 16 vaccines do not cause cancer, auto-immune disease, and brain injury" (Wolfson, 2014).

This simply is not how science works. It makes as much sense as demanding studies that show turmeric does not cause brain cancer and that coconut oil does not cause multiple sclerosis. When studies are presented showing no link between vaccines and autoimmune disease, for example, the anti-vaccine contingent merely claims these studies are bought or unreliable.

Anecdotal Fallacy: Using a Personal Experience or Examples to Extrapolate Without a Statistically Significant Number of Cases That Could Provide Scientifically Compelling Evidence

The way this fallacy is used incorporates two others. First, appeal to emotion: where an argument is made due to the manipulation of emotions, rather than the use of valid reasoning. Second, *post hoc, ergo propter hoc*: after this, therefore because of this. The fallacy that, because one thing happens after another, the first event must have caused the second.

Since anti-vaccine advocates cannot depend on science to support their case, they instead must rely on anecdotes. A large part of the problem is that correlation does not imply causation. In other words, the rooster crowing does not cause the sun to rise. There is no question that tragedies have befallen individuals at some point after getting a vaccine, though of course, the role the vaccine played in that tragedy is often far from clear. Anti-vaccine websites and movies are awash with stories, often

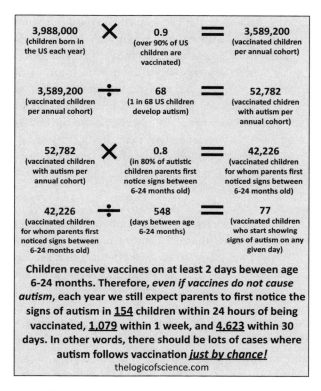

Figure 8.3
A mathematical explanation of why autism symptoms are bound to emerge after vaccination in some children.

accompanied by dramatic pictures and videos, designed to appeal to the fears and emotions of their viewers. When millions of babies are vaccinated multiple times at the same age when many conditions naturally emerge, temporal connections are unavoidable: some bad things are going to happen right after vaccines. One blogger provided a simple mathematical demonstration to demonstrate that a large number of children will develop signs of autism shortly after receiving a vaccine, by chance alone (Logic of Science, 2016).

That is why just a temporal link is not enough. Let's look more closely at two examples. On July 2014, twelve-year-old Meredith Prohaska was found dead in her home. She had received the HPV vaccine that day. Her understandably distraught mother blamed the vaccine, and the claim was repeated both in the regular media (Johnson, 2014) and on anti-vaccine sites (Health Impact News, 2014). From the start there was no evidence of a vaccine connection. That did not stop the rumors, and the story is still circulated as a vaccine-related death. But Meredith did not die from the vaccine. The medical examiner found that what killed her was an overdose of an antihistamine, though it's not clear how the overdose happened—if the girl herself accidentally took too much, or her mother gave it to her (Johnson, 2014). The incident is still regularly shared as an example of a vaccine injury. The strong evidence against the alleged vaccine link makes this case a great example of how problematic these stories are. The mother's initial belief about the vaccine had no evidence behind it, but it was strong and sincere. So are the beliefs of others claiming a range of vaccine harms. Examine the vast majority of these Internet claims, and either there's no evidence or the evidence is pointing the other way. Temporal connections are not good enough to claim something is a vaccine injury—but that's what those using anecdotes tend to rely on.

The claim that the MMR causes autism relied on an article that was, in essence, a collection of stories (Wakefield et al., 1998). As is well known, the article was also later found to be fraudulent—with the dates of MMR administration relative to the symptoms of autism manipulated—and Wakefield was found to have hid the fact that several of the children were recruited from parents engaged in litigation against MMR, and that he himself was paid as an expert witness in that litigation. (There were also other ethical issues.) (Deer, 2011).

Even more important, when the question was studied thoroughly, the claim was found untrue for large numbers of children: rates of autism were similar whether or not a child got MMR (Honda, Shimizu, and Rutter, 2005; Jain et al., 2015; Taylor, Swerdfeger, and Eslick, 2014). In other words, the stories of autism after MMR, however sincere the belief of the parents telling them, were simply mistaken.

Examination of specific cases also showed stories of autism caused by vaccines were incorrect. For example, Dr. Brian Hooker, an anti-vaccine activist, blamed vaccines for his son's autism—but on examination by the National Vaccine Injury Compensation program, it was found that his son showed developmental problems from very early on, rather than sudden regression after the vaccines as claimed. In fact, the medical records contained no indication of serious, sudden problems right after the vaccines (Reiss, 2016).

Furthermore, memories are rarely reliable, especially about emotional events. A temporal connection may not be there; even when it is, that connection may not show causation. For these reasons, anecdotes, the main tool of anti-vaccine movements, are not a good way to argue. Having said that, there is nothing wrong with sharing emotional anecdotes per se, *as long as those stories are consistent with, rather than contradict the science.* Anecdotes can be extremely persuasive at an emotional level and are not necessarily wrong by themselves. After all, Edward Jenner popularized vaccination when he heard anecdotes that milkmaids were immune from smallpox. Pro-vaccine sources often share stories of those who suffered and died due to vaccine-preventable diseases. Indeed, vaccine-hesitant persons were more likely to become favorable if presented with scientific facts and a paragraph by a mother whose ten-month-old son suffered a life-threatening case of measles, as opposed to doubters who were presented with facts alone. However, anecdotes not backed up by science are simply stories, and as many others have said, "The plural of anecdote is not data" (Byrne, n.d.; Horne, Powell, Hummel, and Holyoak, 2015).

False Authority (Single Authority): Using an Expert of Dubious Credentials

There are several anti-vaccine doctors, some of whom have impeccable training and credentials. They have a strong social media presence with numerous blogs and YouTube videos extolling their anti-vaccine views. They also appear in anti-vaccine movies such as *Vaxxed*, *Bought*, and *The Greater Good*. These relatively few doctors are held up as luminaries by the anti-vaccine movement. This is despite the fact that almost without exception, they have not actually done any original research on vaccines and do not currently work in a position where they might encounter a patient with a vaccine-preventable disease. Many appear not to see patients at all anymore, and those who still work with patients often have expensive cash-only practices that cater to the richest 1 percent.

Consider, for example, Dr. Toni Bark, whose credentials are better than those of most anti-vaccine doctors. Dr. Bark is an actual MD with a background in pediatrics. She graduated from medical school in 1986, and then, according to her site, practiced in actual hospitals until 1993, when "her commitment to natural remedies led her to begin her study of Holistic Medicine" (Bark, n.d.). Since then, Dr. Bark has practiced homeopathy. She also runs a store selling supplements and skin care products (Bark, 2016). In other words, for over twenty years Dr. Bark has not practiced evidence-based medicine, instead providing alternative remedies. There is no indication she has had any experience working with patients with preventable diseases in this time, nor that she worked with severely ill patients. Her publications on vaccines stretch to one letter to the editor (she was a middle author) in *The Journal of Infectious Agents and Cancer*. The letter had no original research, instead trying to raise doubts about the effectiveness and safety of HPV vaccines (Tomljenovic, Wilyman, Vanamee, Bark, and Shaw, 2013). Even this almost nonexistent publication record on vaccines places her above most doctors promoting anti-vaccine claims. Dr. Bark's main vaccine-related activities include participation in radio show and films, as well as social media activities ("Dr. Bark in the Media," n.d.).

Other anti-vaccine doctors—including Larry Palevsky, Paul Thomas, Joseph Mercola, Suzanne Humphries, Sherri Tenpenny, or Kelly Brogan—suffer from similar limitations as vaccine experts. In fact, Dr. Bark, badly credentialed as she is, is more credentialed than most of them. Almost without exception, these doctors have no scientific publications related to vaccines, they practice in ways that make them very unlikely to have to confront victims of vaccine-preventable diseases, or otherwise severely ill patients. For most of them, the amount of time they spend on a stage or in front of a movie camera passing themselves off as vaccine experts vastly exceeds the amount of time they spend actually caring for sick children or researching vaccines in a lab.

Despite this lack of expertise anti-vaccine doctors often claim to have done extensive vaccine research. For example, on her Facebook page, Dr. Brogan claims to have done "more than 10,000 hours of personal research" on vaccines. Despite her claim, implausible as it may be, that she has done "research" on vaccines for nearly ten hours every day for three straight years (that's what it would take get to 10,000 hours), she has no peer-reviewed publications on vaccines, outside of a single opinion piece linking vaccines and depression that was published in the journal *Alternative Therapies, Health and Medicine*. She does not have a lab where she could conduct basic experiments in immunology, nor has she been involved in clinical trials of vaccines. She has not made any new discoveries, and therefore cannot be said to have done "research" on vaccines in any meaningful sense of the word. In the same Facebook post where she boasted of 10,000 hours of research, she wrote that vaccines are "recommended for every human on the planet, cradle to grave, one-size-fits all." In fact, the inability of some people to receive vaccines is one of the reasons herd immunity is so important. This information takes literally seconds to find by going to the CDC webpage.

Another striking feature of such doctors tends to be the way they sell alternative treatments and supplements, all the while accusing their pro-vaccine colleagues of profiting from vaccines. While they routinely excoriate vaccines as being "untested" ways for pharmaceutical companies to profit, they promote other treatments, such as coffee enemas, for which there is no evidence of benefit. Dr. Brogan sells coffee enema supplies and has multiple other corporate connections through affiliate marketing on her website. Dr. Thomas sells a wide range of "Metagenic," (the name of his brand) supplements on his website. Dr. Palevky's practice at the Northport Wellness Center also sells a wide range of supplements and products. Dr. Mercola runs an online sales empire that according to one estimate earned him nearly $7 million in 2010. These supplements can be sold to strangers whom the practitioner has never met or examined, and whom they have no responsibility for treating (Brogan, 2016a; Metagenetics.com, 2016; "Welcome to the Northport Wellness Center store," 2016; Smith, 2012).

In contrast to their skepticism on vaccines, many anti-vaccine doctors not only easily accept untested alternative treatments, but they embrace far-fetched conspiracy theories. Dr. Brogan has shared her belief that outbreaks of various diseases such as Ebola, Zika, and even HIV/AIDS are "scripted" to force vaccines on the masses. She even shared an article on her Facebook page entitled "Research: Insulin Can Kill Diabetics; Natural Substances Heal Them." Dr. Tenpenny once shared a phony news item that a neo-Nazi rally was really "ran by Jews." Anti-vaccine doctor Lawrence Palevsky shared the story "Busted Pilot Forgets to Turn Off CHEMTRAILS While Landing" on his Facebook page. This phenomenon is sometimes termed "crank magnetism," the condition

Figure 8.4

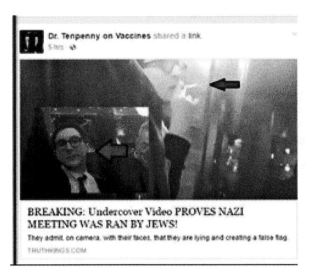

Figure 8.5

where people become attracted to multiple crank ideas at the same time. One could imagine an alternative universe where pharmaceutical companies, government bodies, and scientific organizations all touted the dangers of vaccines, anti-vaccine doctors would fervently embrace and sell them as a suppressed miracle cure.

Another point: consider the behavior of these doctors on social media. They claim to be "awake" and "fearless," and call on their followers to question expert opinion. But any challenge to any antivaccine statement made by one of these doctors, almost always results in the comment being deleted and the commenter being banned. Maintaining an echo chamber is extraordinarily important to anti-vaccine doctors and anti-vaccine pages in general.

Moving the Goalposts (Raising the Bar): After Evidence Is Presented in Response to a Specific Claim, Some Other (Often Greater) Evidence Is Demanded

Anti-vaccine advocates have been blaming vaccines for autism for years and are perfectly comfortable moving from one theory of autism causation to another any time an existing theory runs into difficulties. The case of thimerosal, an organomercury compound used as a preservative in many vaccines, is a good example of how anti-vaccine advocates move the goalposts. Though there was no evidence thimerosal caused autism, it was removed from childhood vaccines in 1999 because of negative publicity and an "abundance of caution." Many anti-vaccine advocates had predicted the autism rate would fall. When it did not, they blamed other vaccines or vaccine ingredients, and even ingredients that aren't in vaccines (like the weed-killer glyphosate) (Gerber and Offit, 2009). The anti-vaccine crusade is also resourceful about what it blames on the vaccines. The list is seemingly endless: autoimmune diseases, SIDS, diabetes, cancer, ADHD, depression, Alzheimer's disease, and many others.

Other common examples of moving the goalposts include blaming the vaccinated for spreading diseases, claiming that preventable diseases can be easily cured with "natural" substances, or even claiming that vaccine-preventable diseases are beneficial. For example, an article published on GreenMedInfo in 2015 was entitled "Measles Transmitted by The Vaccinated, Gov. Researchers Confirm." Yet the same site also published an article entitled "The Unreported Health Benefits of Measles." If measles is both spread by the vaccinated and beneficial, then this a unique (though completely flawed) pro-vaccine argument. Similar an article on GreenMedInfo entitled "Natural Herbal HPV 'Cure' Discovered" was followed two weeks later by an article entitled "HPV Vaccine Maker's Study Proves Natural HPV Infection Beneficial, Not Deadly" (GreenMedInfo, n.d.). It obviously makes no sense to tout a natural "cure" for a "beneficial" disease.

These obvious, blatant contradictions do not bother antivaccine activists. As long as vaccines are deemed ineffective or dangerous, anything goes. Trying to counter anti-vaccine claims is like playing whack-a-mole. As soon as one claim is debunked, another will sprout in its place.

This shifting of goalposts is also very evident when it comes to vaccine safety where impossible standards are often demanded. One common claim of the anti-vaccine movement is that vaccines have never been properly tested. Tens of thousands of studies exist on vaccines' safety and effectiveness, and vaccines are, in fact, tested more thoroughly than most medications. But the anti-vaccine movement claims that these studies are invalid because the entire vaccine schedule has not been tested together over the entire human life span. Obviously, conducting such experiments would be a logistical and ethical impossibility, something some in the anti-vaccine movement are well aware of. This complaint is particularly ironic, given that many members of the anti-vaccine movement readily embrace alternative medicine treatments for which there is no evidence of efficacy (such as "earthing" mats) or treatments for which there is strong evidence of inefficacy (such as homeopathy). In addition, this demand is inconsistent with their approach to other issues. There has never been a double-blind study proving that smoking causes lung cancer, for example, yet anti-vaccine advocates have no trouble accepting that smoking is dangerous.

Another tactic of anti-vaccine advocates is to try to shift the debate into an ethical/legal discussion about freedom and rights. When they have given up trying to argue that science is on their side, they will instead say that should be free to make health care decisions for themselves and their families. The rhetoric used by many anti-vaccine advocates suggests that they believe that squads of gun-toting vaccine police might break down their door at any moment, take their children, and vaccinate them against their will. This is largely a strawman argument, as the vaccination of groups of children against the wishes of their parents is exceedingly rare in American history. It last occurred in Philadelphia in 1991, when nine children were vaccinated during a measles outbreak centered upon an anti-vaccine church. As it happens, the outbreak in question ultimately took the lives of nine children and hundreds of others were sickened (Kelto, 2015).

When all else fails, some anti-vaccine activists will instead frame their opposition to vaccines as part of a broader rejection of science and reason in general. In her essay, "Sacred Activism: Moving Beyond the Ego," Dr. Brogan advises readers to "use logic and reason sparingly" when considering vaccines, and instead "sit quietly with yourself and feel for your truth. Choose what Story will be yours" (Brogan, 2016b). This effectively ends any conversation on vaccines (or any topic at all). As Sam Harris said:

> Water is two parts hydrogen and one part oxygen. What if someone says, "Well, that's not how I choose to think about water"? All we can do is appeal to scientific values. And if he doesn't share those values, the conversation is over. If someone doesn't value evidence, what evidence are you going to provide to prove that they should value it? If someone doesn't value logic, what logical argument could you provide to show the importance of logic? (Harris, 2013)

Genetic Fallacy: Where a Conclusion Is Based Solely on Something or Someone's Origin, With No Regard to Current Meaning or Context

Vaccines are made by pharmaceutical companies, and pharmaceutical companies are out to earn a profit. These facts, do not make vaccines unsafe or ineffective. The profit picture is underscored by the fact that vaccines are some of the least profitable drugs produced by pharmaceutical companies. According to PharmaCompass.com, no vaccines were among the top twenty drugs sold in 2015. However, if vaccines were not profitable, they would not be produced. And several modern vaccines have had high revenue (though, as a reminder, profits mean the difference between revenue and costs, not just revenue). Vaccines may have the advantage of stable demand as well, barring unexpected scandals or problems—when a vaccine is recommended to most babies, there is a clear and reasonably stable market. They are not, however, in the same category as products that are top earners, including medications taken daily, or medications that cost hundreds of thousands of dollars, for example. In spite of this and the fact that vaccine science comes from multiple sources, including governments, nonprofits, and research centers, to many in the anti-vaccine movement, the fact that vaccines are made by pharmaceutical companies is enough to taint the entirety of vaccine-science (PharmaCompass, 2016).

Legitimate scandals involving the pharmaceutical industry, often in fields completely unrelated to vaccines, are often held as examples as to why any science that supports the safety and efficacy of vaccines cannot be trusted. For example, a common theme in an anti-vaccine discussion is reference to Vioxx. Merck makes Vioxx; Merck makes vaccines. Therefore, goes the logic, all vaccines are bad. Strangely, this logic is not applied in other fields of medicine and business. Scandals involving faulty brakes and air bags are not used to discredit the fundamental science behind the internal combustion engine, for example.

Appeal to Motive: Where a Premise Is Dismissed by Calling into Question the Motives of Its Proposer

No one who has advocated publicly for vaccines for even a short amount of time has escaped the accusation of shill. In the minds of many anti-vaccine advocates, no one could possibly defend vaccines without being paid to do so. Most people are unlikely to find it strange that people are willing to speak up for preventing disease and protecting children and community, but the idea is unacceptable to anti-vaccine speakers. This is particularly ironic since the financial motivation of companies may well go the other way. After all, people getting preventable diseases may require medication and treatment sold by pharmaceutical companies, and those costs may be substantially higher than the income from vaccines. As already pointed out, vaccines reduce costs, not the other way around (Zhou et al., 2014).

Cherry Picking: Pointing to Individual Cases or Data That Seem to Confirm a Particular Position, While Ignoring a Significant Portion of Related Cases or Data That May Contradict That Position

As of the writing, searching Pubmed for vaccines yields 279,252 results. If even 1 percent of these papers pointed to safety concerns about vaccines, or to limited effectiveness of vaccines, this would mean an anti-vaccine advocate could find a substantial number of papers to bolster their position. Lists such as "67 Research Papers Showing That Vaccines Can Cause Autism" are common on the Internet, though the papers often do not show what the lists purport (Pierce, n.d.).

Any paper that does show a legitimate flaw in vaccines is touted as infallible by anti-vaccine advocates, whereas any paper that contradicts their position shows that science is flawed and corrupt. A skit called "If Google Was a Guy" by the website collegehumor.com illustrates this point well. In the skit, a woman asks the character representing Google if "vaccines cause autism?" Google responds "Well, I have one million results that says they don't, and one result that says they do." At this point, the woman grabs that one result and declares, "I knew it!" (CollegeHumor, 2014).

One slightly different form of cherry picking is called quote mining. This tactic involves taking phrases out of context to make them appear to agree with the quote miner's agenda. An example of this can be seen in the essay "Pregnancy-Friendly

Protection? The Truth About Whooping Cough Vaccine" by Dr. Brogan. In this essay, she correctly quotes a CDC Morbidity and Mortality Weekly Report (though one that was seven months out of date at the time) as saying: "In prelicensure evaluations, the safety of administering a booster dose of Tdap to pregnant women was not studied" (Brogan, 2013). However, reading the complete CDC report reveals that pregnancy registries were established to monitor safety of the vaccine, and "these studies did not suggest any elevated frequency or unusual patterns of adverse events in pregnant women who received Tdap and that the few serious adverse events reported were unlikely to have been caused by the vaccine" (Advisory Committee on Immunization, 2011). By omitting these additional sentences, Dr. Brogan presents an incorrect picture of the article, using it to support the opposite of its main point. Examples like this abound in anti-vaccine blogs.

Bad Science or Math

Anti-vaccine activists often use bad science or bad math to make their points. Serious errors are included, for example, when anti-vaccine advocates try to use statistics. Consider this statement by Mr. Sayer Ji regarding a measles outbreak of 159 cases that originated in Disneyland in 2015.

18% of the measles cases occurred in those who had been vaccinated against it—hardly the vaccine's claimed "99% effective." (Ji, 2015)

Let's do some math. Let us suppose 10,000 people were exposed to the measles and that 95 percent of them were vaccinated. Let us further suppose that all unvaccinated people will get the disease, while only 1 percent of vaccinated people will. This means that 500 unvaccinated people will get the measles, while 90 vaccinated people will get the measles. This means exactly 18 percent of the people who got the measles were vaccinated. This is to be expected, even with a vaccine that works 99 percent of the time, given that the population of vaccinated people is much larger than the population of unvaccinated people. It is up for debate whether such statistical misstatements are the result of intentional deception or a failure to understand sixth grade math. (It is also worth noting that one of the authors of this chapter was banned from the Facebook page of an anti-vaccine doctor for pointing out the mathematical error in Mr. Ji's statement.)

Similarly, the papers most highly touted in the anti-vaccine community are often of extremely poor quality, such as nonrandomized Internet surveys that claim to show vaccinated children are healthier than unvaccinated children. Many are published in journals which, despite legitimate-sounding names, are predatory publishers that charge high publication fees to authors, without providing legitimate peer review and quality control (Beall, 2018). A recent example is an abstract published, and then retracted, in the journal *Frontiers in Public Health,* and later republished in another predatory journal, *The Journal of Translational Science*. This was, in essence, an anonymous Internet survey of a small number of homeschooling mothers in four states, with strong indications that the survey population was biased against vaccines, and many other methodological flaws. However, anti-vaccine groups uncritically accepted and lauded this flawed abstract as proving a crucial point (McGovern, 2017).

Middle Ground: Saying That the Middle Point between Two Extremes Must Be the Truth

If one person says the earth is round and another person says with equal fervor that the earth is flat, it does not follow that the truth is somewhere in the middle of these claims. Similarly, it does not follow that there is truth on both sides of the vaccine "debate." Yet many of the people we identify as anti-vaccine activists claim not to be anti-vaccine, but rather "pro-safe vaccines." As Jenny McCarthy, the biggest celebrity associated with the anti-vaccine movement wrote:

> I am not "anti-vaccine." This is not a change in my stance nor is it a new position that I have recently adopted. For years, I have repeatedly stated that I am, in fact, "pro-vaccine" and for years I have been wrongly branded as "anti-vaccine." (Berman, 2014)

People such as Ms. McCarthy and Dr. Bob Sears—whatever they themselves think—are fairly described as anti-vaccine. These people consistently and systematically misrepresent the risks of vaccines as greater than they are—including attributing to vaccine risks they do not have (like autism, which Ms. McCarthy attributed to vaccines) and overstating the frequency of vaccine adverse events, consistently understate the benefits of vaccines (for example, by understating the risks of preventable diseases) and reject all or part of vaccine research by drawing on conspiracy theories. That combination puts them clearly in the anti-vaccine category.

Escape to the Future: Claiming That an Idea Will Soon Prevail Because the Emerging Evidence Is Just Around the Corner. "Science Was Wrong Before": Science Has Been Wrong in the Past, Therefore Science Cannot Be Trusted Now

These fallacies seek to cast doubt on vaccine science by drawing on historical cases where advances in science were the result of persistent work by those rejecting the consensus. Science is clearly continuously advancing, and there is much to learn. But this fact does not by itself cast doubt on the 200-year-old science of vaccines. Vaccines are supported by decades of studies on millions of people, and the science supporting them is robust. Yet anti-vaccine advocates are undeterred. Consider James Maskell, who writes, "we have precious little understanding of how vaccines affect microbial balance, but we do have a history of making interventions that seem like a good idea at the time, but backfire spectacularly" (Maskell, 2013). The recognition of the importance of our gut bacteria, the microbiome, in the importance of human health is without doubt an important, emerging field of science, and may lead to changes in many areas. But it does not change the abundant data that show that vaccines work and their risks are small.

Finally, No Discussion of the Anti-Vaccine Movement Would Be Complete Without a Discussion of Conspiracy Theories and Deception

As Neil DeGrasse Tyson tweeted: "Conspiracy theorists are those who claim coverups whenever insufficient data exists to support what they're sure is true" (Tyson, 2011). Conspiracy theories are the bread and butter of the anti-vaccine movement. The number of conspiracy theories promulgated by the anti-vaccine movement could be a book by itself. There are long-standing claims of conspiracies by the Centers for Disease Control and Prevention (CDC) to hide a link between vaccines and autism. Yet the data involved come from around the world, and therefore from outside the CDC's control.

Vaccines are thought by some to be part of a depopulation agenda. When two Mexican children died after receiving a contaminated vaccine, Mike Adams of the pseudoscience site Natural News wrote an article entitled "Depopulation Test Run? 75% of Children who Received Vaccines in Mexican Town Now Dead or Hospitalized" (Adams, 2015). As described above, emerging diseases for which there is no vaccine, such as Ebola and Zika, are thought to be scripted for the purpose of creating vaccines. As Mike Adams proclaimed in 2014: "It's all scripted! Ebola outbreak and impossibly rapid vaccine response clearly scripted; U.S. govt. patented Ebola in 2010 and now owns all victims" (Adams, 2014). Others believe that vaccines are being delivered via chemtrails to the unwitting masses. At times, these can take a much darker turn, with anti-vaccine and anti-Semitic graphics and comments often mingling on the Internet.

It is unfortunately almost impossible to argue against such conspiracies, as evidence against them is seen as evidence of how deep and clever the conspiracy actually is. Research has also shown that conspiracy theorists are not bothered by contradictions. Dr. Karen Douglas found, for example, "The more people were likely to endorse the idea Princess Diana was murdered, the more they were likely to believe that Princess Diana is alive" (Parry, 2012). Similarly, many anti-vaccine conspiracy theorists are equally comfortable with the idea that vaccines are part of a depopulation agenda and that vaccines are merely a way for pharmaceutical companies to profit. This makes little sense unless one considers dead people to be good customers. Anti-vaccine advocates who claim that the MMR vaccine causes autism are not offended by colleagues who claim thimerosal causes autism (the MMR never contained thimerosal). As long as vaccines are blamed for autism, the contradiction in the purported causative agent is of little concern.

Perhaps not surprisingly, only studies showing the benefits of vaccines are felt to be part of a massive conspiracy. When an HIV vaccine trial failed in 2013, there were no cries from the anti-vaccine community that the science on vaccines couldn't be trusted (Miner, 2016). When the pertussis vaccine was shown to have waning immunity, there were no screams from anti-vaccine advocates about the pharmaceutical industry controlling all science (Klein, Bartlett, Fireman, Rowhani-Rahbar, and Baxter, 2012). When vaccinated baboons were shown in a FDA study to be capable of transmitting pertussis, without showing symptoms of it, this study was widely promoted by the same anti-vaccine advocates who previously said nothing the FDA produces can be trusted (Haelle, 2014). The pattern soon becomes clear: there are infallible studies that show what the anti-vaccine advocate wants to be true and corrupt, deceptive studies that show vaccines are safe and effective.

Deception[1]

A good example of deception by an anti-vaccine activist occurs in a document entitled "200 Evidence-Based Reasons NOT to Vaccinate" by Sayer Ji. Though the list purports to be a list of scientific publications supporting his anti-vaccine stance, Mr. Ji simply rearranged the titles for many articles to make them appear anti-vaccine. For example, one of the papers Mr. Ji uses is entitled "Measles Outbreak in a Highly Vaccinated Population, San Diego, 2008: Role of the Intentionally Undervaccinated." The paper concludes:

. . . despite high community vaccination coverage, measles outbreaks can occur among clusters of intentionally undervaccinated children, at major cost to public health agencies, medical systems, and families. Rising rates of intentional undervaccination can undermine measles elimination. (Sugerman et al., 2010; quoted in Ji, 2015a)

Yet in Mr. Ji's hands, the words are rearranged, and the paper reads, "A measles outbreak was reported in a highly vaccinated population, San Diego, 2008." This makes it appear as if the paper documents vaccine failure, and is pointing to a limitation of vaccines, when in fact it is showing the opposite: that even in a highly vaccinated population, intentional nonvaccination can cause harm. A similar example occurs with a paper examining the risk of miscarriage with the HPV vaccine. The paper concludes simply, "There is no evidence overall for an association between HPV vaccination and risk of miscarriage." However, in Mr. Ji's document, the article is titled "The risk of miscarriage increases following HPV vaccination," which of course is the exact opposite of its real conclusion.

Many of the other articles, when given even minimal scrutiny, also fail to back up his anti-vaccine claims (Ji, 2015a). Additionally, many of them are about potential vaccine side effects in cows, mice, sheep, and pigs and likely have little or no relevance to people.

When informed that one of Mr. Ji's articles, entitled "Study Calls into Question Primary Justification for Vaccines," was based on his research, the lead author of the original study on which it was based, Dr. Ulrich von Andrian, responded to a query by saying:

The online article you referenced below misrepresents our paper. Our work in no way calls into question the utility of vaccines, which in my personal view are among the most impactful and cost-effective accomplishments of modern medicine. (Skeptics' College, 2015).

Despite this response being posted in the comments to his article, Mr. Ji has not issued a correction and continues to periodically post his original article on Facebook.

Another unfortunate example of outright lying sometimes occurs after the death of a child. In a since removed blog entitled "Michigan Baby Dies, Pathologists Confirm Vaccines Responsible," the ironically named website VacTruth discussed the tragic death of the baby Elijah Daniel French. The child's own mother, Rachel French, who temporarily was active in the anti-vaccine movement and is now a staunch vaccine defender, wrote:

Unfortunately vactruth [sic] only told parts of the story, embellished other bits, and really added a bunch of propaganda bullshit that is not representative of my views. I didn't realize it until today when I answered ten million questions. (Respectful Insolence, 2015)

It is worth noting that such blatant lying is almost certainly the exception. Many well-meaning parents legitimately believe their children have been injured by vaccines, even if the science does not support their belief.

Conclusion

Anti-vaccine activists draw on different strands of thoughts and may have differing motives when they promote misinformation that may scare people from vaccinating. However, many common themes run through the movement, all of which reinforce the basic reality that anti-vaccine claims have little basis in fact. They appeal to the way people think and can mislead even rational, well-intentioned people into putting their children at risk.

Note

1. Dr. Ulrich von Andrian and Rachel French both confirmed the accuracy of their quotes with the authors directly.

References

Adams, M. (2014). It's All Scripted! Ebola Outbreak and Impossibly Rapid Vaccine Response Clearly Scripted; U.S. Govt. Patented Ebola in 2010 and Now Owns All Victims' Blood. Retrieved from http://www.naturalnews.com/046946_Ebola_outbreak_vaccines_patents.html

Adams, M. (2015). Depopulation Test Run? 75% of Children Who Received Vaccines in Mexican Town Now Dead or Hospitalized. Retrieved from http://www.naturalnews.com/049669_vaccine_injury_depopulation_agenda_deadly_side_effects.html

Advisory Committee on Immuniza. (2011). Updated Recommendations for Use of Tetanus Toxoid, Reduced Diphtheria Toxoid and Acellular Pertussis Vaccine (Tdap) in Pregnant Women and Persons Who Have or Anticipate Having Close Contact with an Infant Aged <12 Months. *Morbidity and Mortality Weekly Report*, *60*(41): 1424–1426. Retrieved from http://www.cdc.gov/mmwr/preview/mmwrhtml/mm6041a4.htm

Anderson, P. A., Rivara, F. P., Maier, R. V, and Drake, C. (1991). The Epidemiology of Seat belt-Associated Injuries. *Journal of Trauma*, *31*: 60–67. Retrieved from http://www.ncbi.nlm.nih.gov/pubmed/1986134

Antivaccinationists Promote a Bogus Internet "Survey." Hilarity Ensues as It's Retracted. (2016). Retrieved from http://scienceblogs.com/insolence/2016/11/29/antivaccinationists-promote-a-bogus-internet-survey-hilarity-ensues-as-its-retracted/

Bark, T. (n.d.). About the Center for Disease Prevention & Reversal. Retrieved from http://www.disease-reversal.com/about-us

Bark, T. (2016). Skin and Chocolate. Retrieved from https://www.skinandchocolate.com/category_s/1822.htm

Berman, M. (2014). Jenny McCarthy Says She Isn't Anti-Vaccine. Here Are Some Other Things She Has Said About Vaccinations. *Washington Post*, April 16. Retrieved from https://www.washingtonpost.com/news/post-nation/wp/2014/04/16/jenny-mccarthy-says-she-isnt-anti-vaccine-here-are-some-other-things-she-has-said-about-vaccinations/?utm_term=.e9cfe7d53bfd

Brogan, K. (2013). Pregnancy-Friendly Protection? The Truth about Whooping Cough Vaccine. Retrieved from http://kellybroganmd.com/pregnancy-friendly-protection-truth-about-whooping-cough-vaccine-pertussis/

Brogan, K. (2016a). Coffee Enema Products. Retrieved from http://kellybroganmd.com/resources/personal-care-products/coffee-enema-products/

Brogan, K. (2016b). Sacred Activism: Moving Beyond Ego. Retrieved from http://kellybroganmd.com/sacred-activism-moving-beyond-ego/

Brogan, K., and Ji, S. (2014). Why Vaccines Aren't Paleo. Retrieved from http://www.greenmedinfo.com/blog/why-vaccines-arent-paleo

Byrne, J. (n.d.). The Plural of Anecdote Is Not Data. Retrieved from https://sites.google.com/site/skepticalmedicine//the-plural-of-anecdote-is-not-data

Centers for Disease Control and Prevention (US). (2011). Rotavirus Vaccine (RotaShield®) and Intussusception. Retrieved from http://www.cdc.gov/vaccines/vpd-vac/rotavirus/vac-rotashield-historical.htm

Centers for Disease Control and Prevention (US). (2016). Seasonal Influenza Vaccine Effectiveness, 2005–2017. Retrieved from http://www.cdc.gov/flu/professionals/vaccination/effectiveness-studies.htm

CollegeHumor. (2014). Vaccines Cause Autism—If Google Was a Guy. Retrieved from https://www.youtube.com/watch?v=77GGn-E607E

Deer, B. (2011). How the Case against the MMR Vaccine Was Fixed. *British Medical Journal, 342*: c5347. Retrieved from http://doi.org/10.1136/bmj.c5347

Dr. Bark in the Media. (n.d.). Retrieved from http://www.disease-reversal.com/dr-bark-in-the-media

Dvorsky, G. (2014). What If Natural Products Came with a List of Ingredients? Retrieved from http://io9.gizmodo.com/what-if-natural-products-came-with-a-list-of-ingredient-1503320184

Gerber, J. S., and Offit, P. A. (2009). Vaccines and Autism: A Tale of Shifting Hypotheses. *Clinical Infectious Disease, 48*(4): 456–461.DOI: doi.org/10.1086/596476

GreenMedInfo. (n.d.) The Science of Natural Healing. Retrieved from http://www.greenmedinfo.com/

GreenMedInfo. (2015). The Unreported Health Benefits of Measles. Retrieved from http://www.greenmedinfo.com/blog/unreported-health-benefits-measles

Haelle, T. (2014). Baboon Study Reveals New Shortcoming of Pertussis Vaccine. Retrieved from https://www.scientificamerican.com/article/baboon-study-reveals-new-shortcoming-of-pertussis-vaccine/

Harris, S. (2013). Sam Harris Quotes. Retrieved from http://www.goodreads.com/quotes/818485-water-is-two-parts-hydrogen-and-one-part-oxygen-what

Health Impact News. (2014). Gardasil Vaccine: One More Girl Dead. Retrieved from https://healthimpactnews.com/2014/gardasil-vaccine-one-more-girl-dead/

Honda, H., Shimizu, Y., and Rutter, M. (2005). No Effect of MMR Withdrawal on the Incidence of Autism: A Total Population Study. *Journal of Child Psychology and Psychiatry, 46*(6): 572–579. DOI: doi.org/10.1111/j.1469-7610.2005.01425.x

Horne, Z., Powell, D., Hummel, J. E., and Holyoak, K. J. (2015). Countering Antivaccination Attitudes. *Proceedings of the National Academy of Sciences of the United States of America, 112*(33): 10321–4.DOI: doi.org/10.1073/pnas.1504019112

Howard, Jacqueline. (2017). Minnesota Measles Outbreak Exceeds Last Year's Nationwide Numbers. Retrieved from http://www.cnn.com/2017/06/02/health/minnesota-measles-outbreak-bn/index.html

Immunization Action Coalition, Personal Belief Exemptions for Vaccination Put People at Risk. Examine the Evidence for Yourself. (n.d.). DOI: doi.org/10.1111/jrh.12019.Epub

Jain, A., Marshall, J., Buikema, A., Bancroft, T., Kelly, J. P., Newschaffer, C. J., . . . TL, L. (2015). Autism Occurrence by MMR Vaccine Status among US Children with Older Siblings with and without Autism. *JAMA, 313*(15): 1534. DOI: doi.org/10.1001/jama.2015.3077

Ji, S. (2014). Should You Trust "The Daily Beast" about Vaccines? Retrieved from http://m.green medinfo.com/blog/should-you-trust-daily-beast-about-vaccines

Ji, S. (2015a). 200 Evidence-Based Reasons NOT to Vaccinate. Retrieved from http://www.greenmedinfo.com/blog/200-evidence-based-reasons-not-vaccinate-free-research-pdf-download

Ji, S. (2015b). Measles Transmitted by the Vaccinated, Gov. Researchers Confirm. Retrieved from http://www.greenmedinfo.com/blog/measles-transmitted-vaccinated-gov-researchers-confirm

Johnson, M. (2014). A Day That Started with a Routine Doctor Visit Ends in Girl's Death. Retrieved fr -visit-ended-in-her-death-b99326702z1-270410121.html

Kelto, A. (2015). Why a Court Once Ordered Kids Vaccinated against Their Parents' Will. Retrieved from http://www.npr.org/sections/health-shots/2015/02/19/386040745/why-a-court-once-ordered-kids-vaccinated-against-their-parents-will

Klein, N. P., Bartlett, J., Fireman, B., Rowhani-Rahbar, A., and Baxter, R. (2012). Waning Protection after Fifth Dose of Acellular Pertussis Vaccine in Children. *New England Journal of Medicine*, *367*(11): 1012–1019. DOI: //doi.org/10.1056/NEJMoa1200850

Logic of Science. (2016). Why Are There So Many Reports of Autism Following Vaccination? A Mathematical Assessment. Retrieved from https://thelogicofscience.com/2016/06/28/why-are-there-so-many-reports-of-autism-following-vaccination-a-mathematical-assessment/

Logic of Science. (2017). Another Terrible Anti-vaccine Study Bites the Dust. Retrieved from https://thelogicofscience.com/2017/05/10/another-terrible-anti-vaccine-study-bites-the-dust/

Maglione, M. A., Das, L., Raaen, L., Smith, A., Chari, R., Newberry, S., . . . Gidengil, C. (2014). Safety of Vaccines Used for Routine Immunization of US Children: A Systematic Review. *Pediatrics*, *134*(2): 325–337. DOI: doi.org/10.1542/peds.2014-1079

Maskell, J. (2013). 8 Reasons I Haven't Vaccinated My Daughter. Retrieved from http://www.mindbodygreen.com/0-11532/8-reasons-i-havent-vaccinated-my-daughter.html

McGovern, Celeste. (2017). Vaccinated vs. Unvaccinated: Mawson Homeschooled Study Reveals Who Is Sicker. Retrieved from http://info.cmsri.org/the-driven-researcher-blog/vaccinated-vs.-unvaccinated-guess-who-is-sicker

Metagenics Genetic Potential through Nutrition. (2016). Retrieved from http://integrative.metagenics.com/store

Mina, M. J., Metcalf, C. J. E., de Swart, R. L., Osterhaus, A. D. M. E., and Grenfell, B. T. (2015). Long-Term Measles-Induced Immunomodulation Increases Overall Childhood Infectious Disease Mortality. *Science*, *348*(6235): 694–699. DOI: doi.org/10.1126/science.aaa3662

Miner, M. (2016). HVTN 505 Phase 2b HIV Vaccine Trial Showed No Efficacy to Reduce HIV Infection Risk. Retrieved from https://www.fredhutch.org/en/labs/vaccine-and-infectious-disease/news/publication-spotlight/hvtn_505_phase_2b_hiv_vaccine_trial.html

Northport Wellness Center. (2016). Welcome to the Northport Wellness Center Store. (2016). Retrieved from https://www.northportwellnesscenter.com/store/

Offit, P. (2010). *Deadly Choices: How the Anti-vaccine Movement Threatens Us All*. New York: Basic Books.

Offit, Paul A. (2017). Did Anti-Vaxxers Spark a Measles Outbreak in an Immigrant Community? Retrieved from http://www.thedailybeast.com/did-anti-vaxxers-spark-a-measles-outbreak-in-an-immigrant-community

Palfreman, J., and McMahon, K. (2010). *Frontine: The Vaccine War*.

Parry, W. (2012). Contradictions Don't Deter Conspiracy Theorists. Retrieved from http://www.livescience.com/18171-contradicting-conspiracy-theories-mistrust.html

Perry, R. T., and Halsey, N. A. (2004). The Clinical Significance of Measles: A Review. *Journal of Infectious Disease*, *189*: S4–S16. Retrieved from http://doi.org/10.1086/377712

PharmaCompass. (2016). Top Drugs by Sales Revenue in 2015: Who Sold the Biggest Blockbuster Drugs? Retrieved from http://www.pharmacompass.com/radio-compass-blog/top-drugs-by-sales-revenue-in-2015-who-sold-the-biggest-blockbuster-drugs

Pierce, B. (n.d.). 67 Research Papers Showing That Vaccines Can Cause Autism. Retrieved from http://circleofdocs.com/67-research-papers-showing-that-vaccines-can-cause-autism/

Reid, A. B., Letts, R. M., and Black, G. B. (1990). Pediatric Chance Fractures: Association with Intra-abdominal Injuries and Seat belt Use. *Journal of Trauma*, *30*(4): 384–391. Retrieved from http://www.ncbi.nlm.nih.gov/pubmed/2325168

Reiss, D. R. (2016). Brian Hooker's Vaccine Injury Claim Denied by NVICP. Retrieved from http://www.skepticalraptor.com/skepticalraptorblog.php/brian-hooker-vaccine-injury-claim-denied/

Respectful Insolence. (2015). No, Vaccines Almost Certainly Did Not Kill Elijah Daniel French. Retrieved from http://scienceblogs.com/insolence/2015/05/18/no-vaccines-almost-certainly-did-not-kill-elijah-daniel-french/

Scutti, Susan. (2017). How Countries around the World Try to Encourage Vaccination. Retrieved from http://www.cnn.com/2017/06/06/health/vaccine-uptake-incentives/

Skeptics' College. (2015). Experience with Dr. Sherri Tenpenny on Facebook. Retrieved from https://skepticscollege.wordpress.com/2015/07/28/experience-with-dr-sherri-tenpenny-on-facebook/

Smith, B. (2012). Dr. Mercola: Visionary or Quack. Retrieved from http://www.chicagomag.com/Chicago-Magazine/February-2012/Dr-Joseph-Mercola-Visionary-or-Quack/

Sugerman, D. E., Barskey, A. E., Delea, M. G., Ortega-Sanchez, I. R., Bi, D., Ralston, K. J., . . . LeBaron, C. W. (2010). Measles Outbreak in a Highly Vaccinated Population, San Diego, 2008: Role of the Intentionally Undervaccinated. *Pediatrics*, *125*(4): 747–755.

Taylor, L. E., Swerdfeger, A. L., and Eslick, G. D. (2014). Vaccines Are Not Associated with Autism: An Evidence-Based Meta-Analysis of Case-Control and Cohort Studies. *Vaccine*, *32*(29): 3623–3629. DOI: doi.org/10.1016/j.vaccine.2014.04.085

Tomljenovic, L., Wilyman, J., Vanamee, E., Bark, T., and Shaw, C. (2013). HPV Prevention Series. *Infectious Agents and Cancer*, *8*(6): 37. DOI: doi.org/10.1186/1750-9378-7-37

Tyson, N. deGrasse. (2011). Conspiracy Theorists . . . Retrieved from https://twitter.com/neiltyson/status/56010861382336513

van Panhuis, W. G., Grefenstette, J., Jung, S. Y., Chok, N. S., Cross, A., Eng, H., . . . Burke, D. S. (2013). Contagious Diseases in the United States from 1888 to the Present. *New England Journal of Medicine*, *369*(22): 2152–2158. Retrieved from http://doi.org/10.1056/NEJMms1215400

Wakefield, A., Murch, S., Anthony, A., Linnell, J., Casson, D., Malik, M., . . . Walker-Smith, J. (1998). RETRACTED: Ileal-Lymphoid-Nodular Hyperplasia, Non-specific Colitis, and Pervasive Developmental Disorder in Children. *Lancet*, *351*(9103): 637–641. Retrieved from http://doi.org/10.1016/S0140-6736(97)11096-0

Wendorf, K., Winter, K., Harriman, K., Zipprich, J., Schechter, R., Hacker, J., . . . Glaser, C. (2016). "Subacute Sclerosing Panencephalitis: The Devastating Measles Complication Is More Common Than We Think." In *Vaccines, Vaccine Preventable Disease, and Their Impact*.

Whale.to.(n.d.). Vaccine Disease. Retrieved from http://www.whale.to/vaccines/diseases.html

Wolfson, J. (2014). Case Dismissed. Retrieved from https://www.wolfsonintegrativecardiology.com/casedismissed/#sthash.c76umI8w.3siF0Iug.dpuf

Yang, T., Delamater, P. L., Leslie, T. F., and Mello, M. M. (2016). Sociodemographic Predictors of Vaccination Exemptions on the Basis of Personal Belief in California. *American Journal of Public Health*, *106*(1): 172–177. http://doi.org/10.2105/AJPH.2015.302926

Zhou, F., Shefer, A., Wenger, J., Messonnier, M., Wang, L. Y., Lopez, A., . . . Rodewald, L. (2014). Economic Evaluation of the Routine Childhood Immunization Program in the United States, 2009. *Pediatrics*, *133*(4): 577–585. DOI: doi.org/10.1542/peds.2013-0698

III Scientific (or Pseudoscientific) Soundness

9 Understanding Pseudoscience Vulnerability through Epistemological Development, Critical Thinking, and Science Literacy

Arnold Kozak

Introduction

As a culture, we are vulnerable to the claims of pseudoscience. Students come out of high school, even college, with insufficient epistemological development, critical thinking skills, and science literacy. This makes the public susceptible to dubious or premature scientific claims. An equally epistemologically unsophisticated media that sensationalizes science exacerbates this susceptibility.

To demonstrate, in radical fashion, the public's susceptibility to unsubstantiated science claims, journalist John Bohannon intentionally conducted a poor methodological scientific trial on the benefits of eating chocolate. The study included a very small sample and a very large number of dependent measures (increasing the likelihood that something significant would be found). Sure enough, the study found that the group eating chocolate (the rest of the diet was not recorded) had lost a pound within the three-week trial, which was statistically but not clinically significant because it was an insignificant amount due to normal fluctuation in weight. Once the study was written up, he immediately bought into a pay-per-publish journal (*The International Archives of Medicine*) for the price of 600 euros. Once the paper was published and a press release concocted, the story was picked up by numerous tabloid newspapers and then television news. "German scientists found out that your diet works better, faster if you combine it with dark chocolate," the public was told. And: "This is real, adding chocolate, you'll lose weight, I'm not kidding!" (Gladstone, 2015). After the study went viral in the media, it was pulled from the publisher's website (Oransky, 2015). Unfortunately, readers and viewers are not equipped to be more discriminating consumers of science claims.

The health and science news food chain that feeds the consuming public is contaminated by different players at different stages along the way. The problem is much worse than a publishing prank that fools millions. It's a credibility crisis for science and for journalism. And consumers at the end of the food chain will be—indeed, already are—poisoned. (Oransky, 2015)

This chapter will outline four factors that contribute to pseudoscience vulnerability: a general lack of science literacy (e.g., not knowing how to distinguish science from pseudoscience), 2) pedagogy that emphasizes memorization of facts over critical thinking, 3) insufficient epistemological development (e.g., not knowing how to evaluate truth claims), and 4) media distortions of science. Discussion of this last topic will include a critical look at the science claims made by the mindfulness movement that is taking place in America (Wilson, 2014). While the mindfulness movement is firmly established as a legitimate scientific enterprise, there are aspects of its presentation to the public that bear no resemblance to pseudoscience.

Science Literacy

The scientific literacy of the American public could withstand substantial improvement. An article in *MIT Technology Review* by David Ewen Duncan (2007) proclaims, "216 million Americans are scientifically illiterate."

This level of science illiteracy may explain why over 40 percent of Americans do not believe in evolution and about 20 percent, when asked if the earth orbits the sun or vice versa, say it's the sun that does the orbiting—placing these people in the same camp as the Inquisition that punished Galileo almost 400 years ago. It also explains the extraordinary disconnect between scientists and much of the public over issues the scientists think were settled long ago—never mind newer discoveries and research on topics such as the use of chimeras to study cancer, or pills that may extend life span by 30 or 40 percent. (Duncan, 2007)

In addition to the findings Ewen cites, only 73 percent of Americans can correctly distinguish between astronomy and astrology. This, among other chilling results, was found in the Pew Research Center's American Trends Panel on over 3000 US adults (Funk and Goo, 2015).

Most of the questions regarded science facts and knowledge, with only one pertaining to scientific methodology. This question asked participants to interpret a scatterplot. Only 63 percent were able to do so correctly. In sum, about a quarter of people don't know the difference between astrology and astronomy and a third cannot make a very basic (i.e., what is a correlation) interpretation of a graph. How could people such as these be expected to reasonably evaluate data-based claims in the media?

The National Science Foundation Science and Engineering Indicators report (2014) presents other lowlights of science literacy, including:

- 26 percent believe the sun orbits the Earth
- 61 percent cannot identify the Big Bang
- 53 percent know an electron is smaller than an atom
- 50 percent cannot identify the theory of evolution

Also troubling was the rising trend of astrology acceptance:

In 2012, slightly more than half of Americans said that astrology was "not at all scientific," whereas nearly two-thirds gave this response in 2010. The comparable percentage has not been this low since 1983. (National Science Foundation, 2014)

One potential cause of these deficits in scientific literacy is the poverty of science education in American schools. Another 2015 Pew Research report found that the general population and members of the American Association for the Advancement of Science (AAAS) both see US K–12 education in science, technology, engineering, and mathematics (STEM) fields as "average" or "below average" compared with other industrialized countries. The report states:

75 percent of AAAS scientists say too little STEM education for grades K–12 is a major factor in the public's limited knowledge about science. An overwhelming majority of scientists see the public's limited scientific knowledge as a problem for science. (Funk and Rainie, 2015)

It appears that a large percentage of the population is ignorant of basic science facts and lacks the ability to reason about science. This stems, perhaps, from poor science education that emphasizes rote learning instead of critical thinking.

Critical Thinking

Possession of a body of facts does not ensure the ability to reason about these facts. Robert Ennis defined *critical thinking* as "reflective and reasonable thinking that is focused on deciding what to believe or do . . . raising questions, formulating hypotheses, questions, alternatives, and plans for experiments" (Ennis, 1985). Critical thinking is activated by a discrepancy between personal beliefs and a new external learning event that initiates the following steps: motivating, information seeking, information-relating, evaluating, expressing, and integrating (Kasschau, 1986). Motivation requires paying attention and having curiosity to dig into the question at hand. Information seeking involves acute observation, understanding of basic concepts, and an ability to organize and use information. Information-relating necessitates convergent and divergent thinking, and the ability to identify patterns, make links, and draw logical inferences. Evaluating results seeks to resolve the initial discrepancy by deciding and then evaluating that decision based on relevance, accuracy, sufficiency, and quality. Metacognitive strategies are employed in this step of the process. In the expressing step, there is a willingness to put the decision out for critical review from others or to submit it to external standards. At the integration stage, learners will resolve the discrepancy by expanding their knowledge base and revising their personal theories.

One study found that at lower levels of epistemological development, critical thinking performance was lower (Ouellette-Schramm, 2015). At earlier levels of cognitive and epistemological development the data may be selected, biased, and misconstrued to fit (assimilate) within firmly held personal theories (e.g., creationism) rather than accommodate to new knowledge structures (e.g., evolution). At low levels of epistemological development, individuals may not even have the requisite curiosity to begin asking questions or to notice the knowledge discrepancy that gives rise to critical thinking. Along the way through the steps outlined above, these individuals will not be able to employ the strategies, such as metacognition, logical inference, and divergent thinking, which are required for critical thinking. While attempts have been made to foster critical thinking within science (e.g., Carson, 2015), their success must be limited unless critical thinking skills are viewed within the context of life-span developmental models.

Epistemological Development

The psychologist Jean Piaget did the groundbreaking work on cognitive development but stopped at the acquisition of formal operations, which may arise in early adolescence but is by no means guaranteed (Kitchener, 1986). Epistemological models cover ground beyond the logical-mathematical operations that Piaget articulated. The initial model was by Perry (1970) and subsequent theorists (Baxter Magolda, 1992; Belenky, Clinchy, Goldberger, and Tarule, 1986; King and K. Kitchener, 1994). Across these different formulations, epistemological development can be understood as how "college students' thinking progresses from a state of simple, absolute certainty into a multifaceted, evaluative system and the capacity of the thinker to bring increasingly complex combinations of abstractions to bear in the solution of problems" (West, 2004, p. 61). Unlike critical thinking, which is viewed as a collection of skills, epistemological development is a developmental stage that increases through maturity and exposure to higher education.

West (2004) has synthesized the available epistemological models into a four-stage model. Stage 1, *absolute knowledge*, is certain and concrete. There are right and wrong answers to all questions and authorities are trusted without question. For students at this level, knowledge claims do not need justification (i.e., truth is self-evident). An example of absolute knowledge would be the contention that the world was created in six days less than six thousand years ago because the Bible says it is so.

At a certain level of education and development learners realize that authorities and experts can disagree therefore the absolutistic stance is abandoned and the student transitions into Stage 2, the *personal level* of thinking. Here, personal subjective knowledge is seen to be just as valid as authoritative knowledge (because nothing can be known for certain). "Justification of knowledge is egocentric during this stage; 'all of us are equally entitled to our own opinions' becomes the battle cry. Personal knowing is a closed system that only allows the knower to decide 'I'm right'" (West, 2004, p. 64). For an example, consider this argument: not every single scientist believes in climate change, so that means anyone's opinion is good enough because the truth is not yet known with absolute certainty.

In Stage 3, *rules-based knowledge*, "students recognize the power of discipline-specific rules for comparing and judging knowledge claims (e.g., replication and hypothesis testing in science, sample size in statistics)" (West, 2004, p. 64). Knowledge within stage 3 functions within a specific knowledge domain and recognizes that knowledge claims have differential quality. Rules-based knowing does not provide the knower with a process for solving problems for which the rules do not apply; using rules-based knowing it is not possible to judge which of two expert opinions from different disciplines is better" (West, 2004, p. 64). For example, a student at this stage would have trouble determining if nuclear power was safe because it involves cross-discipline knowledge.

"To resolve most of the emerging problems in the real-world, students must develop a way of knowing that is capable of evaluating not only the specific problem situation but also of evaluating the ways in which that situation can be thought about" (West, 2004, p. 64). This accomplishment brings the student to Stage 4, *evaluative*. "People using Stage 4 knowing also recognize that they must continue to evaluate all available evidence, experience, and priorities, which may require them to reconsider their decisions" (West, 2004, p. 65). The percentage of adults, even those graduating from college, who function seamlessly within higher stages of epistemological development is low (King and Kitchener, 1994; Richardson, 2013). This makes many of us vulnerable to spurious or premature scientific claims. In the next section, we'll examine media distortions of science.

Media Distortions of Science: The Spread of Pseudoscience

The public's lack of science literacy, critical thinking skills, and epistemological development makes them vulnerable to sensationalized, superficial, and even spurious media presentations of science. For example, the American obsession with "orthorexia," or being healthy (Pollan, 2009), the use of useless or even dangerous supplements never approved by the US Food and Drug Administration (Offit, 2013), the presentation of simplified reductionistic claims by biological psychiatry (Ross and Pam, 1995; Whittaker, 2010), and non–drug and food related interventions such as mindfulness-based interventions (Dimidjian and Segal, 2015). Next, we'll focus on the mindfulness movement to see how the promoters of mindfulness (media, the scientists themselves, and others who profit from the findings) can fall prey to misapprehending or overstating what the experimental data support (Heuman, 2014a, 2014b).

The Mindfulness Movement

The growth in popularity of mindfulness has seen a burgeoning of programs, books, and scientific research studies (Boyce, 2011; Wilson, 2014). Mindfulness has affected nearly all sectors of society including health care, education, law, and the business world (Ryan, 2013). Each day on the Internet you can find blogs touting the benefits of mindfulness for everything from chronic pain to better orgasms. Behind many of the claims lie a lack of adequately rigorous trials and a wealth of overzealous interpretation of data. The media and our craving for the ontological security of certainty conspire to overrepresent what is known about many medical and scientific topics, including mindfulness. The mindfulness movement may be subject to the dynamics of hype that have afflicted other scientific programs, such as the human genome project. The hype pipeline consists of publication pressure, commercialization, institutional press releases, media practices, public interests and expectations, marketing, and the scientific bandwagon effect (Caulfield and Condit, 2012). This same pipeline seems to be active within the mindfulness community. Since the inception of the clinical application of mindfulness in 1979, research data has been collected. But there appears to be an unwillingness or inability on the part of advocates, media, and consumers to discriminate between different levels of evidence. Many of the studies that form the basis of mindfulness' popularity have been simply prepost studies without controls; those with controls frequently employ wait-list controls. Active control conditions are less frequent (Goyal, Singh, et al., 2014). There is more of an emphasis on showing the benefits of multifaceted mindfulness interventions, rather than seeking to understand the exact mechanisms that drive study findings, and that emphasis may ultimately undermine research efforts. "Underemphasizing links to basic research and precise specification of for whom and how a treatment works risks situating the study of MBI [mindfulness-based interventions] less as science and more as pseudoscience in which mindfulness is seen as a panacea for all problems" (Dimidjian and Segal, 2015).

Lilienfeld, Lynn, and Lohr (2014) outline the features of pseudoscience: 1) an overuse of ad hoc hypotheses to immunize claims against falsification, 2) absence of self-correction, 3) evasion of peer review, 4) emphasis on confirmation rather than refutation, 5) reversed burden of proof, 6), absence of connectivity (to established paradigms), 7) overreliance on anecdotal and testimonial data, 8) use of obscurantist language, 9) absence of boundary conditions, and 10) the mantra of holism. As Dimidjian and Segal (2015) caution, the research program of the mindfulness movement touches on aspect 4 (emphasis on confirmation over refutation) and 9 (absence of boundary conditions, which encourages a panacea perception). These problems will be explored in more detail below.

I have been a practitioner of mindfulness meditation since 1989, long before this movement was making its widespread impact on society. I have also been teaching mindfulness, practicing mindfulness-based psychotherapy, and writing about mindfulness for years (e.g., Kozak, 2009, 2015). Wilson (2014) in his book *Mindful America: The Mutual Transformation of Buddhist Meditation and American Culture* devoted over a page to my book, *Wild Chickens and Petty Tyrants: 108 Metaphors for Mindfulness* (Kozak, 2009), as an example of the mainstreaming of mindfulness. My familiarity with the mindfulness community as a clinician, personal practitioner, and promoter makes me aware that there is a hype effect happening currently and that some of this has been created by overzealous media interpretations of preliminary mindfulness-related findings (Heuman, 2014b). Indeed, "At times, it appears that we are witnessing the development of a 'cult of mindfulness' that, if not appropriately recognized and moderated, may result in an unfortunate backlash against it" (Brendel, 2015). I value the practice and the potential value it has for others. At the same time, I want to be *mindful* of not getting ahead of the data and not contributing to the hyping of mindfulness. I confess that, at times, I take shortcuts that are not epistemologically justified. I've issued hopeful, promissory note-type claims on the efficacy of mindfulness because it's compelling, convenient, and lazy to do so. It's harder to spell out the critiques, to unpack the science and embed it in a tutorial on basic epistemological principles, such as how to interpret evidence, critique methodology, and understand the limitations of statistics. In addition to media outlets, the overselling of mindfulness research has been promulgated by leading figures. The next section will present the findings of the most rigorous meta-analysis focusing on mindfulness to date.

The State of the Research

In 2014 Goyal and colleagues published a rigorous meta-analysis of the mindfulness research for the Agency of Healthcare Research and Quality (AHRQ). While there are thousands of citations and hundreds of studies that have been published, many of them in peer-reviewed journals, the criteria imposed for inclusion in the review excluded the vast majority of them. Out of the 18,753 mindfulness citations, only forty-seven met the inclusion criteria for the AHRQ study. Studies of DBT (Dialectical Behavior Therapy) and ACT (Acceptance and Commitment Therapy), which both have a mindfulness component and their own research support, were excluded from the review. Only clinical populations were included. The major inclusion criterion was "RCTs in which the control group was matched in time and attention to the intervention group for the purpose of matching expectations of benefit" (Goyal et al., 2014). These forty-seven trials included 3,515 participants. This high rate of exclusion reveals the methodological laxity of most mindfulness studies.

For those studies included, moderate effect sizes were found for anxiety, depression, and pain. Despite "stress reduction" being in the title of the signature mindfulness-based intervention (Mindfulness-Based Stress Reduction, or MBSR, founded by Jon Kabat-Zinn), low effect sizes were found for stress/distress and mental health-related quality of life. The review also found "low evidence or no effect or insufficient evidence of any effect of meditation programs on positive mood, attention, substance use, eating habits, sleep, and weight. We found no evidence that meditation programs were better than any active treatment (i.e., drugs, exercise, and other behavioral therapies)"

(Goyal et al., 2014). To be clear, it wouldn't be proper to say that mindfulness is *not* useful for stress reduction, only that rigorous evidence is not yet available. Pre-post studies don't control for nonspecific effects like taking questionnaires and attention from researchers. Wait-list controls are more rigorous but still uncontrolled for social support, inspiring teachers, and other factors not specific to meditation. It's astonishing and humbling to see that after thirty-five years of mindfulness research, mostly conducted on clinical populations, only forty-seven of these studies made the grade. This points up the need for important methodological development to take place within the field.

Media Claims by High-Profile Mindfulness Figures

Leading figures within the mindfulness movement such as Jon Kabat-Zinn (e.g., Kabat-Zinn, 2010) and Dan Siegel (e.g., Siegel, 2007, 2010a) have presented mindfulness as proven. The term "proven" is problematic, as clarified by the Understanding Science Project at the University of California, Berkeley:

> Journalists often write about "scientific proof" and some scientists talk about it, but in fact, the concept of proof—real, absolute proof—is not particularly scientific. Science is based on the principle that *any* idea, no matter how widely accepted today, could be overturned tomorrow if the evidence warranted it. Science accepts or rejects ideas based on the evidence; it does not prove or disprove them. (UndSci, n.d.)

The average person is likely ill-equipped to understand a claim of "scientifically proven." In a compelling article, it is argued that the majority of scientific studies will turn out to be false. "Simulations show that for most study designs and settings, it is more likely for a research claim to be false than true. Moreover, for many current scientific fields, claimed research findings may often be simply accurate measures of the prevailing bias" (Ioannidis, 2005). Hence, much of what occurs in the context of biomedical science will later be recognized as not true. Scientists, let alone a scientifically illiterate public, have difficulties comprehending this methodological state of affairs. Siegel does state, "Further research will be needed to verify the repeated studies affirming that long-term improvements are correlated with the mindfulness a practice, and are not just the effect of gathering in a reflective way as a group" (Siegel, 2010b). But the horse, as it were, is already out of the barn. The possibility that study findings are the result of nonspecific group support or other study-related factors (such as charismatic leaders or visiting a university research setting) is precisely the reason that AHRQ review excluded studies without active controls.

Dan Harris has used his high-profile status of co-anchor of ABCs *Nightline* to promote mindfulness in his bestselling book, *10 Percent Happier* (Harris, 2014). In an interview in *Mindful Magazine* he says, "As it turns out, there's all this science that says it can boost your immune system, reduce your blood pressure, and rewire key parts of your brain" (Harris, cited in Mindful Staff, 2014). Harris is likely referring to the oft-cited (682 times) study by Davidson et al. (2003). This study would not have met the inclusion criteria for the AHRQ study because it employed a wait-list control design. Other studies that looked at immune function (also blood pressure) were noncontrolled or nonrandomized, (e.g., Carlson et al., 2007; Robinson, 2003).

Getting Beyond the Hype

Mindfulness researcher Catherine Kerr, assistant professor of medicine and family medicine at Brown University, is one of two strong voices at Brown concerned about the hyping of mindfulness. She says:

> I think we are all going to need to take responsibility and do something so that the coverage looks slightly more balanced. Otherwise, when the inevitable negative studies come, this whole wave will come crashing down on us . . . The *Huffington Post* features mindfulness a lot and tends to represent only the positive findings (and in the most positive light imaginable) rather than offering a balanced reading of the science. They use that approach to justify the idea that every person who has any mental abilities should be doing mindfulness meditation. I don't think the science supports that. The *Huffington Post* has really done mindfulness a disservice by framing it in that way . . . This problem of overestimation is ubiquitous. (Kerr, cited in Heuman, 2014b)

For example, one HuffPo article claims, "The deep, calming breathing you use to meditate has been proven to lower blood pressure and release the bad stress we experience on a regular basis in our lives" (Lauren, 2015). Setting aside the problem that instructions for mindfulness practice do not emphasize deep breathing, here again is a misappropriation of "proven" from what appears to be a preliminary scientific finding. Kerr explains that "People don't really know how to hear a story that a scientist is telling *as* a hypothesis. They don't know how to gauge that. The hypothesis somehow registers as 'already proven'" (Kerr, cited in Heuman, 2014b). What scientists produce is not the same as what the media presents. Kerr goes on to say, "In my brain science course, I bring in examples of what a scientific abstract says and also a news article that reports on it. They are very disconnected from one another. People want ways to reduce suffering and stress and they have grabbed onto mindfulness like a life jacket. I find that very moving, and I want to take it seriously." (Kerr, cited in Heuman, 2014b)

It appears that the public and professional community cannot distinguish between heuristic and definitive studies. Furthermore, there is a prevailing sense that mindfulness has been "proven" without any sense of the epistemological irony of that claim. Since the science literacy of the public is so poor, how much could anyone other than dedicated researchers understand falsifiability, p values, and statistical power? The public assumes (and I would include clinicians in the body public) that publication in a peer-review journal vouchsafes the validity of a study. What gets neglected is a more nuanced appreciation for the different levels of evidence and the types of conclusions that can be drawn from each.

Willoughby Britton, another Brown University professor of psychiatry and mindfulness researcher, echoes Kerr's cautions:

Tricyle: Have the claims for the scientific evidence supporting the efficacy of meditation been overstated by proponents of meditation?
Willoughby Britton: Definitely. Because they take all those studies that I was just describing (like pre-post studies) as evidence. You really shouldn't cite those as evidence. Our natural bias to confirm our own worldview is very much at work. People are finding support for what they believe rather than what the data is actually saying. Ironically, we need a lot of mindfulness to "see clearly" the science of mindfulness. (Heuman, 2014a)

Where the Field Needs to Go

The *American Psychologist*, the flagship journal of the American Psychological Association, recently dedicated an entire issue to mindfulness. The appearance of this special issue is a powerful statement on the rapidly progressing developments of the field. Within this special issue, a paper by Sona Dimidjian and Zindel Segal (one of the developers of Mindfulness-Based Cognitive Therapy [MBCT], one of the best researched mindfulness-based interventions, presents seven critical and corrective recommendations for the methodological state of affairs that confront the field to help it to move to firmer ground. Stages 1—4 are presented below. For the later stages, refer to Dimidjian and Segal (2015).

Recommendation 1: Attend to the Basics, Meaning Specify Intervention Targets and Populations This recommendation focuses on boundary conditions—what will MBI not work for? What is the mechanism of action? For example, Segal's research team has found that MBCT appears to work via the "regulation of dysphoric mood states in ways that inhibited the activation of habitual mood-linked mental content (i.e., ruminative)" (Dimidjian and Segal, 2015, p. 604).

Recommendation 2: Do Not Conflate Promise With Efficacy The field needn't shy away from studies that either don't show an effect or don't show that MBI is superior to a comparison treatment. "Such findings help to inform the "boundary conditions" necessary for scientific progress . . . in which failures in one context create fertile ground in the other. The field will be well served by frank acknowledgement of failure rather than obscuring such findings with multiple or ambiguous primary and secondary outcomes or falling victim to the 'file drawer' problem in which failed trails simply are not published" (Dimidjian and Segal, 2015, p. 605). The field has also ignored adverse effects (Farias and Wikholm, 2015), but that issue is beyond the scope of this chapter.

Excitement is valid but does not constitute efficacy. "The sheer quantity of promising uncontrolled studies cannot substitute for later stages studies; researchers, practitioners, and the public must be cautious not to conflate the fact that many studies exist at Stage I (intervention generation/refinement) with indications of efficacy of effectiveness" (Dimidjian and Segal, 2015). This recommendation has been recognized by other researchers:

The meditative traditions provide a compelling example of strategies and techniques that have evolved over time to enhance and optimize human potential and well-being. The neuroscientific study of these traditions is still in its infancy but the early findings promise to reveal the mechanisms by which such training may exert its effects as well as underscore the plasticity of the brain circuits that underlie complex mental functions. (Lutz, Dunne, and Davidson, 2007)

Their caveat is often overlooked by the media. Less is known about the brain than not known (Ascoli, 2015), so caution is warranted.

Recommendation 3: Engage the Thorny Question of Clinical Training Without standardized clinical trainings for the clinicians that deliver MBIs, studies will lack fidelity. Presently, there is a lack of guidelines for clinician training (e.g., what type of personal meditation practice is required in addition to other training).

Recommendation 4: It's Time to Get Specific about the Specific Effects of MBI This recommendation asks questions such as: Is meditation the active ingredient in MBIs? As discussed already, this question has not yet been answered; "findings are equivocal." If meditation is the active ingredient, is there a dose response relationship? How much is needed for effectiveness? (Dimidjian and Segal, 2015).

Addressing Pseudoscience Vulnerability

For society at large, the way to shift the current trends of science illiteracy and to promote critical thinking and epistemological development, greater attention must be devoted to the quantity and quality of science education. Possession of a more impressive body of facts won't be enough. The pioneering work of Clyde Herreid on case method learning in science (e.g., Herreid, 1998, 2000, Herreid and Kozak, 1995; Herreid, this volume) offers a promising alternative to the current system.

Earlier stages of epistemological development are certainty-focused. The world is reduced to black-and-white terms, and ontological comfort can be secured by choosing a side. The rote learning of facts can reinforce this desire for certainty. The quest for certitude averts existential questions and the ineluctable uncertainty that actually characterizes the world. What is indefensible from an intellectual standpoint may be necessary from an emotional place. For example, Ludwig Wittgenstein, widely considered the most influential philosopher of the twentieth century and the founder of logical positivism, took personal comfort in a belief in God (Monk, 1990).

As one progresses through stages of epistemological development, an appreciation for the tentative and relative grows. Dogmatism, however, represents either being stuck within or regressing to earlier levels of development in a way that provides refuge against the existential realities of uncertainty, change, and loss. Ironically, such a regression can be seen in the history of Buddhism, the religion that is the source for mindfulness meditation. A close reading of the Buddha's lectures shows that he practiced and taught mindfulness in a secular fashion and emphasized the existential realities of life at a macro level (sickness, old age, and death) and micro level—the "trauma of everyday life" (Epstein, 2014). After his death, followers of his teachings sought refuge in dogmatic and sectarian interpretations of this secular psychology. Indian Buddhism took a turn from the raw existential project that the Buddha initiated and constructed religious contexts of belief (Batchelor, 2015). Likewise, the mindfulness movement today seeks an amelioration of these existential challenges and is quick to jump out of the ambiguity of early research data into the certainty provided by the bandwagon.

We live in a world of polarized dualities. At one time or another, loyalties and outlooks have been divided between East and West, red states and blue states, Sunni and Shia. At bottom everyone's reality is divided between *is* and *is not* (in the metaphysical, ontological sense contemplated and argued in most sects of Buddhism). We crave ontological security, and clinging to it retards our intellectual capabilities. Surely our brains, the most complex things in the known universe, are capable of more than dogmatic epistemology. Ironically, mindfulness invites us to experience in direct experiential fashion the existential questions we are situated within. To fully realize the liberating potential of mindfulness, which may help us to mature in our epistemological development by increasing tolerance for uncertainty, ambiguity, and adversity, we have to avoid the trap of *it is* or *it is not*. There is a human tendency to skip existential uncertainty by grabbing onto a belief, often a belief defined by an extreme or binary split between black and white, right and wrong, and so on. As epistemological development progresses, so may the capacity to tolerate unpleasant affects (Kozak, 1992) and the need to avoid such existential vagaries may reduce. Returning to the scientific project of the mindfulness movement, Kerr uses the encouraging neuroscience research on mindfulness as an example of how to move forward from dogmatic certainty to a probabilistic conversation:

It's fair to say that there are some clues from brain science that meditation might help enhance brain function. That is an evidence-based statement. The mistake is investing 100-percent certainty in a result and not holding a probabilistic view of scientific truth or risk and benefit. When people are making decisions for their own well-being, they need to be able to hold that uncertainty in mind. And they need to understand that the scientific context in which they are making their decisions could be different five years from now. (Kerr, cited in Heuman, 2014b)

Fortunately, mindfulness meditation can help us develop the skills to look into our experience and notice the contingency of the existential realities we are embedded within (e.g., impermanence, the constructed nature of self through concept and language). To get this benefit, however, we must relinquish the soteriological hype that surrounds mindfulness—as the salvation for the angst and stress of the Information Age—and take a patient look beyond its surface appeal to its greater promise.

References

Ascoli, G. A. (2015). *Trees of the Brain, Roots of the Mind.* Cambridge: MIT Press.

Batchelor, S. (2015). *After Buddhism.* New Haven, CT: Yale University Press.

Baxter Magolda, M. B. (1992). *Knowing and Reasoning in College: Gender-Related Patterns in Students' Intellectual Development.* San Francisco: Jossey-Bass.

Belenky, M. F., Clinchy, B. M., Goldberger, N. R., and Tarule, J. M. (1986). *Women's Ways of Knowing: The Development of Self, Voice and Mind.* New York: Basic Books.

Bohannon, J. (2015, May 27). I Fooled Millions into Thinking Chocolate Helps Weight Loss. Here's How. Retrieved from http://io9.gizmodo.com/i-fooled-millions-into-thinking-chocolate-helps-weight-1707251800

Boyce, B. (Ed.). (2011). *The Mindfulness Revolution.* Boston: Shambhala.

Brendel, D. (2015). There Are Risks to Mindfulness at Work. Retrieved from https://hbr.org/2015/02/there-are-risks-to-mindfulness-at-work

Carlson, L. E., Speca, M., Faris, P., and Patel, K. D. (2007). One Year Pre-Post Intervention Follow-Up of Psychological, Immune, Endocrine and Blood Pressure Outcomes of Mindfulness-Based Stress Reduction (MBSR) in Breast and Prostate Cancer Outpatients. *Brain, Behavior, and Immunity,* *21*(8): 1038–1049. DOI: doi.org/10.1016/j.bbi.2007.04.002

Carson, S. (2015). Targeting Critical Thinking Skills in a First-Year Undergraduate Research Course. *Journal of Microbiology and Biology Education,* *16*(2): 148–156.

Caulfield, T., and Condit, C. (2012). Science and the Sources of Hype. *Public Health Genomics,* *15*(3–4): 209–217. DOI: doi.org/10.1159/000336533

Davidson, R. J., Kabat-Zinn, J., Schumacher, J., Rosenkranz, M., Muller, D., Santorelli, S. F., . . . Sheridan, J. F. (2003). Alterations in Brain and Immune Function Produced by Mindfulness Meditation. *Psychosomatic Medicine,* *65*(4): 564–570. DOI: doi.org/10.1097/01.PSY.0000077505.67574.E3

Diamond, S. L., and Kozak, A. (1994). A Course on Biotechnology and Society. *Journal of Chemical Engineering Education*: 140–144.

Dimidjian, S., and Segal, Z. V. (2015). Prospects for a Clinical Science of Mindfulness-Based Intervention. *American Psychologist,* *70*(7, SI): 593–620.

Duncan, D. E. (2007). 216 Million Americans Are Scientifically Illiterate (Part I). MIT Technology Review. Retrieved from http://www.technologyreview.com/view/407346/216-million-americans-are-scientifically-illiterate-part-i/

Ennis, R. H. (1985). A Logical Basis for Measuring Critical Thinking Skills. *Educational Leadership*, *43*(2): 44–48.

Epstein, M. (2014). *The Trauma of Everyday Life*. New York: Penguin.

Farias, M., and Wikholm, C. (2015). *The Buddha Pill: Can Meditation Change You?* New York: Watkins.

Funk, C., and Goo, S. K. (2015). A Look at What the Public Knows and Does Not Know about Science. Retrieved from http://www.pewinternet.org/2015/09/10/what-the-public-knows-and-does-not-know-about-science/

Funk, C., and Rainie, L. (2015). Public and Scientists' Views on Science and Society. Retrieved from http://www.pewinternet.org/2015/01/29/public-and-scientists-views-on-science-and-society/

Gladstone, B. (2015). A Skeptic's Guide to Health News and Diet Fads. Retrieved from http://www.onthemedia.org/story/skeptics-guide-health-news-and-diet-fads/

Goyal, M., Singh, S., Sabinga, E. M., et. al. (2014). Meditation Programs for Psychological Stress and Well-Being, (124). Rockville, MD: Agency for Health Care Quality.

Harris, D. (2014). 10% Happier: How I Tamed the Voice in My Head, Reduced Stress without Losing My Edge, and Found Self-Help That Actually Works: A True Story. New York: It Books.

Herreid, C. F. (1998). Why Isn't Cooperative Learning Used to Teach Science? *Bioscience*, *48*(7): 553–559. DOI: doi.org/10.2307/1313317

Herreid, C. F. (2000). Teaching in the Year 2061. *Advances in Physiology Education*, *24*(1): 2–7.

Herreid, C. F., and Kozak, A. (1995). Using Students as Critics in Faculty Development. *Journal on Excellence in College Teaching, 6:* 17–29.

Heuman, L. (2014a). Meditation Nation: How Convincing Is the Science Driving the Popularity of Mindfulness Meditation? Retrieved from http://www.tricycle.com/blog/meditation-nation

Heuman, L. (2014b). Don't Believe the Hype. Retrieved from http://www.tricycle.com/blog/don't-believe-hype

Ioannidis, J. P. A. (2005). Why Most Published Research Findings Are False. *PLoS Medicine*, *2*(8): e124. DOI: doi.org/10.1371/journal.pmed.0020124

Kabat-Zinn, J. (2010, May). Greater Good. Retrieved from http://greatergood.berkeley.edu/gg_live/science_meaningful_life_videos/speaker/jon_kabat-zinn/the_science_of_mindfulness/

Kasschau, R. (1986). "A Model for Teaching Critical Thinking in Psychology." In J. S. Halonen and L. S. Cromwell (Eds.), *Teaching Critical Thinking in Psychology*. Milwaukee: Alverno Productions.

King, P. M. and Kitchener, K. S. (1994). *Developing Reflective Judgment: Understanding and Promoting Intellectual Growth and Critical Thinking in Adolescents and Adults*. San Francisco: Jossey-Bass.

Kitchener, R. F. (1986). *Piaget's Theory of Knowledge: Genetic Epistemology and Scientific Reason*. New Haven, CT: Yale University Press

Kozak, A. (1992). Epistemological Development and Adaptation: Reflective Judgement and Stressful Affect and Their Relationship to Affect-Tolerance and Attributions of Pragmatic Structure. Buffalo, New York: State University of New York at Buffalo. (ProQuest Dissertations Publishing, 1992. 9617879.)

Kozak, A. (2009). *Wild Chickens and Petty Tyrants: 108 Metaphors for Mindfulness*. Boston: Wisdom.

Kozak, A. (2015). *Mindfulness A to Z: 108 Insights for Awakening Now*. Boston: Wisdom.

Lauren, L. (2015). How Mindfulness Meditation Can Improve Your Health. Retrieved from http://www.huffingtonpost.com/linda-lauren/how-mindfulness-through-m_b_8520962.html

Lilienfeld, S. O., Lynn, S. J. and Lohr, J. M. (2014). "Science and Pseudoscience in Clinical Psychology: Initial Thoughts, Reflections, and Considerations." In Lilienfeld, Lynn, and Lohr (Eds.), *Science and Pseudoscience in Clinical Psychology* (2nd ed.), pp. 1–18. New York: Guilford.

Lutz, A., Dunne, J. D., and Davidson, R. J. (2007). "Meditation and the Neuroscience of Consciousness." In Zelazo, Philip David, Moscovitch, Morris, and Thompson, Evan (Eds.), *Cambridge Handbook of Consciousness*, pp. 499–555. Cambridge University Press.

Mindful. (2014). Meditation Can Make Us 10% Happier. Retrieved from http://www.mindful.org/dan-harris-meditation-10-percent-happier/

Monk, R. (1990). *Ludwig Wittgenstein: The Duty of Genius*. New York: Free Press.

National Science Foundation (2014). Science and Engineering Indicators: Science and Technology: Public Attitudes and Understanding. Science and Engineering Indicators: 1–53. Retrieved from http://www.nsf.gov/statistics/seind14/index.cfm/chapter-7/c7h.htm

Offit, P. A. (2013). *Do You Believe in Magic? The Sense and Nonsense of Alternative Medicine*. New York: Harper Collins.

Oransky, I. (2015). Should the Chocolate-Diet Sting Study Be Retracted? And Why the Coverage Doesn't Surprise a News Watchdog: Retraction Watch. Retrieved from http://retractionwatch.com/2015/05/28/should-the-chocolate-diet-sting-study-be-retracted-and-why-the-coverage-doesnt-surprise-a-news-watchdog/

Ouellette-Schramm, J. (2015). Epistemological Development and Critical Thinking in Post-Secondary. *Australian Journal of Adult Learning*, 55(1): 114–134. Retrieved from http://search.proquest.com/docview/1675860095?accountid=14679

Perry, W. G. (1970). *Forms of Intellectual and Ethical Development in the College Years*. New York: Holt, Rinehart and Winston.

Pollan, M. (2009). *In Defense of Food: An Eater's Manifesto*. New York: Penguin.

Richardson, J. T. E. (2013). Epistemological Development in Higher Education. *Educational Research Review*, 9: 191–206.

Robinson, F. P., Mathews, H. L., and Witek-Janusek, L. (2003). Psycho-Endocrine-Immune Response to Mindfulness-Based Stress Reduction in Individuals Infected with the Human Immunodeficiency Virus: A Quasiexperimental Study. *Journal of Alternative and Complementary Medicine*, *9*(5): 683–694. DOI: doi.org/10.1089/107555303322524535

Ross, C. A. and Pam, A. (1995). *Pseudoscience in Biological Psychiatry: Blaming the Body.* New York: Wiley.

Ryan, T. (2013). *A Mindful Nation.* New York: Hayes House.

Siegel, D. (2007). *Mindful Brain.* New York: Norton.

Siegel, D. (2010a). *Mindful Therapist.* New York: Norton.

Siegel, D. (2010b). The Science of Mindfulness. Retrieved from http://www.mindful.org/the-science-of-mindfulness/

UndSci. (n.d.) Tips and Strategies for Teaching the Nature and Process of Science. Retrieved from http://undsci.berkeley.edu/teaching/misconceptions.php#b10

West, E. J. (2004). Perry's Legacy: Models of Epistemological Development. *Journal of Adult Development*, *11*(2): 61–70. DOI: doi.org/10.1023/B:JADE.0000024540.12150.69

Whitaker, R. (2010). Mad in America: Bad Science, Bad Medicine, and the Enduring Mistreatment of the Mentally Ill. New York: Basic Books.

Wilson, Jeff. (2014). Mindful America: The Mutual Transformation of Buddhist Meditation and American Culture. New York: Oxford University Press.

10 Scientific Failure as a Public Good: Illustrating the Process of Science and Its Contrast with Pseudoscience

Chad Orzel

Introduction

Recent years have seen numerous examples of politically motivated and pseudoscience-fueled attacks on the modern scientific consensus across a wide range of fields. The most highly publicized examples are climate change denialism and the anti-vaccine movement. Climate change denialism rejects the findings of an overwhelming majority of climate science studies that show the earth is warming due to greenhouse gas emissions produced by human industrial activity. The anti-vaccine movement has led large numbers of parents to refuse vaccination for their young children due to fears of a link between childhood vaccines and the onset of autism, despite decades of medical research showing vaccines are safe and effective.

Neither of these positions finds significant support in the mainstream scientific community, but both will claim to be supported by research. On closer inspection, however, much of this research turns out to be pseudo-scientific. For example, one of the key sources for the anti-vaccination movement, a study led by Andrew Wakefield, was retracted by its publishing journal for extreme irregularities in its analysis, and Wakefield was stripped of his license to practice medicine.

Despite their lack of mainstream scientific support, both of these movements have been extraordinarily successful in getting publicity for their views, thanks to powerful political connections and celebrity adherents. And both movements have very real costs, most visibly in the recent resurgence of measles outbreaks in the United States (Fiebelkorn et al., 2015). The costs of climate change denialism are less concrete, but may be vastly greater in the long term, as the failure to act to reduce greenhouse gas emissions may already have made it impossible to avoid catastrophic consequences due to sea level rise and changes in weather patterns.

Given this context, it is understandable that many public advocates of science would adopt a combative approach to both climate-change denialism and the anti-vaccine movement. Celebrity science communicators such as Bill Nye and Neil deGrasse Tyson have taken to aggressively promoting and defending the scientific consensus on television and in popular books. Scores of less famous scientists have done the same on blogs, in newspaper op-eds, and in testimony before agencies at many different levels of government.

An interesting side effect of this polarization and politicization has been a shift in the way scientists and science advocates react to scientific failures; that is, scientific results generated through normal scientific processes and disseminated through mainstream scientific channels that nevertheless turn out to be incorrect. These generally involve spectacular findings attracting a large amount of publicity that are directly contradicted (or at least unable to be replicated) by follow-up experiments and observations. A tiny fraction of these turn out to be the result of active scientific fraud, but the majority are simply honest errors in analysis or interpretation. Two recent examples come from the world of physics: the announcement by the OPERA collaboration in 2011 that neutrinos in their experiment appeared to be moving faster than the speed of light (which was later shown to be due to a faulty fiber-optic connection), and the 2014 claim by the BICEP2 collaboration that they had found evidence of primordial space-time ripples predicted by inflationary cosmology (which had to be walked back when it turned out they had underestimated the effect of foreground dust on their signal).

In both cases, some commentators denounced the failures with nearly the same vehemence as toward more directly harmful forms of pseudoscience. In the combative context of public science advocacy, these scientific failures are seen as providing aid and comfort to the enemy by raising public doubts about the reliability of mainstream science. In this view, public failures harm science because they create room for pseudoscientists to profit by sowing doubt. If much-publicized revolutionary results turn out to be false, these commentators argue, it reduces general public trust in mainstream arguments that disprove pseudoscientific claims.

In this chapter, I will argue that, on the contrary, high-profile scientific failures present an opportunity to enhance public understanding of science. Failure is an inevitable part of any human enterprise, and there is no conceivable procedure that could completely eliminate the risk of publicizing scientific results that do not pan out. The stark contrast between the way these failures of genuine science play out and the way pseudoscientific claims linger endlessly, however, allows public advocates of science to draw clear and sharp distinctions between the processes employed in reputable mainstream science and those that are hallmarks of pseudoscience. Recognizing these processes, in turn, can give members of the general public greater appreciation for the self-correcting nature of mainstream science and a better ability to identify pseudoscience.

The Structure of Scientific Failures

The problem of demarcating the precise line between science and pseudoscience is beyond the scope of this chapter (see Haack, 2007, for an attempt on this problem), but there are a set of behavioral norms that are relatively uncontroversial as hallmarks of "real" science. In the normal process of science, researchers investigate some question by various means (laboratory experiments, field observation, theoretical calculation, or computer simulation), and reach some conclusions. The results of this investigation are tested as rigorously as possible against alternative explanations (either individually or in discussion with collaborators on the research project), and then shared with a wider community of researchers studying similar topics. The exact means of disseminating results vary somewhat between disciplines. For most fields, peer-reviewed journal articles are the standard, but computer scientists accord papers at conferences greater prestige, and many physicists regard publication on the Arxiv preprint server as the essential step, with the eventual publication in a journal treated as old news.

The dissemination to the wider community is often regarded as the end point of a scientific project, but in a deeper sense it is merely the beginning of the most important part of science. The true test of a scientific result is not whether it passes the often somewhat cursory standards of prepublication peer review, but what it contributes to a larger discussion within the discipline. Science advances as new results are replicated or extended and become incorporated into the body of knowledge associated with a particular discipline.

Postpublication attempts to replicate or extend new results are a far more stringent test of a new result than is generally possible within the time frame allowed for prepublication peer review, and this is the stage where scientific frauds are usually uncovered. This is also where the high-profile scientific failures of interest for this chapter take place. Most such cases follow a broadly similar trajectory, characterized by three elements:

1) The result in question comes from researchers who are *connected to the relevant research community*.

2) Debate over the validity of the original claim is *conducted via appropriate professional channels*.

3) The final resolution involves *self-correction by the original researchers*.

High-profile failures of genuine science generally start with a prominent initial announcement by researchers who have connections to the research community in their field. These connections typically involve either employment at an established research institution (college, university, government, or industrial research lab) or at least a relevant professional credential (advanced degree in the field of research, or the equivalent). Scientific outsiders are fond of citing the example of Einstein, who wrote his first papers on relativity while employed as a patent clerk. But he was not truly an outsider to the scientific community of his day. Contrary to myth Einstein was a successful student and held a PhD from a well-respected university.

The announcement of what will turn out to be a failure of genuine science is greeted with an initial period of excitement gradually displaced by growing controversy and criticism. This criticism comes from other researchers in the same field and is expressed through the exchange of technical scientific papers distributed by the normal channels for the discipline, often beginning very rapidly. In the case of the controversial December 2010 announcement of a bacterium that seemed to incorporate arsenic in place of phosphorus, detailed responses began appearing online within a few days of the initial press conference (most notably from microbiologist Rosie Redfield on her blog [Refield, 2010]). The eventual print publication of the article the following June (Wolfe-Simon et al., 2011) was accompanied by eight technical comments raising various issues with the original claim, plus a response from the authors of the original paper. The paper generated several additional follow-up articles in the next couple of years, many published in the same journal that ran the original paper. (See Drahl, 2013, for an overview of this incident.)

The other crucial element of these high-profile failures of genuine science is self-correction: the researchers involved in the original result engage directly with their critics and address the criticism in future work. Again, this can begin very early in the process—the 1996 NASA press conference announcing the controversial claim to have found microbial life on Mars included paleobiologist Dr. William Schopf, who expressed skepticism about the results as they were being announced (Wilford, 1996). The original researchers often play a decisive role in resolving the controversy once a source of error is discovered. In 2004, the research group of Moses Chan at Penn State announced that they had observed the formation of an exotic quantum state of helium dubbed a "supersolid," but other researchers raised questions about the results. Controversy about the supersolid continued until 2012, when Chan's group revisited the original experiment with an improved apparatus and identified the systematic effect that had led to their initial mistaken report (Voss, 2012).

These elements—connection to the research community, debate conducted through normal disciplinary channels, and self-correction by the original researchers—are shared by all genuine scientific failures, though the elements manifest in slightly different ways for each case. Pseudoscientific controversies, on the other hand, generally fail on at least one of these counts.

While some pseudoscientific controversies take their initial inspiration from scientific research, most are sustained by proponents well outside of mainstream science. The anti-vaccination movement took much of its impetus from a 1998 paper in *The Lancet* by British medical researcher Andrew Wakefield, but the most influential advocates of anti-vaccination views are celebrities like Jenny McCarthy and Jim Carrey. And in this case, even the initial research was revealed to be outside the bounds of normal science—a series of investigations into Wakefield's research culminated in the retraction of the original article by *The Lancet* (Dyer, 2010), and Wakefield being barred from practicing medicine in the United Kingdom for ethical lapses in his research and in other activities (Kmietowicz, 2010).

Similarly, doubts about human contributions to global climate change are almost exclusively promoted from outside the mainstream scientific community. Multiple studies have shown that an overwhelming majority of peer-reviewed research published in scientific journals—in excess of 90 percent by some measures (Cook et al., 2013)—supports the consensus that human activity contributes to climate change. Doubts about this consensus are promoted almost exclusively by political and industry groups with a strong financial interest in opposing government action on climate change. Their criticisms appear on web sites, in policy publications, and in campaign speeches, not in the research literature.

Finally, self-correction is almost completely absent from pseudoscientific controversies. Despite the retraction of his paper and the loss of his license to practice medicine, Andrew Wakefield continues to publicly advocate against childhood vaccination (Roser, 2015). Previously debunked arguments against climate change continue to circulate, nearly unchanged, sometimes for years after the initial errors have been pointed out.

The stark difference between failures of genuine science and pseudoscientific activity presents an opportunity to draw clear lines between them. Advocates for science can make a sharp distinction between the behavior of scientists who have made genuine errors and that of nonscientists promoting pseudoscience. This, in turn, can enable members of the public to more readily distinguish between real science and pseudoscience without needing to follow technical arguments in great detail, simply by observing the behavior of participants in a controversy.

In the following sections, I will explore in detail the examples of OPERA and BICEP2 to show how they illustrate the proper operation of science. As a contrast to these, I will then discuss the history of "hydrino" physics as an illustration of a pseudoscientific process that clearly violates these norms.

The OPERA Experiment and Faster-Than-Light Neutrinos

One of the most exciting physics stories of 2011 was the September announcement that an experiment in Italy had observed neutrinos traveling faster than the speed of light. If true, this would directly contradict one of the cornerstone predictions of the theory of relativity. A comprehensive revision of modern physics would be needed.

The excitement meeting the initial announcement was quickly matched by skepticism from many parts of the physics community. And a few months later, in March 2012, the results were traced to a bad fiber-optic connection in the timing system. A new and improved measurement a year later confirmed that neutrinos do, in fact, obey the conventional laws of physics.

The faster-than-light neutrino saga evolved very rapidly, with the whole issue completely resolved within nine months. It was also extensively documented at every step, making it an excellent demonstration of the trajectory of a high-profile scientific failure.

Timeline

The Oscillation Project with Emulsion-tRacking Apparatus (OPERA) collaboration involves around 180 physicists from twenty-eight institutions, mostly in Europe. Its goal is to study how the fundamental particles called neutrinos transform as they travel from an accelerator at CERN outside Geneva to a detector at Gran Sasso in Italy. (See appendix A for a more detailed explanation of the experiment.)

In 2011, while analyzing data from its first three years of operation, the OPERA team spotted an anomaly. The neutrino beam produced at CERN is pulsed on and off in a regular way, and neutrinos are detected in real time at Gran Sasso, enabling a measurement of how much time the particles take to travel the 730 kilometers between accelerator and detector. The OPERA data showed particles being detected some sixty nanoseconds sooner than would be expected if they were moving at the speed of light. This would directly contradict Einstein's theory of relativity, which forbids matter or information being transmitted at speeds greater than that of light.

Detailed analysis of the data within the OPERA collaboration provided a very good measurement of this anomaly: $(60.7 \pm 6.9 \pm 7.4)$ ns (the two quoted uncertainties are statistical and systematic, respectively), corresponding to a speed of $1.0000248 \pm 0.0000028 \pm 0.000030$ times the speed of light (OPERA collaboration, 2011). Despite extensive checks of the data, they were unable to find an explanation of this anomaly; everything they could readily test was consistent with a faster-than-light speed for the neutrinos in their particle beam.

Given the potential importance of the result, and their inability to find an error, the OPERA collaboration put together a preprint of a paper describing their experiment, and scheduled a seminar at CERN for September 23, 2011. Rumors circulated on social media before the seminar created a huge sensation surrounding the announcement, which was widely reported in international news media.

The initial public claims of the OPERA collaboration were relatively restrained. The technical article posted to the Arxiv preprint server at the time of the announcement referred only to an "anomaly" and explicitly disavowed "any theoretical or phenomenological interpretation of the results" (OPERA collaboration, 2011). OPERA's physics coordinator, Dario Auterio, concluded the seminar at CERN with "Therefore, we present to you today this discrepancy, this anomaly" (Vastag, 2011), and collaboration coordinator Antonio Ereditato emphasized the preliminary nature of the result, saying, "We made a measurement and we believe our measurement is sound. Now it is up to the community to scrutinize it. We are not in a hurry. We are saying, tell us what we did wrong, redo the measurement if you can" (Sample and Jha, 2011). He also noted that "a result is never a discovery until other people confirm it" (Sample, 2011).

Initial reactions from the physics community were generally positive, with physicists who watched the initial presentation praising it as "very carefully done" and "a very serious job" (Vastag, 2011). But the general sense of excitement was tempered with scientific skepticism. For example, Susan Cartwright of Sheffield University was quoted saying, "Neutrino experimental results are not historically all that reliable, so the words 'Don't hold your breath' do spring to mind" (Sample, 2011), and Lisa Randall of Harvard commented, "If you had to bet, you'd bet on some experimental error" (Vastag, 2011). The most colorful expression of doubt came in the form of an actual bet, when Professor Jim Al-Khalili from the University of Surrey stated that if the OPERA result held up, "I will eat my boxer shorts on live TV" (Sample and Jha, 2011).

The OPERA result also generated a furious burst of activity in the technical literature, with more than forty articles citing the OPERA preprint posted to the Arxiv in the first week after the announcement, and around 180 by the end of 2011 (according to the INSPIRE database tracking citations of high-energy physics papers). These featured a mix of theoretical papers offering mechanisms by which neutrinos might violate the relativistic speed limit (or at least appear to do so), theoretical papers demonstrating the impossibility of faster-than-light neutrinos, and attempts to find errors in the OPERA collaboration's analysis of their anomalous data.

The various attempts to analyze the experiment from a distance were mostly unimpressive—as John Learned of the University of Hawaii presciently noted, "If a screw-up, it is probably in the details not accessible to outsiders" (Overbye, 2011)—but two theoretical papers quickly emerged as a solid foundation for opposition to the result. The first, by Gian Giudice of CERN and two colleagues (Giudice, Sibiryakov, and Strumia, 2011) was posted to the Arxiv on September 26 (just three days after the initial announcement). The paper argued that the speed measured by OPERA would imply other violations of relativity at a level ruled out by previous experiment. The second, by Andrew Cohen and Sheldon Glashow just a few days later (Cohen and Glashow, 2011), described a mechanism by which the OPERA neutrinos should have rapidly lost energy and slowed to speeds below that of light. These were widely taken by theorists as proof that OPERA's initial results must be an error, and stronger public complaints began to appear. For example, Lawrence Krauss of the University of Arizona worried that the whole issue "could wind up embarrassing the profession" (Achenbach 2011).

On the experimental side, the late months of 2011 were a mixed bag for the OPERA collaboration, with a follow-up experiment using shorter pulses of neutrinos (one of the potential systematic problems frequently raised by critics) giving results consistent with their initial measurement. However, the ICARUS collaboration, which operates another detector in the same neutrino beam, analyzed its own data to show that there was no sign of the energy loss required by the Cohen and Glashow mechanism (ICARUS collaboration, 2011), a significant blow to the experimental finding.

In early 2012, the OPERA finding unraveled completely. While the results were being debated in the literature, the experimenters were going over every inch of their apparatus looking for possible sources of error. In late February, they announced a devastating discovery: one of the fiber-optic cables carrying the time signal from a GPS receiver on the surface to the underground laboratory was not correctly connected. The misconnected cable combined with their timing electronics to delay their clock by an amount that was almost identical to the anomaly they originally reported. In mid-March, the ICARUS experiment reported on its own measurement of the velocity of neutrinos from CERN to Gran Sasso, finding a value completely consistent with the speed of light (ICARUS collaboration, 2012).

Finally, in June the OPERA collaboration completed their own follow-up measurement with the fiber-optic connection fixed (OPERA collaboration, 2012). The team found a speed consistent with both the light-speed limit and results from three other experiments using the same neutrino beam (Jha 2012). From start to finish, the whole story took a bit under nine months, at the end of which physics was returned to the status quo.

Commentary

The unraveling of the faster-than-light neutrino anomaly was unquestionably an embarrassment for the OPERA collaboration, with the initial burst of excitement and international media attention vanishing in a wave of recrimination. Collaboration spokesperson Antonio Ereditato and physics coordinator Dario Auterio resigned their positions at the end of March 2012, both blaming media hype for their downfall (Ouellette, 2012).

Plenty of other commentators blamed OPERA for the unfolding of events. From very early in the story, Lawrence Krauss publicly called the incident "an embarrassment," and worried that it was "very unfortunate—for CERN and for science" (Matson, 2011). And Nima Arkani-Hamed of Princeton declared the whole thing a waste of time, as "No really decent theoretical physicist took this seriously from the very start" (Vastag, 2012).

While it is easy for disappointed physicists to point fingers, closer inspection suggests that the OPERA collaboration did little that was wrong. The team's result appeared statistically sound and had some support from an earlier measurement by the MINOS experiment at Fermilab, which had seen a similar timing anomaly in 2007 but without enough statistical power to claim it was a genuine measurement (MINOS collaboration, 2007). And the experimental flaw, when it was uncovered after several months of extensive searching, turned out to be much subtler than the "loose cable" that was often reported in the media. A small gap in a fiber-optic line would only be expected to attenuate the timing signal, not delay it by sixty nanoseconds. The delay came from an unfortunate combination of the attenuation and the particular electronics used; in the words of Belgian physicist Pierre Vilain, "only a very deep knowledge of the electronic chip which treats the signal can explain why it leads to such a long and fatal delay" (Ouellette, 2012).

In fact, the superluminal neutrino saga stands as an excellent demonstration of the process of science. Throughout the process, the OPERA team was careful to emphasize the preliminary nature of their results, and they were very forthright about the measurements they had made and the procedures they followed. They were accused of "science by press conference," but the media attention their anomaly generated was largely a function of a changed media landscape (Orzel, 2011), with reporters and bloggers closely monitoring the Arxiv and CERN calendar for rumors. The common suggestion that they could have quietly sought outside assistance displayed what Dennis Overbye drily called "a quaint faith in the ability of the Twitterverse to keep secrets" (Overbye, 2012).

The superluminal neutrino story also demonstrates the deep engagement with the scientific community that characterizes genuine science and distinguishes it from pseudoscience. In the wake of the initial announcement, there was a real dialogue carried out within the literature. The original report by the OPERA team has been cited in more than 300 papers on the Arxiv and in physics journals, both by theorists proposing models for superluminal neutrinos, and others attacking the original analysis. The Cohen and Glashow paper has itself been cited more than 150 times, and not only by papers arguing against superluminal neutrinos. Many of those citations are in articles proposing new physics, attempting to work around or explain away the mechanism identified by Cohen and Glashow.

Finally, the ultimate resolution of the issue demonstrates the self-correction that is the hallmark of genuine science. The discovery of the faulty connection was publicly announced by OPERA, and the final nail in the coffin was a corrected measurement by the very same collaboration that kicked the whole business off nine months earlier.

In the end, perhaps the best summation of the whole issue was given by Sergio Bertolucci of CERN in announcing the combined results in June 2012:

Although this result isn't as exciting as some would have liked, it is what we all expected deep down. The story captured the public imagination, and has given people the opportunity to see the scientific method in action—an unexpected result was put up for scrutiny, thoroughly investigated and resolved in part thanks to collaboration between normally competing experiments. That's how science moves forward. (Jha, 2012)

BICEP2 and Primordial Gravitational Waves

In March 2014, the BICEP2 collaboration announced the discovery of evidence unmistakably supporting the cosmological theory of inflation. According to inflationary cosmology, the universe expanded exponentially for a fleeting instant not long after the Big Bang. This rapid inflation should lead to a characteristic variation in the cosmic microwave background radiation that would be the modern relic of the hot Big Bang. The BICEP2 experiment claimed to have detected just this pattern.

Just three months later, the BICEP2 discovery unraveled completely. The collaboration turned out to have greatly underestimated the amount of dust in the patch of sky their telescope observed; with a revised estimate of the dust's contribution to their signal, the result was no longer as convincing. In February 2015, the BICEP2 team released another analysis, combining their data with that of a rival experiment, the European Space Agency's Planck satellite. The joint analysis completely retracted the earlier finding.

The story of BICEP2's vanishing evidence for inflation is broadly similar to that of OPERA's faster-than-light neutrinos, with an initial burst of excitement followed by growing skepticism. It differs in a few key respects, however, and sheds additional light on issues of rivalry and replication in genuine scientific failures.

Timeline

The BICEP2 experiment was, as the name suggests, an upgrade of the earlier BICEP1 experiment (both names are an acronym for Background Imaging of Cosmic Extragalactic Polarization), and consists of a large microwave telescope located near the South Pole. It operated from 2009 to 2012, making a detailed map of the faint radiation left over from about 300,000 years after the Big Bang (see Appendix B for more technical detail about the experiment). This "cosmic microwave background" is our best source of information about the conditions of the very early universe.

On March 17, 2014, the BICEP2 collaboration announced, via a press conference and a pair of technical papers (BICEP2 collaboration, 2014a, 2014b), that they had detected a particular pattern of polarization in this background radiation, a pattern characteristic of the primordial gravitational waves produced during inflation. BICEP2's analysis showed that the strength of this pattern corresponded to a value of r=0.20 for the parameter measuring this phenomenon. Such a large value of r would be consistent with the predictions of inflationary cosmology, in which the early universe went through a period of extremely rapid expansion. Most importantly, BICEP2's value of r would be incompatible with most of inflationary cosmology's rivals.

Such a discovery would be a revolutionary breakthrough in cosmology, and it triggered an absolute avalanche of technical articles—according to the Harvard-Smithsonian Astrophysical Data Service, the BICEP2 paper had been cited 1,147 times as of early 2016 (about three times as many citations as the OPERA faster-than-light neutrino paper). While the more esoteric nature of the discovery made for slightly less coverage in mainstream media than in the case of OPERA, the discovery was hailed as heralding "a whole new era in cosmology and physics as well" (Clark, 2014), "allowing us to turn previously metaphysical questions about our origins into scientific ones" (Achenbach, 2014a), and potentially "one of the greatest discoveries in the history of science" (Overbye, 2014a). Even scientists who were loudly skeptical about the OPERA announcement were generally positive, like Lawrence Krauss (Achenbach, 2014a), Martin Rees, and Jim Al-Khalili of the boxer shorts wager (Sample, 2014a).

As is characteristic of genuine science, the early reactions included notes of caution. John Kovac of the BICEP2 team noted in an early interview that "science can never actually prove a theory to be true. There could always be an alternative explanation that we haven't been clever enough to think of" (Achenbach, 2014a). Jocelyn Bell

Burnell, a noted British astrophysicist, went so far as to declare, "I see this announcement as a placeholder, and wait for independent confirmation" (Sample, 2014a)

One striking difference between the OPERA anomaly and the BICEP2 announcement is that BICEP2 was competing with several other rival experiments trying to make the same measurement. In particular, early stories note that the Planck experiment was expected to announce its own results that summer. While the BICEP2 analysis was widely praised even by rival scientists—Adrian Lee of PolarBear called it "very likely to be correct" (Achenbach, 2014a) and Jon Carlstrom of the South Pole Telescope said the analysis was "beautiful and very convincing" (Overbye, 2014a)—scientists associated with rival experiments made a point of calling for independent confirmation. Lee expressed this most clearly, noting that "it's such a hard measurement that we really would like to see it measured with different experiments, with different techniques, looking at different parts of the sky, to have confidence that this is really a signal from the beginning of the universe" (Achenbach, 2014a).

By mid-May, the notes of caution had taken on a sharper edge, thanks largely to an analysis by Raphael Flauger, Colin Hill, and David Spergel of New York University and Princeton, which was posted to the Arxiv by the end of the month (Flauger, Hill, and Spergel, 2014). Flauger argued that the BICEP2 analysis greatly underestimated the effects of dust, in part because the team had used a preliminary data slide from a conference presentation by the Planck experiment to make their estimate. Flauger remained cautiously optimistic, saying, "I'm still hopeful that there's a signal there" (Achenbach, 2014b). Spergel offered a stronger condemnation, saying, "What they have seen is what you would get from dust alone" (Sample, 2014b).

The BICEP2 team responded to this criticism very directly, acknowledging that "new information from Planck makes it look like pre-Planckian predictions of dust were too low" (Overbye, 2014b) and incorporating it into their analysis. The final peer-reviewed version of their initial paper (BICEP2 collaboration, 2014c) removed the Planck data slide from the analysis, thereby increasing the experimental uncertainty. They also pledged to join together with the Planck collaboration, whose data were better suited to determining the effect of dust, to reach a definitive analysis (Overbye, 2014b).

Unfortunately for BICEP2, when that joint analysis was completed several months later, the effect was to completely undermine the original claim (BICEP2 collaboration and Planck collaboration, 2015). What they had thought was clear and unambiguous evidence for inflation was, in fact, badly contaminated by dust. Clem Pryke, who presented the joint analysis, said that while they still hoped to tease out evidence of gravitational waves, "If there is a gravitational wave component, it's less than about half the total signal" (Overbye, 2015). Their original measurement of $r=0.2$ was reduced to an upper limit of $r<0.12$. This smaller value implies a strength of gravitational waves that would be compatible with any number of early-universe theories. The "new era in cosmology" heralded by the BICEP2 announcement lasted all of ten months before returning to the status quo.

Commentary

The chief difference between OPERA's neutrinos and BICEP2's gravitational waves is that where the OPERA collaboration was careful to present their results as an "anomaly" in need of explanation, the BICEP2 team was making an affirmative claim of a dramatic discovery. OPERA launched its story with a technical seminar at CERN, but BICEP2's announcement was in fact a press conference. In a very 2014 touch, it also came with a viral video (https://www.youtube.com/watch?v=ZlfIVEy_YOA, which had 2.9 million views as of January 2016), a clip of BICEP2's Chao-Lin Kuo paying a visit to his Stanford colleague Andrei Linde, a pioneer of inflationary theory, with a bottle of champagne to celebrate the discovery.

This active courting of publicity fueled a lot of the criticism of BICEP2's behavior. The strongest denunciation came in a *Nature* opinion column by Paul Steinhardt of Princeton, saying of future measurements that "If there must be a press conference, hopefully the scientific community and the media will demand that it is accompanied by a complete set of documents, including details of the systematic analysis and sufficient data to enable objective verification" (Steinhardt, 2015).

There is, however, no small amount of self-interest on the part of Steinhardt, who has long been a strident critic of inflationary cosmology—Steinhardt's preferred early-universe model would have been ruled out had the BICEP2 result held up (Overbye, 2014a). Indeed, he devoted the last portion of his *Nature* piece to reiterating his position that inflation is fundamentally unscientific (Steinhardt, 2015). Steinhardt's criticism also ignores the fact that BICEP2's press conference was, in fact, accompanied by the sharing of technical documents, including the very information that allowed Flauger et al. to perform the reanalysis (Flauger, Hill, and Spergel, 2014) that undermined the result.

Despite their much stronger claim, there is little to suggest that the BICEP2 team violated any professional scientific norms. They made their claim based on a careful statistical analysis of the best information available to them; the information about dust that they would have needed to do a better job was available only to their competitors in the Planck collaboration. As with OPERA, the BICEP2 team was actively engaged with the community throughout the process, directly responding to criticism and incorporating it into the final published paper as a result of peer review. And again, BICEP2 demonstrated an admirable commitment to self-correction, joining with Planck for the joint analysis that provided a definitive end to the controversy.

To the extent that there is a negative lesson about science to take from this, it is about the pressures of competition. BICEP2 is guilty of "overinterpreting" their results (Achenbach 2015), but they were pushed to that in part because there were competing experiments working on the same problem. The data they needed to better interpret their results was collected by researchers at Planck who were pursuing the same gravitational-wave measurement. Had BICEP2 waited for Planck to publish its analysis

Scientific Failure as a Public Good

of the dust signal, they might well have missed out on the chance to get credit for a momentous discovery. In light of that competition, they made a decision to go with the data they had, and it's only in hindsight that this was clearly a mistake.

Contrary to claims like that by Marcelo Gleiser of Dartmouth that the BICEP2 saga "harms science because it's an attack on its integrity" (Hall, 2015), the BICEP2 incident is, in fact, an excellent example of the process of genuine science. The result was generated and announced through normal scientific channels, the ensuing debate arose and was carried out in the mainstream disciplinary literature, and the original researchers actively participated in the resolution of the controversy. This presents a stark contrast to the behavior of pseudoscientists, as we will see with our final example.

Hydrino Physics and the Structure of Pseudoscience

In the same way that genuine scientific failures follow a similar trajectory, there are a set of common elements that recur in a wide range of pseudoscience. These are mostly in direct opposition to the characteristic behaviors of genuine science: as we shall see, pseudoscience is conducted by researchers outside the mainstream scientific community, is distributed and publicized largely through nonstandard channels, and most importantly shows a near total absence of the self-correction that we see in failures of genuine science (such as the OPERA and BICEP2 incidents).

While there are numerous examples of public pseudoscience available for study, the physics of "hydrinos" serves as a useful case study in the same way that the OPERA and BICEP2 incidents do. Those scientific failures are extensively documented and went through the full cycle from initial excitement to final retraction in a relatively compact time frame, making it easy to identify the signature features. Hydrino physics has been around since the early 1990s, but it is a relatively small and self-contained area, which facilitates tracing the characteristic behaviors and drawing a contrast with genuine scientific failures like OPERA and BICEP2.

A Brief History of Hydrino Physics

One of the earliest successes of quantum mechanics was Niels Bohr's quantum model of hydrogen in 1913, in which the electron is permitted to occupy only certain special orbits, each with a particular energy. Bohr's model provided the first conceptual explanation of the discrete frequencies of light absorbed and emitted by hydrogen (and other atoms) in terms of "quantum jumps" between these allowed energy states. The development of the complete theory of quantum physics through the 1930s replaced Bohr's planetary-type electron orbits with quantum wavefunctions, but the core concept remains the same: the electron in hydrogen can take on only certain specific values of energy. In particular, there is a "ground state," which is the lowest possible energy for an electron in a hydrogen atom.

Hydrino physics posits the existence of states in hydrogen with energies *below* this ground state. This idea was introduced in the early 1990s by Dr. Randell Mills (who holds a medical degree) as an explanation for the cold fusion phenomena that were then attracting interest[1] (though he rejects the term, as his explanation does not involve fusion) (Broad, 1991). He claimed that the anomalously large energy releases seen in these experiments were given off by hydrogen atoms making a transition to hydrino states. The full explanation of the hydrino is presented in Mills's *Grand Unified Theory of Classical Quantum Physics* (Mills, 2000).

Mills has been promoting the use of hydrinos in various contexts ever since, promising ultracheap power generation from the excess heat generated by hydrinos, lasers, and photovoltaic electricity generation powered by the light emitted in transitions to hydrino states, and even hydrino catalysts to aid the formation of thin-film coatings on various surfaces. The name of Mills's company has changed over the years, from Hydro-Catalysis to BlackLight Power and most recently (in November 2015) to Brilliant Light Power, but the goal of promoting commercial uses of hydrinos has remained consistent.

Hydrinos as Pseudoscience
The most obvious indicator of the pseudoscientific status of hydrino physics is that it remains entirely a fringe phenomenon. Not only does Mills lack relevant professional credentials—he holds a medical degree from Harvard, but no physics training (Broad 1991)—he and his company are essentially the only source of hydrino research. Literature searches find only one hydrino researcher not affiliated with Brilliant Light Power: Jonathan Phillips, a mechanical engineer and former coauthor with Mills. Mills and his collaborators have produced numerous publications, mostly in low-impact journals, journals known for publishing wild speculation and unrefereed conference proceedings. His occasional articles in more mainstream journals are usually quickly followed by critical responses, as discussed at greater length below. Citations of Mills's work are primarily found in other papers by Mills et al.—forty-two of the fifty-seven references in one 2009 paper (Mills, et al., 2009) feature Mills as a coauthor. The ones found in papers from authors outside his immediate circle are mostly critical of his claims.

It probably goes without saying that the mainstream physics community does not take the hydrino idea seriously, and the model has drawn no small amount of mockery (Aaronson, 1999). This does not, however, mean that the various claims made by Mills have not been subjected to proper scientific scrutiny. A 2005 paper by Andreas Rathke (Rathke, 2005) takes a detailed look at Mills's *Grand Unified Theory of Classical Quantum Physics* and finds it riddled with mathematical errors. Rathke shows that the hydrino states Mills describes are mathematically inconsistent even with the equations of his own theory.

Also in 2005, Jan Naudts (a mathematical physicist in Antwerp) published a paper on the Arxiv (Naudts, 2005) identifying a hydrino-like state with an energy below the normal ground state that appeared to be a valid solution for the equations of quantum physics applied to hydrogen. These were quickly investigated by other researchers in mathematical physics, who reported that while one such state could be identified mathematically, it was not consistent with basic principles of physics. The hydrino-like solution exists only for the simplified case where the nucleus is a true point, but vanishes for the more realistic case of a nucleus with a finite but very small size (Dombey, 2006). The solution found by Naudts also does not satisfy the requirement that it be orthogonal to other allowed states of hydrogen (de Castro, 2007), one of the central requirements for a theory to match reality.

Defenders of hydrino physics point to the empirical effects Mills claims to have measured in a variety of experiments, and critical analyses by theoretical physicists generally do not address these. But the claimed experimental evidence for hydrino physics is regularly and vigorously questioned in the literature by experts in experimental physics. One of the most recent experimental articles reporting the observation of X-ray emission attributed to hydrinos (Mills and Lu, 2011) was accompanied by an editorial emphasizing that the paper's publication was "in no way an endorsement of the authors' 'hydrino' hypothesis by the Editors of this journal" (Becker, Mason, and Fabre, 2011). This was followed over the next several months by the publication of two rebuttals in the same journal, one a comment showing that the proposed explanation by Mills was inconsistent with a wide range of other experimental observations (Lawler and Goebel, 2012), and the other a full article demonstrating that the "extra" x-rays could easily be explained by ordinary plasma physics (Phelps and Clementson, 2012). It is useful to compare these responses to Giudice, Sibiryakov, and Strumia (2011), which takes a similar approach to attacking OPERA's superluminal neutrinos.

The most striking contrast between failures in genuine science and hydrino physics comes in the response to criticism. Where OPERA and BICEP2 self-corrected in response to criticism of their claims, the response to criticisms of hydrino physics is notable for its absence. Rathke's detailed picking apart of hydrino physics (Rathke, 2005) has been cited nine times in the past decade (according to the Harvard-Smithsonian Astrophysical Data Service); only one of those citations is in a paper (Naudts, 2005) supporting the existence of hydrino-like states. In fact, the only acknowledgement by hydrino researchers of criticism in the mainstream scientific literature over the past decade comes in the form of a single reply (Phillips, 2005) to a comment (Phelps, 2005) critical of an earlier paper (Phillips, Mills, and Chen, 2004). The reply might fairly be characterized as taking an angry tone, addresses only a few of the concerns raised in the comment, and largely misses the point of one of the issues it does address.

This lack of response does not reflect a lack of opportunity. In the years since Rathke's paper, Mills and his colleagues produced at least sixteen new papers on hydrino physics (according to the Harvard-Smithsonian Astrophysical Data Service), any one of which might have addressed or at least acknowledged the criticism. Their primary focus, however, was not participation within the broader community of scientific research, but the generation of publicity and venture capital funding. Over the same period, the Brilliant Light Power website hosted links to forty news stories and press releases about their activities (though this list omitted some of the less complimentary coverage, for example Lynch, 2015). Many of these, dating back to at least 1999 (Baard, 1999), promise the imminent release of a commercial product, though this has yet to emerge. Media reports about Mills and his company do contain the occasional comment about his critics—he dismissed Rathke's paper as "riddled with mistakes" in one story (Jha, 2005), and on an earlier occasion he threatened legal action against physicists critical of his theory (Reichhardt, 2000). But many years after Mills's theory was first criticized, a detailed response has yet to appear in the mainstream scientific literature.

Thus, there is a clear and striking contrast between the genuine scientific failures on the part of the OPERA and BICEP2 collaborations and the pseudoscience of hydrinos. The OPERA and BICEP2 errors originated with researchers based in the field, were distributed and debated via normal disciplinary channels, and ultimately corrected with the active participation of the original researchers. Hydrino physics, on the other hand, originates with researchers working outside their professional expertise, is largely promoted via nonstandard channels, and almost completely ignores critical responses.

Conclusions

Concerns that the widely publicized failure of a particular result in science will prove damaging to the credibility or institutional prestige of science as a whole implicitly assume that the general public will have trouble distinguishing genuine science from pseudoscience. The case studies presented here demonstrate a sharp distinction that should allay those concerns. Far from damaging the credibility of science in general, failed discoveries like OPERA's superluminal neutrinos or BICEP2's gravitational waves show the process of science at its best.

From the other direction, pseudoscientific attacks on mainstream science are usually predicated on the notion that the whole scientific establishment is mistaken and engaged in a grand conspiracy to cover up the truth. Again, the OPERA and BICEP2 incidents provide vivid counterexamples of what it *really* looks like when scientists are mistaken. The contrast could hardly be clearer. The OPERA and BICEP2 scientists actively participated in fixing their mistakes at the cost of public embarrassment, and pseudo-scientists refuse to engage with criticism in a forthright manner.

Thus, science communicators and public advocates of science should view these episodes not as lasting blots on the discipline, but as an opportunity for public education. The active engagement, widespread debate, and most importantly self-correction on display in these incidents show research scientists as passionate and committed to finding the truth, even at the cost of their own egos. That commitment reflects well on the entire profession. By using these examples to educate the public about the proper procedures of science, we can help bolster confidence in the scientific consensus around the most important challenges facing modern society.

Appendix A: The OPERA Experiment

The Oscillation Project with Emulsion-tRacking Apparatus (OPERA) collaboration involves around 180 physicists from twenty-eight institutions, mostly in Europe. Its goal is to study the way that the fundamental particles called neutrinos transform between their three different "flavors" (electron, muon, and tau). The experiment uses a beam of muon neutrinos generated by an accelerator at CERN near Geneva, which is detected with a 600-ton collection of electronic detectors and photographic emulsion located in the underground laboratory at Gran Sasso in Italy, 730 kilometers away. A tiny fraction of the initial muon neutrinos should transform into tau neutrinos during the trip, and the OPERA experiment aims to detect these and determine the rate of such transformations, which will provide useful insights into the fundamental physics involved.

Neutrinos are extremely elusive particles, and tau neutrinos even more so—from its initial activation in 2008 through 2015, OPERA has detected only five tau neutrinos (OPERA collaboration, 2015). Confirming such rare events requires a wide range of "sanity checks" to ensure that the detected particles are genuinely neutrinos, and not some less exotic and more readily detected particles. To shield against cosmic rays, the source and detector are located deep underground: the Laboratori Nazionali del Gran Sasso is in a set of experimental halls off a freeway tunnel between L'Aquila and Teramo, under about 1400 meters of solid rock. To help exclude other sources of background radiation, the particle beam that creates the neutrinos is switched on and off in a regular way. Particles detected when the beam is off should reflect the natural background, and excess particles detected when the beam is on are presumably due to neutrinos making the trip from CERN.

This pulsed-beam operation requires the precise synchronization of clocks separated by about 730 kilometers. This is accomplished in part using atomic clock signals from the Global Positioning System's network of satellites, detected by antennas on the surface and transmitted down to the underground laboratory via a fiber-optic cable, one of a huge number of cables and wires going to the apparatus. Unfortunately, this particular cable was improperly connected, leaving a small gap between the fiber and the detector. This by itself should have created only a slight attenuation in the signal, but the particular detection electronics used in the experiment converted this into a delay of some sixty nanoseconds.

Appendix B: The BICEP2 Experiment

The BICEP2 experiment consisted of a large microwave telescope that operated from 2009 to 2012. Located near the South Pole, the telescope was used to make a detailed map of the cosmic microwave background in a small patch of the southern sky. The cosmic microwave background radiation is light left over from about 300,000 years after the Big Bang, when the universe first cooled enough to allow the formation of neutral atoms. This radiation was initially in the visible and ultraviolet region of the spectrum, reflecting the extremely high temperature when it was created, but has been stretched out to microwave wavelengths by the last 13 billion years of the universe's expansion and cooling.

Past measurements (such as those by the COBE and WMAP satellites) have found small variations in the temperature of the background radiation from different parts of the sky, which correspond to differences in the density of matter in the early universe. Analysis of these temperature variations provides some of the tightest constraints on the amount and types of matter and energy present in the early universe. These primordial density variations are the seeds around which stars and galaxies formed, so the temperature measurements also constrain models of the evolution of everything we see.

BICEP2 expanded on previous measurements, which only measured the frequency spectrum of the microwaves, by also recording the polarization of the light (loosely speaking, the direction of the oscillating electric field associated with the electromagnetic waves detected by the telescope). This polarization should be sensitive to another type of variation in the density of the universe, a variation caused by gravitational waves created during the period of extremely rapid expansion known as "inflation." The contribution of gravitational waves is distinguished from other hot spots by a characteristic swirling pattern (the jargon term for this is a "B-mode") in the polarization; analysis of this pattern could reveal key properties of the gravitational waves in this very early era, and thus offer evidence for or against inflationary cosmology or one of its rival theories.

The light polarization can also be affected by much more recent events, most importantly the scattering of light by dust in our galaxy. BICEP2 was pointed at a region of sky believed to contain relatively little dust, but the project had no way of directly measuring the amount of dust in the field of view. In analyzing their data, the team had to rely on preliminary estimates that suggested the dust would have only a small effect. The Planck satellite, covering a wider field of view and different range of frequencies, could more directly measure the effects of dust, which turned out to be much larger than anticipated by BICEP2.

Appendix C: "Hydrino" Physics

A "hydrino" is a hypothesized state of hydrogen with an energy below the ground state predicted by conventional quantum mechanics. The textbook treatment of the hydrogen atom—that is, a single electron interacting with a single proton by means of the electrostatic interaction—involves solving the Schrödinger equation for a single electron attracted to a single proton (see, for example, Shankar 2011), and leads to a set of allowed wave functions for the electron whose energies have the mathematical form:

$$E_n = -\frac{E_0}{n^2}$$

where n is a positive integer and E_0 is the binding energy of hydrogen, 13.6 eV. There is an infinite number of such states, whose energy approaches $E=0$ as n increases (by convention, the zero of energy is taken to be an electron and proton at infinite distance; thus, the negative total energy indicates a state where the electron is bound to the proton), but there is a definite lowest energy state, when $n=1$. More advanced treatments replace the Schrödinger equation with the relativistic Klein-Gordon equation or the Dirac equation, but these do not fundamentally change the outcome: there is a single, definite lowest-energy state for a hydrogen atom, and its energy can be calculated (and experimentally measured) with great precision.

The hydrino theory promoted by Mills and his collaborators posits the existence of states with an energy *below* the ground state. Mathematically, this is accomplished by allowing n to take on fractional values, leading to much lower energies—taking $n=1/2$, for example, would give a state with an energy of –54.4 eV. Ordinary ground-state hydrogen making a transition to this hydrino state would release 40.8eV of energy, either as an x-ray photon or a burst of heat.

Such solutions are mathematically incompatible with the mathematical treatment using the Schrödinger equation, and thus are incompatible with well-understood and experimentally confirmed modern physics. As a matter of pure mathematics, a state with an energy below the standard ground state can be found using the relativistic Klein-Gordon equation for a very particular set of assumptions (Naudts, 2005), but this solution does not meet the requirements for a physically relevant solution (de Castro, 2007) and vanishes when using a more realistic description of the nucleus (Dombey, 2006).

Note

1. "Cold fusion" refers to the 1989 announcement at a press conference by Martin Fleischmann and Stanley Pons that their electrochemical experiments were generating more heat than could be explained by conventional physics. They attributed the excess to nuclear fusion taking place inside the palladium electrodes, at implausibly low temperatures for such reactions. These cold fusion reactions were touted as a potential source for generating vast amounts of clean energy.

Many labs attempted to reproduce their results without success, and today most reputable scientists regard the whole episode as a cautionary tale showing the dangers of "science by press conference," though a small fringe community of cold fusion enthusiasts continues to work on electrochemical cells to this day.

References

Aaronson, S. (1999). Hydrino Theory, which Overturns Quantum Theory, Is in Turn Overturned by Doofusino Theory. Retrieved from http://www.scottaaronson.com/writings/doofusino.html

Achenbach, J. (2011). Einstein, the Sky Is Falling! Or Not. *Washington Post*, Nov. 15, p. E01.

Achenbach, J. (2014a). Instant Cosmos. *Washington Post*, March 18, p. A01.

Achenbach, J. (2014b). A Cosmological Dust-up Over Big Bang Findings. *Washington Post*, May 17, p. A01.

Achenbach, J. (2015). Analysis Dispels Big-Bang Revelation. *Washington Post*, Jan. 31, p. A03.

Baard, E. (1999). Quantum Leap. *Village Voice*, Dec. 21. Retrieved from https://www.villagevoice.com/1999/12/21/quantum-leap/.

Becker, K. H., Mason, N. J., and Fabre, C. (2011). Editorial by the Editors-in-Chief Regarding the Highlighted Paper "Time-Resolved Hydrino Continuum Transitions with Cutoffs at 22.8 nm and 10.1 nm" by R.L. Mills and Y. Lu. *European Physical Journal D* 64(1): 63–63.

BICEP2 collaboration. (2014a). BICEP2 I: Detection of B-mode Polarization at Degree Angular Scales. arXiv:1403.3985v1 [astro-ph.CO]

BICEP2 collaboration. (2014b). BICEP2 II: Experiment and Three-Year Data Set. arXiv:1403.4302v1 [astro-ph.CO]

BICEP2 collaboration. (2014c. Detection of B-Mode Polarization at Degree Angular Scales by BICEP2. *Physical Review Letters*. 112, 241101, DOI: 10.1103/PhysRevLett.112.241101

BICEP2 collaboration and Planck collaboration. (2015) Joint Analysis of BICEP2/Keck Array and Planck Data. *Physical Review Letters*. 114, 101301, DOI: 10.1103/PhysRevLett.114.101301

Broad, W. J. (1991). 2 Teams Put New Life in "Cold" Fusion Theory. *New York Times*, April 26.

Clark, S. (2014). Primordial Gravitational Wave Discovery Heralds "Whole New Era" in Physics. *Guardian*, March 17.

Cohen, A., and Glashow, S. (2011). New Constraints on Neutrino Velocities. arXiv:1109.6562 [hep-ph]. (Later published in *Physical Review Letters* DOI: 10.1103/PhysRevLett.107.181803)

Cook, J., et al. (2013). Quantifying the Consensus on Anthropogenic Global Warming in the Scientific Literature. *Environmental Research Letters*. 8 024024 DOI: 10.1088/1748-9326/8/2/024024

de Castro, A. S. (2007). Orthogonality Criterion for Banishing Hydrino States from Standard Quantum Mechanics. *Physics Letters A* 369. 380.

Dombey, N. (2006). "The Hydrino and Other Unlikely States." *Physics Letters A* 360. 6.

Drahl, C. (2013). The Arsenic-Based-Life Aftermath. *Chemical & Engineering News, 90*(5):42–47. DOI: 10.1021/cen-09005-scitech1

Dyer, C. (2010). Lancet Retracts Wakefield's MMR Paper. BMJ. 340. DOI: http://dx.doi.org/10.1136/bmj.c696

Fiebelkorn, A., et al. (2015). A Comparison of Postelimination Measles Epidemiology in the United States, 2009–2014 versus 2001–2008. *Journal of Pediatric Infectious Diseases,* December 13. Retrieved from https://academic.oup.com/jpids/article/6/1/40/2957338/A-Comparison-of-Postelimination-Measles. DOI: 10.1093/jpids/piv080

Flauger, R., Hill, J. C., and Spergel, D. N. (2014). Toward an Understanding of Foreground Emission in the BICEP2 Region, May 28. arXiv:1405.7351v2 [astro-ph.CO].

Giudice, G. F., Sibiryakov, S., and Strumia, A. (2011). Interpreting OPERA Results on Superluminal Neutrino. arXiv:1109.5682v1 [hep-ph]

Haack, S. (2007). *Defending Science—Within Reason: Between Scientism and Cynicism*. New York: Prometheus Books.

Hall, S. (2015). BICEP2 Was Wrong, but Publicly Sharing the Results Was Right. Retrieved from http://blogs.discovermagazine.com/crux/2015/01/30/bicep2-wrong-sharing-results/#.VqegnPkrIVQ

ICARUS collaboration. (2011). "A Search for the Analogue to Cherenkov Radiation by High Energy Neutrinos at Superluminal Speeds in ICARUS." arXiv:1110.3763v2 [hep-ex]

ICARUS collaboration. (2012). Measurement of the Neutrino Velocity with the ICARUS Detector at the CNGS Beam. arXiv:1203.3433v3 [hep-ex].

Jha, A. (2005). Science: Fuel's Paradise? Power Source That Turns Physics on Its Head: Scientist Says Device Disproves Quantum Theory: Opponents Claim Idea Is Result of Wrong Maths. *Guardian*, Nov. 4, p. 9.

Jha, A. (2012). Neutrino Researchers Admit Einstein Was Right. *Guardian*, June 8.

Kmietowicz, Z. (2010). Wakefield Is Struck Off for the "Serious and Wide-Ranging Findings against Him." BMJ. Retrieved from http://www.bmj.com/content/340/bmj.c2803. 340 DOI: http://dx.doi.org/10.1136/bmj.c2803

Lawler, J. E., and Goebel, C. J. (2012). Comment on "Time-Resolved Hydrino Continuum Transitions with Cutoffs at 22.8 nm and 10.1 nm." *European Physical Journal D* 66:29.

Lynch, M. (2015). Warning Signs for Energy Technology Investors 3: Yes, They Can Be That Stupid. Forbes.com. Retrieved from http://www.forbes.com/sites/michaellynch/2015/06/01/warning-signs-for-energy-technology-investors-3-yes-they-can-be-that-stupid/

Matson, J. (2011). Faster-Than-Light Neutrinos? Physics Luminaries Voice Doubts. *Scientific American*. Retrieved from http://www.scientificamerican.com/article/ftl-neutrinos/ (retrieved 1/25/2016)

Mills, R. L. (2000). *The Grand Unified Theory of Classical Quantum Mechanics*. Cranbury, NJ: BlackLight Power.

Mills, R., Good, W., Jansson, P., and He, J. (2009). Stationary Inverted Lyman Populations and Free-Free and Bound-Free Emission of Lower-Energy State Hydride Ion Formed by an Exothermic Catalytic Reaction of Atomic Hydrogen and Certain Group I Catalysts. *Open Physics* 8(1): 7–16. ISSN (Online) 2391-5471, DOI: 10.2478/s11534-009-0052-6, November 2009.

Mills, R. L., and Lu, Y. (2011). Time-Resolved Hydrino Continuum Transitions with Cutoffs at 22.8 nm and 10.1 nm. *European Physical Journal D* 64: 65–72.

MINOS collaboration. (2007). Measurement of Neutrino Velocity with the MINOS Detectors and NuMI Neutrino Beam. *Phys. Rev. D* 76, 072005

Naudts, J. (2005). On the Hydrino State of the Relativistic Hydrogen Atom. arXiv:physics/0507193 [physics.gen-ph]

OPERA collaboration. (2011). Measurement of the Neutrino Velocity with the OPERA Detector in the CNGS Beam. arXiv:1109.4897v1 [hep-ex]

OPERA collaboration. (2012). Measurement of the Neutrino Velocity with the OPERA Detector in the CNGS Beam Using the 2012 Dedicated Data. arXiv:1212.1276v2 [hep-ex]

OPERA collaboration. (2015). Discovery of τ Neutrino Appearance in the CNGS Neutrino Beam with the OPERA Experiment. *Phys. Rev. Lett.* 115, 121802 DOI: 10.1103/PhysRevLett.115.121802

Orzel, C. (2011). The Brave New-Media World. *Physics World* 24(11): 19.

Ouellette, J. (2012). "Faster-Than-Light" Neutrino Team Leaders Resign. *Discovery News*. Retrieved from http://news.discovery.com/space/opera-leaders-resign-after-no-confidence-vote-120404.htm

Overbye, D. (2011). Particles Faster than the Speed of Light? Not So Fast, Some Say. *New York Times*, Oct. 25; section D, p. 3.

Overbye, D. (2012). The Trouble with Data That Outpaces a Theory. *New York Times*, March 2; section D, p. 1.

Overbye, D. (2014a). Space Ripples Reveal Big Bang's Smoking Gun. *New York Times*, March 18; section A, p. 1.

Overbye, D. (2014b). Astronomers Stand by Their Big Bang Finding, but Leave Room for Debate. *New York Times*, June 20; section A, p. 20.

Overbye, D. (2015). Speck of Interstellar Dust Obscures Glimpse of Big Bang. *New York Times*, Jan. 31; section A, p. 11.

Phelps, A. V. (2005). Comment on "Water Bath Calorimetric Study of Excess Heat Generation in Resonant Transfer Plasmas" [J. Appl. Phys.96, 3095 (2004)]. Journal of Applied Physics 98, 066108 (2005); doi: http://dx.doi.org/10.1063/1.2010616

Phelps, A. V., and Clementson, J. (2012). Interpretation of EUV Emissions Observed by Mills et al. *European Physical Journal D*, May 2012, 66:120.

Phillips, J. (2005). Response to "Comment on 'Water Bath Calorimetric Study of Excess Heat Generation in Resonant Transfer Plasmas'" [*J. Appl. Phys.* 96, 3095 (2004)]. *J. Appl. Phys.* 98, 066109; http://dx.doi.org/10.1063/1.2010617

Phillips, J., Mills, R. L., and Chen, X. (2004). Water Bath Calorimetric Study of Excess Heat Generation in "Resonant Transfer" Plasmas." *J. Appl. Phys.* 96, 3095 (2004). DOI: http://dx.doi.org/10.1063/1.1778212

Rathke, A. (2005). A Critical Analysis of the Hydrino Model. *New J. Phys.* 7, 127.

Redfield, R. R. (2010). Arsenic-Associated Bacteria (NASA's Claims). Retrieved from http://rrresearch.fieldofscience.com/2010/12/arsenic-associated-bacteria-nasas.html

Reichhardt, T. (2000). New Form of Hydrogen Power Provokes Scepticism. *Nature* 404: 216.

Roser, M. J. (2015). Discredited Autism Guru Andrew Wakefield Takes Aim at CDC. *Austin American Statesman*, June 27. Retrieved from http://www.mystatesman.com/news/news/local/discredited-autism-guru-andrew-wakefield-takes-aim/nmk5p/

Sample, I. (2011). Was Einstein Wrong? Lab Particles Travel Faster than Light: Scientists Find Particles Going Faster than Light. *Guardian*, September 23, p. 1.

Sample, I. (2014a)."We Have a First Tantalising Glimpse of the Cosmic Birth." *Observer*, March 23, p. 20.

Sample, I. (2014b). Gravitational Wave Triumph Turns to Dust after Flawed Analysis Revealed. *Guardian*, June 4.

Sample, I., and Jha, A. (2011). "Faster than Light" Neutrinos Met with Shock, Awe and Scepticism: Results Appear to Question Cornerstone of Physics: Scientists Urge Caution before Accepting Findings. *Guardian*, Sept. 24, p. 1.

Shankar, R (2011). *Principles of Quantum Mechanics* (2nd ed.) New York: Plenum Press.

Steinhardt, P. (2015). Big Bang Blunder Bursts the Multiverse Bubble. *Nature* 510: 9. DOI: 10.1038/510009a

Vastag, B. (2011). Faster than Light: Revolution or Error? *Washington Post*, Sept. 24.

Vastag, B. (2012). Not so Fast: Neutrinos Clocked at Light Speed. *Washington Post*, March 17, p. A05.

Voss, D. (2012). Focus: Supersolid Discoverer's New Experiments Show No Supersolid. *Physics* 5: 111. Retrieved from https://physics.aps.org/articles/v5/111

Wilford, J. N. (1996). Replying to Skeptics, NASA Defends Claims About Mars. *New York Times*, August 8; section A, p.1.

Wolfe-Simon, F., Switzer Blum, J., Kulp, T. R., Gordon, G. W., Hoeft, S. E., Pett-Ridge, J., Stolz, J. F., Webb, S. M., Weber, P. K., Davies, P. C. W., Anbar, A. D., and Oremland, R. S. (2011). A Bacterium That Can Grow by Using Arsenic Instead of Phosphorus. *Science* 332: 1163–1166. DOI: 10.1126/science.1197258

11 Evidence-Based Practice as a Driver of Pseudoscience in Prevention Research

Dennis M. Gorman

Introduction

Many of the subject matters discussed under the topic of pseudoscience (such as parapsychology) can be rather easily distinguished from science proper, and there are few individuals with any serious scientific training who would be fooled into thinking these are science-based disciplines. Harder to identify and distinguish are those disciplines that began as genuine science but have transformed into pseudoscience primarily through their use of irregular data analysis practices in the pursuit of results that support a favored hypothesis. Elsewhere I have argued that drug prevention research has largely evolved into a pseudoscience, in that it is based on consensual thinking, rarely rejects hypotheses that programs are effective, is hostile to criticism, explains away unfavorable results, and employs a variety of questionable analytic practices (such as one-tailed tests of statistical significance and post hoc subgroup analysis) to manufacture positive results (Gorman, 2003, 2010).

There are a number of drivers of the decline of drug prevention research into a pseudoscience. Some of these are influences and incentives that are common to the research cultures of a variety of disciplines. Conflict of interest, for example, is associated with use of questionable research practices and the production of positive results in medicine and public health (Ioannidis et al., 2014a), psychology (Ioannidis, Munafo, Fusar-Poli, Nosek, and David, 2014b; Nosek, Spies, and Motyl, 2012), public policy analysis (MacCoun, 2005), and addiction studies (Holder, 2010). In drug prevention research, conflict of interest manifests itself in the form of program developers evaluating the interventions they developed and financially profit from (Gorman and Conde, 2007). Such evaluations have been shown to produce more positive results than evaluations conducted by independent evaluators, as well as employing questionable data analysis practices such as multiple subgroup analysis and post hoc exclusion of subject and selection of outcome variables (Eisner, 2009; Gorman, 2016). Another common driver of the use of questionable analytic practices is that academic journals prefer positive results, and academics must publish in journals for the sake of their careers (Fanelli,

2010a, 2010b). Drug prevention researchers primarily come from psychology, a discipline so predisposed towards publishing positive results that we hear of its "aversion" to the null hypothesis (Ferguson and Heene, 2012).

In this chapter I focus on a driver of pseudoscience that is found mainly among applied disciplines, most notably those that seek to identify interventions to prevent or treat health and social problems (for example, public health, health education, social work, medicine, criminology, and clinical psychology). In particular, I focus on the incentive to use questionable data analysis practices to produce positive results that is created by the existence of evidence-based practices. The evidence-based practice movement has played a prominent role in the drug prevention field since the early 1990s, and resulted in the generation of a number of lists of "best practices." Here I focus on what is probably the most influential list, one developed by the US Substance Abuse and Mental Health Services Administration. I argue that this particular list has had a detrimental effect on the quality of drug prevention research and has contributed substantially to it becoming a pseudoscience.

The Substance Abuse and Mental Health Services Administration's National Registry

The impetus for the development of what have come to be termed evidence-based practice lists came in large part from criticism that was directed at government-funded drug prevention activities of the early 1990s; many of these activities were supported only by anecdotal evidence (Gorman, 1998; US General Accounting Office, 1997). As a result, a spate of "best practice" or "science-based" or "model" lists of alcohol, drug, and violence prevention programs were produced by federal and private agencies (Petrosino, 2003). The best known of these were the Center for Substance Abuse Prevention's (CSAP) list of science-based programs (Center for Substance Abuse Prevention, 2001, 2002), the US Department of Education's Exemplary and Promising programs (US Department of Education Safe, Disciplined, and Drug-Free Schools Expert Panel, 2001), the University of Colorado's *Blueprints for Violence Prevention* (Mihalic, Irwin, Elliott, Fagan, and Hansen, 2001, and the National Institute on Drug Abuse's (NIDA) *Research-Based Guide* (National Institute on Drug Abuse, 2003). The programs that appeared on these lists were recommended for widespread use and dissemination on the basis of apparently scientific research demonstrating their efficacy. Prevention practitioners, it was argued, must abandon the use of interventions that had been shown to be ineffective (e.g., the Drug Abuse Resistance Education [DARE] program[1]) and use instead programs that appeared on these lists.

The Substance Abuse and Mental Health Services Administration's (SAMHSA) efforts to identify evidence-based prevention programs began in 1999 with the establishment of the National Registry of Effective Prevention Programs (NREPP), described by the agency that administered it as "a process to screen and identify intervention programs that because of their scientific support and practical findings warrant national dissemination and replication" (Center for Substance Abuse Prevention, 2002, p. 2). Candidate programs for NREPP review came from four main sources: existing scientific literature, effective programs assessed by other rating processes (such as that of NIDA), final reports of SAMHSA grantees, and general solicitations to the field. Members of a team of twenty-seven doctoral-level scientists conducted the reviews, and assessed programs using fifteen criteria (including theory, process evaluation measures, outcome measures, and analysis). Although each candidate program was assessed on all fifteen of the NREPP criteria, only two were used to designate the program as "model" (Center for Substance Abuse Prevention, 2002, pp. 16–17). These criteria were integrity and utility, both of which had a somewhat global quality to them. Integrity, which determined the scientific rigor of the evaluation, referred to "the quality of the intervention's implementation, the evaluation study design, and the actual conduct of the study," and required "reviewers to rate the merits of the science that guided the evaluation" (Center for Substance Abuse Prevention, 2002, p. 16). Utility pertained to "whether, and to what degree, a program produces a consistent pattern of results and is usable and appropriate for widespread application and dissemination" (Center for Substance Abuse Prevention, 2002, p. 16). In order to be designated a "model" program by NREPP, the research literature pertaining to a program had to be rated at least 4.0 on each of the 5-point scales of integrity and utility, and the program developers had to be capable of providing "quality materials, training and technical assistance to practitioners who wish to adopt their programs" (Center for Substance Abuse Prevention, 2002, p. 17). The original NREPP review process resulted in relatively few intervention programs being identified as "model." The 2001 NREPP annual report contained descriptions of thirty-eight model programs (Center for Substance Abuse Prevention, 2001), and this increased to forty-four in the 2002 report (Center for Substance Abuse Prevention, 2002).

In March 2006, SAMHSA revised its rating criteria and selection procedure and renamed the *National Registry of Effective Prevention Programs* the *National Registry of Evidence-based Programs and Practices* (with the acronym, NREPP, remaining) (US Department of Health and Human Services, 2006). The revisions were in response to comments that had been elicited from the public and that focused on limiting the system to interventions that had demonstrated effects on behavioral outcomes, training reviewers with specialized knowledge of the interventions under review, providing descriptions of programs that detailed the programs' readiness for dissemination, their cultural diversity, and their implications for practice, and recognizing multiple streams of evidence. The strength of evidence for programs was presented as a pyramid, with case, pilot, and observational studies at the base and meta-analysis and expert reviews at the apex. Study designs without comparison conditions were not to be included in the NREPP system, so the focus remained on randomized trials and quasi-experiments. There was also a continued focus on data that were collected using valid and reliable measures and on analyses that took into account confounding and missing data. In the new review procedures, the only requirement for potential inclusion in the NREPP review process was "for an intervention to have demonstrated one or more significant behavioral change outcomes" (US Department of Health and Human Services, 2006, p. 13135). This "one behavioral effect" requirement was a fairly low threshold to cross, especially as there was no consideration given as to how many analyses were conducted to produce it (Gorman, 2015a). Not surprisingly, the number of programs included on the NREPP list began to increase, and by July 2009, 140 interventions were included (SAMHSA, 2009).

NREPP was revised once again in the summer of 2015 (US Department of Health and Human Services, 2015). The minimal requirements for review were again that there be at least one evaluation (using an experimental or quasi-experimental design) of the intervention that had assessed mental health, substance use, or other behavioral health-related outcomes pertaining to individuals, communities, or populations with or at risk of mental health or substance use problems, and that the results be published in a peer-reviewed journal, other professional publication, or comprehensive evaluation report (US Department of Health and Human Services, 2015, p. 38717). Implementation material, training, and support resources were no longer required as part of the review. The review procedure was also revised such that the studies and outcomes to be reviewed would be identified through "standardized screening criteria" and more accurately reflect the evidence base of the intervention rather than simply highlighting positive outcomes. Ideally two reviewers would assess each intervention, and rate its "effectiveness" on each outcome based on the methodological rigor of the study in which the outcome was reported and the magnitude and direction of the effect. But it was also noted that inclusion on NREPP did not indicate that all of the evidence pertaining to the intervention had been reviewed; as in the 2006 iteration, it was stated that inclusion did not constitute an endorsement of a program or practice.

The new NREPP review guidelines described in the Federal Register were implemented on November 23, 2015. At this point the number of programs included in the NREPP list had risen to 356, and the revisions proposed that these "legacy" programs and practices be re-reviewed according to the new criteria between 2015 and 2018 (US Department of Health and Human Services, 2015). By the end of February 2016, a further thirty-six programs had been added to the NREPP list. The number of programs stood at 394 programs by July 5, 2016 (SAMHSA, 2016a).

A Systems Approach to Understanding Pseudoscience in Drug Prevention Research

System research has shown that interventions can produce iatrogenic effects (that is, effects that are detrimental and exacerbate a problem rather than solving it) due to the mismatch between the complexity of the systems we seek to improve and the static and simplistic nature of many of our solutions, as well as the fact that the agents whose behavior we seek to change react to interventions in an attempt to restore the balance of the system in their favor (Sterman, 2006). It is quite possible, although rarely given much thought, that the introduction of measures intended to improve the quality of practice and research can create a self-reinforcing feedback loop that actually has the opposite effect to the system-correcting one intended. I have suggested that the introduction of evidence-based lists in prevention research is an example of such a feedback mechanism (Gorman, 2015a).

Figure 11.1 presents two ways in which researchers have reacted to (and essentially gamed) the system in the form of causal loop models, one involving increased use of flexible data analysis practices (the top loop) and the other involving minimal adherence to study design criteria (the bottom loop). In each case, researchers (many of whom are developers of the programs they evaluate) promote their programs further by using the evidence-based review system that has been put in place to improve the quality of practice. The result is that interventions that are largely ineffective continue to be used by practitioners; indeed, they are more likely to be used as they now carry the stamp of approval of some government or private agency. In addition, rather than the science improving practice, the quality of the science in the field is degraded as it uses questionable data analysis and presentation practices as a means of generating results that can be used to promote intervention programs. The net result is a field comprised of pseudoscientific research practices and ineffective interventions.

Loop 1: Flexible Data Analysis and Selective Reporting

There are numerous examples in the literature of the first of the loops shown in figure 11.1, that is, evaluations that use flexible data analysis practices and selective reporting to produce isolated statistically significant effects that meet the very minimal criteria for inclusion on the NREPP list. These include the evaluations of the ATLAS Program (Gorman, 2002), Eye Movement Desensitization and Reprocessing (Herbert et al., 2000), the Life Skills Training program (Gorman, 2005, 2011), Multisystemic Therapy (Littell, Popa, and Forsythe, 2005), Project ALERT (Gandhi, Murphy-Graham, Petrosino, Chrismer, and Weiss, 2007; Gerstein and Green, 1993; Gorman and Conde, 2010), and the Strengthening Families Program (Gorman and Conde, 2009; Gorman, Conde, and Huber, 2007).

A key issue raised in these critiques is that the results presented in support of the efficacy of the programs evaluated are the result of data dredging and selective reporting. Thus, the selection and review criteria of NREPP allows further cherry picking of what are already cherry-picked results. Consider Project Towards No Drug Abuse (Project TND). Elsewhere, I have reviewed all of the seven evaluation studies of Project TND and concluded that the evidence in support of the program is very weak and

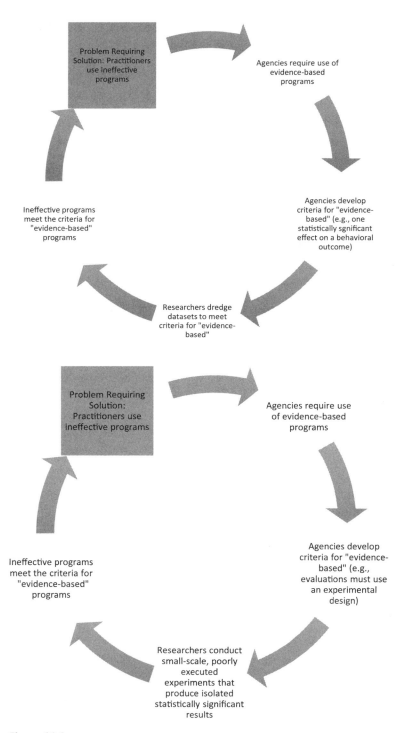

Figure 11.1
Causal Loops Depicting How Researchers Manipulate the Evidence-Based Review System. Use of flexible data analysis practices (top loop) and minimal adherence to study design criteria (bottom loop)

inconsistent (Gorman, 2014). Although isolated statistically significant effects emerge from the evaluations, especially when subgroup analyses are conducted, the preponderance of evidence supports the null hypothesis of no effect.

The results presented in the NREPP review to support the inclusion of Project TND as a drug prevention intervention are summarized in table 11.1. These are said to come from four papers, pertaining to three studies, published between 1998 and 2006 (Dent, Sussman, and Stacy, 2001; Sussman et al., 1998, 2003; Sun et al., 2006). However, a number of the specific results cited in the NREPP summary come from an additional paper that reports one-year prevalence data for three of the studies (Sussman et al., 2002). Sussman et al. (1998) and Sun et al. (2006) contain data from the same study that assessed two versions of the TND program (a nine-session curriculum and the nine-session curriculum plus a school as community component), the former reporting data from the one-year follow-up and the latter from a combined two- and three-year follow-up and a combined four- and five-year follow-up. Both of these studies report data from a thirty-day frequency measure, as does Dent et al. (2001). Sussman et al. (2003) reports thirty-day prevalence data from a two-year follow-up of a revised twelve-session TND curriculum, delivered both in classroom setting and through self-instruction.

Table 11.1.
"Key Findings" from the Project TND Evaluations Cited in the NREPP Review of the Program

Drug	Follow-Up	Curriculum	Key Finding: TND versus Control Group	Subgroup Analysis	Studies
Alcohol	1 year	9-session and 12-session	7%–12% reduction across 3 studies in use among baseline users	Yes	Sussman et al. (1998) Sussman et al. (2002) Dent et al. (2001)
Tobacco	1-year	12-session	27% reduction in use	No	Sussman et al. (2002)
Tobacco	2-year	12-session	Half as likely to use	No	Sussman et al. (2003)
Marijuana	1-year	12-session	22% reduction in use	No	Sussman et al (2002)
	2-year	12-session	Male baseline non-users one-tenth as likely to use	Yes	Sussman et al. (2003)
Hard Drugs	1-year	9-session and 12-session	25% reduction across 3 studies in use	No	Sussman et al. (2002)
	2-year	12-session	One-fifth less likely to use	No	Sussman et al. (2003)
	4–5-year	9-session	Less likely to use	No	Sun et al. (2006)

Sussman et al. (2002) contains thirty-day prevalence results for the one-year follow-up data reported in Sussman et al. (1998) and Dent et al. (2001). And it is the only published account of one-year follow-up data from the study also described in Sussman et al. (2003). Thus, the 25 percent reduction in prevalence of hard-drug use across three studies at the one-year follow-up reported in the NREPP summary is found in Sussman et al. (2002), and not in Sussman et al. (1998), Dent et al. (2001), or Sussman et al. (2003).

Table 11.2 locates the results selected for inclusion in the NREPP review of Project TND within the context of all of the results pertaining to main effects reported in the

Table 11.2.
Statistically Significant Main Effects in the Project TND Evaluations Reported in the NREPP Review

Publication	TND Program	Follow-Up	Alcohol	Tobacco	Marijuana	Hard Drugs
Sussman et al. (1998)	9-session curriculum	1 year	-	-	-	-
	Curriculum+SAC	1 year	-	-	-	-
Sussman et al. (2002)	9-session curriculum	1 year	-	-	-	✓
	Curriculum+SAC	1 year	-	-	-	✓
Dent et al. (2001)	9-session curriculum	1 year	-	-	-	*
Sussman et al. (2002)	9-session curriculum	1 year	-	-	-	✓
Sussman et al. (2002)	12-session curriculum	1 year	-	✓	✓	✓
	Self-instruction	1 year	-	-	-	-
Sussman et al. (2003)	12-session curriculum	2 years	-	✓	-	✓
	Self-instruction	2 years	-	-	-	-
Sun et al. (2006)	9-session curriculum	2–3 years	-	-	-	-
	Curriculum+SAC	2–3 years	-	-	-	-
	9-session curriculum	4–5 years	-	-	-	✓
	Curriculum+SAC	4–5 years	-	-	-	✓

Key:
- = No Statistically Significant Difference
* = Statistically Significant Difference Not Reported in NREPP
✓ = Statistically Significant Difference Reported in NREPP
SAC = School as Community Version of Project TND

five publications from which these were selected. It can be seen from the table that there were no statistically significant main effects for alcohol reported in any of the studies cited by NREPP in support of Project TND. As noted in table 11.1, the "key findings" pertaining to alcohol use are all from subgroup analyses conducted on students who had used alcohol at the baseline assessment in the three studies. No mention is made in the NREPP review of the fourteen analyses of main effects that found no statistically significant differences between the Project TND and the control conditions in these studies. Moreover, subsequent evaluations of Project TND have not replicated the effect pertaining to baseline alcohol users (Gorman, 2014). With regard to cigarette use, only two of the fourteen comparisons made between the TND conditions and the control conditions in the five papers were statistically significant. Both are reported in the NREPP review, while no mention is made of the twelve null results. The only result reported for marijuana use in the NREPP key findings is the sole statistically significant main effect finding reported in the five publications. There is no discussion of the thirteen null findings.

Eight of the fourteen results for hard drugs are statistically significant, and seven of these are reported in the NREPP summary. The validity of these findings is questionable, however, for a number of reasons (Gorman, 2014). First, the procedure used to transform the hard-drug use measurement scale may have artificially inflated the scores of the control group in the Sun et al. (2006) study and these may be driven by a few extreme scores among control group subjects. Second, the studies contained significant selection bias that was introduced at the recruitment phase of each and exacerbated over the course of the follow-ups. For example, 30 percent of eligible subjects refused to participate in the Sussman et al. (1998) study, and by the final follow-up almost two-thirds of subjects had dropped out. Of those eligible to participate in the Sussman et al. (2003), study 40 percent refused to participate and 45 percent dropped out by the two-year follow-up. Despite these serious methodological problems, the quality of the research on which the results pertaining to hard-drug use are extracted is rated 3.4 on the 4-point scale used by NREPP.

In summary, the positive results that have emerged from the Project TND studies cited by NREPP are the exception rather than the rule and they suffer from serious methodological problems that undermine their validity. Yet the NREPP review makes no mention of the plethora of null findings nor the threats to the validity of the positive results that it highlights.

Loop 2: Minimal Adherence to Study Design Criteria

With regard to the second loop shown in figure 11.1, an extreme example of this phenomenon is the Moment Program which was added to NREPP on February 3, 2016. This "classroom-based, mindfulness education program" is said to improve cognitive functioning and reduce disruptive behavior disorders and antisocial behavior (SAMHSA,

2016b; see also Kozak, in volume, for further discussion of mindfulness). The "Evaluation Findings by Outcome" section of the NREPP program description of the Moment Program makes reference to pretest-posttest changes that were found in a study that compared program participants to a wait-list control group. The only reference presented for this study is an undated, unpublished manuscript produced by the organization that sells the Moment Program (Parker and Kupersmidt, n.d.). Similarly, only one unpublished preliminary report is cited in support of the inclusion of An Apple a Day (O'Neill, 2008; SAMHSA, 2016c), which was added to the list in June 2011, and two unpublished reports in support of AMIkids Personal Growth Model (Early, Blankenship, and Hand, 2011a; Early, Hand, Blankenship, and Chapman, 2011b; SAMHSA, 2016d), which was added December, 2011.

Other examples of programs on NREPP that have been evaluated using small, methodologically weak evaluation designs that nonetheless meet its research design requirements include Accelerated Resolution Therapy (ART) and Internal Family Systems (IFS). The former was added to NREPP on November 22, 2015 (a day before the move to the latest review guidelines) and the latter on November 20, 2015. ART is a brief (two to five sessions of sixty to seventy-five minutes each) exposure-based therapy designed to treat combat-related post-traumatic stress disorder, depression, anxiety, phobias, obsessive-compulsive disorder, and substance use. It was developed in 2008 by Laney Rosenzweig (Rosenzweig Center for Rapid Recovery, 2016), and has been employed in numerous military, community-based, and private settings in the USA since 2011 (SAMHSA, 2016e). ART participants visualize traumatic events or scenes, using clinician-directed eye movement to help replace these with positive imagery, sensations, and emotional reactions. The NREPP webpage lists two publications under its evaluation findings section, both of which come from a single study (Kip et al., 2013, 2014). Kip et al. (2013) found statistically significant differences at follow-up that favored the ART group on symptoms of PTSD, depression, and guilt and distress (termed personal resilience/self-concept on the NREPP webpage).

Thus, in line with the NREPP criterion, the Kip et al. (2013) results show that ART produces significant change on one or more mental health outcomes. The study also clearly meets the NREPP research design criterion, as it was a trail that randomly allocated subjects to either the ART group or a control group (comprised of two one-hour sessions of either fitness assessment and planning or career assessment and planning). Further, the evaluation included certain methodological refinements, such as the use of intent-to-treat analysis. However, the trial is, at best, a pilot study as just fifty-seven subjects were identified as eligible for inclusion and the follow-up period was a mere three months. At this point, only twenty-one of the twenty-nine subjects assigned to the ART group, and only seventeen of the twenty-eight assigned to the control group, completed the follow-up (see figure 1 of Kip et al., 2013). It is difficult to see how the results from a study with so few subjects and so brief a follow-up can be used as the sole basis from which to designate ART an "evidence-based" practice.[3]

The IFS program is a psychotherapeutic intervention delivered in small groups and designed to teach "patients to attend to and interact with their internal experience mindfully" (Shadick et al., 2013, p. 1832). It was developed by Richard C. Schwartz in 1995, and by 2013 more than 2,200 therapists worldwide had received training in its use (Shadick et al., 2013). In the only research study listed on the NREPP webpage, the program was evaluated with patients diagnosed with rheumatoid arthritis recruited from a hospital in Boston. Thirty-nine subjects were randomly allocated to the IFS program, and forty to the control group that received a rheumatoid arthritis education program. Subjects were assessed at baseline, three-, six-, nine-, and twenty-one-month follow-up. The primary outcomes were improvement in disease activity (assessed using two instruments), depression, anxiety, and physical function. The secondary outcomes were self-compassion and self-efficacy. At the twenty-one-month follow-up there were statistically significant differences between the IFS group and the control group on one of the disease activity measures, the measure of depression, and the self-compassion measure.

Shadick et al. (2013) point to a number of limitations of their study and are cautious in their interpretation of its results, stating, for example, that the IFS intervention *may* be helpful to patients and *may* complement medical management of the disease. However, no such caution is displayed in the IFS announcement on its webpage that the program is included in NREPP. According to this, the NREPP listing affirms "the vast potential of IFS Therapy for advancing emotional healing and mental well-being" (Center for Self Leadership, 2013). Further, it is said to represent "an indirect acknowledgment of the work of Richard Schwartz, PhD, who developed the modality 30-some years ago, and the longstanding efforts of this growing community of IFS trainers, practitioners, psychotherapy clients, and individuals for whom the model has offered a new way of being in the world" (Center for Self Leadership, 2013). So much for the Department of Health and Human Services' insistence that inclusion on NREPP does not constitute an endorsement of a program or practice.

As with Kip et al.'s (2013) evaluation of ART, Shadick et al.'s (2013) evaluation of IFS is, at best, a pilot study. Moreover, interpretation of the results is severely limited by the selection bias introduced at the start of the study. Initially, 857 potential subjects were identified, of whom 290 were not contacted and 209 declined to be contacted. Of the remaining 358 assessed for eligibility, twenty were determined to be ineligible and 259 declined to participate. By the final follow-up, sixty-eight of the seventy-nine subjects randomly allocated to the study condition provided data. This represents just 20 percent of those eligible to take part in the study (this does not include the 468 subjects who declined to participate at earlier stages of the evaluation, many of whom would have likely met the eligibility criteria). Thus, IFS appears on the NREPP list on the basis of a single, small-scale study in which considerable selection bias was present. The study also included the program developer among its research team and as an author on the manuscript describing the results of the study, although no mention is made in the Shadick et al. (2013) publication of this conflict of interest.

Discussion

The above discussion of the development of the evidence-based list known as NREPP suggests that this list has done little to improve the practice of drug prevention and has led to a degradation of the science of drug prevention. Regarding the latter, the development of NREPP has provided an incentive to evaluators of drug prevention programs to employ study designs and data analysis and presentation practices that capitalize on the production or manufacture of chance trivial findings. In some cases, NREPP allows program developers to conduct evaluations of very low quality and use the findings generated in these in support of their case that their programs are "evidence-based." In other cases, NREPP allows program developers to conduct numerous analyses of their datasets and to cherry pick those in support of their case that their programs are evidence-based, while ignoring the vast majority of results that support the null hypothesis of no effect. Thus, rather than improving prevention practice through science, NREPP has largely functioned to turn prevention science into a pseudoscience.

With regard to its primary function of improving the quality of drug prevention practice, NREPP has also been unsuccessful. Figure 11.2 presents stock and flow models contrasting the role envisioned for NREPP with how it actually functions. Prior to NREPP, the view was that there were too many programs of very low quality in use among practitioners (figure 11.2A). The role of NREPP, and indeed all lists of evidence-based practices, was seen as filtering out the low-quality interventions from high-quality interventions. This filtering would be performed through the critical application of standards developed by scientists to distinguish valid and reliable information from invalid and unreliable information. Thus, the flow from the stock of existing programs into the stock of programs widely in use would be reduced; there would be fewer programs in use but these would be of higher quality (figure 11.2B). The problem is that while the basic trappings of the scientific method were applied in the filtering process (e.g., the evidence had to meet basic standards of having been produced in a study with a comparison condition and having a statistically significant effect on one behavioral outcome), the key feature of the scientific method (namely a critical-rational approach to evidence) was never applied (Gorman, 2005, 2015a). Accordingly, the pseudoscientific approach to filtering out programs that has been employed by NREPP has done little to affect the flow of low-quality programs into the stock of programs in use (figure 11.2C).

Evidence-Based Practice as a Driver of Pseudoscience in Prevention Research

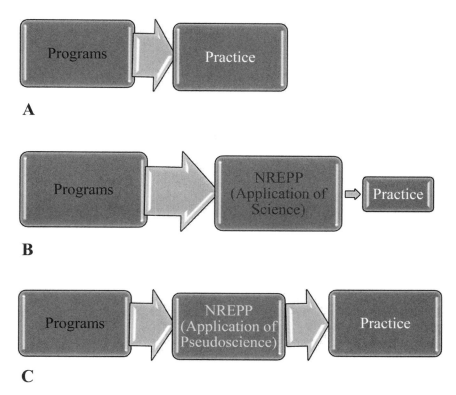

Figure 11.2
Program Flow. A: Pre-NREPP; B: NREPP as envisioned; C: NREPP in practice.

The various attempts made by SAMHSA to revise and improve the NREPP selection criteria and review process suggest that it is aware of the severe limitations of these criteria and the manner in which they are operationalized by reviewers in their assessments of candidate programs. It is likely, however, that NREPP does not have the organizational capacity to improve its review process and conduct a thorough, independent assessment of the many programs that it receives for review. If this is the case, and NREPP cannot ensure the quality of the programs that appear on its webpage, then it would be in the interests of both science and practice for it to desist in its efforts. Giving a stamp of approval to programs that appear effective only through the application of pseudoscientific approaches to knowledge generation serves neither the scientific community nor those in need of effective drug prevention services.

An alternative would be for NREPP to rely on a third party to ensure the quality of the programs it recommends. That is, rather than conducting its own review of material largely submitted by program developers, it could ask developers to only submit materials that have been vetted and approved by some independent party. Study registration with a third party such as ClinicalTrials.gov has been proposed as a means of improving the quality of prevention research (e.g., Olds, 2009). The problem with this approach is that it allows registration of even weak study designs (for example, the ART study is registered with ClinicalTrial.gov), permits vague descriptions of outcome variables and measures, and is limited as a means of deterring analytic flexibility and selected reporting of findings (Fleming, Koletsi, Dwan, and Pamdis, 2015; Gorman, 2015b). A more rigorous mechanism for ensuring quality would involve requiring any program that is submitted for inclusion in NREPP to have been evaluated in a study whose results have been published in the form of a registered report. A registered report moves beyond simply registering a study with a third party in that journals review research proposals prior to data collection and analysis based on methodological soundness and make a commitment to publish study results (provided the analysis plan is adhered to) regardless of whether these are positive or negative or support the null (Nosek and Lakens, 2014). Thus, a registered report is potentially more effective in limiting the use of flexible data analysis practices and distorted reporting than simple registration of a project as it directly removes the incentives to use such practices from the publication process and provides a mechanism for detecting their use prior to publication.

Adopting registered reports as an essential requirement for submission of program evaluation materials to NREPP would likely reduce the current list to nothing, and program evaluators would need some time to meet the new entry criterion. And it may be that few evaluators wish to subject their programs to the rigors of the registered reports mechanism, in which case NREPP would simply wither on the vine. But both of these options—an NREPP slowly repopulated by quality programs that meet real scientific standards of quality, or an NREPP whose quality standards cannot be met by any program evaluation—are better than the current NREPP, whose pseudoscientific approach so poorly serves drug prevention practice.

Notes

1. It should be noted that use of flexible data analysis practices can make the DARE Program appear to be an effective drug prevention intervention (see Gorman and Huber, 2009).

2. Some of the Project TND evaluations do not state whether one-tailed or two-tailed tests of statistical significance were used, although the default in these studies is to use one-tailed tests. Other studies state that one-tailed tests were used but do not report a specific p value, and therefore it is impossible to know if the result would be statistically significant had a two-tailed test been used. In both of these cases, it was assumed that the test would be statistically significant with a two-tailed test and a ✓ appears in the table.

3. The data and analyses presented in Kip et al. (2014) are from the same study with the same subjects and therefore suffer from the same methodological shortcomings.

References

Begley, C. G., and Ioannidis, J. P. A. (2015). Reproducibility in Science: Improving the Standard for Basic and Preclinical Research. *Circulation Research*, *116*: 116–126.

Center for Self-Leadership (2013). IFS, an Evidenced Based Practice. Retrieved from http://selfleadership.org/evidence-based-practice.html Retrieved March 8, 2016.

Center for Substance Abuse Prevention (US). (2001). *Annual Report of Science-Based Prevention Programs*. Rockville, MD: Center for Substance Abuse Prevention, Substance Abuse and Mental Health Services Administration.

Center for Substance Abuse Prevention (US). (2002). *Science-Based Prevention Programs and Principles, 2002*. Rockville, MD: Center for Substance Abuse Prevention, Substance Abuse and Mental Health Services Administration..

Dent, C. W., Sussman, S., and Stacy, A. W. (2001). Project Towards No Drug Abuse: Generalizability to a General High School Sample. *Preventive Medicine*, *32*(6): 514–520.

Department of Health and Human Services (US). (2006). Changes to the National Registry of Evidence-Based Programs and Practices (NREPP). *Federal Register*, *71*(49): 13132–13155.

Department of Health and Human Services (US). (2015). National Registry of Evidence-Based Programs and Practices. *Federal Register*, *80*(127): 38716–38718.

Early, K. W., Blankenship, J. L., and Hand, G. A. (2011a). *Evaluation of AMIkids Alternative School and Juvenile Justice Program Educational Outcomes: An Examination of Pre/post Test Academic Change*. Tallahassee, FL: Justice Research Center.

Early, K. W., Hand, G. A., Blankenship, J. L., and Chapman, S. F. (2011b). Experiential Community-based Interventions for Delinquent Youth: An Evaluation of Recidivism and Cost-Effectiveness. Unpublished manuscript.

Eisner, M. (2009). No Effect in Independent Prevention Trials: Can We Reject the Cynical View? *Journal of Experimental Criminology*, *5*: 163–183.

Fanelli, D. (2010a). "Positive" Results Increase Down the Hierarchy of Science. *PLOS ONE*, *5*(4), e10068. DOI: org/10.1371/journal.pone.0010068

Fanelli, D. (2010b). Do Pressures to Publish Increase Scientists' Bias? An Empirical Support for US States Data. *PLOS ONE*, *5*(4), e10271. DOI: 10.1371/journal.pone.0010271

Ferguson, C. J., and Heene, M. (2012). A Vast Graveyard of Undead Theories: Publication Bias and Psychological Science's Aversion to the Null. *Perspectives on Psychological Science*, *7*: 555–561.

Fleming, P. S., Koletsi, D., Dwan, K., and Pamdis, N. (2015). Outcome Discrepancies and Selective Reporting: Impacting the Leading Journals? *PLOS ONE*, *10*(5), e0127495. DOI: 10.1371/journal.pone.0127495

Gandhi, A. G., Murphy-Graham, E., Petrosino, A., Chrismer, S. S., and Weiss, C. H. (2007). The Devil Is in the Details: Examining the Evidence for "Proven" School-Based Drug Abuse Prevention Program. *Evaluation Review*, *31*: 43–74.

General Accounting Office (US). (1997). *Safe and Drug-Free Schools: Balancing Accountability with State and Local Flexibility* (GAO/HEHS-98-3). Washington, DC: General Accounting Office.

Gerstein, D. R., and Green, L. W. (1993). *Preventing Drug Abuse: What Do We Know?* Washington, DC: National Academy Press.

Gorman, D. M. (1998). The Irrelevance of Evidence in the Development of School-Based Drug Prevention Policy, 1986–1996. *Evaluation Review*, *22*: 118–146.

Gorman, D. M. (2002). Defining and Operationalizing "Research-Based" Prevention: A Critique (with Case Studies) of the US Department of Education's Safe, Disciplined and Drug-Free Schools Exemplary Programs. *Evaluation and Program Planning*, *25*: 295–302.

Gorman, D. M. (2003). Prevention Programs and Scientific Nonsense. *Policy Review*, *117*: 65–75.

Gorman, D. M. (2005). Drug and Violence Prevention: Rediscovering the Critical Rational Dimension of Evaluation Research. *Journal of Experimental Criminology*, *1*: 39–62.

Gorman, D. M. (2010). Commentary: Understanding Prevention Research as a Form of Pseudoscience. *Addiction*, *105*: 582–583.

Gorman, D. M. (2011). Does the Life Skills Training Program Prevent Use of Marijuana? *Addiction Research and Theory*, *19*: 470–481.

Gorman, D. M. (2014). Is Project Towards No Drug Abuse (TND) an Evidence-Based Drug and Violence Prevention Program? A Review and Reappraisal of the Evaluation Studies. *Journal of Primary Prevention*, *35*: 217–232.

Gorman, D. M. (2015a). "Everything Works": The Need to Address Confirmation Bias in Evaluations of Drug Misuse Prevention Interventions for Adolescents. *Addiction*, *110*: 1539–1540.

Gorman, D. M. (2015b). Analytic Flexibility in the Evaluation of the Drug Education in Victoria Schools (DEVS) Programme. *International Journal of Drug Policy*, *26*: 719–720.

Gorman, D. M. (2016). Can We Trust Positive Findings of Intervention Research? The Role of Conflict of Interest. *Prevention Science*, Epub ahead of print: PMID: 27106694

Gorman, D. M., and Conde, E. (2007). Conflict of Interest in the Evaluation and Dissemination of "Model" School-Based Drug and Violence Prevention Programs. *Evaluation and Program Planning*, *30*: 422–429.

Gorman, D. M. and Conde, E. (2009). Further Comments on the Path to Drawing Reasonable Conclusions about Prevention. *Addiction*, *104*: 152–154.

Gorman, D. M., and Conde, E. (2010). The Making of Evidence-Based Practice: The Case of Project ALERT. *Children and Youth Services Review*, 32(2): 214–222.

Gorman, D. M., Conde, E., and Huber, Jr., J. C. (2007). The Creation of Evidence in "Evidenced-Based" Drug Prevention: A Critique of the Strengthening Families Program Plus Life Skills Training Evaluation. *Drug & Alcohol Review*, 26: 585–593.

Gorman, D. M., and Huber, Jr., J. C. (2009). The Social Construction of "Evidence-Based" Drug Prevention Programs: A Reanalysis of Data from the Drug Abuse Resistance Education (DARE) Program. *Evaluation Review*, 33: 396–414.

Herbert, J. D., Lilienfeld, S. O., Lohr, J. M., Montomery, R. W., O'Donohue, W. T., Rosen, G. M., and Tolin, D. F. (2000). Science and Pseudoscience in the Development of Eye Movement Desensitization and Reprocessing: Implications for Clinical Psychology. *Clinical Psychology Review*, 20: 945–971.

Holder, H. D. (2010). Prevention Programs in the 21st Century: What We Do Not Discuss in Public. *Addiction*, 105: 578–581.

Ioannidis, J. P. A., Greenland, S., Hlatky, M. A., Khoury, M. J., Macleod, M. R., Moher, D., Schulz, K. F., and Tibshirani, R. (2014a). Increasing Value and Reducing Waste in Research Design, Conduct, and Analysis. *Lancet*, 383: 166–175.

Ioannidis, J. P. A., Munafo, M. R., Fusar-Poli, P., Nosek, B. A., and David, S. P. (2014b). Publication and Other Reporting Biases in Cognitive Sciences: Detection, Prevalence, and Prevention. *Trends in Cognitive Sciences*, 18: 235–241.

Kip, K. E., Rosenzweig, L., Hernandez, D. F., Shuman, A., Diamond, D. M., Girling, S. A., . . . and McMillan, S. C. (2014). Accelerated Resolution Therapy for Treatment of Pain Secondary to Symptoms of Combat-Related Posttraumatic Stress Disorder. *European Journal of Psychotraumatology*, 5. DOI: 10.3402/ejpt.v5.24066

Kip, K. E., Rosenzweig, L., Hernandez, D. F., Shuman, A., Sullivan, K. L., Long, C. J., . . . and Diamond, D. M. (2013). Randomized Controlled Trial of Accelerated Resolution Therapy (ART) for Symptoms of Combat-Related Post-Traumatic Stress Disorder (PTSD). *Military Medicine*, 178: 1298–1309.

Littell, J., Popa, M., and Forsythe, B. (2005). Multisystemic Therapy for Social, Emotional, and Behavioral Problems in Youth Aged 10–17. *Campbell Systematic Reviews*. DOI: 10.4073/csr.2005.1

MacCoun, R. (2005). "Conflicts of Interest in Public Policy." In D. A. Moore, D. M. Cain, G. Loewenstein, and M. Bazerman (Eds.), *Conflicts of Interest: Challenges and Solutions in Business, Law, Medicine, and Public Policy*, pp. 233–262. London: Cambridge University Press.

Mihalic, S., Irwin, K., Elliott, D., Fagan, A., and Hansen, D. (2001). Blueprints for Violence Prevention. *OJJDP Bulletin*. Washington, DC: US Department of Justice, Office of Justice Programs, Office of Juvenile Justice and Delinquency Prevention.

National Institute on Drug Abuse. (2002). *Preventing Drug Use among Children and Adolescents: A Research-Based Guide*. Second Edition. Rockville, MD: National Institute on Drug Abuse.

Nosek, B. A., and Lakens, D. (2014). Registered Reports: A Method to Increase the Credibility of Published Results. *Social Psychology, 45*: 137–141.

Nosek, B. A., Spies, J. R., and Motyl, M. (2012). Scientific Utopia II: Restructuring Incentives and Practices to Promote Truth over Publishability. *Psychological Science, 7*: 615–631.

Olds, D. L. (2009). In Support of Disciplined Passion. *Journal of Experimental Criminology, 5*: 201–214.

O'Neill, S. (2008). An Evaluation of ACCA's An Apple A Day Substance Abuse Prevention Program: Preliminary Report. Unpublished manuscript.

Open Science Collaboration. (2015). Estimating the Reproducibility of Psychological Science. *Science, 349*. DOI: 10.1126/science.aac4716

Parker, A. E., and Kupersmidt, J. B. (n.d.). *The Moment Program.* Durham, NC: Innovation Research and Training.

Petrosino, A. (2003). Standards for Evidence and Evidence for Standards: The Case of School-Based Drug Prevention. *Annals of the American Academy of Political and Social Science, 587*: 180–207.

Rosenzweig Center for Rapid Recovery. (2016). *About Laney.* Retrieved from http://acceleratedresolutiontherapy.com/web/about-laney/

Safe, Disciplined, and Drug-Free Schools Expert Panel (US Department of Education). (2001). *Exemplary and Promising Safe, Disciplined, and Drug-Free Schools Programs 2001.* Jessup, MD: US Department of Education.

Shadick, N. A., Sowell, N. F., Frits, M. L., Hoffman, S. M., Hartz, S.A., Booth, F. D., . . . and Schwartz, R. C. (2013). A Randomized Controlled Trial of an Internal Family Systems-Based Psychotherapeutic Intervention on Outcomes in Rheumatoid Arthritis: A Proof-of-Concept Study. *Journal of Rheumatology*, 40: 11; DOI: 10.3899/jrheum.121465

Sterman, J. D. (2006). Learning from Evidence in a Complex World. *American Journal of Public Health, 96*: 505–514.

Substance Abuse and Mental Health Services Administration (US). (2009). NREPP: Find Results. Retrieved from http://www.nrepp.samhsa.gov/listofprograms.asp?textsearch=Search+specific+word+or+phrase&ShowHide=2&Sort=1

Substance Abuse and Mental Health Services Administration (US). (2016a). NREPP: All Programs. Retrieved from http://nrepp.samhsa.gov/AllPrograms.aspx

Substance Abuse and Mental Health Services Administration (US). (2016b). Moment Program: Program Description. Retrieved from http://nrepp.samhsa.gov/ProgramProfile.aspx?id=62#hide3

Substance Abuse and Mental Health Services Administration (US). (2016c). An Apple a Day. Retrieved from http://legacy.nreppadmin.net/ViewIntervention.aspx?id=165#std522

Substance Abuse and Mental Health Services Administration (US). (2016d). AMIkids Personal Growth Model. Retrieved from http://legacy.nreppadmin.net/ViewIntervention.aspx?id=252

Substance Abuse and Mental Health Services Administration (US). (2016e). Accelerated Resolution Program: Program Description. Retrieved from http://nrepp.samhsa.gov/ProgramProfile.aspx?id=7#hide3

Sun, W., Skara, S., Sun, P., Dent, C. W., and Sussman, S. (2006). Project Towards No Drug Abuse: Long-Term Substance Use Outcome Evaluation. *Preventive Medicine, 42*: 188–192.

Sussman, S., Dent, C. W., and Stacy, A. W. (2002). Project Towards No Drug Abuse: A Review of the Findings and Future Directions. *American Journal of Health Behavior, 26*: 354–365.

Sussman, S., Dent, C. W., Stacy, A. W., and Craig, S. (1998). One-Year Outcomes of Project Towards No Drug Abuse. *Preventive Medicine, 27*: 632–642.

Sussman, S., Sun, P., McCuller, W. J., and Dent, C. W. (2003). Project Towards No Drug Abuse: Two-Year Outcomes of a Trial That Compares Health Educator Delivery to Self-Instruction. *Preventive Medicine, 37*: 155–162.

12 Scientific Soundness and the Problem of Predatory Journals

Jeffrey Beall

Introduction

The gentleman's agreement, or social contract, of science publishing has been breached. A horde of scholarly publishers no longer feels bound to publish valid research and only valid research. Genuine researchers see their work devalued, and a crucial gatekeeper has been removed that protected students who lack the credentials to differentiate between real science and junk science.

This chapter examines the role that predatory journals and publishers have played in enabling the increased publication and distribution of pseudoscience. It also looks at some of the consequences of this increased publication of junk science, including the mingling of junk and authentic science in scholarly data bases. I believe that more junk science is being published than ever before, published in journals that appear at first to be authentic scientific journals, but upon close analysis are really only fake journals, publications bearing only the window dressing of authentic scholarly journals.

Setting the Scene

Social and technological changes dating back to the 1990s made the decline in scholarly publishing possible. Prior to the advent of the Internet, most scholarly journals were print-based and available only through subscriptions. Libraries subscribed to the print journals and made them available to their users, either directly or through interlibrary loan. Moreover, many researchers belonged to scholarly societies and received copies of one or more of their societies' journals though the mail.

At that time, it was more difficult to publish pseudoscience in scholarly journals. Subscription journals did not want to risk losing subscriptions from libraries and aimed to accept and publish only the highest-quality, most novel, and most important scientific manuscripts from among those submitted. The subscription publishing model has always incentivized quality, for higher-quality journals gained more subscribers, and lower-quality journals risked losing them. Elsevier's *Homoeopathy*, to take one example, shows that a reputable publisher can devote a subscription journal to pseudoscience. But such journals remain the exception in the subscription world.

In the late 1990s, online journals began to appear, and traditional publishers began making their journals available over the Internet, sometimes as a bonus included in the print subscription cost, sometimes as a feature available at an additional cost. In the early 2000s, most scholarly publishers went through the process of "flipping their model," making the online subscription the basic one, with print copies available optionally for an additional cost. This is when online journals became legitimate scientific journals.

At the same time, a vocal, demanding, and militant social movement gained momentum: the open-access movement. The more militant members of the movement declared that any organization putting scholarly journals behind a paywall, were enemies to subscription publishers and scholarly societies alike.

Among the publishing models the movement promoted was the gold open-access model, also called the author-pays model. The model shifts the costs of publishing from the subscribers to the authors, with the advantage that the papers are made open-access. Though this model started in the early 2000s, it didn't gain traction until about 2008 (Beall, 2013). Author payments were new for most. Some nonprofit scholarly societies had in the past implemented "page charges" to authors to help subsidize the publishing costs, but most commercial publishers had abandoned these charges long ago. Author fees were in fact unheard of in many fields, such as library science, and in the social sciences and humanities.

But because researchers in the biomedical sciences have often won grant funding to support their research, the idea of author fees didn't come as a shock to them as it would in other fields, especially fields in which research grants were rare, such as the arts and humanities. The gold open-access model slowly gained acceptance as a model for financing the publishing of scholarly journals, albeit those without grant funding were reluctant to pay.

This gradual acceptance of the gold open-access model led to the creation of predatory publishers, those that exploit the model for their own profit. Slowly realizing the potential income that easy acceptance of scholarly articles could bring, several publishers appeared in South Asia, including several in Faisalabad, Pakistan, in the middle of this century's first decade.

Under the subscription model, journals often issued a "Call for papers" as a traditional method of soliciting manuscripts. Prior to the Internet, these calls were printed, generally on a single sheet of paper, and mailed to individuals and to university departments. Often, they were mentioned at faculty meetings and posted on bulletin boards in departmental offices. They also appeared on a page at the back of journal issues.

Adapting this practice, online journals began issuing calls for papers, but they did it through email—specifically, spam emails sent directly to authors. Many scholarly authors in the West began to receive such emails around 2008. The spam emails were poorly composed and contained prominent grammatical errors. Many boldly used the ungrammatical phrase "Call for paper."

One of the main things that distinguishes print scholarly publishers from online scholarly publishing is the high barrier to starting up a publishing operation. Under the print model, launching a new journal—or indeed even a new publishing house—required a large investment. The costs, such as for printing, mailing, marketing, and managing subscriptions, were high.

Online journals—especially open-access journals—have relatively low start-up costs, especially for journals that don't adhere to industry practices, such as digital preservation and copyediting. Newcomers may lack experience in publishing, but they can easily copy the look and feel of the legitimate websites, including creating names for publishing operations that mimic those of legitimate publishers. Examples include Academic and Scientific Publishing and the Global Institute for Research & Education. Journal titles, too, can be mimicked and copied, and even the format of a publisher's PDF can be copied and applied to a new "publisher's" journal. Moreover, dozens of journals can easily be launched using templates.

Thus in 2008, we saw the birth of what would later come to be called predatory publishers and journals. The first few appeared in Pakistan, but opportunists around the world quickly realized they could easily copy the model and make almost instant money with a minimal investment. The number of predatory publishers began to proliferate rapidly as the recipe for the predatory open-access publishing model became known and popularized.

I coined the term "predatory publisher" in 2010 and realize that at this time it may not be the perfect term to describe the concept, but it is now a widely used and conventional term (Shen and Björk, 2015). I also use the term "questionable publisher." Either term means journals and publishers that exploit the gold open-access model for their own profit.

Along with scamming money from young academics in developing countries, one of the most negative aspects of predatory journals is their failure to carry out a standard, science-based peer review. Because predatory publishers and journals want to generate as much revenue as possible, the practice of rejecting papers for publication is contrary to their interests. It means money walking away. Therefore, they often perform a fake peer review, they fail to perform any peer review but claim to, or they go through the motions of peer review, but ignore "reject" recommendations from peer reviewers. In short, they corrupt the system so that all or most papers are accepted for publication and the authors are invoiced for the article processing fee.

The more articles that predatory publishers accept and publish, the more money they make. This practice has led to the increasing use of the term "pay-to-publish journals" to describe them.

Peer review is traditionally carried out privately, with those involved generally unaware of the identities of the reviewers, so it is the hardest component of scholarly publishing to analyze and evaluate. Thus when evaluating scholarly publishers, one has to examine the product of their peer-review processes: the published papers. A run of unscientific or low-quality papers probably indicates that peer review is not being done according to established standards and practices. Additionally, some authors of papers submitted have shared the peer-review reports they received from predatory publishers, and the reports appear as mere templates, using general statements and without any specific references to the paper in question (Beall, 2014). Finally, journalists and others have carried out sting operations on questionable publishers, testing whether they would accept for publication papers that report obviously unsound science.

The initial shock to the scientific community of the egregious practices and enormity of the predatory journals seems to have worn off. These journals now routinely publish junk science, with perhaps insufficient concern being expressed by the scientific community. Quackery in scholarly communication is expanding and proliferating, enabled by the predatory journals and their fake peer review processes.

Predatory Journals are Enabling the Publishing and Distribution of Much Pseudoscience

Consider the publisher called Science Publishing Group. Its URL is http://www.sciencepublishinggroup.com/home/index.aspx. I blogged about this publisher when I first became aware of it in December 2012 (Beall, 2012), soon after it launched with fifty-two new open-access journals. After three years the publisher had a fleet of 203 such journals.

The publisher accepts pretty much anything that is submitted; it is like a vanity press, but worse (Millard, 2013). It's worse because it publishes pseudoscience without concern for the damage and confusion this may cause.

Science Publishing Group gives this headquarters address on its website:

548 Fashion Avenue
New York, NY 10018
U.S.A

The address is bogus. It may be used gratuitously, or it may be the address of a virtual office company. It certainly is not the real address of this publisher, which probably operates out of Pakistan.

Here are three articles published by this publisher that verify its status as a predatory publisher that publishes pseudoscience without hesitation.

1. Modification of Einstein's $E=MC^2$ to $E=1/22\ MC^2$ (Marek-Crnjac, 2013).

This article, published in Science Publishing Group's *American Journal of Modern Physics*, claims that Einstein's famous mass–energy equivalence equation is inaccurate. The article provides a corrected equation, originally posited by another author, Mohamed S. El

Naschie. The corrected equation is given in the article's title. The article is indexed in Google Scholar, which indicates it has been cited eleven times, but most of these citations are by the author, Leila Marek-Crnjac, or by Mohamed S. El Naschie. No mainstream scientists have cited the article, and it is complete quackery. Einstein's original equation stands, and this purported correction is bogus.

2. The Case against Educated Mathematical Dogma (Precise Mathematical Pi Value by Finite Equation = 3.14159292035) 3+(1/(7+1/16)) Precise Finite Value (Cameron, 2013).

In this article, published in Science Publishing Group's *Pure and Applied Mathematics Journal*, Wisconsin-based author Vinoo Cameron explains that pi, contrary to what we've all been taught, is a rational number with a precise value. This is a clear example of pseudoscience, one that any competent high-school student could identify. No honest scholarly journal would ever consider publishing such an article; most would reject it without even sending it out for peer review. Articles such as this are evidence that some predatory journals do not carry out peer review. Instead, they accept all articles submitted, dress them up as scholarly articles in PDF format, and publish them in exchange for the author fee.

3. Novel Strategy of a Method to Cure from the Cancer [sic] (Rojeab, 2015).

This article appears in Science Publishing Group's *Cancer Research Journal*. The article lists the author's affiliation as "Electrical and Electronic Engineering Department, The London College UCK." We have a barely literate article written by an electrical engineering faculty member, an article that claims to have discovered a cancer cure.

Motivations for Authoring Pseudoscience and Submitting It to Scholarly Publications

What motivates authors to submit bogus science to scholarly journals? These reasons and factors bear on epistemology, namely questions such as, "What is the truth?" and "What does it mean to know something?" Do authors of pseudoscience know that what they're authoring and submitting is bogus?

In the long run, it matters less what the authors know or what their motivations are; peer review is supposed to police the submission of journal articles, book chapters, and monographs by preventing unscientific works from being published in the first place. But because predatory journals have largely abandoned peer review, the system is broken; pseudoscience is flooding and saturating the scholarly record. Therefore, leaving aside the question of whether authors of junk science know it is junk, let's examine instead the apparent reasons why pseudoscience is published at all.

- To promote monetary gain.

An as-yet unapproved drug, Gc protein-derived macrophage activating factor (GcMAF), is marketed as an anticancer drug in Europe, and a group of scientists has been successfully submitting and publishing articles demonstrating the drug's efficacy in numerous predatory journals, exploiting the easy acceptance the journals offer and the scientific facade they provide. Several of these articles (Smith et al., 2013; Thyer et al., 2013; Ward et al., 2014) have appeared in the *American Journal of Immunology*, a journal published by a firm called Science Publications (which is not the same as Science Publishing Group, mentioned earlier).

Note that to someone unfamiliar with predatory publishing, such as a prospective buyer of an anticancer drug, a publisher named Science Publications and a journal entitled *American Journal of Immunology* will likely sound legitimate. However, I think both the publisher and the journal are highly questionable. If you look at the "Contact" page on the Science Publications website, there is no location listed. I believe that the publisher is really based somewhere in Asia, so the use of the geographical term "American" in its journal titles is misleading and gratuitous. Science Publications is really just a pay-to-publish publisher, and authors essentially use it as a scientific vanity press.

- To support a political, religious, or social agenda.

The fake peer review and easy article acceptance that predatory publishers offer benefit those with political agendas who want to make their views seem grounded in science. If you disagree with global warming, you can publish articles showing that the earth is actually cooling or that sea levels are falling.

Antinuclear activists have used the easy publishing offered by the China-based publisher Scientific Research Publishing (SCIRP), specifically its *Open Journal of Pediatrics*, to report the fallout from the Fukushima Daiichi nuclear disaster as being worse than it really was (Beall, 2014b), an attempt to tarnish all nuclear power.

- To document and prove conspiracy theories.

I authored a blog post (Beall, 2015) about an article published in MDPI's *International Journal of Environmental Research and Public Health*. The article, entitled "Evidence of Coal-Fly-Ash Toxic Chemical Geoengineering in the Troposphere: Consequences for Public Health," was a manifestation of the chemtrail conspiracy theory. The author, based in San Diego, shot pictures of contrails above his house and included the pictures in the article. He claimed the contrails were part of a government operation to spread coal fly ash in the atmosphere to combat global warming, and said that the ash, when it reached the ground, was toxic to humans.

Soon after the article's publication, several conspiracy theory blogs ran stories on the article's publication and used the article to show that the conspiracy theory was "proven" in a "scientific" journal. Soon after my blog post was published, however, MDPI quickly retracted the article. One wonders whether MDPI, the publisher, was testing the waters to see how much pseudoscience it could get away with.

- To promote the author as the one who first answered open scientific questions.

Ramzi Suleiman is an associate professor of psychology at the University of Haifa in Israel. Venturing into cosmology in ambitious fashion, he wrote the article "The Dark Side Revealed: A Complete Relativity Theory Predicts the Content of the Universe," a piece published in a questionable open-access journal called *Progress in Physics* (Suleiman, 2013). The journal is largely ignored by mainstream cosmologists and physicists, except to warn colleagues about the junk science it publishes.

Suleiman's article purports to "correct" Einstein and answer most all of the outstanding questions in cosmology, including the nature of dark energy and dark matter. Ignored by legitimate researchers, the article has deservedly had no impact on current cosmological theory. Even so, the paper is indexed in Google Scholar.

Cosmology is a favorite field of those seeking fame and glory. According to Gardner (1953, p. 9), "Often the quickest road to fame is to overturn a firmly-held belief." Many have tried to disprove Einstein and other cosmologists, hoping to hack out a shortcut to fame. One of the reasons cosmology may be a popular target among those seeking fame is that it's easy to write statements about cosmology that are not easily disproved. Pseudoscientists exploit some of the field's nebulous aspects to posit their wild theses, ideas that are time-consuming to disprove and unworthy of attention anyway.

- To conduct sting operations on publishers.

Journalists and others have hoodwinked dubious publishers, submitting scientifically flawed manuscripts and documenting their acceptance. The most famous of these sting operations was the one conducted by *Science* reporter John Bohannon (2013). He submitted a made-up paper containing obvious unscientific methodologies and conclusions to randomly selected open-access journals and recorded how many accepted the bogus manuscript for publication. A significant proportion did accept the manuscript (upon acceptance, he asked that the fake paper be withdrawn).

The Imprimatur of Science

In her book *The AIDS Conspiracy: Science Fights Back*, South African author and researcher Nicoli Nattrass (2012) uses the term "the imprimatur of science" to describe science's seal of approval, which is, of course, peer review. Defending peer review, she says, "For all its faults, peer review remains an essential mechanism for the allocation of trust in the results of others" (p. 139).

Incorporating ideas posited by Gieryn (1983), Nattrass discusses demarcation and boundary work in the context of AIDS denialism and its appearance, for a time, as validated science in peer-reviewed journals. The role of demarcation is to identify that which is valid science and that which is not. It requires making decisions and acting on the decisions, especially in the context of peer review, as to what qualifies as science. Boundary work is that work carried out by the scientific community to defend demarcation. The article called "Science and Pseudo-Science" in the *Stanford Encyclopedia of Philosophy* sums up the reality well:

> Since science is our most reliable source of knowledge, in a wide variety of areas, we need to distinguish scientific knowledge from its look-alikes (Hansson, 2015; p. 2).

Unfortunately, predatory journals are increasingly skilled at "looking like" authentic peer-reviewed and respected journals.

I think these two practices—demarcation and boundary work—have become front and center in the context of the problems created by predatory publishers. To fool authors, their prey, they pretend to go through the motions of an authentic peer review. One component of peer review is the report the author receives from the journal. By its nature, peer review is blind, and much of the process is managed in secret, with the identity of the reviewers, the authors, or both, hidden from the others. A few authors, however, have shared peer-review reports with me, reports that appear to be boilerplate documents, with each author receiving the same report for different manuscripts. They are written in such a way that they do not comment on any specific aspects of any submission but merely contain general comments that could apply to any accepted paper. Figure 12.1 shows one of two components of a peer-review report that was shared with me recently.

The other component of this peer-review report was a textual report that followed the graded evaluation results shown in figure 12.1. This text is copied here:

Suggestions for the Author:
The manuscript possesses interesting accounts on the topic. Overall, it is a noteworthy effort and the reviewers provided following comments about the manuscript.

EVALUATION RESULTS	(Grade: 5, 4, 3, 2, 1 – Highest to Lowest)
Organization and Presentation:	4
Soundness of the Methodology:	3
Evidence Supports Conclusion:	3
Adequacy of Literature Review:	4
Contribution to Existing Knowledge:	3
Overall Evaluation on the Paper:	**3.5**

Figure 12.1
One of two evaluation sections of a peer review report from a questionable publisher. (Quoted from an anonymous peer review shared with the author.)

Abstract covers all the important aspects of the subject and is well written in general summarizing information on methodology also. Introduction also clarifies essence of the manuscript ahead. Adequate amount of Literature is consulted and it is appreciative that the consulted literature on the topic is comparatively recent. Rest of the manuscript is also very well structured representing the dedication and knowledge of the researcher about the topic and skill on research.

Sufficient discussion have been used and conclusions are adequately portrayed. Technical quality of the manuscript is reasonable and reviewers find no major modification or any other reason to decline the manuscript from publication.

Therefore, manuscript shall be accepted for publication as it is.

Decision: ACCEPTED

It's easy to conclude that this peer report is completely bogus. The text contains grammatical and idiomatic errors, it is overly general, and it displays a positive and even triumphal tone throughout. The text panders to authors, especially those who have suffered rejections in the past, or those needing a publication to bolster their tenure dossier or to augment their annual inventory of accomplishments, such as those lists evaluated by supervisors and evaluation committees.

Unfortunately, peer-review reports such as this one are increasingly common, and they constitute the imprimatur of science for individuals, academic departments, universities, and even governments. English is the language of science, but in countries and regions where English is not the native language, the grammatical errors in the peer-review report may fail to ring any warning bells. For an emerging researcher who receives a peer-review report such as this one as his or her first-ever peer review, there's nothing to compare it to. They may incorrectly assume that all peer-review reports look like this one.

Predatory publishers and journals counterfeit the imprimatur of science in other ways. They have articles in PDF, they have editorial boards, and they have volumes and issues. They have impact factors assigned by fake impact-factor companies. And they mimic the titles of legitimate journals. One example is *The Journal of Depression & Anxiety*, a name suspiciously close to that of the established journal *Depression and Anxiety* (Beall, 2015b). The copycat journal sent out numerous spam emails, perhaps hoping to confuse recipients into believing that the messages originated from the legitimate journal and publisher.

Payments from Authors and the Breakdown of Demarcation

The system of payments from authors is corrupting scholarly publishing and damaging the integrity of published science. Open-access publishers understand that, to appear credible and to continue to attract submissions from authors, they have to have at least a facade of peer review, going through the motions so that authors can get academic credit for publications that appear in their journals.

In many cases they accept and publish papers despite receiving "reject" recommendations from peer reviewers. They publish flawed science in return for payments from authors and against professional advice. The scientific record becomes polluted with fringe science, and researchers' time is wasted because the work they devote to peer reviewing is essentially disregarded.

One publisher that has been criticized for abusing researchers is Switzerland-based Frontiers. Frontiers has used an in-house journal management software that does not give reviewers the option to recommend the rejection of manuscripts they have reviewed. The publisher's systems are set up to make it almost impossible to reject papers, perhaps to keep potential revenue from jumping to a rival publisher. Increasingly, journal management software is designed to optimize a publisher's revenue. One blogger described Frontiers' system this way:

> What I learned is that even the associate editors often find their power limited: once a manuscript has been sent out for peer review, Frontiers editors have hardly any option to reject it. This may explain how controversial papers came to be published in Frontiers, e.g. one denying that HIV is the cause of AIDS, or another suggesting that vaccinations cause autism (Schneider, 2015)

Gold open access has engendered a shift for scholarly journals and monographs. A substantial part of the industry, the open-access part, is driven by authors and not by readers (in the form of library decision makers). Writers are now the customers, and the industry aims to please them. What do they want? They want the quick, easy, and cheap publishing of their works. Unfortunately, there is a subset of authors whose science is junk science, and because the competition is so intense, many of these authors are easily finding publishers that are eager to earn their money and publish their works along with the works of legitimate scientists.

Accordingly, demarcation is severely wounded; if you have a pseudoscientific idea, a manuscript, and a few hundred dollars, you can easily see the publication of your work. In too many cases, the scholarly publishing industry is no longer providing the service of demarcation. The consumers (readers) of a scholarly publisher's products are the ones being ripped off, for the content they read is full of corrupt science. Their voice as consumers is lost because there are no journal subscriptions to cancel; open access serves the authors.

Pseudoscience at the Journal Level

Earlier, I described and listed some pseudoscientific articles published in low-quality or predatory journals. Unfortunately, I have observed what appears to be an increase in the number of so-called scientific journals, each of which is entirely devoted to a single pseudoscience. They are published both by predatory and supposedly legitimate publishers. Most of these focus on one of the various junk sciences that comprise what is euphemistically referred to as complementary and alternative medicine (CAM).

CAM includes "Everything from homeopathy to acupuncture, from ear candles to urine therapy, from herbal medicine to aromatherapy" (Hall, 2015). Capitalizing on both the popularity of CAM and the potential to earn more author fees, numerous scholarly publishers have launched journals devoted to individual CAM fields. Ayurveda, a traditional Indian medicine, is the subject of more than a dozen, with some of these "scholarly" journals devoted to Ayurveda alone (for example, *The International Ayurvedic Medical Journal*), others to Ayurveda and some other pseudoscience (for example, the ambitiously titled *Journal of AYUSH: Ayurveda, Yoga, Unani, Siddha and Homeopathy*). Most current Ayurveda research can be classified as "tooth fairy science," research that accepts as its premise something not scientifically known to exist (Carroll, 2015). Ayurveda is a long-standing system of beliefs and traditions, but its claimed effects have not been scientifically proven. Most Ayurveda researchers might as well be studying the tooth fairy.

The German publisher Wolters Kluwer bought the Indian open-access publisher Medknow in 2011 (Wolters Kluwer India Pvt. Ltd., 2011). It acquired its entire fleet of journals, including those devoted to pseudoscience topics such as *An International Quarterly Journal of Research in Ayurveda*. This journal is indexed in PubMed, a data base published by the US National Center for Biotechnology Information, a center that is part of the National Institutes of Health. What better credentials could a junk science journal wish to earn? The inclusion of the journal—along with other CAM journals included in PubMed—grants it instant legitimacy, essentially a seal of approval from the US government. It's no wonder that CAM is seeing so much resurgence, expansion, and interest in the West.

Information Resources Are Polluted by Research Published in Predatory Journals

One of the advantages of scholarly open-access journals is that their content—the published articles—is freely available to anyone with Internet access. Moreover, there are free academic search engines that index the content of both subscription and open-access journals. The most popular of these is Google Scholar. It aims to be comprehensive and index as much published scholarly content as possible, without regard to the indexed content's quality or adherence to scientific principles. That is to say, in addition to the quality content it indexes, Google Scholar indexes much pseudoscience. It is perhaps the world's largest index of junk science. It is also the most popular and most accessible academic search engine.

Of course, there are proprietary scholarly indexes that compete with Google Scholar. These indexes are licensed by academic libraries and are produced by companies such as Gale, EBSCO, and Thomson Reuters. Another major academic index is Elsevier's Scopus. Proprietary databases such as these have stricter inclusion criteria than Google Scholar. They don't want to waste resources indexing the junk science emanating from predatory publishers, for example, and they don't want their products to be stigmatized by metadata from predatory journals.

Still, these data bases compete with each other. Libraries cannot afford to pay for every scholarly index, so they make choices from among those available. The producers of the indexes, on the other hand, want to license their products to as many libraries as possible. One of the ways the index producers market their products to libraries is by bragging about the number of journals they cover. Here, quantity is a selling point. This pressure for a data base provider to index more journals than its competitors incentivizes even the legitimate indexing agencies to include borderline and even predatory publishers in their journal portfolios.

These borderline and predatory journals then use their inclusion in the indexes as a mark of quality, bragging about it on their websites. In many countries, the earning of an advanced degree is accomplished only after a candidate has a requisite number of publications in journals covered by certain academic indexes. It's the same for academic advancement, such as earning tenure or being promoted from associate to full professor.

All this means that it is easier than ever before for pseudoscience to get indexed in legitimate scholarly indexes. The barrier has been breached, pseudoscience and science that has not been through an honest peer review is mingled with mainstream, vetted science.

The mass media is also taken in by predatory journals. It is the role of science journalists to translate novel and important scientific ideas and discoveries to a popular audience, but the number of competent practicing science journalists seems to be declining (Lucibella, 2009). Many journalists with little or no background in science are now reporting on it. Like the general public, many are unable to differentiate valid science from pseudoscience. The result is that junk science is sometimes reported as authentic science in the mass media.

This idea was tested by *Science* reporter John Bohannon. In his "Misleading chocolate study" (Wikipedia, 2015), Bohannon, along with several collaborators, wrote up and submitted for publication a bogus study that "found" that eating chocolate aids in weight loss. The intent was to get the intentionally bogus study quickly published in a pay-to-publish journal, pitch the story to the mass media, and see whether reporters would take the bait and report on the fake findings. The plan worked. The study (Bohannon, et al. 2015) was quickly accepted and published in the journal *International Archives of Medicine*, whose publisher—iMedPub Journals—has been identified as a potential predatory publisher.

Many media outlets reported on the fake study and its false conclusion. According to Bohannon,

We landed big fish before we even knew they were biting. *Bild* rushed their story out—"Those who eat chocolate stay slim!"—without contacting me at all. Soon we were in the *Daily Star*, the

Irish Examiner, *Cosmopolitan*'s German website, the *Times of India*, both the German and Indian site of the *Huffington Post*, and even television news in Texas and an Australian morning talk show. (Bohannon, 2015).

The sting's success signals a significant failure of science journalism. Anyone can pay to have an article published in a predatory journal and then promote the article as science to naïve reporters and editors, reporters and editors who help promote and publicize the bogus science.

Some of those promoting questionable science have used a relatively new process in scholarly publishing—post-publication peer review (PPPR)—to package fringe science as authentic science. In PPPR, no peer review is done at first, and articles are quickly vetted to eliminate the obviously nonscientific ones. Then they are published on the journal's website and researchers are invited to contribute peer reviews following the articles' publication.

I wrote about one case that dealt with Lyme disease (Beall, 2015c). There is an organization that posits that the disease may be sexually transmitted, despite statements from the Centers for Disease Control and Prevention that there's no scientific evidence for this. The organization submitted a research paper (Middelveen et al. 2015) to the PPPR journal *F1000Research*, and the paper was quickly published on the journal's website.

Immediately after the article appeared on the journal's website, the organization issued a press release entitled "Expanded Study Confirms that Lyme Disease May Be Sexually Transmitted." The press release led readers to believe that the article was a traditional, vetted scientific article and didn't mention that it hadn't gone through peer review. One may conclude that the paper's authors exploited the PPPR model just to be able to grant the imprimatur of science to their paper. They found a way to get their questionable paper published in a scientific journal and then used that publication in press releases to publicize their unvetted findings.

Many with unscientific ideas want to disseminate them as widely as possible, and if the ideas can be packaged as science, all the better. Prior to the Internet, this dissemination was more difficult, as peer review often prevented the ideas from being published in scientific journals. However, the advent of predatory journals and journals using the post-publication peer-review model has changed this process. Any scientific thesis, no matter how outlandish, can be published in a journal that looks like a scientific journal and be promoted as vetted science.

The Rise of Boundary Work

Fortunately, some scientists are increasingly rising to engage in boundary work and demarcate science from the growing body of published pseudoscience, using new tools. Gieryn (1983) characterizes boundary work in this way:

The focus is on boundary-work of scientists: their attribution of selected characteristics to the institution of science (i.e., to its practitioners, methods, stock of knowledge, values and work organization) for purposes of constructing a social boundary that distinguishes some intellectual activities as "non-science."

One of the tools that scientists are using to enforce demarcation is blogging. There are numerous science blogs that comment on recently published articles, especially articles that make questionable claims. For instance, one article that generated much discussion and boundary work was published in the MDPI journal *Entropy* (Seneff, Davidson, and Liu, 2012). Entitled, "Empirical Data Confirm Autism Symptoms Related to Aluminum and Acetaminophen Exposure," the article is one of many coauthored by MIT professor Stephanie Seneff that appear in journals identified as questionable or predatory.

The science blog called Respectful Insolence has done much boundary work involving the research output of Seneff and her coauthors, output that also describes the effects of the herbicide glyphosate. The blog's author is given as Orac, a pseudonym for David Gorski. In a blog post entitled "Oh, No! GMOs Are Going to Make Everyone Autistic!" (Gorski, 2014), Gorski analyzes the work of Seneff and her coauthors, demonstrating that their work is unscientific. The website PubPeer provides a means for researchers to leave comments on published science articles, comments that relate to methodological flaws or to fabricated or falsified data and images. The website deals mostly with scientific misconduct rather than pseudoscience, but those promoting unscientific agendas may engage in misconduct to make their research appear scientific, so PubPeer has much potential in combating future junk science. The site allows anonymous commenting, but it is really set up to allow commenting only on articles that have valid digital object identifiers (DOIs). Most predatory publishers do not assign DOIs to their published articles, so this makes it more difficult to cite them or to refer to them in linking websites such as PubPeer.

Looking Ahead: Scholarly Publishing and the Dissemination of Pseudoscience

The future of scientific soundness in the context of scholarly publishing is grim. Advocates for scholarly open-access publishing promote the pay-to-publish model (gold open access) while averting their gaze from the increasing amount of junk science that is being published in open-access journals that are dressed up as scientific journals.

The system of payments from authors is degrading published science. The consumers of scholarly publishing are increasingly the authors of research works, many of whom have unscientific agendas. Readers are no longer the consumers of scholarly publishing services, and given that they increasingly no longer finance scholarly publishing—either directly or through academic libraries—their voices are silenced. They no longer play a role in determining which journals succeed or fail. Scholarly publishers now increasingly work for authors or for anyone with money to pay the article processing charge. The notion of selectivity in scholarly publishing is disappearing.

Governments grant their approval to pseudoscience by indexing it in the databases they sponsor and create and by funding research into it. The open-access movement has done more to enable the publishing of pseudoscience than any other social movement in history.

If you are a reader or other consumer of scholarly literature, beware. Scholarly data bases are full of junk science. If you are an author, demand a full and standard peer review, and refuse to patronize journals that devalue peer review and act as mere vanity presses.

References

Beall, J. (2012). Medical Publishing Triage: Chronicling Predatory Open Access Publishers. *Annals of Medicine and Surgery 2*(2), 47–49. Retrieved from http://dx.doi.org/10.1016/S2049-0801(13)70035-9

Beall, J. (2014). Peer Review Reports from Questionable Publishers: Three Examples. Retrieved from http://scholarlyoa.com/2014/07/17/peer-review-reports-from-questionable-publishers-three-examples/

Beall, J. (2014b). Fallout from Questionable Article in a Pediatrics Journal. Retrieved from http://scholarlyoa.com/2014/04/29/fallout-from-questionable-article-in-oa-pediatrics-journal/

Beall, J. (2015). More Pseudo-Science from Swiss / Chinese Publisher MDPI. Scholarly Open Access. Retrieved from http://scholarlyoa.com/2015/08/25/more-pseudo-science-from-swiss-chinese-publisher-mdpi/

Beall, J. (2015b). OMICS Group Aims to Trick Researchers with Copycat Journal Titles. Scholarly Open Access. Retrieved from http://scholarlyoa.com/2015/02/19/omics-group-aims-to-trick-researchers-with-copycat-journal-titles/

Beall, J. (2015c). I'm Following a Fringe Science Paper on *F1000Research*. Scholarly Open Access. Retrieved from: http://scholarlyoa.com/2015/01/06/im-following-a-fringe-science-paper-on-f1000research/

Bohannon, J. (2013). Who's Afraid of Peer Review? *Science, 342*(6154): 60–65.

Bohannon, J. (2015). I Fooled Millions into Thinking Chocolate Helps Weight Loss: Here's How. Retrieved from http://io9.gizmodo.com/i-fooled-millions-into-thinking-chocolate-helps-weight-1707251800

Bohannon, J., Koch, D., Homm, P., and Driehaus, A. (2015). Chocolate with High Cocoa Content as a Weight-Loss Accelerator. *International Archives of Medicine, 8*(55): 1–8. Retrieved from http://doi.org/10.3823/1654

Cameron, V. (2013). The Case against Educated Mathematical Dogma (Precise Mathematical Pi Value by Finite Equation = 3.14159292035) 3+(1/(7+1/16)) Precise Finite Value. *Pure and Applied Mathematics Journal, 2*(5): 169. Retrieved from http://doi.org/10.11648/j.pamj.20130205.14

Carroll, R. T. (2015). Tooth Fairy Science and Fairy Tale Science. The Skeptic's Dictionary. Retrieved from http://skepdic.com/toothfairyscience.html

Gardner, M. (1953). *Fads & Fallacies in the Name of Science*. New York: Dover.

Gieryn, T. F. (1983). Boundary-Work and the Demarcation of Science from Non-Science: Strains and Interests in Professional Ideologies of Scientists. *American Sociological Review (48)*6: 781–795.

Gorski, David. (2014). Oh, No! GMOs Are Going to Make Everyone Autistic! Respectful Insolence [Blog]. Retrieved from: http://scienceblogs.com/insolence/2014/12/31/oh-no-gmos-are-going-to-make-everyone-autistic/

Hall, H. (2015). Course Guide for the Video Series *Science-Based Medicine* by Harriet Hall, MD. Retrieved from http://web.randi.org/uploads/3/7/3/7/37377621/course_guide.pdf

Hansson, S. O. (2015). Science and Pseudo-Science. In Edward N. Zalta (ed.), *Stanford Encyclopedia of Philosophy*. Retrieved from http://plato.stanford.edu/archives/spr2015/entries/pseudo-science.

Lucibella, M. (2009). Science Journalism Faces Perilous Times. *APS News 18*(4): 5, 7.

Marek-Crnjac, L. (2013). Modification of Einstein's $E=MC^2$ to $E=1/22\ MC^2$. *American Journal of Modern Physics, 2*(5): 255–263. DOI: dx.doi.org/ 10.11648/j.ajmp.20130205.14

Middelveen, M. J., Burke J., Sapi E., et al. (2015). Culture and Identification of Borrelia Spirochetes in Human Vaginal and Seminal Secretions [Version 3; Referees: 1 Approved, 2 Not Approved]. *F1000Research 3*:309. DOI: dx.doi.org/10.12688/f1000research.5778.3

Millard, W. B. (2013). Some Research Wants to Be Free, Some Follows the Money. *Annals of Emergency Medicine, 62*(2): A14–A20. DOI: doi.org/10.1016/j.annemergmed.2013.06.009

Nattrass, Nicoli. (2012). *The AIDS Conspiracy: Science Fights Back*. Johannesburg: Wits University Press.

Rojeab, A. Y. (2015). Novel Strategy of a Method to Cure from the Cancer. *Cancer Research Journal, 3*(1): 6–10. DOI: doi.org/10.11648/j.crj.20150301.12

Schneider, L. (2015). Is Frontiers a Potential Predatory Publisher?. Retrieved from https://forbetterscience.wordpress.com/2015/10/28/is-frontiers-a-potential-predatory-publisher/

Seneff, S., Davidson, R., and Liu, J. (2012). Empirical Data Confirm Autism Symptoms Related to Aluminum and Acetaminophen Exposure. *Entropy, 14*(12): 2227–2253. DOI: doi.org/10.3390/e14112227

Shen, C., and Björk, B.-C. (2015). "Predatory" Open Access: A Longitudinal Study of Article Volumes and Market Characteristics. *BMC Medicine, 13*(230): 1–15. DOI: dx.doi.org/:10.1186/s12916-015-0469-2

Smith, R., Thyer, L., Ward, E., Meacci, E., Branca, J. J. V., Morucci, G., Gulisano, M. R., Ruggiero, M., Pacini, A. Paternostro, F., Di Cesare Mannelli, L., Noakes, D. J., and Pacini, S. (2013). Effects of Gc-Macrophage Activating Factor in Human Neurons; Implications for Treatment of Chronic Fatigue Syndrome. *American Journal of Immunology, 9*(4): 120–129.

Suleiman, R. (2013). The Dark Side Revealed: A Complete Relativity Theory Predicts the Content of the Universe. *Progress in Physics* 9(4): 34–40.

Thyer, L., Ward, E., Smith, R., Branca, J. J. V., Morucci, G., Gulisano, M., Noakes, D. and Pacini, S. (2013). Therapeutic Effects of Highly Purified De-Glycosylated GcMAF in the Immunotherapy of Patients with Chronic Diseases. *American Journal of Immunology*, 9(3): 78–84.

Ward, E., Smith, R., Branca, J. J. V., Noakes, D., Morucci, G., and Thyer, L. (2014). Clinical Experience of Cancer Immunotherapy Integrated with Oleic Acid Complexed with De-Glycosylated Vitamin D Binding Protein. *American Journal of Immunology*, 10(1): 23–32.

Wikipedia (2015). John Bohannon. Retrieved from https://en.wikipedia.org/wiki/John_Bohannon

Wolters Kluwer India Pvt. Ltd. (2011). Wolters Kluwer Health Acquires Leading Open Access STM Journal Publisher in India. Available from http://www.medknow.com/contactus.asp

13 Pseudoscience, Coming to a Peer-Reviewed Journal Near You

Adam Marcus and Ivan Oransky

In December 2014, the publisher Scientific Research issued a retraction notice for a paper that had appeared in its journal *Health* with the anodyne title "Basic Principles Underlying Human Physiology." According to the notice, the action resulted from "the fact that the contents of this paper need further research and study" (*Health* editors, 2014).

Except that they don't. A quick look at the now-retracted article reveals that it is an effort to promote the false and wholly discredited notion that the human immunodeficiency virus (HIV) does not cause AIDS: "HIV is not etiologically involved in AIDS. It is just a common retrovirus found in AIDS conjuncturally. There is only AIDS that may not be strictly associated neither to a primary immune deficiency nor to an acquired immune deficiency. Actually, heart failure represents the causal factor of AIDS and many other 'primary' immune deficiencies . . ." (Pavel, 2014).

The retraction notice ends with a laughable, and head-scratching, assurance: "*Health* strives to promote the circulation of scientific research, offering an ideal research publication platform to the world with specific regard to the ethical, moral and legal concerns involved. We would like to extend our sincere apologies for any inconvenience it may cause" (*Health* editors, 2014).

What all that means is anyone's guess—which is nicely ironic, given that the article being retracted was a classic example of pseudoscience. (For more, see Kalichman, this volume.)

Pseudoscience by definition is not supposed to find its way into scientific publications. From peer review to layers of ostensibly expert editorial scrutiny, the barriers to entry for nonsense are high—at least in theory. The reality, however, can differ substantially from the theoretical. Peer review can be porous, in that it allows errors, significant methodological problems, and misconduct into the literature. It also is vulnerable to gaming by researchers who exploit sloppy editorial processes to slip pseudoscience into the literature.

Scientific Research and other "predatory publishers"—a term coined by Jeffrey Beall (for more, see Beall, this volume) to describe outfits that claim to be legitimate scientific publishers, but in reality exist only to collect researchers' money—can be an efficient pipeline for peddlers of pseudoscience (Butler, 2013). For a fee, these journals will print virtually anything they receive after arranging for the most cursory of peer reviews. So much is clear from the nature of the articles they are subsequently forced to retract. The publisher Frontiers, for example, is an open-access, all-digital imprint that produces many journals, and which as of 2016 was on Beall's list of predatory publishers (Beall, 2016), although some dispute the classification (Bloudoff-Indelicato, 2015), as some dispute the legitimacy of Beall's list—which was taken down in early 2017 (Oransky, 2017)—altogether (Crawford, 2014). The publisher reportedly accepts for publication nearly 90 percent of the manuscripts it receives but found itself backpedaling after one of its journals published a 2014 article questioning the link between HIV and AIDS. The article, by a researcher at Texas A&M University named Patricia Goodson, was not some Trojan horse with a bland title. Its thrust was perfectly clear from the headline alone: "Questioning the HIV-AIDS hypothesis: 30 Years of Dissent" (Goodson, 2014).

Facing a stiff backlash from readers, Frontiers retreated—but did not surrender. As the publisher explained in a notice:

Frontiers has received several complaints from public health professionals related to the article "Questioning the HIV-AIDS hypothesis: 30 years of dissent," which questions the link between HIV and AIDS. Acknowledging the gravity of these concerns, and the implications that the speculation on the lack of an HIV-AIDS link has on public health in general, an internal investigation was conducted. During the course of the investigation, Frontiers had sought expert input from the Specialty Chief Editors of the HIV and AIDS section of *Frontiers in Public Health* and *Frontiers in Immunology*. Based on the conclusion of the investigation the article type of "Questioning the HIV-AIDS hypothesis: 30 years of dissent" has been changed to an Opinion article, which represents the viewpoint of an individual. In addition, a commentary on the article has been published "Commentary on 'Questioning the HIV-AIDS hypothesis: 30 years of dissent,'" which discusses the concerns and analyzes the viewpoint within a scientific discourse on the topic. (Frontiers publishers, 2015)

On its face, that explanation seems reasonable. After all, science that lacks room for debate is indistinguishable from religious orthodoxy. But opinion pieces that "represent the viewpoint of an individual" and offer hypotheses without testing them are the opposite of science. Goodson's paper, then, would seem to be at odds with Frontiers' ostensible commitment "to producing a high quality scientific journal of interest to researchers and practitioners from many disciplines" (*Frontiers in Public Health* editors, 2016a). Pseudoscience is many things, but a discipline it is not.

Frontiers in Public Health does not appear to have learned much from the experience. In late June 2016, the journal published an article in support of the bizarre and demonstrably false claim that the trails shed by jets consist not of ice crystals, as is the case, but coal fly ash, a harmful pollutant (Herndon, 2016). The theory is a favorite of conspiracy theorists. Outraged readers immediately objected, and within three weeks, the editors issued an expression of concern about the article, soon followed by a retraction—acknowledging that the concerns about the paper were valid and that "the article does not meet the standards of editorial and scientific soundness" for the journal (*Frontiers in Public Health* editors, 2016b).

That would have been a good place to stop, but Frontiers allowed the author, J. Marvin Herndon—who had a previous paper on coal fly ash "chemtrails" retracted—to get the last word: "The author considers the retraction to be unwarranted and therefore does not agree to the statement" (*Frontiers in Public Health* editors, 2016b).

Lest anyone think that pseudoscience is the province solely of predatory publishers, however, even the big, top-tier houses fall victim. Springer and the Institute of Electrical and Electronic Engineers, two leading publishers, in 2014 retracted more than 120 articles that had appeared in conference proceedings after learning that they had been written not by scientists but by a convincing computer text generator called SCIgen (Van Noorden, 2014). The program—a sort of industrialized version of the Sokal hoax (Sokal & Bricmont, 1999)—allows anyone to create a "scientific paper" by simply providing author names. The resulting text and graphics look like a proper scientific paper, but are gibberish. The fact that any were published means that no one peer reviewed the manuscripts.

Although one might assume that journals would hold a strong hand when it comes to ridding themselves of bogus papers, that's not always the case. In 2011, Elsevier's *Applied Mathematics Letters* retracted a paper by Granville Sewell of the University of Texas, El Paso, that questioned the validity of the second law of thermodynamics—a curious position for an article in a mathematics journal, but not so curious for someone like Sewell, who apparently favors intelligent design theories over Darwinian natural selection (Sewell, 2011).

The journal's editor, Ervin Rodin, blamed the appearance of the paper on "hastiness" and acknowledged that the article had no place in the publication. "Please accept our apologies for our erroneous judgement in even considering this paper," Rodin replied to a critic of the Sewell article, which was eventually retracted (Oransky, 2011a).

The affair ought to have ended there. But Sewell sued and Elsevier, the world's largest scholarly publisher, blinked. Not only did it pick up the tab for Sewell's legal fees—a $10,000 hit—but it took the unusual step of apologizing to him (although it did not order the journal to reinstate the article) (Oransky, 2011b). The article was retracted, according to the notice, "because the Editor-in-Chief subsequently concluded that the content was more philosophical than mathematical and, as such, not appropriate for a technical mathematics journal such as Applied Mathematics Letters" (*Applied Mathematics Letters* editors, 2011).

Beyond the financial remuneration, the real value of the settlement for Sewell was the ability to say—with a straight face—that the paper was not retracted because it was wrong. Such stamps of approval are, in fact, why some of those who engage in pseudoscience want their work to appear in peer-reviewed journals.

And it means that the gatekeepers of science—peer reviewers, journal editors, and publishers—need always be vigilant for the sort of "not even wrong" work that pseudoscience has to offer. Online availability of scholarly literature means that more such papers come to the attention of readers, and there's no question there are more lurking. Be vigilant. Be very, very vigilant.

References

Applied Mathematics Letters editors. (2011). A Second Look at the Second Law. *Applied Mathematics Letters*, *24*(11): 1968.

Beall, J. (2016). Beall's List of Predatory Publishers 2016. Scholarly Open Access. Retrieved from https://scholarlyoa.com/2016/01/05/bealls-list-of-predatory-publishers-2016/

Bloudoff-Indelicato, M. (2015). Backlash after Frontiers Journals Added to List of Questionable Publishers. *Nature*, *526*(7575): 613. DOI: 10.1038/526613f

Butler, D. (2013). Investigating Journals: The Dark Side of Publishing. *Nature*, *485*(7442): 433–435. DOI: 10.1038/495433a

Crawford, W. (2014). Ethics and Access 1: The Sad Case of Jeffrey Beall. *Cites & Insights*, *14* (4): 1–14.

Frontiers in Public Health editors. (2016a). About [Journal]. *Frontiers in Public Health*. Retrieved from http://journal.frontiersin.org/journal/public-health#about

Frontiers in Public Health editors. (2016b). Retraction: Human and Environmental Dangers Posed by Ongoing Global Tropospheric Aerosolized Particulates for Weather Modification. *Frontiers in Public Health*, 4:156. DOI: 10.3389/fpubh.2016.00156

Frontiers publishers. (2015). Publisher Statement on "Questioning the HIV-AIDS Hypothesis: 30 Years of Dissent." *Frontiers in Public Health*, 3:37. DOI: 10.3389/fpubh.2015.00037

Goodson, P. (2014). Questioning the HIV-AIDS Hypothesis: 30 Years of Dissent. *Frontiers in Public Health*, *2*(154): 1–12. DOI: 10.3389/fpubh.2014.00154

Health editors. (2014). Announcement from Editorial Board. *Health*, *6*(14): 1816–1821. DOI: 10.4236/health.2014.614213

Herndon, J. M. (2016). Human and Environmental Dangers Posed by Ongoing Global Tropospheric Aerosolized Particulates for Weather Modification. *Frontiers in Public Health*, *4*(139): 1–16. DOI: 10.3389/fpubh.2016.00139

Oransky, I. (2011a). More on Applied Mathematics Letters: Journal Retracted Paper Questioning Second law of Thermodynamics. Retrieved from http://retractionwatch.com/2011/03/16/more-on-applied-mathematical-letters-journal-retracted-paper-questioning-second-law-of-thermodynamics/

Oransky, I. (2011b). Elsevier Apologizes for Applied Mathematics Letters Retraction, Pays Author's Legal Fees. Retrieved from http://retractionwatch.com/2011/06/08/elsevier-apologizes-for-applied-mathematical-letters-retraction-pays-authors-legal-fees/

Oransky, I. (2017). Why Did Beall's List of Potential Predatory Publishers Go Dark? Retrieved from http://retractionwatch.com/2017/01/17/bealls-list-potential-predatory-publishers-go-dark/

Pavel, D. (2014). Basic Principles Underlying Human Physiology. *Health*, *6*(14): 1816–1821. DOI: 10.4236/health.2014.614213.

Sewell, G. (2011). A Second Look at the Law. *Applied Mathematics Letters*, in press. DOI: 10.1016/j.aml.2011.01.019

Sokal, A., and Bricmont, J. (1999). Fashionable Nonsense: Postmodern Intellectuals' Abuse of Science. New York: Picador.

Van Noorden, R. (2014). Publishers Withdraw More Than 120 Gibberish Papers. *Nature*. DOI: 10.1038/nature.2014.14763 Retrieved from http://www.nature.com/news/publishers-withdraw-more-than-120-gibberish-papers-1.14763

1

IV Pseudoscience in the Mainstream

14 "Integrative" Medicine: Integrating Quackery with Science-Based Medicine

David H. Gorski

Most of us take it for granted that medicine should be based on science, because science is generally the best method for determining what causes disease, how to treat disease, and which treatments work best in which patients. This view, however, is a relatively recent development. Indeed, for most of its history, medicine was based far more on religion and superstition than anything resembling science, with empiric uncontrolled observations of individual practitioners serving as the basis for determining what "works." It is not a coincidence that the very earliest known physicians in ancient Egypt also functioned as priests. In fact, one of Hippocrates' greatest advances was to promulgate the principle among his followers that disease is not caused by the gods or by evil spirits, but rather by natural causes that human beings can come to understand. That is not to say that belief in the supernatural as a cause of disease did not persist for many hundreds of years after Hippocrates. Rather it is to acknowledge that before medicine could progress its practitioners had to accept that there are *natural*, physical causes of diseases. Even after that principle became widely accepted, it was still centuries before more scientific means of observation, such as randomized controlled clinical trials, supplanted bias- and error-prone anecdotal observations as the preferred tools for determining which treatments work and which do not. Although James Lind is usually credited with what was arguably the first planned controlled clinical trial in 1747 (he tested whether providing British sailors with citrus fruit could prevent scurvy), what are now recognizable as the first modern, double-blind, randomized clinical trials were not carried out until the 1940s (Bhatt, 2010). Arguably, science-based medicine (SBM) as we know it is, at most, a century old. And what a century it has been in terms of improvements in human life expectancy!

Based on that history, perhaps it should not be so surprising that there still exist forces in medicine that are striving to turn back the clock, reembrace magical thinking, and return to less rigorous methods of investigation as the preferred tools for evaluating medical treatments. These forces have always been there, but since the 1990s or so two related phenomena have arisen in medicine. The first is now known as "integrative medicine," but has also been known by other names, such as "complementary and alternative medicine" (CAM). As will be discussed, many CAM modalities that make up integrative medicine are based on prescientific and, in some cases, superstitious understandings of disease and human physiology, which is why the rise of integrative medicine has led to the infiltration of pseudoscience in academic medical centers, a phenomenon for which a blogger named Robert W. Donnell coined a term, "quackademic medicine" (Donnell, 2008). In essence, at the same time when advances in genomics, proteomics, and metabolomics have led to exponentially increasing improvements in our understanding of the detailed molecular mechanisms of human physiology and disease, there is a growing movement in medicine seeking to go backward and to "integrate" prescientific ideas with SBM.

The speed with which integrative and quackademic medicine has insinuated itself into the medical mainstream has been truly breathtaking. In 1983, Clark Glymour and Douglas Stalker published a scathing editorial in *The New England Journal of Medicine* (NEJM) about "holistic medicine"—what we would now refer to as CAM—that dismissed it as a "pabulum of common sense and nonsense offered by cranks and quacks and failed pedants who share an attachment to magic and an animosity toward reason" (Glymour and Stalker, 1983). But, beginning in the 1980s, a sea change occurred. Indeed, when I published two articles taking CAM to task (Gorski, 2014; Gorski and Novella, 2014), neither of them anywhere near as blunt as Glymour and Stalker's, I endured considerable criticism, some of it from influential practitioners and proponents of integrative medicine. In addition, the NEJM has published studies of CAM (Wang et al., 2010), as well as a case study recommending acupuncture for chronic back pain (Berman, Langevin, Witt, and Dubner, 2010). During the same time period, other respectable journals published many other similar studies. What Glymour and Stalker once dismissed as magical thinking has become increasingly mainstream, with no end in sight.

Pseudoscience Insinuates Itself in Medical Academia and Medicine

The twin phenomena of integrative and quackademic medicine have progressed in highly respected medical centers from skepticism to outright embrace. As a result, many academic medical centers have established integrative medicine programs that offer everything from potentially useful modalities, such as nutritional counseling, meditation, and lifestyle alterations, to treatment modalities that can only be described from a strictly scientific viewpoint as pure pseudoscience, such as "energy medicine" (Mease, 2011), reflexology (Ernst, 2009), acupuncture (Colquhoun and Novella, 2013; Ernst, 2006), and homeopathy (Smith, 2012a, 2012b). Nor is this phenomenon restricted to medical academia. For example, a prominent private hospital chain, the Cancer Treatment Centers of America, is based on a business model offering "integrative oncology" and "naturopathic oncology" alongside conventional cancer treatments, despite the lack of evidence for naturopathy (Atwood, 2004).

More examples abound. For example, an NBC News report in 2012 by Nancy Snyderman described how treatment modalities such as acupuncture, traditional Chinese medicine, and even reflexology are increasingly offered at cancer centers as venerable as Memorial Sloan Kettering Cancer Center (MSKCC). Included along with Dr. Snyderman's report on the NBC website was a video of an acupuncturist explaining how acupuncture "stimulates the flow of chi." Chi is basically the "life energy," a prescientific vitalistic concept that a special mystical "force" or "energy" is what makes matter living as opposed to nonliving—and is completely "natural." It is the same concept at the root of "energy medicine" like healing touch, in which practitioners claim to be able to manipulate human life energy fields to healing effect. In her report, Dr. Snyderman voiced her enthusiastic approval of this development, even going so far as to state bluntly about integrative medicine, "Quite frankly, if a doctor doesn't know, I think it's time to ask for a referral to someone who does. Certainly in my world of cancer treatment, the really good cancer centers know the difference." She even lamented how most insurance companies do not cover many of the CAM modalities considered part of "integrative medicine"!

Another major academic medical center that has gone "all in" embracing prescientific medicine is the Cleveland Clinic Foundation (CCF). For many years, the CCF has had a "wellness institute" offering unscientific and pseudoscientific treatment modalities such as acupuncture, chiropractic, Reiki, and Chinese herbal therapies. Indeed, I often use the mystical description of Reiki featured on the CCF website in talks about quackademic medicine, because it describes "energy" from the "universal source" being channeled through the Reiki master and being neither created nor destroyed. As bad as that is, in 2014, the CCF went beyond this, when *The Wall Street Journal* reported that it had opened a traditional Chinese herbal clinic, complete with a naturopath carrying out "tongue diagnosis," a system of diagnosis that maps body organs to areas on the tongue in much the same way that reflexology maps them to areas on the palms of the hand and soles of the feet (Reddy, 2014). Naturopathy, it should be noted, is an alternative medicine "discipline" that claims to treat diseases naturally and consists of a mishmash of some sensible advice about diet and exercise mixed with outright quackery, such as homeopathy, applied kinesiology, and "energy medicine."

Later that same year, the Cleveland *Plain Dealer* reported that integrative medicine guru Mark Hyman, arguably the most famous practitioner of "functional medicine," would be opening a functional medicine clinic under the auspices of the CCF (Townsend, 2014). Functional medicine is a poorly defined branch of integrative medicine that claims, as does integrative medicine itself, to "address the underlying causes" of diseases, particularly chronic diseases. On the surface, functional medicine sounds harmless enough, but if one digs deeper, one finds a vaguely defined "discipline" in which it is claimed that measuring a whole laundry list of metabolic factors and other lab values will lead to a "holistic" approach to disease. Often this involves running a number of expensive and generally unnecessary tests in order to "diagnose" various disorders. In addition to addressing the "underlying causes" of disease, Hyman also claims that functional medicine focuses "on the whole person rather than an isolated set of symptoms" (sound familiar?) and "looking at the interactions among genetic, environmental, and lifestyle factors that can influence long-term health and complex, chronic disease" (Hyman, n.d.). Unfortunately, the specific recommendations made by functional medicine practitioners are rarely rooted in strong science. Demonstrating his susceptibility to pseudoscience, in 2015 Hyman himself coauthored a book with Robert F. Kennedy Jr. entitled *Thimerosal: Let the Science Speak: The Evidence Supporting the Immediate Removal of Mercury—a Known Neurotoxin—from Vaccines*, a book that promotes the long-discredited link between thimerosal and autism and other neurodevelopmental conditions. This is the person who founded a major medical clinic at a major academic medical center—and quite successfully, as well. In 2016, Patrick Hanaway, who oversees the day-to-day operation of the CCF Center for Functional Medicine, reported that it would be doubling its patient care capacity in response to "unbelievable pent up demand for this kind of care" (Goldman, 2016).

This embrace of pseudoscience has consequences, as well. Medical evidence is overwhelming that vaccines are safe and effective and, contrary to claims by antivaccine activists, do not cause autism. Yet in 2017 Dr. Daniel Neides, director and CEO of the Cleveland Clinic's Wellness Institute, came under heavy criticism and attracted national attention for publishing an antivaccine screed on a local website that was sympathetic to the discredited claims that vaccines cause autism and railed against "toxins" in adjuvants used in vaccines (Neides, 2017). Of note, Dr. Neides consistently won teaching awards. One has to wonder what Dr. Neides taught students and residents about vaccines and how many other physicians in "integrative medicine" centers harbor antivaccine views.

There are many other institutions besides the CCF embracing pseudoscience as well. For example, the Center for Integrative Medicine at the George Washington University Medical Center (GWCIM) boasts a list of services that includes acupuncture, chiropractic, craniosacral therapy, infrared light therapies, glutathione infusions, Myers' Cocktail, naturopathy, Reiki, intravenous high-dose vitamin C, and unvalidated medical tests that include "customized interpretation" of 23andme.com genetic profile results with a specific accent on methylation and "detoxification" profiles (http://www.gwcim.com/). Almost none of this has any resemblance to SBM, and includes outright quackery like Reiki. Not to be outdone, the University of Kansas Medical Center Integrative Medicine Program is led by Jeanne A. Drisko, whose titles include the Riordan Endowed Professor of Orthomolecular Medicine. Orthomolecular medicine is a form of alternative medicine that was popularized by Linus Pauling decades ago, when he claimed that high-dose vitamin C was an effective treatment for cancer and the common cold. Its main precepts consist of the idea that there is an optimum nutritional environment in the body and that most disease reflects "imbalances" or deficiencies in this environment. In orthomolecular medicine, treatment of disease involves supplementation to address these "deficiencies" with vitamins, amino acids, minerals, trace elements, and fatty acids. In practice, this usually involves supplements and megadoses of vitamins, such that the orthomolecular approach has sometimes been referred to as "megavitamin therapy." Yet another example is the University of Maryland at Baltimore, whose Center for Integrative Medicine offers acupuncture, Reiki, reflexology, and rolfing, a system of "bodywork" that involves painful deep massage of fascia. It even offers inpatient Reiki services for trauma patients.

The embrace of alternative medicine in medical academia can even reach the truly bizarre. The University of Michigan, where I received both my chemistry degree and my MD, offers a Holistic Family Practice Medicine Program. There, you can find the now depressingly usual assortment of physicians and practitioners offering alternative medicine modalities, such as functional medicine and naturopathy. However, this program also includes a physician named Ricardo R. Bartelme, who offers anthroposophic medicine in collaboration with the Rudolf Steiner Health Center in Ann Arbor. Steiner was a late-eighteenth-century philosopher who founded an esoteric spiritual movement known as anthroposophy, whose goal was to find a synthesis between science and spirituality and resulted in the founding of Waldorf Schools. In terms of science, the best that can be said about anthroposophy is that it is pure pseudoscience. For example, biodynamic farming, which is based on anthroposophy, can involve bizarre practices, such as stuffing yarrow blossoms into the bladder of red deer and burying the bladders during the winter, stuffing chamomile blossoms into cattle intestines and burying them in the fall, and placing oak bark into the skull of a domesticated animal and burying it surrounded with peat "in a place where lots of rain water runs by." As for anthroposophic medicine, Simon Singh and Ezard Ernst characterize it thusly in their book *Trick or Treatment: The Undeniable Facts about Alternative Medicine*:

Applying his philosophical concepts to health, he [Rudolf Steiner] founded, together with Dr. Ita Wegman, an entirely new school of medicine. It assumes metaphysical relations between planets, metals, and human organs, which provide the basis for therapeutic strategies. Diseases are believed to be related to actions in previous lives; in order to redeem oneself, it may be best to live through them without conventional therapy. Instead, a range of other therapeutic modalities is employed in anthroposophic medicine: herbal extracts, art therapy, massage, exercise therapy, and other unconventional approaches.

To get an idea of what mystical nonsense anthroposophic medicine is, I like to quote straight from the horse's mouth, namely the Physician's Association for Anthroposophic Medicine, in its pamphlet for patients:

Medicine based purely on material science is limited to explaining an illness solely on the basis of the laws of physics and chemistry. Anthroposophic medicine is more ambitious. It takes into account additional factors, both general and individual, that may affect the patient's life, mind, and soul, and their physical manifestation: in growth, regeneration, microcirculation, fluid retention in the skin, muscle tone, biorhythms, head distribution, posture, uprightness, gait, mental focus, speech. When illness occurs, examination of the above may reveal deviations, imbalances, and extremes—additional diagnostic parameters that need to be considered when selecting a therapy.

Elsewhere in the pamphlet, homeopathy is recommended as a part of anthroposophic medicine:

In addition, other substances tailored to the patient's unique characteristics are administered. These are frequently homeopathic substances designed to stimulate the organism and its powers of self-healing.

It should be clear to the average reader that the above passage is utter nonsense, particularly the claim about going beyond the laws of physics and chemistry to explain illness. Yet this is a form of "medicine" being offered at a respected medical school.

It is not just medical academia, either. There are now at least two US government entities that, combined, spend approximately $250 million a year funding research and education in CAM and integrative medicine (Mielczarek and Engler, 2012). The first is the National Center for Complementary and Integrative Health (NCCIH, formerly known as the National Center for Complementary and Alternative Medicine), whose budget for fiscal 2015 was $124.1 million. The NCCIH is relatively well known and has been at times during its two-decade history the subject of rancorous scientific and political debate over its mission and even continued existence (Boyle, 2011; Marcus and Grollman, 2006). Second and not as well known, the National Cancer Institute (NCI) spends approximately as much on CAM research and education through its Office of Cancer Complementary and Alternative Medicine (OCCAM, the most inappropriate acronym ever). Its average yearly budget over the last decade has been roughly equal to that of NCCAM.

Even fewer physicians appear to know of the Academic Consortium for Integrative Medicine and Health (ACIMH), which consists of sixty-nine academic medical centers in North America (sixty-two in the United States, five in Canada, and two in Mexico). After being relatively stable for a number of years, the number of ACIMH members has increased dramatically, up from fifty in 2014 (Gorski, 2014). Of these academic medical centers, fifty-five are medical-school based, and the rest are affiliated with various medical schools. Sadly, the roster of ACIMH members includes some of the most prestigious medical schools in the United States, including the Stanford University School of Medicine, the Johns Hopkins University School of Medicine, five of the University of California campuses, Harvard Medical School, Yale University School of Medicine, Columbia University Medical Center, Duke University, and a number of prestigious cancer centers, such as MSKCC and the UT-M.D. Anderson Cancer Center. Thus, the question is now not so much where to find integrative medical programs, but which academic medical centers still lack such programs.

What Is "Integrative Medicine"?

Despite its popularity, both in academia and increasingly in community hospitals, one of the most vexing problems that complicates any discussion of integrative medicine, regardless of the specific medical specialty, is defining what constitutes CAM or integrative medicine. In general, existing definitions (Department of Health and Human Services, 2007; 2012) tend to stress that CAM encompasses treatments that are "outside of the mainstream," although they rarely make clear what constitutes "mainstream." This definition is particularly problematic for a subset of CAM modalities, such as nutrition and exercise, interventions that are routinely claimed by CAM but, depending on how they are applied, can clearly be "mainstream" or part of SBM (Grosso et al., 2013; Zou, Yang, He, Sun, and Xu, 2014), thus bringing into question why a separate CAM category is needed in the first place. (I argue that for nutrition and lifestyle, no separate CAM category is required.) In addition, although some CAM interventions are so unlikely to cause harm that it is difficult to have too strong an objection to them in general, it is also highly questionable to ascribe anything other than nonspecific effects to them. For instance, massage can make patients feel better, and yoga can be viewed as a system of exercise that, properly done, can be as beneficial to general health as many other forms of exercise. The confusion arises when massage becomes "massage therapy," a billable, reimbursable medical *treatment*, rather than support, and when the religious and mystical underpinnings of some forms of yoga, such as kundalini energy, are represented as essential for its benefits. Taken this way, it is difficult not to conclude that in CAM and integrative medicine there has developed a "medicalization" of practices not previously considered medical—even spirituality.

This "blurring of the lines" in CAM between medicine and nonspecific supportive measures coupled with the appropriation by CAM of "mainstream" science-based modalities such as exercise results in serious problems determining which treatments are CAM and which are mainstream, which interventions are actually used with therapeutic intent, and which are, like massage, simply helpful in helping patients cope. Consider a very common disease treated by "conventional medicine," type II diabetes. Is prescribing weight loss and exercise as the first intervention for preventing the progression from prediabetes to type II diabetes, as most primary care doctors routinely do and as the American Diabetes Association recommends (American Diabetes Association, 2014), mainstream medicine or CAM? What, if anything, is outside of the mainstream about it? Nothing. Yet dietary interventions, particularly probiotics and supplements, are often labeled as "integrative" for a variety of conditions.

"Official" definitions of CAM (Department of Health and Human Services, 2007; 2012) fail to clarify this issue. For example, the NCCIH defines "alternative" medicine as "using a non-mainstream approach in place of conventional medicine," and "complementary" medicine as "using a non-mainstream approach together with conventional medicine"

(*What's in a . . .*, 2016). Again, what is "mainstream" and what is not? Using quackery like energy medicine or homeopathy would clearly be outside of the mainstream, but there is nothing outside of the mainstream about many dietary interventions, although it does behoove me to point out that integrative medicine often embraces dietary interventions resting on foundations of pseudoscience, like orthomolecular medicine. Also, according to the NCCIH, "integrative" medicine refers to combining "treatments from conventional medicine and CAM for which there is some high-quality evidence of safety and effectiveness" (Department of Health and Human Services, 2012). However, very few CAM modalities have sufficient high-quality evidence supporting their efficacy and safety to justify their routine integration into mainstream medicine.

To confuse matters further, there are conflicting definitions of CAM/integrative medicine from different government agencies, and the definitions change. For example, the NCCIH 2011–2015 strategic plan (National Center for Complementary and Alternative Medicine, 2011) redefined what were formerly "mind-body" interventions as "mind *and* body" interventions and broadened the category to include virtually any CAM treatment that isn't diet or herbal, including "energy medicine," acupuncture, meditation, yoga, craniosacral therapy, and even reflexology (table 14.1). In the NCCIH strategic plan for 2016–2021 (https://nccih.nih.gov/about/strategic-plans/2016) and on the NCCIH webpage on research priorities (https://nccih.nih.gov/about/researchfocus), NCCIH appears, for research purposes, to recognize only two broad categories of CAM: natural products and everything else. In contrast, OCCAM subdivides CAM into eight areas (table 14.2) (Office of Cancer Complementary and Alternative Medicine, 2011), separating out nutritional therapies such as macrobiotic diets ("Questionable Methods of Cancer Management: 'Nutritional' Therapies," 1993) and the Gerson protocol ("Unproven Methods of Cancer Management: Gerson Method," 1990) for cancer from "biologically based" therapies, such as herbal medicines, and adding a category for exercise therapy. OCCAM's definition also encompasses "spiritual therapies" completely without basis in science. These include intercessory prayer, which does not work (Masters, Spielmans, and Goodson, 2006), and energy medicine such as Reiki or therapeutic touch, whose purported mechanism rests on the ability of human beings to manipulate patient life "energy" fields that have never been shown to exist (Baldwin and Hammerschlag, 2014; Rosa, Rosa, Sarner, and Barrett, 1998).

Arguably, there is no scientific reason why biologically based therapies should be considered "alternative" or "integrative," making their inclusion as CAM problematic. Nor is there a reason why integrative medicine interventions with some biological plausibility cannot be tested in randomized controlled trials (RCTs) just like any other intervention. But what of alternative whole medical systems, like homeopathy or naturopathy, the latter of which is a hodgepodge of interventions that include homeopathy, traditional Chinese medicine, and dietary interventions (Atwood, 2003)? It is also

Table 14.1.
CAM Subtypes as Defined by NCCIH after 2011

CAM subtype	Definition	Examples
Manipulative and body-based practices	Practices involving manipulation of musculoskeletal structures to affect physiology	Osteopathy, chiropractic, craniosacral therapy, massage therapy*
Mind and body practices	Practices that focus on the interactions among the brain, mind, body, and behavior, with the intent of using the mind to affect physical functioning and promote health	Exercise, meditation, acupuncture, yoga, massage therapy, tai chi, progressive relaxation, hypnotherapy, guided imagery
Natural products	Practices involving the use of substances found in nature, including diet	Herbal medicines, dietary supplements, probiotics, nutrition and diet manipulation
Other CAM practices	Practices that do not fall into the above NCCAM definitions	Energy medicine (Reiki, chi gong, healing touch*), alternate whole medical systems (homeopathy, naturopathy, traditional Chinese medicine, Ayurvedic medicine)

Source: Adapted from *What Is Complementary and Alternative Medicine?* U.S. Department of Health and Human Services, National Institutes of Health and National Center for Complementary and Alternative Medicine, Washington, DC, 2012.
*Note that in the NCCAM Third Strategic Plan (National Center for Complementary and Alternative Medicine, 2011) massage and healing touch are considered part of mind and body practices.

difficult not to wonder whether the inclusion of lifestyle interventions, such as diet, exercise, and relaxation techniques (which arguably have been underutilized in medicine) in the same category as homeopathy, naturopathy, and spiritual interventions lend by association the appearance of scientific plausibility to interventions that, from a basic science viewpoint, are incredibly implausible at best, while "rebranding" more plausible nonpharmacological and nonsurgical treatments as "alternative" or "integrative." Unfortunately, the rationales for many of the treatments falling under the rubric of "respectable" integrative medicine (i.e., the sort practiced in "evidence-based" integrative medicine programs in academic medical centers) are based on many of the very same prescientific, vitalistic concepts as a lot of disreputable alternative medicine. Therein lies the key problem with integrative medicine. The less "alternative" the intervention, the more it resembles "conventional" medicine; the more "alternative" the intervention, the more it resembles quackery. Yet it is quite possible to find homeopathy being

Table 14.2.
Subtypes of CAM as Defined by OCCAM

CAM subtype	Definition	Examples
Alternative medical systems/whole medical systems	*Alternative medical systems* are built upon complete systems of theory and practice. Often these systems have evolved apart from and earlier than the conventional medical approach used in the United States.	Ayurveda, homeopathy, traditional Chinese medicine, Tibetan medicine
Energy therapies	There are two types:	
	Biofield therapies are intended to affect energy fields that purportedly surround and penetrate the human body. The existence of such fields has not yet been scientifically proven.	Chi gong, Reiki, therapeutic touch
	Electromagnetic-based therapies involve the unconventional use of electromagnetic fields, such as pulsed fields, magnetic fields, or alternating current or direct current fields.	Pulsed electromagnetic fields, magnet therapy
Exercise therapies	*Exercise therapies* include health-enhancing systems of exercise and movement.	Tai chi, yoga
Manipulative and body-based methods	*Manipulative and body-based methods* in CAM are based on manipulation and/or movement of one or more parts of the body.	Chiropractic, therapeutic massage, osteopathy, reflexology
Mind-body interventions	*Mind-body medicine* uses a variety of techniques designed to enhance the mind's body to affect bodily function and symptoms.	Meditation, hypnosis, art therapy, biofeedback, imagery, relaxation therapy, music therapy, cognitive-behavioral therapy, aromatherapy
Nutritional therapeutics	*Nutritional therapeutics* are an assortment of nutrients and non-nutrients, bioactive food components used as chemopreventive agents, and specific foods or diets used as cancer prevention or treatment strategies.	Macrobiotic diet, vegetarianism, Gerson therapy, Kelley/Gonzalez regimen, vitamins, soy

Table 14.2.
(continued)

CAM subtype	Definition	Examples
Pharmacological and biological treatments	*Pharmacological and biologic treatments* include the off-label use of certain prescription drugs, hormones, complex natural products, vaccines, and other biological interventions not yet accepted in mainstream medicine.	Antineoplastons, low-dose naltrexone, immunoaugmentative therapy, Laetrile
	Subcategory: Complex natural products are an assortment of plant samples (botanicals), extracts of crude natural substances, and unfractionated extracts from marine organisms used for healing and treatment of disease.	Herbs and herbal extracts, mistletoe, mixtures of tea polyphenols
Spiritual therapies	*Spiritual therapies* are therapies that focus on deep, often religious beliefs and feelings, including a person's sense of peace, purpose, connection to others, and beliefs about the meaning of life.	Intercessory prayer, spiritual healing

Source: Adapted from Office of Cancer Complementary and Alternative Medicine's Annual Report on Complementary and Alternative Medicine, 2011.

used in academic medical centers, and Reiki is particularly prevalent. Both of these can be described most appropriately as the rankest quackery.

The Problem with Integrative Medicine

Integrative medicine, as a specialty, has a huge problem that makes it very difficult indeed to ever consider it as science-based. The reason is simple. Many of the "outside the mainstream" treatments being "integrated" rest on principles that, from a strictly basic science standpoint, range from highly implausible to virtually impossible, even violating well-established laws of physics and chemistry; prime examples include homeopathy and Reiki, both of which are based on ideas rooted in prescientific vitalism. There are CAM modalities that are based on anatomical structures or abnormalities that do not exist, such as acupuncture "meridians" or chiropractic "subluxations." Other CAM modalities seek to correct abnormalities in nonexistent physiological functions.

Particularly silly examples of this phenomenon include craniosacral therapy, which seeks to correct the "craniosacral rhythms" of the cerebrospinal fluid by manipulating joints in the skull (Ernst, 2012), and reflexology, which claims a nonexistent connection between organs and specific points on the soles of the feet or palms of the hands (White, Williamson, Hart, and Ernst, 2000). Yet one can find reflexology offered in academic medical centers as more than just a pleasant foot massage, and craniosacral therapy is offered in academic medical centers as varied as Beth Israel Deaconness Medical Center, the University of Pittsburgh, the Cleveland Clinic, and the Beaumont Health System.

Perhaps the most disturbing development in integrative medicine is how whole medical systems based on pseudoscience or prescientific ideas are becoming not just respected, but perceived as essential parts of an academic medical center. Consider the case of traditional Chinese medicine, clearly one of the most popular whole medical systems, if not the most popular. While it is certainly possible that herbal medicines used by TCM practitioners might have substances with useful pharmacological activity that might be used as drugs, many arguments in favor of TCM cite its ancientness, as though being hundreds or thousands of years old somehow indicates that a medicine works. Appeals to antiquity aside, simply put, TCM postulates links between specific organs and anatomic structures similar to the links between locations on the soles of the feet and hands and specific organs that form the basis of reflexology. For example, "tongue diagnosis" links areas on the tongue to specific organs (Anastasi, Currie, and Kim, 2009). In addition, TCM ascribes illness to the six "pernicious influences" of wind, cold, heat, dampness, dryness, and summer heat (Chan, 1995) and/or imbalances in the "five elements," earth, fire, metal, wood, and water. One notes that these "influences" strongly resemble ideas at the core of ancient European medicine dating back to Hippocrates, who postulated imbalances in the "four humors" as the cause of disease. Integrative medicine practitioners enthusiastically accept TCM's ancient concept of "imbalances" in the five elements, while at the same time being appropriately skeptical of the ancient concept of imbalances in blood, yellow bile, black bile, and phlegm. Yet the Cleveland Clinical Foundation (CCF) is far from alone in having TCM practitioners in its integrative medicine program. Many, if not most, academic medical centers offer at least acupuncture, and many offer more TCM modalities than that.

Another problematic "whole medical system" is homeopathy, which is among the most obvious quackeries. Originating in Germany and more popular in Europe than in the United States, homeopathy is still a trans-Atlantic presence. For instance, in those states that license naturopathy, homeopathy is part of the Naturopathic Physicians Licensing Examinations. In fact, homeopathy is built into naturopathy, with many hours of training included in the curriculae of naturopathy schools. Unfortunately, homeopathy rests on two laws that are pure pseudoscience. The first is the law of similars, which states that to relieve a symptom requires the use of a substance that causes that symptom in healthy individuals, and the law of infinitesimals, which states that serial dilution of a remedy, with vigorous shaking between dilutions, "potentizes" it, thereby making it stronger rather than weaker. Neither law has any scientific basis. Yet that doesn't stop homeopaths from, as Richard Dawkins put it, paddling "boldly further up the creek of pseudoscience." Indeed, most homeopathic dilutions surpass 10^{60}-fold; that is more than 10^{36}-fold greater than Avogadro's number (Ernst, 2008), meaning that it is incredibly unlikely that there is a single molecule of original substance left. Yet homeopathy is not only legal, with remedies included in the Homeopathic Pharmacopeia of the United States and exempt from requiring premarket approval by the Food and Drug Administration, but even offered at academic medical centers, usually by naturopaths. The NCCIH has even funded grants to study homeopathy, including one that examined what is the best method to "succuss" homeopathic remedies. (Succussion is the word homeopaths use to describe the shaking step between each serial dilution of a homeopathic remedy.)

The Blind Spot of Evidence-Based Medicine

Advocates of integrative medicine often argue that science is constantly changing based on evidence and that the development of medical treatments has long largely been empirical. Both points are true but irrelevant. Although much of evidence-based medicine (EBM) does remain primarily empirical, as with the aggregated results of high-quality Randomized Clinical Trials (RCTs) as the very highest form of evidence (Seshia and Young, 2013a, 2013b), while ranking basic science considerations at or near the bottom, this appeal to the empiricism of EBM is a red herring and largely irrelevant to scientific criticisms of highly implausible CAM claims. Accepting that chemotherapy, a drug, or surgery "works" based on empirical observation and clinical studies does not require scientists to accept a mechanism that violates laws of physics or chemistry, as homeopathy does, or the existence of anatomic structures or abnormalities that clearly do not exist, as reflexology or craniosacral therapy does. Indeed, the physics and chemistry that conclude that CAM treatments such as homeopathy and Reiki are virtually impossible are so well established that, were healing phenomena in humans due to homeopathy or Reiki to be unequivocally demonstrated beyond a shadow of a doubt, then what we understand about physics and chemistry would be cast into serious doubt.

What has allowed integrative medicine to progress so far and quackademic medicine to become so respected is a massive blind spot in EBM. That blind spot assumes that treatments do not reach the stage of RCTs without having amassed sufficient high-quality preclinical evidence supporting biological plausibility to justify the effort, time, and expense of RCTs, and, even more importantly, to justify the use of human subjects. Indeed, so integral to this process are biological plausibility and preclinical evidence that they are enshrined in the Declaration of Helsinki governing the ethical treatment of human research subjects. In the era of RCTs, treatments without biological plausibility and compelling preclinical evidence rarely, if ever, progress to the stage of RCTs. CAM has thrown that assumption completely on its head. Indeed, it is rare to find laboratory and animal experimentation supporting CAM modalities other than natural products—and acupuncture-related studies. Such preclinical studies, even when they exist, are often over-interpreted and/or have little relevance to humans. Many clinical trials are being done of incredibly implausible treatments, such as homeopathy, with depressingly predictable false positive results. Remember, setting $p < 0.05$ as the level for statistical "significance" guarantees that, even if all clinical trials were conducted perfectly, 5 percent of studies would appear to be "positive" through random chance alone. Given the imperfections in the clinical trial process, we know that this "false positive" number is likely far higher (Ioannidis, 2005).

There is a problem inherent in doing clinical trials of CAM. In RCTs testing modalities with very low pretest probability (i.e., low plausibility), any confounding effects are vastly magnified, producing many "false positives" (Goodman, 1999; Ioannidis, 2005). When tested in RCTs, CAM treatments (such as homeopathy or Reiki) that have prior probabilities based on scientific plausibility that can only be characterized as very close to zero, if not zero, will in essence do little more than identify the "noise" in the RCT process. In the interest of scientific rigor, one must admit that this is by no means a problem confined only to RCTs of CAM treatments (Ioannidis, 2005). For example, biomarker (Ioannidis, 2012) and omics (Ioannidis and Khoury, 2011) studies are prone to it as well—but not to the same degree, because they have a higher pretest probability. In any case, the infinitesimally low prior probabilities associated with CAM modalities such as Reiki or homeopathy enormously amplify this problem in clinical trials of CAM (Gorski and Novella, 2014). Which is more likely, that homeopathy is placebo medicine supported by a few bias- and error-prone clinical trials, or that physicists and chemists are wrong about fundamental principles of physics and chemistry? It is, of course, remotely possible that existing scientific theories addressing such a question, built up over hundreds of years and supported by vast evidence, might be wrong. That possibility must always be conceded. But think of it this way. To show that homeopathy works would require evidence of at least the same quality and quantity as the evidence that concludes that it cannot work. As yet, that has not happened.

Integrative Medicine: Harms versus Questionable Benefits

CAM, as a term, lumps together the highly implausible with the plausible, and "integrative medicine" seeks to integrate those highly implausible treatments with established science- and evidence-based medicine. But what CAM treatments might be plausible and therefore potentially useful? Integrative treatments that might have a basis in science tend, from a scientific perspective, to be mundane. These include interventions that tend to fall under either the "mind and body" category (NCCIH) or exercise therapies (OCCAM), such as yoga and tai chi, or the "biologically based" (NCCIH) or "pharmacological and biologically based treatments or nutritional therapeutics" (OCCAM), where there are scientific studies identifying specific mechanisms of action of diet or specific chemicals in herbal remedies. Unfortunately, plausibility does not equal efficacy, something that is well known in standard drug development, where the vast majority of experimental therapeutics fail to make it through the gauntlet of phase I, II, and III testing to win FDA approval. Not surprisingly, many supplements tested for their effect on cancer prevention have failed, such as selenium and vitamin E for prostate cancer (Klein et al., 2011) and β-carotene, which actually increased the risk of lung cancer (Virtamo et al., 2003). Even though prior plausibility is no guarantee of positive results in clinical trials, that does not mean that considering prior plausibility is "preselecting" what is likely to work and that we therefore need to test every implausible health claim, as some CAM advocates claim. The reason is simple. Prior probabilities as close to zero as those of homeopathy are an excellent indicator that clinical trial results will be negative.

But what's the harm of adopting integrative medicine? It's not an unreasonable question given that most CAM "integrated" into integrative medicine is used for symptom relief rather than to cure disease. There are two harms, one scientific and one ethical, both intimately related. The scientific argument rests on the observation that the hard-won improvements in life expectancy and the curing of diseases since the 1930s have come about not from diluting the scientific basis of medicine, but rather from the evermore rigorous application of evidence and science. Certainly, no scientist denies that the process is nowhere near as linear and neat as that, although that is sometimes a caricature of the process used by CAM advocates to question why one scientific study will show one result for a CAM treatment but another will show a very different result. Clinical observations cross-pollinate basic science observations and vice versa.

Unfortunately, the integration of alternative medicine into medicine risks both degrading the rigorous science behind cancer clinical trials and compromising the ethics of clinical trials by subjecting subjects to interventions that are so implausible that it is reasonable to conclude there is no realistic probability of a positive result. The most infamous example of this phenomenon was a clinical trial of chelation therapy for cardiovascular disease, which had close to zero pretest probability. Yet the National Institutes of Health (NIH) spent $32 million on an RCT known as TACT (trial to assess chelation therapy) to test whether chelation therapy decreased morbidity and mortality due to atherosclerotic cardiovascular disease. Its results were essentially negative (Lamas et al., 2013), except in one subgroup (diabetic patients), a result that, upon closer examination, appears to have been almost certainly spurious (Nissen, 2013). Worse, even if this trial's results were taken at face value and the most optimistic spin put on it, the best that could be said about chelation therapy is to consider it for diabetics but not for anyone else with heart disease. Predictably, no such recommendation was issued. Chelation therapy continues to be widely used for heart disease by integrative medicine practitioners, and the TACT principal investigator is busy promoting his trial as evidence that chelation therapy shows great promise as a nonsurgical treatment for atherosclerotic heart disease while lobbying for more money for another clinical trial.

Another particularly egregious example of an NIH-funded trial to test a protocol for a highly implausible alternative medicine treatment involved the Gonzalez protocol for pancreatic cancer. Named after the integrative physician in New York who invented it based on earlier similar protocols, Nicholas Gonzalez, the protocol involves extreme dietary modifications, consumption of large quantities of vegetable juices and supplements (eighty-one capsules per day), skin brushing, salt and soda baths, and twice-daily coffee enemas. (Yes, you read that correctly.) After many years and abandonment of the RCT format for an unblinded "patient's choice" design, the results showed that one-year survival rates of patients undergoing the dietary protocol were nearly fourfold worse than those of patients receiving standard-of-care chemotherapy (Chabot et al., 2010). Survival rates were also worse than expected based on historical controls, and subjects in the experimental group had poorer quality of life scores. By any reasonable measure, this trial was a disaster for science and, worse, a disaster for the patients who were enrolled.

All clinical trials, not just RCTs, should be based on scientifically well-supported preclinical observations that justify them, preferably with biomarkers to guide patient selection and follow up. Until specific CAM modalities achieve that level of preclinical evidence in oncology, RCTs testing them cannot be scientifically or ethically justified. Unfortunately, strong scientific and clinical evidence appear no longer to be prerequisites for the widespread adoption of alternative medicine therapies as part of "integrative medicine." Acupuncture, Reiki, and, less frequently, homeopathy are increasingly practiced at academic medical centers, despite little or no convincing scientific evidence (Colquhoun and Novella, 2013) for specific effects distinguishable from placebo for any condition. Indeed, the evidence that most nonpharmacological or dietary CAM modalities, such as acupuncture, are little more than "theatrical placeboes" (Colquhoun and Novella, 2013) is so compelling that some proponents of acupuncture have, in essence, conceded this point by advocating the "harnessing of placebo effects" or developing "meaningful placeboes" (Moerman, 2011).

Recently, the situation has even gotten to the point where clinical guidelines in pseudoscientific treatment are being promoted by organizations like the Society for Integrative Oncology, which has issued clinical guidelines for the integrative oncological care of breast cancer patients coauthored by physicians and naturopaths (Greenlee et al., 2014). One notes that the society accepts as members basically all practitioners of medicine and alternative medicine related to cancer, including acupuncturists, naturopaths, homeopaths, reflexologists, and more. A well-attended session on integrative oncology at the 2014 meeting of the American Society of Clinical Oncology featured approving discussion not only of diet and yoga, but also of acupuncture and "mind-body" interventions.

"Integrating" Quackery: The Future of Medicine?

The pseudoscience at the heart of so many CAM treatments that are being "integrated" into integrative medicine is so pervasive that the scientific "wheel" is being reinvented and sent rolling backward. Medical practitioners are being encouraged to favor crude plant extracts over pure drugs, and to celebrate what can only be described as magical thinking (energy medicine and acupuncture). At the same time, science is telling us more than it ever has before about how the human body functions and malfunctions. Unfortunately, integrative medicine, even as full as it is with such vitalism and the questionable treatments discussed above, has not only found an increasingly respectable niche in medical academia, but is now taught in the undergraduate medical curriculum and in primary care residencies as a recommended core competency (Locke, Gordon, Guerrera, Gardiner, and Lebensohn, 2013). Such programs tend not to be skeptical and science-based, but rather strongly biased toward CAM; the underlying assumption is that the fairy dust being taught actually has value and works (Marcus and McCullough, 2009). In the United States, thanks to integrative medicine popularizer Andrew Weil, there is board certification in integrative medicine for primary care specialties. Although this board certification is granted by a much less widely recognized board than the American Board of Medical Specialties, which oversees board certifications of core medical specialties such as surgery, internal medicine, and pediatrics, given the trajectory of integrative medicine it is not beyond the pale to imagine a time in the not-too-distant future when integrative medicine becomes a fully accepted board certification granted by the ABMS, just like all the other major specialties.

Unfortunately, integrative medicine "integrates" far too much pseudoscience and downright bad science with science-based oncology. I would, however, like to end this chapter on a hopeful note. Returning to Glymour and Stalker (1983), practicing truly holistic medicine does not require rejecting science and embracing pseudoscience. It is possible to introduce scientifically supportable elements of CAM, such as certain dietary and lifestyle interventions, into medicine as science- and evidence-based modalities, and thereby reverse the rebranding of what should be considered evidence- and science-based interventions as somehow "alternative," a rebranding at which integrative medicine has been so successful. There should only be medicine, medicine with strong evidence supporting efficacy and safety. There should be no such thing as "alternative" or "integrative" medicine. We in the medical field can make this happen by embracing critical thinking and teaching it to our medical and nursing students.

References

American Diabetes Association. (2014). Standards of Medical Care in Diabetes—2014. *Diabetes Care, 37*(15): 123–127. Supplement 1, S14–80. DOI: 10.2337/dc14-S014

Anastasi, J. K., Currie, L. M., and Kim, G. H. (2009). Understanding Diagnostic Reasoning in TCM Practice: Tongue Diagnosis. *Alternative Therapies in Health and Medicine, 15*(3): 18–28.

Atwood, K. C. (2003). Naturopathy: A Critical Appraisal. *MedGenMed: Medscape General Medicine, 5*(4): 1–8.

Atwood, K. C. (2004). Naturopathy, Pseudoscience, and Medicine: Myths and Fallacies vs Truth. *MedGenMed: Medscape General Medicine, 6*(1): 33.

Baldwin, A. L., and Hammerschlag, R. (2014). Biofield-Based Therapies: A Systematic Review of Physiological Effects on Practitioners During Healing. *Explore, 10*(3): 150–161. DOI: 10.1016/j.explore.2014.02.003

Berman, B. M., Langevin, H. M., Witt, C. M., and Dubner, R. (2010). Acupuncture for Chronic Low Back Pain. *New England Journal of Medicine, 363*(5): 454–461. DOI: 10.1056/NEJMct0806114

Bhatt, A. (2010). Evolution of Clinical Research: A History Before and Beyond James Lind. *Perspectives in Clinical Research, 1*(1): 6–10.

Bopp, A., & Schürholz, J. (2003). *Anthroposophic medicine: Its nature, its aims, its possibilities*. Dornach, Switzerland.

Boyle, E. W. (2011). The Politics of Alternative Medicine at the National Institutes of Health. *Federal History, 1*(3): 16–32.

Chabot, J. A., Tsai, W. Y., Fine, R. L., Chen, C., Kumah, C. K., Antman, K. A., and Grann, V. R. (2010). Pancreatic Proteolytic Enzyme Therapy Compared with Gemcitabine-Based Chemotherapy for the Treatment of Pancreatic Cancer. *J Clin Oncol, 28*(12): 2058–2063. DOI: 10.1200/JCO.2009.22.8429

Chan, K. (1995). Progress in Traditional Chinese Medicine. *Trends Pharmacol Sci, 16*(6): 182–187.

Colquhoun, D., and Novella, S. P. (2013). Acupuncture Is Theatrical Placebo. *Anesth Analg, 116*(6): 1360–1363. DOI: 10.1213/ANE.0b013e31828f2d5e

Department of Health and Human Services (US). (2007). *What Is Complementary and Alternative Medicine? CAM Basics*. Washington, DC: US Department of Health and Human Services, National Institutes of Health, and National Center for Complementary and Integrative Health.

Department of Health and Human Services (US). (2012). *What Is Complementary and Alternative Medicine? CAM Basics*. Washington, DC: US Department of Health and Human Services, National Institutes of Health, and National Center for Complementary and Alternative Medicine.

Donnell, R. (2008). Exposing Quackery in Medical Education. Retrieved from http://doctorrw.blogspot.com/2008/01/exposing-quackery-in-medical-education.html

Ernst, E. (2006). Acupuncture—A Critical Analysis. *J Intern Med, 259*(2): 125–137. DOI: 10.1111/j.1365-2796.2005.01584.x

Ernst, E. (2008). The Truth About Homeopathy. *Br J Clin Pharmacol, 65*(2): 163–164. DOI: 10.1111/j.1365-2125.2007.03007.x

Ernst, E. (2009). Is Reflexology an Effective Intervention? A Systematic Review of Randomised Controlled Trials. *Medical Journal of Australia, 191*(5): 263–266.

Ernst, E. (2012). Craniosacral Therapy: A Systematic Review of the Clinical Evidence. *FACT, 17*(4): 197–201. DOI: 10.1111/j.2042-7166.2012.01174.x

Glymour, C., and Stalker, D. (1983). Engineers, Cranks, Physicians, Magicians. *New England Journal of Medicine, 308*(16): 960–964. DOI: 10.1056/NEJM198304213081611

Goldman, E. (2016). Facing Huge Demand, Cleveland Clinic Doubles Its Functional Medicine Center. Retrieved from https://www.holisticprimarycare.net/topics/topics-a-g/functional-medicine/1772-facing-huge-demand-cleveland-clinic-doubles-its-functional-medicine-center.html

Goodman, S. N. (1999). Toward Evidence-Based Medical Statistics. 2: The Bayes Factor. *Annals of Internal Medicine, 130*(12): 1005–1013.

Gorski, D. H. (2014). Integrative Oncology: Really the Best of Both Worlds? *Nat Rev Cancer, 14*(10): 692–700. DOI: 10.1038/nrc3822

Gorski, D. H., and Novella, S. P. (2014). Clinical Trials of Integrative Medicine: Testing Whether Magic Works? *Trends Mol Med, 20*(9): 473–476. DOI: 10.1016/j.molmed.2014.06.007

Greenlee, H., Balneaves, L. G., Carlson, L. E., Cohen, M., Deng, G., Hershman, D., . . . Tripathy, D. (2014). Clinical Practice Guidelines on the Use of Integrative Therapies as Supportive Care in Patients Treated for Breast Cancer. *J Natl Cancer Inst Monogr, 50*: 346–358. DOI: 10.1093/jncimonographs/lgu041

Grosso, G., Buscemi, S., Galvano, F., Mistretta, A., Marventano, S., La Vela, V., . . . Biondi, A. (2013). Mediterranean Diet and Cancer: Epidemiological Evidence and Mechanism of Selected Aspects. *BMC Surg, 13 Suppl 2*, S14. DOI: 10.1186/1471-2482-13-S2-S14

Hyman, M. (n.d.) About Functional Medicine. Retrieved from http://www.ultrawellnesscenter.com/becoming-a-patient/about-functional-medicine/

Ioannidis, J. P. (2005). Why Most Published Research Findings Are False. *PLOS Medicine, 2*(8): e124. DOI: 10.1371/journal.pmed.0020124

Ioannidis, J. P. (2012). Biomarker Failures. *Clin Chem, 59*(1): 202–204. DOI: 10.1373/clinchem.2012.185801

Ioannidis, J. P., and Khoury, M. J. (2011). Improving Validation Practices in "Omics" Research. *Science, 334*(6060): 1230–1232. DOI: 10.1126/science.1211811

Klein, E. A., Thompson, I. M., Jr., Tangen, C. M., Crowley, J. J., Lucia, M. S., Goodman, P. J., . . . Baker, L. H. (2011). Vitamin E and the Risk of Prostate Cancer: The Selenium and Vitamin E Cancer Prevention Trial (SELECT). *JAMA, 306*(14): 1549–1556. DOI: 10.1001/jama.2011.1437

Lamas, G. A., Goertz, C., Boineau, R., Mark, D. B., Rozema, T., Nahin, R. L., . . . Investigators, T. (2013). Effect of Disodium EDTA Chelation Regimen on Cardiovascular Events in Patients with Previous Myocardial Infarction: The TACT Randomized Trial. *JAMA, 309*(12): 1241–1250. DOI: 10.1001/jama.2013.2107

Locke, A. B., Gordon, A., Guerrera, M. P., Gardiner, P., and Lebensohn, P. (2013). Recommended Integrative Medicine Competencies for Family Medicine Residents. *Explore, 9*(5): 308–313. DOI: 10.1016/j.explore.2013.06.005

Marcus, D. M., and Grollman, A. P. (2006). Science and Government. Review for NCCAM Is Overdue. *Science, 313*(5785): 301–302. DOI: 10.1126/science.1126978

Marcus, D. M., and McCullough, L. (2009). An Evaluation of the Evidence in "Evidence-Based" Integrative Medicine Programs. *Acad Med, 84*(9): 1229–1234. DOI: 10.1097/ACM.0b013e3181b185f4

Masters, K. S., Spielmans, G. I., and Goodson, J. T. (2006). Are There Demonstrable Effects of Distant Intercessory Prayer? A Meta-Analytic Review. *Ann Behav Med, 32*(1): 21–26. DOI: 10.1207/s15324796abm3201_3

Mease, M. R. (2011). Reiki at University Medical Center, Tucson, Arizona, a Magnet Hospital: Mega R. Mease Is Interviewed by William Lee Rand. *Holist Nurs Pract, 25*(5): 233–237. DOI: 10.1097/HNP.0b013e31822a0291

Mielczarek, E. V., and Engler, B. D. (2012). Measuring Mythology: Startling Concepts in NCCAM Grants. *Skeptical Inquirer, 36*(1): 36–43.

Moerman, D. E. (2011). Meaningful Placebos—Controlling the Uncontrollable. *New England Journal of Medicine, 365*(2): 171–172. DOI: 10.1056/NEJMe1104010

National Center for Complementary and Alternative Medicine (US). (2011). *Third Strategic Plan 2011–2015: Exploring the Science of Complementary and Alternative Medicine.* (NIH Publication No. 11-7643). Washington, DC: US Department of Health and Human Services. Retrieved from http://nccam.nih.gov/about/plans/2011

National Center for Complementary and Integrative Health. (2016). *Complementary, Alternative, or Integrative Health: What's In a Name?* Retrieved from https://nccih.nih.gov/health/integrative-health

Neides, D. (2017). Make 2017 the Year to Avoid Toxins (Good Luck) and Master Your Domain: Words on Wellness. Retrieved from http://www.cleveland.com/lyndhurst-south-euclid/index.ssf/2017/01/make_2017_the_year_to_avoid_to.html

Nissen, S. E. (2013). Concerns About Reliability in the Trial to Assess Chelation Therapy (TACT). *JAMA, 309*(12): 1293–1294. DOI: 10.1001/jama.2013.2778

Office of Cancer Complementary and Alternative Medicine (US). (2011). *Annual Report on Complementary and Alternative Medicine.* Retrieved from http://cam.cancer.gov/cam_annual_report_fy11.pdf

Questionable Methods of Cancer Management: "Nutritional" Therapies. (1993). *CA Cancer J Clin, 43*(5): 309–319.

Reddy, S. (2014). A Top Hospital Opens Up to Chinese Herbs as Medicines, News. *Wall Street Journal*, April 21. Retrieved from http://online.wsj.com/news/articles/SB10001424052702303626 804579509590048257648

Rosa, L., Rosa, E., Sarner, L., and Barrett, S. (1998). A Close Look at Therapeutic Touch. *JAMA, 279*(13): 1005–1010.

Seshia, S. S., and Young, G. B. (2013a). The Evidence-Based Medicine Paradigm: Where Are We 20 Years Later? Part 1. *Can J Neurol Sci, 40*(4): 465–474.

Seshia, S. S., and Young, G. B. (2013b). The Evidence-Based Medicine Paradigm: Where Are We 20 Years Later? Part 2. *Can J Neurol Sci, 40*(4): 475–481.

Singh, S., & Ernst, E. (2009). *Trick or Treatment*. New York: Norton.

Smith, K. (2012a). Against Homeopathy—A Utilitarian Perspective. *Bioethics, 26*(8): 398–409. DOI: 10.1111/j.1467-8519.2010.01876.x

Smith, K. (2012b). Homeopathy Is Unscientific and Unethical. *Bioethics, 26*(9): 508–512. DOI: 10.1111/j.1467-8519.2011.01956.x

Townsend, A. (2014). Cleveland Clinic to Open Center for Functional Medicine; Dr. Mark Hyman to be director, News. *Plain Dealer*, September 22. Retrieved from http://www.cleveland.com /healthfit/index.ssf/2014/09/cleveland_clinic_to_open_cente.html

United States Department of Health and Human Services (2007). *CAM basics*.

United States Department of Health and Human Services (2012). *CAM basics*.

Unproven Methods of Cancer Management: Gerson Method. (1990). *CA Cancer J Clin, 40*(4): 252–256.

Virtamo, J., Pietinen, P., Huttunen, J. K., Korhonen, P., Malila, N., Virtanen, M. J., . . . Group, A. S. (2003). Incidence of Cancer and Mortality Following Alpha-tocopherol and Beta-carotene Supplementation: A Postintervention Follow-Up. *JAMA, 290*(4): 476–485. DOI: 10.1001/jama.290.4.476

Wang, C., Schmid, C. H., Rones, R., Kalish, R., Yinh, J., Goldenberg, D. L., . . . McAlindon, T. (2010). A Randomized Trial of Tai Chi for Fibromyalgia. *New England Journal of Medicine, 363*(8): 743–754. DOI: 10.1056/NEJMoa0912611

White, A. R., Williamson, J., Hart, A., and Ernst, E. (2000). A Blinded Investigation into the Accuracy of Reflexology Charts. *Complement Ther Med, 8*(3): 166–172. DOI: 10.1054/ctim.2000.0380

Zou, L. Y., Yang, L., He, X. L., Sun, M., and Xu, J. J. (2014). Effects of Aerobic Exercise on Cancer-Related Fatigue in Breast Cancer Patients Receiving Chemotherapy: A Meta-Analysis. *Tumour Biol*. DOI: 10.1007/s13277-014-1749-8

15 Hypnosis: Science, Pseudoscience, and Nonsense

Steven Jay Lynn, Ashwin Gautam, Stacy Ellenberg, and Scott O. Lilienfeld

Hypnosis, long regarded as a mysterious or even occult phenomenon, has slowly but surely made inroads into the scientific mainstream. Starting in the late twentieth century, numerous researchers have conducted (a) rigorous studies to elucidate the determinants of hypnotic phenomena, and (b) clinical trials to examine the value of hypnosis as an adjunctive treatment for an impressive array of psychological and medical conditions, ranging from acute stress disorder to dermatological problems.

Still, the media, popular lore, and even some members of the professional community latch onto views of hypnosis with no scientific backing and advance claims about hypnosis that are pseudoscientific and unsubstantiated at best, and patent nonsense at worst. These claims are often at face improbable. Some of them resound with the bells and whistles of science, but are based largely on anecdotes and testimonials. Nevertheless, because they appeal to people's wishes, hopes, and fantasies, these claims have acquired a strong foothold within the popular imagination. As we will discover, many of the claims for hypnosis rest on popular myths. In this chapter, we will present evidence garnered from the scientific literature to counter or refute these prevalent yet mistaken beliefs.

Our discussion expands from the root myth about hypnosis, from which other "mythlets" branch: the idea that hypnosis is a special or unique state of consciousness that can confer on the hypnotized participant's special or unique abilities or powers. This idea has populated everyday culture since the time of Franz Anton Mesmer (1734–1815), when dramatic alterations in hypnotized participants' actions, emotions, and appearance persuaded credulous onlookers that there was something special about hypnosis. By the later eighteenth century, some scholars advanced claims that hypnosis empowered individuals to perform astral travel, see through the skin without the use of the eyes, and experience paranormal and supernatural occurrences.

In 1784, Benjamin Franklin headed a commission that found the effects of "mesmerism" were the products of belief and imagination. (Franklin was the American ambassador to France at the time, and his fellow commission members included the chemist Antoine Lavoisier and the physician Joseph-Ignace Guillotin, who would become famous for his own unique solution to the mind-body problem.) The debunking was complete by the mid-nineteenth century. Nevertheless, the idea that hypnosis inculcates a special state of consciousness has survived. Consider the words of John Gruzelier (1996) who contended, "We can now acknowledge that hypnosis is indeed a 'state.'" In the remainder of this chapter, we challenge this perspective and cast a light on corollary myths and misconceptions about hypnosis.

Popular Media and Hypnosis

Media dramatizations illustrate the staying power of the myth of hypnosis as producing a trancelike state, often depicted as one in which people are powerless to resist suggestions and must behave in a mindless, robotic way. Lilienfeld, Lynn, Ruscio, and Beyerstein (2010) cited the following movies, in which everyday people will: (a) commit an assassination (*The Manchurian Candidate*), (b) commit suicide (*The Garden Murders*), (c) disfigure themselves with scalding water (*The Hypnotic Eye*), (d) assist in blackmail (*On Her Majesty's Secret Service*), (e) perceive only a person's internal beauty (*Shallow Hal*), (f) steal (*Curse of the Jade Scorpion*), (g) access past lives (*Dead Again*), and (h) fall victim to brainwashing by alien preachers who use hypnotic messages in sermons (*Invasion of the Space Preachers*).

The first inkling of the power of such representations of hypnosis to capture the public imagination can be traced to George du Maurier's novel, *Trilby* (1894), in which Svengali—whose name today connotes a ruthless manipulator—uses hypnosis to dominate the young singer Trilby. The novel popularized the idea of hypnosis as associated with powerlessness, loss of control, special abilities, amnesia, and a sleep-like trance state—myths that endure in many current depictions of hypnosis.

Today, the Internet is a thriving medium for the propagation of pseudoscience and balderdash of various guises, and hypnosis is no exception. Some websites make preposterous and totally improbable claims. One site, entitled Hypnosis Black Secrets, touts a course entitled "7 Steps of Feminization Hypnosis." Readers are told, "Feminization of Body Induction helps to transform each part of your body, hair, your figure, voice and so on." For example, "You successfully transform your body into a desired feminine shape," or "You are changing your voice into a woman's voice." Other sites are no less restrained in their promises. The site Mindfit Hypnosis is sponsored by Dr. Dobson (PhD) who avers that he is a "real psychologist" and offers MP3 hypnosis sessions in breast and penis enlargement, "get high hypnosis" (like on drugs) and "think and grow rich hypnosis." For those looking to balance their chakras, hypnotize animals, and explore alien abduction experiences "help" in achieving these goals and many more can be found on the Internet.

Stage hypnosis, also called performance hypnosis, capitalizes on a variety of misconceptions about hypnosis. Viewing raucous and often hilarious shenanigans of people onstage who bark like dogs, disrobe, fall asleep on command, or do a second-rate impersonation of a strutting Mick Jagger, to name a few examples, can convey the impression that hypnosis compels total control of a person's actions. Yet audiences are unaware of the fact that onstage the performance is not the product of a trance state. Rather, it is the product of one or more of the following: (a) social and compliance demand characteristics to put on a compelling, entertaining performance, in front of an expectant audience, (b) ordinary waking suggestibility and selection of highly suggestible people based on their responses to a few waking suggestions provided by the hypnotist, (c) outright trickery (e.g., bringing one or more shills to the show who are paid to do exactly what the hypnotist suggests), and (d) straightforward exhortations to play along (e.g., whispering in the person's ear, "Let's not disappoint them and give them a good show."). In short, despite appearances, stage hypnosis does not demonstrate that hypnosis engenders a special state of consciousness.

Is Hypnosis a Trance State?

Before we address the question of whether hypnosis produces a trance state, as media representations and stage hypnosis would have us believe, we first need to define what hypnosis is. In our view, hypnosis is a situation in which imaginative suggestions for changes in thoughts, feelings, and actions are provided to a person in a context defined as "hypnosis" (Lynn, Laurence, and Kirsch, 2015). Imaginative suggestions are requests to experience an imaginary state of affairs as if it were real (Kirsch and Braffman, 2001). Imaginative suggestions can be provided with or without the formal induction of hypnosis. Hypnotic inductions consist of any number of suggestions or procedures that individually or in total imply, directly or indirectly, that the person will experience or "enter" hypnosis, thereby establishing the context as "hypnotic."

Many people are surprised to learn that the induction of hypnosis confers little advantage in suggestibility compared with the same imaginative suggestions administered without a prior induction. As the famed American psychologist Clark Henry Hull (1933) commented that the effect of hypnosis over and above waking suggestions is unimpressive, "probably far less than the classical hypnotists would have supposed had the question ever occurred to them" (p. 298). When differences between the same suggestions administered in a hypnotic and nonhypnotic state are found, they are typically small, hovering around 1.5 suggestions out of 12 (Barber, 1969; Hilgard, 1965; Kirsch and Lynn, 1995). Accordingly, hypnosis does not induce a state of greatly heightened suggestibility beyond waking suggestibility, as would be expected if (a) hypnosis produced an altered state marked by enhanced responsiveness to suggestions, and (b) the induction of an altered state constituted a sufficient or adequate explanation for (or precursor to) dramatically enhanced hypnotic suggestibility.

Depictions in the movies and on the Internet push relentlessly the notion that there is something unique about hypnotic inductions, and that some inductions are more special than others. For example, the site Hypnosis Black Secrets offers instructions on "How to Hypnotize Someone to Have Sex" and "Street Hypnosis" (hypnosis performed in the street), and boast of a "Method that Hypnotists Use to Hypnotize in 3–6 Seconds." Inductions ranging from the classic pocket watch dangling from a chain to the hypnotist staring directly into the participant's eyes and intoning suggestions in a commanding voice are by now seamlessly stitched into our cultural psyche.

Yet research shows that the nature of the induction has little or no bearing on hypnotic responsiveness. It does not matter whether the induction contains suggestions worded in a permissive versus authoritative manner, or whether they are direct and traditionally worded or indirect and open ended (Gibbons and Lynn, 2010). Interestingly, when the induction presents hypnosis as an altered state of consciousness, participants are *less* likely to respond to suggestions than when it is presented as involving cooperation (Lynn, Vanderhoff, Shindler, and Stafford, 2002).

One of the special properties that many have ascribed to hypnosis is the feeling of involuntariness that often (but not always) accompanies suggestion-related responses. Inductions often include suggestions with the implication that responses will occur involuntarily or automatically. For example, the hypnotist might say, "You will respond to the suggestions that follow easily and effortlessly." Moreover, the suggestions that follow the induction often include wording that includes imagery that calls for the response to be experienced with a sense of involuntariness or automaticity, as in the following suggestion: "Imagine that your hand is attached to helium balloons, and the helium balloons are lifting your hand off the resting surface, as the helium balloons rise up and up in the air, your hand rises up and up, quite automatically on its own, right off the resting surface." These sorts of suggestions provide an imaginative strategy to facilitate responsiveness and clear cues or demands for involuntary responding.

Not surprisingly, participants who respond to such suggestions often report that their hand lifted effortlessly into the air. Yet when participants receive the same imaginative suggestions in the absence of an induction, they still report suggestion-related involuntariness. Researchers typically find a substantial correlation between subjective and behavioral hypnotic responsiveness (number of suggestions passed, $r=.82$ for subjective scores and $r=.67$ for behavioral scores; Braffman and Kirsch, 1999) with respect to the same imaginative suggestions (e.g., hand levitation, visual hallucination) administered across hypnotic and nonhypnotic (i.e., no prior induction) contexts. In short, despite media hype and dramatizations, there is nothing particularly special, let alone unique, about responding to hypnotic suggestions, even in terms of subjective experience.

Although hypnotic inductions can elicit a wide range of changes in cognition, emotion, memory, sensation, and perception, these changes are largely the by-product of three variables, shown to account for as much as 50 percent of variability in hypnotic responsiveness: motivation, imaginative suggestibility, and expectancies (Braffman and Kirsch, 1999; Meyer and Lynn, 2011). From the time of the Franklin Commission, skeptics have argued that the seemingly miraculous powers of hypnosis were largely attributable to participants' expectations regarding hypnosis. Research continues to accord with this view. The relatively small shifts in hypnotic suggestibility that hypnotic induction produces may be due to (a) widely prevalent culturally based expectancies that hypnosis boosts responsivity and (b) enhanced motivation to respond, which is often implicitly or explicitly (e.g., "Do your very best to respond to the suggestions") conveyed by the induction and the hypnotic context more broadly.

Researchers have systematically studied the role of expectancies in establishing when a particular response occurs or does not occur, as well as the nature of the response to suggestions. One of the myths about hypnosis, which is the linchpin of stage hypnosis shows, is that most people cannot resist suggestions. In fact, the ability to resist suggestions can be altered greatly by what people are told about that ability (Lynn, Nash, Rhue, Frauman, and Sweeney, 1984; Silva and Kirsch, 1987; Spanos, Cobb, and Gorassini, 1985).

Researchers have shown that individuals informed that hypnotized people can resist suggestions succeed in resisting suggestions, whereas individuals told that hypnotized people cannot resist suggestions report an inability or diminished ability to resist (Lynn et al., 1985). Similarly, "spontaneous" amnesia for the experience of hypnosis, a familiar fixture of popular media depictions, is limited to people who expect to be amnesiac (Young and Cooper, 1972). Henry (1985) reported that peoples' preconceptions about hypnosis mediated whether they reported time slowing down or speeding up, logical thought becoming easier or more difficult, the hypnotist's voice sounding closer or farther away, sounds being clearer or more muffled, and so forth. It seems that in hypnosis, you get what you expect (Lynn and Kirsch, 2006).

Rather than the ability to achieve a trance state, research supports the following factors as influential in determining the outcome of a hypnotic induction: (a) participants' prehypnotic expectations, attitudes, beliefs, and intentions regarding hypnosis; (b) their motivation and ability to become absorbed in suggestions and to think and imagine along with suggested activities; (c) their ability to establish rapport with the hypnotist; (d) their ability to interpret suggestions appropriately and view their responses as successful; (e) their ability to discern task demands and cues; and (f) the nature of the hypnotist's and participant's ongoing interaction (Barber, 1985, Gibbons and Lynn, 2010; Lynn et al., 1996).

Hypnotic Phenomena

A corollary of the view that hypnosis is a special state is the idea that certain phenomena typify this state, which is marked by definable properties. To date, investigators have provided no compelling evidence for such a state. In addition, they have generated alternative explanations for "hypnotic" phenomena that do not depend on the instantiation of an "altered" or unique hypnotic state. Although suggestion can indeed produce dramatic alterations in consciousness, these alterations are evident in nonhypnotic as well as hypnotic contexts and/or are demonstrably the product of implicit or explicit expectancies. We now consider these findings in more detail in light of a number of alleged markers of a hypnotic state.

Hypnosis Produces a Sleep-Like State

Surely one of the most popular images of hypnosis is the slack-jawed subject slumping in his or her chair, deeply immersed in a sleep-like trance yet still responsive to suggestion. By the time that Braid (1843) coined the term neurohypnosis (from the Greek word *hypno*, meaning sleep), which was shortened to hypnosis, the view of hypnosis as a sleep-like trance phenomenon was firmly entrenched in popular consciousness. Today, researchers have discredited the idea that hypnosis produces a sleep-like state, as inductions that call for mind expansion and alertness (called "alert hypnosis," Gibbons, 1973; Gibbons and Lynn, 2010; Wark, 2006) are just as effective as are sleepy-drowsy inductions in enhancing hypnotic suggestibility. Moreover, regardless of whether the induction centers on relaxation or suggestions are delivered while the participant pedals a stationary bicycle, the effect on suggestibility is the same (Bányai, 1991). Perhaps most tellingly, the pattern of electrophysiological activity in the brain during hypnosis varies in tandem with suggested images and tasks and does not mirror patterns of activity typically observed during sleep (Lynn and Kirsch, 2006). Indeed, electroencephalogram (EEG) data clearly demonstrate that hypnotized participants are awake.

Literalism Is a Marker of Hypnosis

Some theoreticians (Pattie, 1956; Shor, 1962; Weitzenhoffer, 1957) have argued that the hypnotic state is characterized by a penchant for literal responding. Milton Erickson (1980), a famed clinical hypnotist, stated that he assessed literalism in 1,800 hypnotized individuals and 3,000 responses in nonhypnotized persons by asking questions, such as "Do you mind telling me your name?" A literal response would be scored as "no" to the question. Erickson claimed that 95 percent of individuals responded nonliterally by stating their names, whereas between 80 and 97 percent of hypnotized individuals, depending whether they were in a "light" or "deep trance," responded literally (e.g., replied "no" or exhibited a negative shaking of the head).

Nevertheless, research conducted in our laboratory (Green et al., 1990) found that less than 30 percent (29.2 percent) of individuals who tested in the "highly suggestible" range on two measures of hypnotic suggestibility responded literally to a series of questions, whereas low-suggestible role-playing participants, asked to simulate the performance of what they believed to be an excellent hypnotic subject, exhibited twice the rate of literal responding (58.3 percent). Moreover, contrary to Erickson's claim that literal responses are rarely displayed outside the context of hypnosis, we found that more than a fifth (21.78 percent) of students approached in or outside the campus library responded in a literal manner, a rate comparable to that of our highly suggestible subjects tested in the context of hypnosis.

In a second study (Lynn et al., 1990), we found that highly hypnotizable individuals who received "waking" instructions to try their best to respond (i.e., task-motivating instructions) performed on a par with highly hypnotizable individuals tested during hypnosis: interestingly, only 12.5 percent of hypnotizable participants' responses were nonliteral. In terms of explaining the genesis of literal responses, such responses were associated not only with participant perceptions that literal responses (e.g., responding "yes" or "no" to a question) were appropriate, but also with feelings of passivity such that they did not want to expend effort to respond in a nonliteral manner.

Hypnosis and Trance Logic
For many years, it was taken as a virtual given that one of the key features of the hypnotic state was the ability to tolerate or not be fazed by logical inconsistencies that would not be tolerated in everyday life (e.g., following a suggestion to experience hypnotic deafness, responding "no" in response to the question "Can you hear me now?"). In his classic paper on the nature of hypnosis, Orne (1959) defined *trance logic* (TL) as the ability to mix reality-based perceptions with those that stem from imagination in a manner that ignores everyday rational standards. The image transparency response is the index of TL that has garnered the most research attention. Image transparency is captured in Orne's (1959) recounting of a participant's report of transparency as follows: "This is very peculiar; I can see Joe sitting in the chair and I can see the chair through him." Although it is understandable that such reports might be accounted for in terms of TL, researchers have proffered a more parsimonious explanation, supported by numerous studies (e.g., Spanos, deGroot, and Gwynn, 1987; Stanley, Lynn, and Nash, 1986): transparency reports merely reflect the fact that even high-hypnotizable individuals are not able to maintain a realistic, lifelike hallucination of a person or object. In short, there is nothing illogical about the failure to maintain a solid, compelling hallucination, whether during hypnosis or not. Indeed, nonhypnotized participants report image transparency just like hypnotized individuals (see Lynn and Rhue, 1991, for a review). Again, what is purported to be a unique feature of hypnosis is not limited to the hypnotic context and can be explained in mundane terms.

The Hidden Observer
In the case of the *hidden observer* phenomenon, what initially appeared to be a feature of hypnosis was shown to be the product of suggestion-induced expectancies. According to Ernest Hilgard (1991), a dissociated or concealed part of the person, which is aware of memories or experiences that are unavailable to the person under ordinary

circumstances, can be contacted during hypnosis. For example, in a pain study in which analgesia suggestions are provided, the purportedly dissociated part of the personality contacted by the experimenter—the "hidden observer"—typically reports higher levels of pain than does the "hypnotized" part.

Researchers have mounted a strong challenge to Hilgard's interpretation of such findings as supporting a dissociation theory of hypnosis by showing that hidden observer reports vary markedly as a function of interpretations of suggestions or instructions provided by the experimenter. For example, Spanos and Hewitt (1980) found that participants' pain reports of the hidden observer depended on whether they were told that: (a) the hidden part would continue to feel high levels of pain while their hypnotized part experienced diminished pain, or (b) their hidden part was so deeply hidden that it would be even less aware of what had been experienced than their hypnotized part. Participants responded in keeping with suggestions that their hidden part was either more or less aware of pain than their hypnotized part. This and another study (Spanos, Gwynn, and Stam, 1983), which varied the explicitness of cues regarding the nature of the hidden observer, provide strong evidence that the hidden observer phenomenon is not an intrinsic feature of hypnosis, but is the product of suggestions and expectancies, much like many other hypnotic phenomena and everyday experiences (Kirsch and Lynn, 1998; Spanos, 1991).

Hypnosis and the Stroop Effect

More recent investigations have provided further serious challenges to the "special" nature of hypnosis. The well-known Stroop interference effect (SIE), also known as "Stroop conflict," arises when participants are asked to identify the color of a word (RED) that is printed in an incompatible color (BLUE). In the presence of this conflict, color-naming speed (and sometimes accuracy) will be typically impaired relative to when participants are asked to identify the color of a word (RED) that is printed in a congruent color (RED).

Raz, Shapiro, Fan, and Posner (2002) found that a posthypnotic suggestion to see the words in a foreign (nonnative or previously learned) language completely eliminates the SIE among highly suggestible participants. Performance on the Stroop test has been described as the "gold standard" for studying attentional processes (see Raz and Shapiro, 2002). Attempts to eliminate Stroop conflict, which appears to arise from automatic processing of words at the semantic level (MacLeod, 1991) and response competition (Raz et al., 2002), have generally met with little success, and when they have been reported, have generally failed to conform to the standard Stroop procedure (MacLeod and Sheehan, 2003). Accordingly, Raz interpreted his findings to mean that highly suggestible subjects have "extreme attentional abilities," and MacLeod and Sheehan (2003) noted that Stroop results "suggested a genuine ability to ignore the words . . . something that people cannot do ordinarily, even after extensive practice and given considerable effort" (p. 349).

In a second study, providing the same suggestions used by Raz et al., but not presented in the context of hypnosis, also significantly reduced Stroop conflict (Raz, Kirsch, Pollard, and Nitkin-Kaner, 2006). What initially appeared to be a "hypnosis effect," appears, on closer inspection, to be a "suggestion effect." Suggestions to see the words in a foreign language, whether presented during hypnosis or not, provide participants with a workable strategy to overcome the SIE.

Hypnosis Eliminates the Optokinetic Reflex
The most recent high-profile claim for hypnotic distinctiveness or uniqueness centers on research with a single highly responsive hypnotic subject who displayed a glassy-eyed "hypnotic stare," reminiscent of media portrayals of a person in a trance (Kallio et al., 2011). The researchers reported that this individual could inhibit normally occurring automatic eye movements (i.e., the optokinetic reflex) when she was asked to focus her gaze at a fixation point at the center of a computer screen while lines moved vertically down the screen. Among typical participants, this setup produces automatic eye movements away from the fixation point. The researchers argued that they had produced conclusive evidence that hypnosis causes an altered state of consciousness.

We (Lynn, Inhoff, Maxwell, and Lemons, 2016) tested six highly hypnotizable participants screened on two measures of hypnotizability in an attempt to replicate these intriguing findings. In a direct replication attempt, we found that none of our six participants could inhibit the optokinetic reflex. But when we provided them with a suggestion that provided a strategy to potentially produce the desired effect, (i.e., "Imagine that there is an object, a ball for instance, that is hovering somewhere in between you and the computer monitor. Fixate your eyes on this ball and ignore the screen, while keeping your eyes straight.") one participant was able to successfully inhibit automatic eye movements. These findings do not support the hypothesis that a special state induced by hypnosis is responsible for largely vitiating eye movements; rather, strategy use may be key to producing the dramatic effect reported by Kallio and his colleagues (Kallio et al., 2011). Even if the participant in the original study somehow managed to produce the response without using this strategy, we failed to replicate the effect in all the participants tested under the initial testing conditions with no strategy provided. In summary, researchers have failed to identify a single consistent marker of a special or altered state of hypnosis.

Myths and misconceptions regarding hypnosis and the attribution of a special state to hypnosis can have serious personal and social consequences. For example, when used for memory recovery to bolster eyewitness testimony, hypnosis can lead to grave miscarriages of justice. Moreover, in the context of psychotherapy, hypnosis can be misused to the possible detriment of patients and their family members.

Hypnosis and Recent Memory

Scholars have long claimed that hypnosis improves memory; that is, that it exerts a hypermnesic effect such that it enhances recall above and beyond nonhypnotic conditions (Pintar and Lynn, 2009). Yet scientific evidence provides a robust challenge to this claim. By the late 1970's (Orne, 1979), a small number of reports began to add a tincture of skepticism to claims regarding the ability of hypnosis to enhance recall. For example, Timm (1981) noted that several hypnotized subjects reported with certitude that the mock assassin they were attempting to identify was unkempt and wearing a khaki-colored army jacket and gloves. In reality, the person was immaculately groomed, donned a blue formal jacket, and was not wearing gloves. Beyond such anecdotes, a steady stream of studies that cast doubt on the reliability of hypnotically augmented recall appeared in the 1980's.

By 1994, the courts in most of the forty cases reportable concluded that suggestibility effects (e.g., inaccurate recall and unwarranted confidence in memories independent of accuracy) justified the exclusion of hypnotically elicited testimony (Wakefield and Underwager, 1998). At the present time, twenty-seven states have barred hypnotically elicited testimony from the courtroom, although some states admit such testimony (subject to cross-examination) when certain procedures (e.g., videotaping of the hypnotic proceedings to assess pre-hypnotic knowledge and rule out the possibility of leading questions) are followed scrupulously.

Twenty-three studies have shown inflated confidence rates for hypnotically augmented memories, a phenomenon known as "memory hardening." Nine studies of the studies provided no evidence for enhanced confidence across hypnotic and nonhypnotic conditions. In five of the studies (Putnam, 1979; Ready, Bothwell, and Brigham, 1997; Sanders and Simmons, 1983, Scoboria, Mazzoni, Kirsch, and Milling, 2002; Yuille and McEwan, 1985), hypnosis produced more errors or less accurate information on some or all measures. In all of the remaining studies (Gregg and Mingay, 1987; Mingay, 1986; Scoboria, Mazzoni, Kirsch, and Jimenez, 2006; Spanos et al., 1989), with one exception (Terrance, Matheson, Allard, and Schnarr, 2000), there were no differences in memory accuracy across hypnotic and nonhypnotic conditions. In sum, the findings support Steblay and Bothwell's (1994) conclusion: "Hypnosis is not necessarily a source of accurate information; at worst it may be a source of inaccurate information provided with confident testimony" (p. 649), which can lead to false identifications and wrongful convictions of innocent people as perpetrators of crimes.

Hypnosis and Distant Memories

If memories of recent events recalled through hypnotic interventions can be shown to be demonstrably false, what about memories of the more distant past and even of past lives? Based on his review of more than sixty years of research on *hypnotic age regression* (a technique in which a subject is asked to respond to specific hypnotic suggestions to think, feel, or act like a child at a particular age), Nash (1987) concluded that the behaviors and experiences of age-regressed adults were often different from those of actual children. No matter how compelling "age-regressed experiences" appear to observers, they reflect participants' expectancies, beliefs assumptions, and fantasies about childhood; they rarely, if ever, represent literal reinstatements of childhood experiences, behaviors, and feelings. For example, as Nash (1987; see also Nash, Drake, Wiley, Khalsa, and Lynn, 1986) demonstrated, adults age-regressed to childhood do not exhibit the EEG patterns typical of children, nor do they perform similarly to children on standardized measures of intelligence, thinking, or visual perception, including optical illusions that children, but not adults, tend to experience. Subsequent studies have found that by manipulating expectances about recall of events during early life, it is possible to instantiate memories with hypnosis and imaginative suggestions well before the cutoff of infantile amnesia (age two), a time at which scientists (see Malinoski, Lynn, and Sivec, 1998) agree that memories are highly unlikely to be inaccurate. Researchers have structured expectancies such that most participants who were age-regressed with hypnotic and nonhypnotic procedures reported memories of infancy (Spanos et al., 1999; see also Terrance et al., 2000).

Hypnosis and Past-Life Memories

There is a vibrant cottage industry of therapists who make extravagant claims about the power of past-life therapy to salve current psychological wounds. In 2008, Dr. Mehmet Oz introduced Dr. Brian Weiss, a psychiatrist with a reputable academic pedigree, to viewers of *The Oprah Winfrey Show* (Oz, 2008). Some twenty years prior to his appearance on the show, Weiss (1988) had published a widely publicized series of cases of patients he had hypnotized and age-regressed to "go back to" the origin of a present-day problem. When patients were regressed, with incantations such as "Go through the door and through the light . . . Join the scene where the experience or the person on the other side of the light is in a past life," Weiss interpreted the events they reported as the sources of their current problems in living. He further contended that recalling such memories could alleviate their present-day psychological problems. As of this writing, Weiss actively promotes his past-life therapy not only on national television, but also in large workshops (often at convention centers) on a national and international level. There, participants can supposedly experience past lives and, in between recalling their past stints as Napoleon or Cleopatra, can purchase his books and CDs, and can become certified in past-life regression therapy.

Weiss is far from alone in such promotional efforts. The Internet provides vast opportunities for interested parties to receive training in past-life therapy to "heal the eternal soul" with past-life regression therapy, buy software to facilitate past-life regressions, and, of course, buy copious books, CDs, and DVDs on the subject.

With all the hoopla about past lives so readily available, one might ask, "What does psychological research have to contribute to our understanding of past-life experiences?" Certainly a key question is whether the memories reported during such experiences are accurate or merely chimera of imaginative constructions. In a clever laboratory study, Spanos, Menary, Gabora, DuBreuil, and Dewhirst (1991) found that the information participants provided about specific time periods during their hypnotic past-life regression was almost "invariably incorrect" (p. 137). For example, one participant who was regressed to ancient times claimed to be Julius Caesar, "emperor" of Rome, in 50 BC, even though the designations of BC and AD were not adopted until centuries later, and Julius Caesar died years prior to the first Roman emperor. Spanos et al. (1991) informed some participants that past-life identities were likely to be of a different gender, culture, and race from that of the present personality, whereas other participants received no prehypnotic information about past-life identities. Participants' past-life experiences were elaborate, conformed to induced expectancies about past-life identities (e.g., gender, race), and varied in terms of the pre-hypnotic information participants received about the frequency of child abuse during past historical periods. In sum, Occam's razor (the principle of parsimony) holds that hypnotically induced past-life experiences are most likely to be fantasies constructed from available cultural narratives about past lives and known or surmised facts regarding specific historical periods, as well as cues in the hypnotic situation (Spanos, 1996).

On *The Oprah Winfrey Show*, Dr. Oz suggested that past-life regression, whether accessing true memories or not, "has so much wonderful healing potential." In contrast, we would counter that anecdotal reports are a flimsy basis to contend that an intervention is effective. There have been no empirical tests of the short-term, much less the long-term, effectiveness of past-life therapy. Moreover, even if such therapy generates short-term gains, they are probably attributable to nonspecific or placebo effects, demand characteristics, regression to the mean, or any number of other potential sources of apparent gain apart from age regression (see Lilienfeld, Ritschel, Lynn, Cautin, and Latzman, 2014). Finally, participants who engage in past-life therapies are not only participating in potentially costly and time-consuming interventions lacking empirical support, but they also are forgoing treatment with effective methods, such as behavioral and cognitive-behavioral therapies.

Hypnosis and Psychotherapy

Up to this point, we have dwelled on the dangers and misuses of hypnosis. At the same time, there is accumulating evidence that hypnosis, when properly administered, may be helpful in the treatment of a number of psychological and medical conditions. Narrative reviews and meta-analyses (mathematical syntheses of the literature) document the effectiveness or promise of hypnosis in treating an array of psychological and medical conditions ranging from acute and chronic pain to obesity (see Lynn et al., 2000; Elkins, 2016, for reviews). Furthermore, meta-analyses have shown that hypnosis enhances the effectiveness of both psychodynamic and cognitive behavioral psychotherapies (Kirsch, 1990; Kirsch, Montgomery, and Sapirstein, 1995). That said, researchers have some way to go in terms of identifying a specific effect of hypnosis or any altered state of hypnosis in producing salutary effects, apart from the effects of relaxation, placebo effects, and other nonspecific effects.

Conceptualizing hypnosis primarily in terms of an altered state obscures the real and often profound potential of waking suggestions to increase the pliability of consciousness and behavior. The primacy of the altered-state view also relegates to secondary status efforts to examine the crucial social-cognitive variables (e.g., expectancies, imaginative processes, rapport) that mediate responsiveness to suggestion and how hypnosis can be used as a vehicle to create analogues of psychopathology such as delusions. We suggest that hypnosis can be a viable means of examining involuntary versus controlled processes in cognition, automaticity in human action and thought, the link between imagination and behavior, expectancies and response sets, and the attribution of behaviors to the self versus others. We are skeptical that the view of hypnosis as an altered state of hypnosis will prove to be utilitarian in these investigations.

Conclusions

It may turn out that a case can be made that hypnosis produces an altered state of consciousness at some time in the future. To make progress on the theoretical or investigative front of such a possibility, it will be necessary for investigators to carefully specify in advance what such a state might be or how it would be detected. This has been a problem in the area of brain imaging research, where any indications of differences in activation or brain areas have been attributed to state-like changes in consciousness. Furthermore, studies have typically lacked adequate controls for relaxation, imagination, and demand characteristics (Lynn et al, 2009). Efforts to identify an altered state are particularly challenging in that "ordinary experiences are not static and instead reflect constant and often transient changes in our awareness, emotions, thoughts, sensations, and action tendencies that are typically difficult, if not impossible, to parse into discrete states" (Lynn, Laurence, and Kirsch, 2015, p. 315).

For hundreds of years, hypnosis has struggled to find a respected and valued place in the panoply of science. Indeed, as we have seen, many widespread claims regarding hypnosis have embodied the core features of pseudoscience, such as an overreliance on anecdotal evidence ("anecdata"), extravagant assertions that go well beyond the available data, inadequate linkages with established scientific findings ("connectivity"), and a failure to self-correct in the face of negative evidence (see Lilienfeld, Lynn, and Lohr, 2015 for a discussion).

Despite the myths and misconceptions surrounding hypnotic phenomena, the scientific quest to study hypnosis has advanced beyond hokum and pseudoscience to move hypnosis increasingly into the mainstream of social, cognitive, and behavioral science. Efforts to tease apart science and pseudoscience, and fact from fiction, in the study of hypnosis should help to ensure that hypnosis continues to attract the interest of researchers and clinicians who appreciate its potential to increase our understanding of basic psychological processes and to mitigate human suffering.

References

Bányai, É. I. (1991). "Toward a Social-Psychobiological Model of Hypnosis." In S. J. Lynn and J. W. Rhue (Eds.), *Theories of Hypnosis*, 564–600. New York: Guilford.

Barber, T. X. (1969). *Hypnosis: A Scientific Approach*. New York: VanNostrand Reinhold.

Barber, T. X. (1985). "Hypnosuggestive Procedures as Catalysts for Psychotherapies." In S.J. Lynn, and J. P. Garske (Eds), *Contemporary Psychotherapies: Models and Methods*, 333–375. Columbus, OH: Charles E. Merrill.

Braffman, W., and Kirsch, I. (1999). Imaginative Suggestibility and Hypnotizability: An Empirical Analysis. *Journal of Personality and Social Psychology*, 77(3): 578.

Braid, J. (1843). *Neurypnology; or, the Rationale of Nervous Sleep, Considered in Relation with Animal Magnetism. Illustrated by Cases of Its Application in the Relief and Cure of Disease*. London, UK: John Churchill.

Brown, D. (2007). Evidence-Based Hypnotherapy for Asthma: A Critical Review. *International. Journal of Clinical and Experimental Hypnosis*, 55(2): 220–249.

Corey Brown, D., and Corydon Hammond, D. (2007). Evidence-Based Clinical Hypnosis for Obstetrics, Labor and Delivery, and Preterm Labor. *International Journal of Clinical and Experimental Hypnosis*, 55(3): 355–371.

du Maurier, G. (1894). *Trilby*. (Reprinted 1999, New York: Oxford University Press.)

Elkins, G. (2016). *Clinician's Guide to Medical and Psychological Hypnosis: Foundations, Applications, and Professional Issues*. New York: Springer.

Erickson, M. H. (1980). *The Nature of Hypnosis and Suggestion,* vol. 1. New York: Irvington Press.

Gibbons, D. E. (1973). Beyond Hypnosis: Explorations in Hyperempiria. South Orange, NJ: Power Publishers.

Gibbons, D. E., and Lynn, S. J. (2010). "Hypnotic Inductions: A Primer." In S. J. Lynn, J. W. Rhue, and I. Kirsch (Eds.), *Handbook of Clinical Hypnosis* (2nd ed.), 267–293. Washington, DC: American Psychological Association.

Green, J. P., Lynn, S. J., Weekes, J. R., Carlson, B. W., Brentar, J., Latham, L., and Kurzhals, R. (1990). Literalism as a Marker of Hypnotic "Trance": Disconfirming Evidence. *Journal of Abnormal Psychology, 99*(1): 16.

Gregg, V. H., and Mingay, D. J. (1987). Influence of Hypnosis on Riskiness and Discriminability in Recognition Memory for Faces. *British Journal of Experimental & Clinical Hypnosis, 42*: 65–75.

Gruzelier, J. (1996). The State of Hypnosis: Evidence and Applications. *Quarterly Journal of Medicine-Oxford, 89*: 313–318.

Henry, D. C. (1985). *Subjects' Expectancies and Subjective Experience of Hypnosis* (doctoral dissertation). ProQuest Information & Learning.

Hilgard, E. R. (1965). Hypnosis. *Annual Review of Psychology, 16*(1): 157–180.

Hilgard, E. R. (1991). A Neodissociaiton Interpretation of Hypnosis. In S. J. Lynn, and J. W. Rhue (Eds.). *Theories of Hypnosis,* 83–104. New York: Guilford.

Hull, C. L. (1933). *Hypnosis and Suggestibility*. New York: Appleton-Century-Crofts.

Kallio, S., Hyönä, J., Revonsuo, A., Sikka, P., and Nummenmaa, L. (2011). The Existence of a Hypnotic State Revealed by Eye Movements. Public Library of Science One 6(10): e26374.

Kirsch, I. (1990). *Changing Expectations: A Key to Effective Psychotherapy*. Thomson Brooks/Cole.

Kirsch, I., and Braffman, W. (2001). Imaginative Suggestibility and Hypnotizability. *Current Directions in Psychological Science, 10*(2): 57–61.

Kirsch, I., and Lynn, S. J. (1995). Altered State of Hypnosis: Changes in the Theoretical Landscape. *American Psychologist, 50*(10): 846.

Kirsch, I., and Lynn, S. J. (1998). Dissociation Theories of Hypnosis. *Psychological Bulletin, 123*: 100–115.

Kirsch, I., Montgomery, G., and Sapirstein, G. (1995). Hypnosis as an Adjunct to Cognitive- Behavioral Psychotherapy: A Meta-analysis. *Journal of Consulting and Clinical Psychology, 63*(2): 214.

Lilienfeld, S. O, Lynn, S. J., and Lohr, J. (Eds.). (2015). *Science and Pseudoscience in Clinical Psychology* (2nd ed.). New York: Guilford Press.

Lilienfeld, S. L., Lynn, S. J., Ruscio, J. B., and Beyerstein, B. (2010). *50 Great Myths of Popular Psychology: Shattering Widespread Misconceptions about Human Behavior*. Malden, MA: Wiley-Blackwell.

Lilienfeld, S. O., Ritschel, L. A., Lynn, S. J., Cautin, R. L., and Latzman, R. D. (2014). Why Ineffective Psychotherapies Can Appear to Work: A Taxonomy of Causes of Spurious Therapeutic Effectiveness. *Perspectives on Psychological Science*, 9(4): 355–387.

Lynn, S. J., Boycheva, E., Deming, A., Lilienfeld, S. O., and Hallquist, M. N. (2009). "Forensic Hypnosis: The State of the Science." In J. Skeem, K. Douglas, and S. O. Lilienfeld (Eds.), *Psychological Science in the Courtroom: Controversies and Consensus*, pp. 80–99. New York: Guilford.

Lynn, S. J., Green, J. P., Weekes, J., Carlson, B., Brentar, J., Latham, L., and Kurzhals, R. (1990). Literalism and Hypnosis: Hypnotic versus Task Motivated Subjects. *American Journal of Clinical Hypnosis, 33:* 113–119.

Lynn, S. J., Inhoff, A., Maxwell, R., and Lemons, P. (2016). Hypnosis and Eye Movements as an Index of "Trance": A Failure to Replicate (unpublished manuscript). Binghamton University.

Lynn, S. J., and Kirsch, I. (2006). *Essentials of Clinical Hypnosis: An Evidence-Based Approach*. Washington, D.C.: American Psychological Association

Lynn, S. J., Kirsch, I., Barabasz, A., Cardena, E., and Patterson, D. (2000). Hypnosis as an Empirically Supported Clinical Intervention: The State of the Evidence. *International Journal of Clinical and Experimental Hypnosis, 48*: 343–361.

Lynn, S. J., Laurence, J. R., and Kirsch, I. (2015). Hypnosis, Suggestion, and Suggestibility: An Integrative Model. *American Journal of Clinical Hypnosis, 57*(3): 314–329.

Lynn, S. J., Nash, M. R., Rhue, J. W., Carlson, V., Sweeney, C., Frauman, D., and Givens, D. (1985). "Non-volition and Hypnosis. Reals vs. Simulators: Experiential and Behavioral Differences in Response to Conflicting Suggestions during Hypnosis." In D. Waxman, P. Misra, M. Gibson, and *M. Basker (Eds.), *Modern Trends in Hypnosis*, pp. 109–117. New York: Plenum.

Lynn, S. J., Nash, M. R., Rhue, J. W., Frauman, D., and Sweeney, C. (1984). Nonvolition, Expectancies, and Hypnotic Rapport. *Journal of Abnormal Psychology, 93*: 295–303.

Lynn, S. J., Neufeld, V., Green, J., Rhue, J., and Sandberg, D. (1996). "Daydreaming, Fantasy, and Psychopathology." In R. Kunzendorf, N. Spanos, and B. Wallace (Eds.), *Hypnosis and Imagination*, pp. 67–98. New Jersey: Baywood Press.

Lynn, S. J., and Rhue, J. W. (1991). "An Integrative Model of Hypnosis." In S. J. Lynn, and J. W. Rhue (Eds.), *Theories of Hypnosis*, pp. 397–438. New York: Guilford.

Lynn, S. J., Vanderhoff, H., Shindler, K., and Stafford, J. (2002). Defining Hypnosis as a Trance vs. Cooperation: Hypnotic Inductions, Suggestibility and Performance Standards. *American Journal of Clinical Hypnosis, 44*: 231–240.

MacLeod, C. M. (1991). Half a Century of Research on the Stroop Effect: An Integrative Review. *Psychological Bulletin, 109*(2): 163.

MacLeod, C. M., and Sheehan, P. W. (2003). Hypnotic Control of Attention in the Stroop Task: A Historical Footnote. *Consciousness and Cognition, 12*(3): 347–353.

Malinoski, P., Lynn, S. J., and Sivec, H. (1998). "The Assessment, Validity, and Determinants of Early Memory Reports: A Critical Review." In S. J. Lynn, and K. M. McConkey (Eds.), *Truth in Memory*, 109–136.

Meyer, E., and Lynn, S. J. (2011). Responding to Hypnotic and Nonhypnotic Suggestions: Performance Standards, Imaginative Suggestibility, and Response Expectancies. *International Journal of Clinical and Experimental Hypnosis*, 59(3): 327–34

Mingay, D. J. (1986). Hypnosis and Memory for Incidentally Learned Scenes. *British Journal of Experimental & Clinical Hypnosis*, 3: 173–183.

Nash, M. (1987). What, If Anything, Is Regressed about Hypnotic Age Regression? A Review of the Empirical Literature. *Psychological Bulletin*, 102(1): 42–52.

Nash, M. R., Drake, S. D., Wiley, S., Khalsa, S., and Lynn, S. J. (1986). Accuracy of Recall by Hypnotically Age-Regressed Subjects. *Journal of Abnormal Psychology,* 95(3): 298–300.

Nash, M. R., Lynn, S. J., Stanley, S., and Carlson, V. (1987). Subjectively Complete Hypnotic Deafness and Auditory Priming. *International Journal of Clinical and Experimental Hypnosis*, 35(1): 32–40.

Orne, M. T. (1959). The Nature of Hypnosis: Artifact and Essence. *Journal of Abnormal and Social Psychology*, 58(3): 277–299.

Orne, M. T. (1979). "On the Simulating Subject as a Quasi-control Group in Hypnosis Research: What, Why, and How." In E. Fromm and R. Shor (Eds), *Hypnosis: Developments in Research and New Perspectives*, 519–565. New Brunswick, NJ: Transaction Publishers.

Oz, M. (2008). Is Past-Life Regression Real? Retrieved from http://www.oprah.com/health/Were-You-Here-Before-Dr-Oz-Explores-Past-Life-Regressions_2/5

Pattie, F. A. (1956). "Methods of Induction, Susceptibility of Subjects, and Criteria of Hypnosis." In R. M. Dorcus (Ed.), *Hypnosis and Its Therapeutic Applications,* pp. 1–24. New York: McGraw-Hill.

Pintar, J., and Lynn, S. J. (2009). *Hypnosis: A Brief History*. New York: John Wiley & Sons.

Putnam, W. H. (1979). Hypnosis and Distortions in Eyewitness Memory. *International Journal of Clinical and Experimental Hypnosis*, 27(4): 437–448.

Raz, A., Kirsch, I., Pollard, J., and Nitkin-Kaner, Y. (2006). Suggestion Reduces the Stroop Effect. *Psychological Science*, 17(2): 91–95.

Raz, A., and Shapiro, T. (2002). Hypnosis and Neuroscience: A Cross Talk between Clinical and Cognitive Research. *Archives of General Psychiatry*, 59(1): 85–90.

Raz, A., Shapiro, T., Fan, J., and Posner, M. I. (2002). Hypnotic Suggestion and the Modulation of Stroop Interference. *Archives of General Psychiatry,* 59(12): 1155–1161.

Ready, D. J., Bothwell, R. K., and Brigham, J. C. (1997). The Effects of Hypnosis, Context Reinstatement, and Anxiety on Eyewitness Memory. *International Journal of Clinical and Experimental Hypnosis*, 45(1): 55–68.

Sanders, G. S., and Simmons, W. L. (1983). Use of Hypnosis to Enhance Eyewitness Accuracy: Does It Work? *Journal of Applied Psychology, 68*(1): 70–77.

Scoboria, A., Mazzoni, G., Kirsch, I., and Jimenez, S. (2006). The Effects of Prevalence and Script Information on Plausibility, Belief, and Memory of Autobiographical Events. *Applied Cognitive Psychology, 20*(8): 1049–1064.

Scoboria, A., Mazzoni, G., Kirsch, I., and Milling, L. S. (2002). Immediate and Persisting Effects of Misleading Questions and Hypnosis on Memory Reports. *Journal of Experimental Psychology: Applied, 8*(1): 26–32.

Shor, R. E. (1962). Three Dimensions of Hypnotic Depth. *International Journal of Clinical and Experimental Hypnosis, 10*(1): 23–38.

Silva, C. E., and Kirsch, I. (1987). Breaching Hypnotic Amnesia by Manipulating Expectancy. *Journal of Abnormal Psychology, 96*(4): 325–329.

Spanos, N. P. (1991). *Theories of Hypnosis: Current Models and Perspectives.* New York: Guilford Press.

Spanos, N. P. (1996). *Multiple Identities and False Memories: A Sociocognitive Perspective.* Washington, DC: American Psychological Association.

Spanos, N. P., Burgess, C. A., Burgess, M. F., Samuels, C., and Blois, W. O. (1999). Creating False Memories of Infancy with Hypnotic and Non-Hypnotic Procedures. *Applied Cognitive Psychology, 13*(3): 201–218.

Spanos, N. P., Cobb, P. C., and Gorassini, D. R. (1985). Failing to Resist Hypnotic Test Suggestions: A Strategy for Self-Presenting as Deeply hypnotized. *Psychiatry, 48*(3): 282–292.

Spanos, N. P., deGroot, H., and Gwynn, M. (1987). Hypnosis as Incomplete Responding. *Journal of Social and Personality Psychology, 53*: 911–921.

Spanos, N. P., Gwynn, M. I., Comer, S. L., Baltruweit, W. J., and de Groh, M. (1989). Are Hypnotically Induced Pseudomemories Resistant to Cross-Examination? *Law and Human Behavior, 13*(3): 271.

Spanos, N. P., Gwynn, M. I., and Stam, H. J. (1983). Instructional Demands and Ratings of Overt and Hidden Pain during Hypnotic Analgesia. *Journal of Abnormal Psychology, 92*(4): 479–488.

Spanos, N. P., and Hewitt, E. C. (1980). The Hidden Observer in Hypnotic Analgesia: Discovery or Experimental Creation? *Journal of Personality and Social Psychology, 39*(6): 1201–1214.

Spanos, N. P., Menary, E., Gabora, N. J., DuBreuil, S. C., and Dewhirst, B. (1991). Secondary Identity Enactments During Hypnotic Past-Life Regression: A Sociocognitive Perspective. *Journal of Personality and Social Psychology, 61*(2): 308–320.

Stanley, S. M., Lynn, S. J., and Nash, M. R. (1986). Trance Logic, Susceptibility Screening, and the Transparency Response. *Journal of Personality and Social Psychology, 50*(2): 447–454.

Steblay, N. M., and Bothwell, R. K. (1994). Evidence for Hypnotically Refreshed Testimony: The View from the Laboratory. *Law and Human Behavior, 18*(6): 635–651.

Tart, C. T., and Hilgard, E. R. (1966). Responsiveness to Suggestions under "Hypnosis" and "Waking-Imagination" Conditions: A Methodological Observation. *International Journal of Clinical and Experimental Hypnosis*, 14(3): 247–256.

Terrance, C. A., Matheson, K., Allard, C., and Schnarr, J. A. (2000). The Role of Expectation and Memory-Retrieval Techniques in the Constructions of Beliefs about Past Events. *Applied Cognitive Psychology*, 14: 361–377.

Timm, H. W. (1981). The Effect of Forensic Hypnosis Techniques on Eyewitness Recall and Recognition. *Journal of Police Science and Administration*, 9: 188–194.

Wakefield, H., and Underwager, R. (1998). "The Application of Images in Child Abuse Investigations." In *Image-Based Research: A Sourcebook for Qualitative Researchers*, 176–194. London: Brunner Routledge.

Wark, D. M. (2006). Alert Hypnosis: A Review and Case Report. *American Journal of Clinical Hypnosis*, 48(4): 291–300.

Weiss, B. (1988). *Many Lives, Many Masters*. New York: Simon & Schuster.

Weitzenhoffer, A. M. (1957). *General Techniques of Hypnotism*. New York: Grune & Stratton.

Young, J., and Cooper, L. M. (1972). Hypnotic Recall Amnesia as a Function of Manipulated Expectancy. *Proceedings of the 80th annual convention of the American Psychological Association, 7*: 857–858.

Yuille, J. C., and McEwan, N. H. (1985). Use of Hypnosis as an Aid to Eyewitness Memory. *Journal of Applied Psychology*, 70(2): 389–440.

16 Abuses and Misuses of Intelligence Tests: Facts and Misconceptions

Mark Benisz, John O. Willis, and Ron Dumont

Technical, and at times esoteric, terminology from the field of psychology often finds its way into popular culture, sometimes with little real understanding of the terms and often coupled with dramatic alterations of their meaning. For example, we chat comfortably, if not always accurately, about cognitive dissonance or schizophrenia. Although cognitive dissonance is often believed to be mental distress about a belief or a decision, it really refers to the distress aroused by a specific inconsistent thought, belief, or attitude. Similarly, schizophrenia is often mischaracterized as having multiple personalities, which is an entirely different disorder.

Perhaps some of the most widely used and potentially misused psychological terms are *intelligence and intelligence quotient (IQ)*. An example of this can be found in an interview in *The New York Times Magazine* (Cox, 2016) in which -the IQ was described as being innate, a view that is still hotly debated within the scientific community. The reality is that an IQ is subject to change, defined differently by various authorities, and measured in different ways by different instruments. In fact, many measures of intelligence or cognitive ability do not even use the term *IQ* as a label for their scores.

The concept of IQ in popular usage has expanded to include almost any evaluation of general or highly specific knowledge or skills in such domains as finances (Kiyosaki, 2008), job interviews (Martin, 2012), hospitality (Cox, 2016), opera (Bass, 1997), and soccer (Blank, 2012; Stewart, 2013), to give just a few random examples. As we discuss later, this is not the meaning of IQ as used by psychologists.

For over a century, the cognitive abilities of humans have been measured using a variety of instruments designed by researchers mainly from the field of psychology. These instruments are known colloquially as IQ tests. This abbreviation, which used to stand for "intelligence quotient," long ago entered the vernacular and is generally defined as a single whole number, two or three digits long, that is supposed to be a measure of an individual's overall intelligence. Within the field of psychology, however, the concept of intelligence and the theories and practices of its measurement have evolved so radically and divergently since the appearance of the first IQ tests that the construct of intelligence, as it is measured by contemporary psychologists, does not correspond to the meaning of IQ in idiomatic speech. There are many instruments commercially available today used to measure cognitive or mental ability. In fact, a search of the Buros Mental Measurements Yearbook website (Buros Center for Testing, n.d.), using the terms "intelligence and general aptitude" found well over 200 results. Most of these instruments, and in particular those instruments that are the most widely used, do not necessarily measure intelligence as that construct is understood by many people. Additionally, few, if any, of them yield an intelligence *quotient*. These are among the many fallacies and misconceptions surrounding contemporary IQ testing. For an authoritative and detailed, but very readable discussion of IQ testing, we recommend Alan Kaufman's (2009) *IQ Testing 101*.

The evolution of the measurement of intelligence can be traced to the pseudoscience of the nineteenth century, when phrenology became widely popular in the United States (Walsh, 1976). This pseudoscience posited that specific mental aptitudes could be measured by analyzing bulges on the skull corresponding to enlarged areas of the brain beneath. Other roots lie in psychophysical measurements such as grip strength and the ability to discriminate the weights of small objects (e.g., Galton, 1907), which were thought to assess intelligence. Phrenology was long ago abandoned, and psychophysical measurements came to be used mostly for neuropsychological assessment rather than intelligence testing. Early psychologists trained in the scientific method used it to gradually develop intelligence tests, thereby distancing intelligence testing from pseudoscience. However, some of the historical and contemporary *uses* of the IQ scores derived from these tests have been controversial and unscientific. IQ scores have been the cause of many bitter, hostile, and heated debates within scientific circles and the popular press (Block and Dworkin, 1976; Kamin and Egerton, 1973).

There are several underlying reasons behind such hostilities. IQ scores have often been misused to discriminate against minorities, reject immigrants to the United States if they did poorly on English-language IQ tests, legally authorize the sterilization of women with low IQ scores, deny employment to qualified individuals, place nondisabled students in restrictive special education classrooms, and refuse college admission to prospective students (Ravitch, 2000; Schultz and Schultz, 2015). In the United States, as a result of these misuses of IQ scores, a backlash occurred to the point where, in *Larry P. v. Riles* (1986), the California Supreme Court banned the use of IQ tests with African American children when determining eligibility for special education under the Education for all Handicapped Children Act (1975). Despite the highly questionable premise upon which the case was decided, and disappointing outcomes based upon its implementation, the ban was still in effect decades later (Powers, Hagans-Murillo, and Restori, 2014; Sattler, 2008).

Yet, despite the history of misuse and the continued controversies and concerns surrounding the concept, the IQ or its modern equivalent remains firmly entrenched in our laws, our educational system, and our mental diagnostic system. The use of IQ tests or other cognitive assessment tools is widespread in schools, hospitals, clinics, and the private offices of psychologists. Decades of research on IQ tests and neuroscience have refined the theoretical and statistical underpinnings of these tests (Kaufman, 2009). When used ethically and properly, IQ testing can be an extremely useful tool, aiding in diagnosis and helping to plan targeted rehabilitative and remedial interventions. In this chapter we provide an overview of some of the major misunderstandings and inappropriate uses of intelligence testing, as well as the subjects that continue to generate controversy in the present day.

Can Intelligence Be Defined as a Singular Construct?

Intelligence is a singular noun and the intelligence quotient, also singular, was represented by one number in the early days of intelligence testing. This characterization of intelligence was developed from the work of Alfred Binet (1910) and the research of Charles Spearman (1927). Binet, along with Henri Simon, developed the first test of intelligence that measured the mental age of school children with questions and puzzles, and contrasted that with their actual or chronological age. Lewis Terman (1916) translated the scale into English, expanded and revised it, and normed it on a sample of Americans.

The effort to quantify intelligence progressed from simply comparing mental age with chronological age, to (1) assigning alphabetical labels (e.g., "B"; Yoakum and Yerkes, 1920) across several grades of ability, to (2) assigning verbal labels (e.g., "idiot," "imbecile," and "moron"; Levine and Marks, 1928) to levels attained on tests, to (3) creating a genuine quotient by dividing the mental age by the chronological age, to (4) the current method of creating a standard score. The modern standard score uses sophisticated statistical procedures that base scores on the distance between the individual's scores and the average score for the individual's age, producing the infamous "normal curve" distribution of test scores.

All of these systems assumed that the tests were measuring a unitary quality of intelligence. Spearman (1927) used factor analysis to prove the existence of a general factor of intelligence—something he labeled g (lowercase and italicized). Test developers such as Lewis Terman, Maud Merrill (e.g., Terman and Merrill, 1937), and David Wechsler (1939) believed that the score on their tests, the IQ, was a good estimate of g. For many decades after the appearance of the first intelligence test, the results of the tests were summarized as a single figure, the IQ. Even after some tests were introduced that did yield more than one score (for example, verbal and nonverbal, or "performance," composites), the test developers deemphasized the other scores in favor of the global IQ score.

Despite the popularity of g and its representation as the IQ, some researchers disputed the existence of g from its inception (e.g., Cattell, 1941; Thurstone, 1938). Furthermore, statistical factor analyses showed that the existing IQ tests did much more than measure a single construct. Theories disputing the notion of intelligence as a singular construct slowly became more popular. Neurological research into the function of the brain also suggested the existence of many interconnected but separate components that worked together to process information (e.g., Luria, 1973). The research and acceptance of theory within the scientific community led to a new understanding of intelligence. Many authorities (e.g., Hale and Fiorello, 2001; Horn and Blankson, 2012) argued that intelligence is best understood as the interaction of several different cognitive abilities that may show wide variation within an individual rather than as a single ability (g). These data and opinions influenced the development of intelligence tests and as a result, the majority of modern psychological assessment batteries no longer measure intelligence only as a single construct. For example, one of the most widely used tests, the Wechsler Intelligence Scale for Children (WISC), currently in its fifth edition (WISC-V; Wechsler, 2014a), yields thirteen composite scores, in addition to a global full scale IQ score. One composite includes knowledge of words and verbal reasoning; another distinct composite is working memory; and a third composite is a measure of visual-spatial skills. This approach recognizes intelligence as comprising many distinct components and allows the evaluator to discover patterns of strengths and weaknesses within and between the specific composites.

Many authorities, however (e.g., Canivez, 2008; Gottfredson, 2008), argue that the current IQ tests still primarily measure g, and that the tests should be interpreted only at that level. They believe that any efforts to measure separate abilities, such as vocabulary or visual-spatial skills, should be made with separate tests of the specific ability in question, not by a profile of subtest scores on an IQ test. It is difficult to defend either of the extreme positions. Many individuals demonstrate a consistent range of intellectual abilities, sometimes all within the range of intellectual disability or within the range of intellectual giftedness; such patterns argue for the primacy of g. Furthermore, even when there is variability among different cognitive abilities, the total test score (e.g., full scale IQ or other proxy for g) is often a fairly accurate predictor of academic and vocational success (e.g., Gottfredson, 2008). Correlations have been reported between g and seemingly unrelated variables (e.g., semen quality) (Pierce, Miller, Arden, and Gottfredson, 2009). However, there are also many individuals whose abilities vary widely, from the level of intellectual disability to that of giftedness in different domains. Although their total scores may be fairly accurate predictors of academic and vocational success, that prediction may simply reflect the disabling effects of a narrow, specific cognitive weaknesses on what would otherwise be higher levels of accomplishment expected from the cognitive strengths. There is little scientific evidence to support either the belief that g is the only important cognitive variable or the belief that g is irrelevant.

This difference in opinion as to what intelligence is and, what is more important, the overall IQ (g) or the separate abilities that make up g, can be described as *how* a person is smart versus *whether* a person is smart (Davis, Christodoulou, Seider, and Gardner, 2011). How intelligence and IQ testing are conceptualized often depends upon the question one is interested in answering.

The Definition of Intelligence

When using the scientific method in research, it is necessary to clearly define the variables that are to be measured. This process is known as operationalization and can often mean the difference between a well-executed study and a study that cannot be replicated. It would seem to follow logically that, since there is an abundance of scientifically sounds tests on the market that are assumed to measure intelligence, the concept of intelligence is very well defined. However, nothing could be farther from the truth. In fact, within the psychology community, attempts to arrive at a consensus have failed. Edwin Boring once famously said, "Intelligence as a measurable capacity must at the start be defined as the capacity to do well in an intelligence test. Intelligence is what the test tests," (Boring, 1923). Legg and Hutter (2007) compiled thirty-five different definitions of intelligence from prominent psychologists and eighteen more from dictionaries and other institutional sources.

Wechsler's (1939, p. 3) definition of intelligence, "the global capacity of a person to act purposefully, to think rationally, and to deal effectively with his environment," is often quoted but obviously not universally accepted. Even among the test developers, there are multiple, sometimes conflicting definitions of intelligence. For this reason, many of the different IQ assessments contain tests or subtests that measure dissimilar abilities. As a result, different tests will often yield significantly different scores for the same person. If, for example, one is very strong or very weak in a specific ability, such as working memory, the global test score will be different on tests that do (e.g., WISC-V) or do not (e.g., DAS-II) include that particular ability in the total score. It is likely, for this reason among others, that many contemporary tests have replaced the term intelligence with "cognitive abilities" or "intellectual" or "mental ability." The total scores on various cognitive ability measures include, among others, "Full Scale IQ" (FSIQ; Roid, 2003; Wechsler, 1939, 2008, 2012, 2014a), General Conceptual Ability (GCA; Elliott, 2007), "Full Scale" Composite Index (FS; Naglieri, Das, and Goldstein, 2014), General Intellectual Ability (GIA; Schrank, McGrew, and Mather, 2014), Composite Intelligence Index (CIX; Reynolds and Kamphaus, 2015), and Mental Processing Index (MPI) and Fluid-Crystallized Index (FCI; Kaufman and Kaufman, 2004).

Intelligence as a Quotient

As mentioned above, the Binet-Simon test was designed to yield a mental age. The tasks it contained were grouped by age level. Thus, a four-year-old child who was able to complete tasks at the five-year-old age level was said to have a mental age of five.

Stern (1912, 1914) suggested taking the mental age and dividing it by the chronological age of the child (5 ÷ 4, in the present example) to derive a *mental quotient*. This concept was further refined by multiplying the quotient by 100 and dropping any decimals to yield a whole number. In the example given above, the mental age of five divided by the chronological age of four yields a quotient of 1.25; multiplied by 100, this equals an IQ of 125. Conversely, if the four-year-old child had been able to do only the test tasks at the three-year-old level, the child's IQ would have been 75. A child who scored at age level (i.e., an average child), no matter what the age, would always have a ratio IQ of 100.

This system for calculating ratio IQs had many drawbacks, among them the inconsistency of scores. Using the example above, the four-year-old child who is one year behind has an IQ of 75. However, the thirteen-year-old who is two years behind has a much higher IQ—close to 85. Adult IQs required using an arbitrary number, such as sixteen years, in place of the adult's actual age. Therefore, the representation of intelligence as a quotient was replaced in the Wechsler-Bellevue Intelligence Scale (introduced in 1939) with the statistically derived standard score or deviation IQ, which has continued in use. (Wechsler did not invent this system of test scores, but was the first to apply it to individually administered intelligence tests, two years after Terman and Merrill [1937] considered and rejected the idea because it would have been too confusing for practitioners.) The old ratio intelligence quotient gradually disappeared as new tests and new editions using the deviation IQ were developed. In this system, all scores are considered to be normally distributed with the mean arbitrarily set at 100 and a standard deviation of 15. The score of each test taker is then represented by a number that indicates by how much the obtained score of the test taker deviates from the mean score of 100. Therefore, the term IQ is actually a misnomer, since the score is no longer a quotient. None of the current Wechsler intelligence scale manuals (2008, 2012, 2014a) even mentions the word "quotient."

Is Intelligence a Fixed Construct?

For as long as intelligence has been measured, its malleability has been debated. Is the obtained IQ score of an individual fixed over that person's lifetime, or is the IQ score subject to changes based on environmental variables? Binet (1909, p. 141) deplored what he described as the brutal pessimism of some of his colleagues in their views of IQ as a static, unchanging measurement. This opinion is shared by some modern researchers (Blackwell, Rodriguez, and Guerra-Carrillo, 2015) who believe that changes in IQ score are quite possible. However, many researchers view intelligence as a relatively stable construct over the lifetime of an individual (Mortensen, Andresen, Kruuse, Sanders, and Reinisch, 2003). The practical implications of each of these views can have far-reaching consequences. For example, if the IQ of a six-year-old child is measured and found to be significantly below average, those who believe in IQ as a stable construct might make certain educational recommendations that are less rigorous than the educational demands for children of average to above average intelligence. The child's academic and vocational outcomes can become a self-fulfilling prophecy. If, however, the view of intelligence as being malleable is taken, the child could be enrolled in remedial programs that are overly and unnecessarily taxing.

In fact, this debate oversimplifies the very nature of intelligence. All intelligence tests measure IQ with error and therefore many evaluators of intelligence will often report their *obtained* scores as an interval or range, in which, at a certain level of confidence, the *true* score would be expected to be found. An obtained IQ score of 104 might be reported as a number that falls between 99 and 109. Furthermore, it is highly likely that when using a different measure of intelligence for retesting, a score that is significantly higher or lower than the original IQ score would be a result of the different abilities measured by each test, as explained above. Even if the same test is administered to the same individual (after an appropriate interval of time has elapsed), competent evaluators expect the scores to be different, due to the statistical phenomenon known as regression to the mean. IQ test scores have a tendency to move closer to the mean score (standard score 100) upon second administration of a test.

Broad Theories of Intelligence

Structures of Intelligence

Two research-based theories of intelligence have come to dominate the construction of IQ tests. One is the Cattell-Horn-Carroll (CHC) theory (e.g., Carroll, 2012; Flanagan, Ortiz, and Alfonso, 2013; Schneider and McGrew, 2012). CHC theory has been developed from the work of Raymond Cattell (1941) and Cattell's student and later colleague, John Horn (1988; Horn and Blankson, 2012), and John Carroll (1993). Based primarily on factor-analytic studies (determining how closely different IQ subtests and scales are correlated with each other and with measures of academic achievement), the current version of Cattell-Horn-Carroll theory is a hierarchical model that contains three strata: Stratum III or g (general intelligence), Stratum II (broad abilities), and Stratum I (narrow abilities). Each broad, Stratum II ability includes several narrow abilities (Stratum I). For example, fluid reasoning (Gf) includes inferential reasoning (I), general sequential or deductive reasoning (RG), and quantitative reasoning (RQ). Some theorists emphasize the fact that all of these abilities are at least moderately correlated with each other in the general population and treat Stratum III (g) as the most important aspect of the theoretical structure (e.g., Carroll, 1993, 2012; Gottfredson, 2008). Others focus on the wide range of ability levels in individuals and consider the Stratum III measure of g a statistical artifact that is not very helpful for understanding the cognitive functioning

	Cattell-Horn-Carroll Theory (CHC) Abilities							Planning, Attention, Simultaneous, Successive (PASS) Theory Abilities			
	Verbal Ability	Fluid Reasoning	Visual-Spatial Ability	Short-Term and Working Memory	Long-Term Storage and Retrieval	Processing Speed	Auditory Processing	Attention	Sequential Processing	Simultaneous Processing	Planning
CAS2								X	X	X	X
DAS-II	X	X	X	(x)	(x)	(x)	(x)				
KABC-II FCI	X	X	X	X	X						
KABC-II MPI				(x)	X				X	X	X
RIAS-2	X	X				(x)					
SB5	X	X	X	X							
WISC-V	X	X	X	X	(x)	X					
WAIS-IV	X		X¹	X		X					
WJ IV	X	X	X	X	X	X	X				

Figure 16.1. Cognitive abilities explicitly included in the total scores of several tests.

¹ Fluid reasoning and visual-spatial subtests are combined in a single Perceptual Reasoning composite.

X One or more scored tests or subtests are included in the total score for the battery.
(x) The battery includes one or more scored tests or subtests, but they are not included in the total.

CAS2 = Cognitive Assessment System-Second Edition (Naglieri, Das, & Goldstein, 2014).
DAS-II = Differential Ability Scales-Second Edition (Elliott, 2007).
KABC-II = Kaufman Assessment Battery for Children-Second Edition (Kaufman & Kaufman, 2004).
    FCI = Fluid-Crystallized Composite.
    MPI = Mental Processing Composite.
RIAS-2 = Reynolds Intellectual Assessment Scale-Second Edition (Reynolds & Kamphaus, 2015).
SB5 = Stanford-Binet Intelligence Scales, Fifth Edition (Roid, 2003).
WISC-V = Wechsler Intelligence Scale for Children-Fifth Edition (Wechsler, 2014a).
WAIS-IV = Wechsler Adult Intelligence Scale-Fourth Edition (Wechsler, 2008).
WJ IV = Woodcock-Johnson IV Tests of Cognitive Ability (Schrank, McGrew, & Mather, 2014).

of an individual (e.g., Horn and Blankson, 2012). Many psychologists are agnostic on the issue, agreeing that Stratum III g is real and that individuals do demonstrate varying levels of g, but also focusing on the various broad (Stratum II) and narrow (Stratum I) abilities for practical issues of diagnosis and remediation. (For extended discussions of CHC theory, see Flanagan, Ortiz, and Alfonso [2013] and Schneider and McGrew [2012].) One cognitive ability test, the Woodcock-Johnson (Schrank, McGrew, and Mather, 2014) is explicitly based on CHC theory. Several other prominent tests, although not based on the theory, are closely aligned with it.

Another theory, the Planning, Attention, Simultaneous, and Successive (PASS) theory is based on the work of the Soviet neuropsychologist Alexander Luria (e.g., 1973), J. P. Das (e.g., 1973, 1980), Jack Naglieri (e.g., Das, Naglieri, and Kirby, 1994), and others. It proposes that there are four distinct cognitive processes that interact to determine intelligent behavior. The Cognitive Assessment System—Second Edition (CAS2; Naglieri, Das, and Goldstein, 2014) is explicitly designed to measure the PASS abilities. The Kaufman Assessment Battery for Children—Second Edition (KABC-II; Kaufman and Kaufman, 2004) is constructed to be interpreted by both CHC and PASS theories. Figure 16.1 provides an analysis of several tests by CHC and PASS theories.

Alternative Theories of Intelligence
IQ tests have long been criticized for their narrow focus on verbal, mathematical, and visual/spatial intelligence (e.g., Gardner, 1983, 2011; Sternberg, 1988). Because most IQ test items have one or only a few possible correct answers, they tend to reward convergent rather than divergent thinking and ignore or even penalize creativity (Kaufman, J. C., 2010). Such concerns have led some experts to propose alternative conceptualizations of intelligence. For example, Howard Gardner (1983, 2011) has posited eight "intelligences": linguistic, logic-mathematical, musical, spatial, bodily/kinesthetic, interpersonal, intrapersonal, and naturalistic. A ninth, existential intelligence, has also been discussed (Gardner, 1999). Robert Sternberg (2012) has proposed a "triarchic" theory of "successful intelligence." The three subtheories relate intelligence to the individual's internal world, the individual's experience, and the individual's external world.

Keith Stanovich (2009) agrees that there are many types of behavior that could be described as intelligent, but shares with many psychologists a concern that applying the word "intelligence" to a wide variety of different behaviors adds to the considerable existing confusion. Stanovich (2009) noted that some people, despite having adequate intelligence, fail to think or act in rational manner, such as the sad examples of prominent, successful politicians who make breathtakingly dumb personal decisions. These considerations need to be kept in mind when one is tempted to assume either that IQ means nothing in the real world or that a high IQ guarantees rational behavior and personal and vocational success.

Flynn Effect

Another interesting phenomenon that affects the obtained score of an IQ test is the Flynn effect (Flynn, 2009). Flynn and other researchers (Nisbett, et al., 2012) found that from 1930 onward IQ scores in many parts of the world had increased substantially. For example, in the United States, test scores increase on average at a rate of three points per decade (Flynn, 2009). The cause of these increases is disputed within the scientific community (Kaufman and Weiss, 2010). Some researchers believe that the upward trend of IQ scores reflects a rise in intelligence—we are getting smarter. Others believe that people are more adept at the particular skills that are measured by IQ tests. The Flynn effect has practical implications for test developers. Test developers must renorm the tests when their original norming data become outdated. They must also review the subtests and subtest items and update them as appropriate.

One other problematic use of test scores in light of the Flynn effect is when a cutoff score is used to apply a punitive consequence or to determine eligibility for a benefit. People with intellectual disabilities are eligible for various government programs, accommodations, and protections. If an outdated test is used and, because of the Flynn effect, an inflated score is obtained, this can result in the denial of benefits, thereby preventing people with genuine disabilities from availing themselves of programs they need to increase their quality of life. These are not theoretical apprehensions, as researchers have documented the use of antiquated tests, especially during periods of transitions to new tests (Sénéchal, Larivée, Audy, and Richard, 2007). Furthermore, in some parts of the United States, an inflated IQ score can have real consequences for the application of the death penalty (Kaufman and Weiss, 2010).

Genetics and Intelligence

One of the most heated controversies in IQ testing is the observed differences between racial and cultural groups on IQ tests (Ortiz, Ochoa, and Dynda, 2012; Sattler, 2008, pp. 134–182). Historically, members of minority groups have obtained IQ scores that are significantly lower than the scores of the dominant culture. This led some of the early developers of IQ tests to decide that these minority groups were intellectually inferior to members of the dominant group. These observations were not limited to groups that have a history of being oppressed in the United States, but included almost any native or immigrant group whose ancestors did not emigrate from northern Europe. The common-sense idea that English-language IQ tests might not be accurate measures of intelligence for persons who did not speak English surprisingly took a long time to come to the attention of the field. This misconceived viewpoint was taken seriously in this county and helped shape policies that curtailed immigration from countries deemed to produce intellectually inferior immigrants (e.g., Brigham, 1923; Goddard, 1913, 1917; Hirsch, 1926; Immigration Act of 1924). Even today, minority children tend to be overrepresented in special education classrooms, in part due to IQ scores (Rhodes, Ochoa, and Ortiz, 2005).

One continuously debated topic related to IQ and intelligence is the old "nature versus nurture" debate. Are we born with a certain level of intelligence or is our intelligence molded and shaped by the environment in which we are raised? In 1969, in an article in the *Harvard Educational Review* entitled "How Much Can We Boost IQ and Scholastic Achievement?" Arthur Jensen (1969) concluded, among other things, that head start programs designed to boost African American IQ scores had failed, largely because, in Jensen's estimation, 80 percent of the variance in IQ in the population studied was the result of genetic factors. The article itself became one of the most highly cited in the history of psychology, but many of the citations were rebuttals of Jensen's arguments or used the paper as an example of controversy. In contrast, Kamin (1974) insisted that, "there exists no data which should lead a prudent man to accept the hypothesis that IQ test scores are in any way heritable" (p. 1). At about the same time, William Shockley, a Nobel Prize winner in Physics (he was the co-inventor of the transistor), became interested in and vocal about the genetics of IQ. He maintained that IQ was largely genetic and that the genetics accounted for racial differences in mean IQs (Pearson, 1992). Shockley went so far as to assert: "Nature has color-coded groups of individuals so that statistically reliable predictions of their adaptability to intellectually rewarding and effective lives can easily be made and profitably be used by the pragmatic man in the street." ("Shockley's Views," 1971).

Herrnstein and Murray (1994), in their controversial book *The Bell Curve*, argued that low IQ—independent of socioeconomic variables—lies at the root of many social problems, such as out-of-wedlock births and unemployment, and that a "cognitive elite" has emerged. The authors made several recommendations including the elimination of welfare policies that encourage the "wrong women" to have babies (Herrnstein and Murray, 1994). They also made other recommendations, including the curtailing of immigration. These recommendations are reminiscent of the arguments made by early eugenicists as mentioned by Kamin (1974). Russell Jacoby and Naomi Glauberman (1995) edited a collection of eighty-one contributions to *The Bell Curve Debate*, many of which argued strenuously against this simplistic conception.

IQ tests are a product of the culture from which they emanate. It is impossible to produce a truly culture-free test. When tests are designed to reduce prerequisite cultural knowledge, the differences between groups are reduced. For example, as a group, whites earn higher IQs, on average, than African Americans. The difference has typically been about 15 IQ points on Wechsler's scales (Kaufman and Lichtenberger, 2005) but is reduced to about half that on the KABC-II (Kaufman and Kaufman, 2004).

Intelligence Tests and Real-Life Outcomes

Intelligence tests and their equivalents are often used to make high-stakes decisions that can have long-term effects on people. Many colleges in the United States require their applicants to take the SAT College Admissions Test (College Board, n.d.) or ACT (ACT, n.d.), using this information along with other criteria to grant or deny admission to their programs. Interestingly, the SAT and ACT, tests of scholastic aptitude, are so highly correlated with intelligence tests that some (e.g., Frey and Detterman, 2004; Koenig, Frey, and Detterman, 2008) suggest that they can be considered measures of *g*. Some companies also require job applicants to take an IQ test such as the Wonderlic (Bell, Matthews, Lassiter, and Leverett, 2002; Wonderlic, 1992) and use that information to help determine the suitability of the applicant. The scores derived for these purposes thus serve as predictors of job performance (Schmidt and Hunter, 2004). It is important to emphasize that the Americans with Disabilities Act Amendments Act (2008) and the various laws and regulations enforced by the US Equal Employment Opportunity Commission (n. d.) restrict the use of tests for selecting employees. If a test has disparate selection rates for one or more protected groups, the employer would have to demonstrate that the test was an accurate measure or predictor of a necessary job skill or of job performance. If, for example, an employer's IQ test selected unequal proportions of applicants from protected groups, the employer would have to be able to show that the IQ accurately predicted performance on skills that were essential for the job and was not being used simply to reject applications from certain groups (Arnold, 2012).

Pseudoscience

Learning Disabilities

For many years specific learning disabilities were identified via a process referred to as the ability-achievement discrepancy. In this model, various formulae were developed in order to compare an IQ score (referred to as "ability") with a score on a standardized academic achievement test. If the two scores (ability vs. achievement) were significantly discrepant, meaning that the achievement score was significantly lower than the IQ score, a diagnosis of learning disability could be applied.

One of the problems inherent in the ability-achievement discrepancy method for determining severe discrepancy is the overrreliance on IQ as the cornerstone for decision making. Kelman and Lester (1997) coined the phrase "IQ Fetishism" to describe the often mindless fixation on IQ-achievement discrepancy despite the lack of empirical support for such distinctions. The relationship between IQ and achievement is not necessarily a strong relationship. Knowing that a person has a high or a low IQ score may tell us something about what they might or should be doing on achievement tests, but that information is typically not very meaningful. Just because we know an IQ score does not mean that we can predict all things from it. For example, if you were told that one person has an IQ score of 150 and that a second has an IQ score of 70, could you predict which one could hold his or her breath longer than the other? You would most likely laugh and say, "Of course not. Knowing one's IQ tells us little about the ability to hold one's breath. The two are not related." This is true. It is also true to some extent when we try to predict the academic ability of a student based upon his or her IQ score. Although related, the relationship is just not good enough to make the prediction.

This approach was also criticized because "significantly discrepant" was poorly defined, leaving many varied, but often rigidly applied, practices in determining exactly what constituted a real discrepancy. For example, for a particular child, would the 20-point difference between the IQ (130) and the achievement (110) be "significant" enough? According to some school districts, a 15-point difference (defined as 1 standard deviation) was enough to be considered significant, while in other districts 22 points (1.5 standard deviations) was the "magic number" required).

Another difficulty with this approach is the mistaken assumption that there is a perfect or near perfect correlation between ability (IQ) and achievement. Contrary to general belief, the correlation between measures of ability and achievement is modest (Wechsler, 2014b). This fact has been known (or at least the information has been available) for almost forty years. Hammill and McNutt (1981) reviewed all correlational studies between reading and other variables in twenty-five journals from 1950 through 1978. Their meta-analysis found, among other things, that the median correlation between the Wechsler scales full scale IQs and reading scores, based on thirty-four coefficients in thirteen studies, was only +0.44. This means that about 19 percent of the variance in reading scores could be accounted for by Wechsler full-scale IQs. The remaining 81 percent of the variance in reading scores was attributable to other factors!

There are other problems inherent in the use of IQ as a determiner of a learning disability. These include that ability-achievement calculations also assume that a single number on either test can summarize a student's abilities. Using this formula does not help inform on how to intervene and help the student. The process tends to overidentify students as being learning disabled when in fact they might not have had adequate education or exposure to the content assessed by the achievement test. And finally, there is often a huge gap between the science of learning disability identification and the social policy involved with the identification of learning disabilities. In the area of reading for example, many studies have demonstrated that the information-processing abilities that are at the root of poor readers' difficulties are the same for poor readers with low and high IQs (Felton and Wood, 1991; Fletcher, Francis, Rourke, Shaywitz, and Shaywitz, 1992; Jimenez-Glez and Rodrigo-Lopez, 1994; Siegel, 1998).

Capital Punishment

Perhaps no other area in psychology has true life-or-death consequences more than the use of IQ tests to determine the application of capital punishment. In 2002, the Supreme Court of the United States ruled in *Atkins v. Virginia* (2002) that applying the death penalty to people with an intellectual disability constituted cruel and unusual punishment, thereby making it unconstitutional. The court did not specify what constituted an intellectual disability, leaving it to the states to adopt standards and regulations. As a result, the states that still maintain a death penalty adopted regulations that differ from each other. These standards were based in part on the criteria used to diagnose an intellectual disability at the time. Until the publication of the *Diagnostic and Statistical Manual, Fifth Edition* (DSM-5) in 2013, heavy emphasis was placed on an obtained IQ score, with an intelligence score of two or more standard deviations below the mean being considered an intellectual disability. Since the most widely used adult intelligence test, the Wechsler Adult Intelligence Scale (Wechsler, 2008), has always reported its scores as standard scores—defined by a mean of 100 and a standard deviation of 15—a score of 70 is two standard deviations ($2 \times 15 = 30$) below the mean ($100 - 30 = 70$). Yet herein lies the problem. As stated above, all scores are measured with error and an obtained score of 75 could deviate—either up or down—from an individual's true score. Depending on the statistical reliability (consistency) of the particular test and the confidence interval set by the examiner (such as 90 or 95 percent confidence), some criminals who obtained an IQ score of 75 might really meet criteria for intellectual disability if their true scores are below 70. (Others obtaining an IQ of 75 might have a true score of 80.) Also, as noted above, different tests will yield different scores for the same individual. There is no way to ever know what a person's true score is. Perhaps it is for this reason, among others, that the DSM-5 de-emphasized the actual score when diagnosing intellectual disability in favor of assessing levels of adaptive behavior. As of this time, state laws have not been adjusted to conform to the updated diagnosis of intellectual disability.

Brain Training

In recent years popular media have reported on a number of for-profit private companies that claim to raise the IQ scores of their customers with a variety of brain training techniques. Some of these companies have publicized data to support their claims of boosted intelligence. The problems inherent with the claims of these companies become apparent when their data are analyzed. Often these companies report their scores as percentile ranks or age equivalents, instead of using standard scores. Small or insignificant increases in standard scores are inflated when percentile ranks or age equivalents (mental ages) are used in their place. A standard score of 90, which is considered average, has a percentile rank of 25. A standard score of 100, also average, has a percentile rank of 50. Thus a 10-point and largely insignificant difference in score is a

25-percentage-point difference. Age-equivalent or mental age scores are almost useless for measuring change. They are simple transformations of raw scores (number of items passed) and are not equal units. Consider the amount of mental growth, for example, between age 3:0 (three years, zero months) and age 4:0 compared to the amount of mental growth between ages 13:0 and 14:0. Since neither the raw scores nor their age equivalencies are equal units, we cannot add, subtract, multiply, divide, or average them. We can be fairly sure that a mental age of 13:0 is higher than one of 9:0, but we cannot know how much higher. On tests with relatively small numbers of items, or tests with large clusters of items at certain levels and sparsely spread-out items across other ages, passing just one additional item may dramatically increase the mental age or age-equivalent score. For example, on the WISC-V (Wechsler, 2014a) Similarities subtest (on which the child attempts to explain how two different words, such as horse and moose or hope and fear, could be alike), a raw score of 28 earns an age equivalent of 11:6. A raw score of 29 has an age equivalent of 11:10. One more point raised the age equivalent by four months. However, a raw score of 30 escalates the mental age to 12:10, a gain of twelve months, and 31 raw-score points yield a mental age of 14:2, a gain of sixteen months! Since each item can be scored 2 points for a high-quality response, 1 point for a marginal response, or 0 points, a good answer to a single item could raise the raw score from 29 to 31 and inflate the age equivalent by twenty-eight months. Scaled scores (a form of standard score with a mean of 10 and standard deviation of 3 used for Wechsler subtests) are much more consistent measurements. If the child's actual age was 12:10, the raw scores of 28, 29, 30, and 31 would give scaled scores of 9, 9, 10, and 10, respectively. The actual differences in ability represented by raw scores of 28 through 31 were insignificant, but the supposed gain of thirty-two months (11:6 to 14:2) would appear to be a miraculous improvement.

There is also the statistical phenomenon of regression to the mean (Galton, 1886), where measurements that are below the mean on the first testing tend to be higher, or closer to the mean, upon second testing. (Similarly, initial scores above the mean tend to be lower on a repeated test unless there has been a practice effect or a genuine improvement in the tested trait.) This regression occurs on its own without any brain training or other intervention.

Some of the testing methodologies employed by these companies are also flawed. If a standardized test is administered to a single person more than once in a short span of time, it is likely that the second score is inflated in part due to practice effects, a well-known confounding variable in scientific research. As these brain training programs are usually no more than six to twelve weeks long, it is likely that the second, higher set of scores obtained are due in part to practice effects. In addition, some of these companies use "modified" versions of standardized tests that are administered by people who would not otherwise be qualified to administer an intelligence test in its standardized version.

It should also be noted that there is to date little evidence published in peer-reviewed journals that can be taken as evidence of real-life improved outcomes for individuals who have gone through brain-training programs. The Federal Trade Commission (2016) recently fined Lumos Labs, "The creators and marketers of the Lumosity 'brain training' program . . . $2 million to settle FTC deceptive advertising charges for its 'Brain Training' program," alleging that the "company claimed [the] program would sharpen performance in everyday life and protect against cognitive decline . . . The proposed stipulated federal court order requires the company and the individual defendants . . . to have competent and reliable scientific evidence before making future claims about any benefits for real-world performance, age-related decline, or other health conditions. The order also imposes a $50 million judgment against Lumos Labs, which will be suspended due to its financial condition after the company pays $2 million to the Commission. The order requires the company to notify subscribers who signed up for an auto-renewal plan between January 1, 2009 and December 31, 2014 about the FTC action and to provide a means to cancel their subscription."

IQ Tests in Magazines and on the Internet
Supposed IQ tests are readily available in magazines, on the Internet (e.g., IQ Check, n.d.; Stanford-Binet test, n.d.), and in paperback books (e.g., Eysenck, 1960). Hans Eysenck was, incidentally, a psychologist with an international reputation who should have known better. The online version of the Stanford-Binet is most emphatically not the actual Stanford-Binet Intelligence Scales, Fifth Edition, published by Houghton Mifflin Harcourt (Roid, 2003).

The most important difference between such easily available tests and scientific IQ tests is that norms matter. Authors and publishers of scientific IQ tests devote extraordinary amounts of time, labor, thought, and money to developing accurate, reliable, and valid tests of intelligence. The tests readily available to the general public are not created with such care.

Conclusions

IQ is no longer a "quotient." It is a statistically derived number based on the difference between an examinee's score and the average (mean) score for the examinee's age. Different intelligence tests measure intelligence in many different ways, based on various theories of intelligence. The best IQ tests give, over short periods of time, fairly consistent results, although these results are not precise. Therefore, nobody has "an" IQ. Scores will vary between tests and over time. Professional IQ tests are developed by large teams of experts over several years, and scores are based on large normative samples of individuals who were administered the test. The tests found in popular books and magazines and on the Internet are not trustworthy.

Although the history of intelligence testing has had its share of controversy, and even today, intelligence test scores can be misused and misinterpreted in ways the test developers—respected members of the scientific community—never intended, they remain useful tools. When used appropriately by evaluators who are adequately trained in their administration, scoring and interpretation, intelligence tests can provide a wealth of detailed information that can guide parents, educators, therapists, and officials to provide targeted interventions and assistance to those who need it the most.

References

ACT. (n.d.). Retrieved from https://www.act.org/products/k-12-act-test/

American Psychiatric Association. (2013). *Diagnostic and Statistical Manual of Mental Disorders* (5th ed.). Arlington, VA: APA.

Americans with Disabilities Act Amendments Act of 2008 (ADAAA). 42 USCA § 12101.

Arnold, D. W. (2012). Cognitive Ability Testing (Wonderlic White Paper). Retrieved from http://www.wonderlic.com/sites/default/files/cognitiveAbilityTesting_LG_4.11.2.pdf

Atkins v. Virginia, 536 U.S. 304 (2002).

Bass, I. (1997). *What's Your Opera IQ?* (updated ed.). New York: Citadel.

Bell, N. L., Matthews, T. D., Lassiter, K. S., and Leverett, J. P. (2002). Validity of the Wonderlic Personnel Test as a Measure of Fluid or Crystallized Intelligence: Implications for Career Assessment. *North American Journal of Psychology, 4*(1): 113–120.

Binet, A. (1909). *Les idées modernes sur les enfants.* Paris: Ernest Flammarion. Retrieved from ftp://ftp.bnf.fr/006/N0067926_PDF_1_-1DM.pdf

Binet, A. (1910). Nouvelles recherches sur la mesure du niveau intellectuel chez les enfants d'école. *L'année psychologique, 17*(1): 145–201. DOI: 10.3406/psy.1910.7275

Blackwell, L. S., Rodriguez, S., and Guerra-Carrillo, B. (2015). "Intelligence as a Malleable Construct." In S. Goldstein, D. Princiotta, and J. Naglieri (Eds.), *Handbook of Intelligence*, pp. 263–282. New York: Springer.

Blank, D. (2012). Soccer IQ: Things That Smart Players Do (vol. 1). Retrieved from http://www.soccerpoet.com/blog/blog/

Block, N. J., and Dworkin, G. (Eds.). (1976). *The IQ Controversy.* New York: Random House (Pantheon Books).

Boring, E. G. (1923). Intelligence as the Tests Test It. *New Republic, 35*: 35–37.

Brigham, C. C. (1923). *A Study of American Intelligence.* Princeton, NJ: Princeton University Press.

Buros Center for Testing. (n.d.). Mental Measurements Yearbook. Retrieved from http://buros.org/test-reviews-information

Canivez, G. L. (2008). Orthogonal Higher Order Factor Structure of the Stanford-Binet Intelligence Scales—Fifth Edition for Children and Adolescents. *School Psychology Quarterly, 23*(4): 533–541. DOI: 10.1037/a0012884

Carroll, J. B. (1993). *Human Cognitive Abilities: A Survey of Factor-Analytic Studies.* Cambridge, UK: Cambridge University Press.

Carroll, J. B. (2012). "The Three-Stratum Theory of Cognitive Abilities." In D. P. Flanagan and P. L. Harrison (Eds.), *Contemporary Intellectual Assessment: Theories, Tests, and Issues* (3rd ed.), pp. 883–890. New York: Guilford Press.

Cattell, R. B. (1941). Some Theoretical Issues in Adult Intelligence Testing. *Psychological Bulletin, 38*(7) 592.

College Board. (n.d.). SAT College Admissions Test. New York: College Board. Retrieved from https://www.collegeboard.org/

Cox, A. M. (2016). Danny Meyer Thinks Tipping Is Socialist. *New York Times Magazine*, February 7, p. 66.

Das, J. P. (1973). Structure of Cognitive Abilities: Evidence for Simultaneous and Successive Processing. *Journal of Educational Psychology, 65*: 103–108.

Das, J. P. (1980). Planning: Theoretical Considerations and Empirical Evidence. *Psychological Research, 41*(2): 141–151.

Das, J. P., Naglieri, J. A., and Kirby, J. R. (1994). *Assessment of Cognitive Processes.* Boston, MA: Allyn & Bacon.

Davis, K., Christodoulou, J., Seider, S., and Gardner, H. (2011). "The Theory of Multiple Intelligences." In R. J. Sternberg and S. B. Kaufman (Eds.), *The Cambridge Handbook of Intelligence*, pp. 485–503. New York: Cambridge University Press.

Education for All Handicapped Children Act of 1975 (EHA), 20 U.S.C. § 1400.

Elliott, C. D. (2007). *Differential Ability Scales* (2nd ed.). San Antonio, TX: Psychological Corporation.

Equal Employment Opportunity Commission (EEOC; US). (n.d.). http://www.eeoc.gov/

Eysenck, H. J. (1960). *Know Your Own IQ.* London: Penguin Books.

Federal Trade Commission. (2016). Lumosity to Pay $2 Million to Settle FTC Deceptive Advertising Charges for Its "Brain Training" Program: Company Claimed Program Would Sharpen Performance in Everyday Life and Protect Against Cognitive Decline. Retrieved from https://www.ftc.gov/news-events/press-releases/2016/01/lumosity-pay-2-million-settle-ftc-deceptive-advertising-charges

Felton, R., and Wood, F. (1991). A Reading Level Match Study of Nonword Reading Skills in Poor Readers with Varying IQ. *Journal of Learning Disabilities, 25*(5): 318–326.

Flanagan, D. P., Ortiz, S. O., and Alfonso, V. C. (2013). *Essentials of Cross-Battery Assessment* (3rd ed.). Hoboken, NJ: Wiley.

Fletcher, J. M., Francis, D. J., Rourke, B. P., Shaywitz, S. E., and Shaywitz, B. A. (1992). The Validity of Discrepancy-Based Definitions of Reading Disabilities. *Journal of Learning Disabilities, 25*(9): 555–561.

Flynn, J. R. (2009). *What Is Intelligence?* (rev. ed.). New York: Cambridge University Press.

Frey, M. C., and Detterman, D. K. (2004). The Relationship between the Scholastic Assessment Test and General Cognitive Ability. *Psychological Science, 15*(6): 373–378.

Galton, F. (1886). Regression towards Mediocrity in Hereditary Stature. *Journal of the Anthropological Institute, 15*(18): 246–263.

Galton, F. (1907). *Inquiries into Human Faculty and Its Development* (2nd ed.). London: J. M. Dent & Co. (Everyman). First electronic edition (G. Tredoux, Ed.), retrieved from http://www.galton.org/books/human-faculty/text/galton-1883-human-faculty-v4.pdf

Gardner, H. (1983). *Frames of Mind*. New York: Basic Books.

Gardner, H. (1999). *Intelligence Reframed: Multiple Intelligences for the 21st Century*. New York: Basic Books.

Gardner, H. (2011). *Frames of Mind* (3rd ed.). New York: Basic Books.

Goddard, H. H. (1913). The Binet Tests in Relation to Immigration. *Journal of Psycho-asthenics, 18*(1): 105–107. Retrieved from https://peboundaries.omeka.net/items/show/12

Goddard, H. H. (1917). Mental Tests and the Immigrant. *Journal of Delinquency, 2*: 243–277. Retrieved from http://harpending.humanevo.utah.edu/Documents/goddard.html

Gottfredson, L. S. (2008). "Of What Value Is Intelligence?" In A. Prifitera, D. Saklofske, and L. G. Weiss (Eds.), *WISC-IV Applications for Clinical Assessment and Intervention* (2nd ed.), pp. 545–563. Burlington, MA: Academic Press.

Hale, J. B., and Fiorello, C. A. (2001). Beyond the Academic Rhetoric of "G": Intelligence Testing Guidelines for Practitioners. *School Psychologist, 55*(4): 113–117, 131–135, 138–139.

Hammill, D. D., and McNutt, G. (1981). *The Correlates of Reading: The Consensus of Thirty Years of Correlational Research*. Austin, TX: Pro-Ed.

Herrnstein, R. J., and Murray, C. (1994) *The Bell Curve*. New York: Free Press.

Hirsch, N. D. M. (1926). A Study of Natio-Racial Mental Differences. *Genetic Psychology Monographs, 1*, 231–406.

Horn, J. L. (1988). "Thinking About Human Abilities." In J. R. Nesselroade and R. B. Cattell (Eds.), *Handbook of Multivariate Psychology* (rev. ed.), pp. 645–685. New York: Academic Press.

Horn, J. L., and Blankson, N. (2012). "Foundation for Better Understanding of Cognitive Abilities." In D. P. Flanagan and P. L. Harrison (Eds.), *Contemporary Intellectual Assessment: Theories, Tests, and Issues* (3rd ed.), pp. 73–98. New York: Guilford Press.

Immigration Act of 1924. Pub. L. No 68-139, 43 Stat. 153. Retrieved from http://library.uwb.edu/static/USimmigration/1924_immigration_act.html

Individuals with Disabilities Education Act of 2004 (IDEA), 20 U.S.C. § 1400.

IQ Check. (n.d.). Retrieved from http://iqcheck.org/

Jacoby, R., and Glauberman, N. (Eds.). (1995). *The Bell Curve Debate: History, Documents, Opinions*. New York: Times Books.

Jensen, A. R. (1969). "How Much Can We Boost IQ and Scholastic Achievement?" *Harvard Educational Review, 39*(1): 1–123.

Jimenez-Glez, J. E., and Rodrigo-Lopez, M. R. (1994). Is It True That Differences in Reading Performance between Students with and without LD Cannot Be Explained by IQ? *Journal of Learning Disabilities, 27*(3): 155–163.

eKamin, L. J. (1974). *The Science and Politics of I.Q.* Potomac, MD: Erlbaum.

Kamin, L., and Egerton, J. (1973). The Misuse of IQ Testing. *Change: The Magazine of Higher Learning, 5*(8): 40–43.

Kaufman, A. S. (2009). *IQ Testing 101*. New York: Springer Publishing.

Kaufman, A. S., and Kaufman, N. L. (2004). *Kaufman Assessment Battery for Children* (2nd ed., Manual). Circle Pines, MN: American Guidance Service.

Kaufman, A. S., and Lichtenberger, E. O. (2005). *Assessing Adolescent and Adult Intelligence*. Hoboken, NJ: Wiley.

Kaufman, A. S., and Weiss, L. G. (Eds.). (2010). Special Issue on the Flynn Effect. *Journal of Psychoeducational Assessment, 28*(5): all pages.

Kaufman, J. C. (2010). Using Creativity to Reduce Ethnic Bias in College Admissions. *Review of General Psychology, 14*(3): 189–203.

Kelman, M., and Lester, G. (1997). *Jumping the Queue: An Inquiry into the Legal Treatment of Students with Learning Disabilities*. Cambridge, MA: Harvard University Press.

Kiyosaki, R. T. (2008). *Rich Dad's Increase Your Financial IQ: Get Smarter with Your Money*. New York: Business Plus (Hachette).

Koenig, K. A., Frey, M. C., and Detterman, D. K. (2008). ACT and General Cognitive Ability. *Intelligence, 36* (2): 153–160.

Larry P. v. Riles. (1986). 343 F. Supp. 1306 (ND Cal. 1972) (preliminary injunction). add 502 F. 2d 963 (9th cir. 1974); 495 F. Supp. 926 (ND Cal. 1979) (decision on merits) add (9th cir. no. 80-427 Jan. 23, 1984). Order modifying judgment, C-71-2270 RFP.

Legg, S., and Hutter, M. (2007). A Collection of Definitions of Intelligence. (IDSIA Technical Report). Lugano, Italy: Istituto Dalle Molle di Studi sull'Intelligenza Artificiale. Retrieved from http://arxiv.org/pdf/0706.3639.pdf

Levine, A. H., and Marks, L. (1928). *Testing Intelligence and Achievement*. New York: Macmillan.

Luria, A. R. (1973). *The Working Brain* (B. Haigh, trans.). London: Penguin Books.

Martin, C. (2012). *Boost Your Interview IQ* (2nd ed.). New York: McGraw-Hill.

Mortensen, E. L., Andresen, J., Kruuse, E., Sanders, S. A., and Reinisch, J. M. (2003). IQ Stability: The Relationship between Child and Young Adult Intelligence Test Scores in Low-Birthweight Samples. *Scandinavian Journal of Psychology, 44*(4): 395–398.

Naglieri, J. A., Das, J. P., and Goldstein, S. (2014). *Cognitive Assessment System* (2nd ed., Technical and Interpretive Manual). Austin, TX: Pro-Ed.

Nisbett, R. E., Aronson, J., Blair, C., Dickens, W., Flynn, J., Halpern, D. F., and Turkheimer, E. (2012). Intelligence: New Findings and Theoretical Developments. *American Psychologist, 67*(2): 130–159.

Ortiz, S. O., Ochoa, S. H., and Dynda, A. M. (2012). "Testing with Culturally and Linguistically Diverse Populations: Moving beyond the Verbal-Performance Dichotomy into Evidence-Based Practice." In D. P. Flanagan and P. L. Harrison (Eds.), *Contemporary Intellectual Assessment: Theories, Tests and Issues* (3rd ed.), pp. 526–552. New York: Guilford Press.

Pearson, R. (Ed.). (1992). *Shockley on Eugenics and Race: The Application of Science to the Solution of Human Problems*. Washington, DC: Scott-Townsend.

Pierce, A., Miller, G., Arden, R., and Gottfredson, L. S. (2009). Why Is Intelligence Correlated with Semen Quality? Biochemical Pathways Common to Sperm and Neuron Function, and Their Vulnerability to P-pleiotropic Mutations. *Communicative & Integrative Biology, 2*(5): 385–387.

Powers, K. M., Hagans-Murillo, K. S., and Restori, A. F. (2014). Twenty-Five Years After *Larry P.*: The California Response to Overrepresentation of African Americans in Special Education. *California School Psychologist, 9*(1): 145–158. First online: January 13, 2014. DOI: 10.1007/BF03340915

Ravitch, D. (2000). *Left Back: A Century of Battles over School Reform*. New York: Simon & Schuster.

Reynolds, C. R., and Kamphaus, R. W. (2015). *Reynolds Intellectual Assessment Scales* (2nd ed.). Lutz, FL: Psychological Assessment Resources.

Rhodes, R. L., Ochoa, S. H., and Ortiz, S. O. (2005). *Assessing Culturally and Linguistically Diverse Students: A Practical Guide*. New York: Guilford Press.

Roid, G. H. (2003). *Stanford-Binet Intelligence Scales* (5th ed.). Itasca, IL: Riverside.

Sattler, J. M. (2008). *Assessment of Children: Cognitive Foundations* (5th ed.). San Diego, CA: Jerome M. Sattler.

Schmidt, F. L., and Hunter, J. (2004). General Mental Ability in the World of Work: Occupational Attainment and Job Performance. *Journal of Personality and Social Psychology, 86*(1): 162–173. DOI: 10.1033/0022-3514.86.1.162

Schneider, J., and McGrew, K. S. (2012). "The Cattell-Horn-Carroll Model of Intelligence." In D. P. Flanagan and P. L. Harrison (Eds.), *Contemporary Intellectual Assessment: Theories, Tests and Issues*, (3rd ed.), pp. 99–144. New York: Guilford Press.

Schrank, F., McGrew, K. S., and Mather, N. (2014). *Woodcock-Johnson IV Tests of Cognitive Ability.* Rolling Meadows, IL: Riverside.

Schultz, D., and Schultz, S. (2015). *A History of Modern Psychology.* Boston: Cengage Learning.

Sénéchal, C., Larivée, S., Audy, P., and Richard, E. (2007). L'effet Flynn et la déficience intellectuelle. *Canadian Psychology/Psychologie Canadienne, 48*(4): 256–270. DOI: 10.1037/cp2007022

"Shockley's Race View Called 'Senile,' 'Fascist.'" (1971). *St. Petersburg Times,* September 8, pp. 1-A, 4-A. Retrieved from https://news.google.com/newspapers?nid=888&dat=19710908&id=sewNAAAAIBAJ&sjid=vnUDAAAAIBAJ&pg=4930,1230689&hl=en

Siegel, L. S. (1998). "The Discrepancy Formula: Its Use and Abuse." In B. K. Shapiro, P. J. Accardo, and A. J. Capute (Eds.), *Specific Reading Disability: A View of the Spectrum,* pp. 123–135. Timonium, MD: York Press.

Spearman, C. (1927). *The Abilities of Man: Their Nature and Measurement.* New York: Macmillan.

Stanford Binet (n.d.) Stanford Binet test. Retrieved from https://www.stanfordbinet.net

Stanovich, K. E. (2009). *What Intelligence Tests Miss.* New Haven, CT: Yale University Press.

Stern, William. (1912, 1914). *Die psychologischen Methoden der Intelligenzprüfung: und deren Anwendung an Schulkindern* [The Psychological Methods of Testing Intelligence and Their Application to School Children]. *Educational Psychology Monographs, 13.* (G. M. Whipple, trans.). Baltimore, MD: Warwick & York. Retrieved from http://babel.hathitrust.org/cgi/pt?id=uc1.$b239630;view=1up;seq=14

Sternberg, R. J. (2012). "The Triarchic Theory of Successful Intelligence." In D. P. Flanagan and P. L. Harrison (Eds.), *Contemporary Intellectual Assessment: Theories, Tests and Issues* (3rd ed.), pp. 156–177. New York: Guilford Press.

Sternberg, R. J. (Ed.) (1988). *Advances in the Psychology of Human Intelligence.* Hillsdale, NJ: Erlbaum.

Stewart, W. (2013). *You're the Basketball Ref: 101 Questions to Test Your IQ.* New York: Skyhorse Publishing.

Terman, L. M. (1916). *The Measurement of Intelligence: An Explanation of and a Complete Guide for the Use of the Stanford Revision and Extension of The Binet-Simon Intelligence Scale.* Boston: Houghton Mifflin.

Terman, L. M., and Merrill, M. A. (1937). *Measuring Intelligence: A Guide to the Administration of the New Revised Stanford-Binet Tests of Intelligence.* Boston: Houghton Mifflin.

Thurstone, L. L. (1938). Primary Mental Abilities. *Psychometric Monographs, 1.* Chicago, IL: University of Chicago Press

Walsh, A. A. (1976). The "New Science of the Mind" and the Philadelphia Physicians in the Early 1800s. *Transactions and Studies of the College of Physicians in Philadelphia, 43*(4): 397–413.

Wechsler, D. (1939). *The Measurement of Adult Intelligence*. Baltimore: Williams and Wilkins.

Wechsler, D. (2008). *Wechsler Adult Intelligence Scale* (4th ed.). San Antonio, TX: Pearson.

Wechsler, D. (2012). *Wechsler Preschool and Primary Scale of Intelligence* (4th ed., Administration and Scoring Manual). Bloomington, MN: Pearson.

Wechsler, D. (2014a). *Wechsler Intelligence Scale for Children* (5th ed., Administration and Scoring Manual). Bloomington, MN: Pearson.

Wechsler, D. (2014b). *Wechsler Intelligence Scale for Children* (5th ed., Technical and Interpretive Manual). Bloomington, MN: Pearson.

Wonderlic, E. F. (1992). Wonderlic Personnel Test. Vernon Hills, IL: Wonderlic. Retrieved from http://www.wonderlic.com/assessments/ability/cognitive-ability-tests/contemporary-cognitive-ability-test

Yoakum, C. S., and Yerkes, R. M. (1920). *Army Mental Tests*. New York: Holt.

17 Reflections on Pseudoscience and Parapsychology: From Here to There and (Slightly) Back Again

Christopher C. French

Introduction

Why do I find it so difficult to start writing this chapter? If I am honest with myself, it is because I cannot really shake off the feeling that personal reflections are not something that "proper scientists" should waste their time on. I was brought up in an era when science was presented as being all about objective facts and observations with absolutely no room for subjective opinions. Indeed, it was actually forbidden to ever write from a first-person perspective in a scientific report. At all costs, the impression had to be maintained that a scientific report was nothing more and nothing less than a completely objective, value-free description of a series of events that had taken place that were relevant to deciding whether or not a particular scientific hypothesis was likely to be true or not. Any sign of the involvement of real people with beliefs, emotions, ambitions, intentions, or any other human quality was to be strictly avoided.

My view of science has changed a lot since then, mainly due to the many decades I have spent actively involved in real research in psychology and parapsychology, often working collaboratively with other scientists. I still strongly believe that science provides our best hope of understanding ourselves and our universe, but I now also recognize that, because scientists are human beings after all, their behavior and beliefs are often strongly influenced by emotional and motivational factors. Science as a process can never be perfect because it is carried out by imperfect human beings. However, the main advantage it has over other approaches to pursuing the truth is that, at its best, it recognizes that personal beliefs can affect the validity of scientific observations, and it provides the means to control for such factors.

Despite the fact that I like to think that I now hold a somewhat more sophisticated view of how science really works than perhaps I did in those early days, I still feel quite uncomfortable sitting down to write this chapter (I mean, "intellectual journey"—how pretentious!). So, I won't be at all offended if you decide to skip this chapter altogether. After all, why should you give a damn what Chris French thinks and how he came to think it? Or you might decide to treat this chapter as an interlude of light entertainment when you are feeling a bit too tired to tackle the intellectually meatier chapters from other contributors. But if you decide to read it, I promise I will try to be as honest as I can in reflecting upon the issues covered herein.

Early Years

For as long as I can remember, I have had an interest in the paranormal. Up until well into my early adulthood, I was not only interested in the paranormal but I was pretty much convinced that a wide range of paranormal claims were true. I would not say that my interest in the weird and wonderful was an all-consuming passion. But, particularly in my teenage years, I read a steady stream of pro-paranormal books and would eagerly watch any TV programs covering such material. It never really occurred to me that it might be possible to account for ostensibly paranormal phenomena in non-paranormal terms or that there were any books or articles out there that took a skeptical position with respect to such claims.

I should point out that I am using the word *paranormal* here in the loose sense of referring to anything "weird and wonderful," as opposed to the rather stricter definition preferred by some parapsychologists. For the latter, the term is reserved for the phenomena of extrasensory perception (ESP, which includes telepathy, precognition, and clairvoyance), psychokinesis, and evidence relating to the possibility of postmortem survival. The Loch Ness Monster, aliens, the Bermuda Triangle, and astrology, among others, would be excluded.

I recall that, as a teenager, I was particularly smitten with the best-selling books of Erich von Däniken, such as *Chariots of the Gods* (von Däniken, 1971), in which he presented his "evidence" that extraterrestrials had often visited the Earth in the past and bestowed upon humanity many examples of advanced technology. I found his ideas fascinating and exciting, and his arguments and "evidence" completely convincing. I naively assumed that it must all be true because it was written in a book. It was not until many years later that I fully appreciated just how wrong I was, thanks to such critical works as Ronald Story's *The Space Gods Revealed* (Story, 1976).

My interest and belief in the paranormal was hugely boosted by the appearance on the scene of a certain Uri Geller back in the 1970s. I was taken in hook, line, and sinker by Geller's presentation of relatively simple conjuring tricks as evidence of paranormal powers. I so *wanted* him to be the real deal and was absolutely delighted when a number of distinguished scientists, perhaps most notably Professor John Taylor of the University of London, declared that they believed that Geller did indeed appear to possess amazing psychic abilities (Taylor, 1975). Taylor subsequently became much more skeptical (Taylor, 1980). I have a vague recollection (this could be a false memory) of occasionally seeing on TV someone who could duplicate all of Geller's feats, most famously of course that of spoon bending, using basic conjuring techniques.

If my memory is genuine, this was undoubtedly James Randi, known as The Amazing Randi. The effects certainly looked the same whether the feats were performed by Geller or Randi. My response to this? I saw it as totally irrelevant that a conjuror could do something that looked the same as Geller's psychic feats because Geller had already told us that he was doing it using psychic powers. As I have heard many conjurors say since, if Uri really is using psychic powers to do that stuff, he's doing it the hard way!

It was at about this time that I remember *New Scientist* magazine ran a survey regarding readers' attitudes toward the paranormal and parapsychology. I was sufficiently interested and motivated to take the time to complete the questionnaire, put it in an envelope, address the envelope, stick on a stamp, and walk to the post box to submit my views. Surveys took some effort to complete in those pre-Internet days. Needless to say, my views were all totally in favor of the reality of the paranormal and in support of more research into such topics.

I began studying for my BSc in Psychology at the University of Manchester in 1974. As I recall, there was little, if any, coverage of the paranormal as part of that program. One exception to this occurred via the tutorial system then operated at Manchester. In those more relaxed days, my tutorial "group" consisted of me and my best friend, Graeme Gillespie. Our tutor asked us what we'd like to discuss in the tutorials. One of the topics that occurred to us was telepathy. "Okay," said our tutor pointing at me. "You go away and prepare the case in support of the existence of telepathy. And you," he said, pointing at Graeme, "prepare the case against." Over the following week, I had a much easier time of it than Graeme. I could find lots of material in the university library in support of the existence of telepathy, perhaps most notably the writings of arguably the most influential British psychologist of the time, the late Professor Hans J. Eysenck (e.g., Eysenck, 1957). Graeme struggled to find any skeptical academic commentary on the topic. I won our "debate" hands down.

I graduated from Manchester in 1977 and, after a brief stint at the University of Bangor, eventually settled down to study for a PhD at the University of Leicester. My PhD research investigated possible differences between the two cerebral hemispheres in terms of electroencephalogram activity. Like many postgraduates, I taught courses in psychology at a local adult education college in order to earn some extra cash. I must have had a lot more time on my hands in those days because my approach was to use my first session with a new class to give a general introductory lecture on psychology and to then ask class members to simply suggest topics that they would like me to prepare a lecture on. I would then go away and spend many hours reading around the topic and present them with a brand-new lecture the following week. One of the topics I covered in this way was parapsychology. I was still unaware at this time that skeptical books on the paranormal even existed. My totally uncritical coverage assured my students that scientists had proven the existence of a range of paranormal abilities beyond all reasonable doubt. I now look back and cringe at the thought that many of my students were probably sitting there thinking, "Well, he's doing a PhD. He must know what he's talking about . . ." I didn't.

Epiphany

I can pinpoint my conversion to skepticism very precisely. Toward the end of my PhD research a friend recommended a newly published book by Canadian social psychologist James Alcock. The book was called *Parapsychology: Science or Magic?* (Alcock, 1981) and I loved it. It was the first in-depth skeptical critique of parapsychology that I had ever read. Among other things, it provided non-paranormal explanations for ostensibly paranormal phenomena, explanations that seemed to be supported by good empirical evidence. It also contained the first discussion that I had ever come across regarding the differences between real science and pseudoscience. According to Alcock, parapsychology was clearly in the pseudoscience category. I would still rate this book as being one of the best I have ever read—and, in terms of its influence on my own subsequent career, it would have to be right at top of the list. At the time that I was reading the book, I would never have dreamed that one day I would know the author personally and consider him a friend, but I am pleased to say that that is what happened.

When I first read Alcock's book, a whole new world opened up to me. Even today, pro-paranormal books, magazines, and newspaper articles, not to mention TV and radio programs, massively outnumber skeptical accounts—but the situation was even worse back then. Alcock's book alerted me to the fact that there were such accounts in existence if one knew where to look. The references listed at the end of the book contained quite a number of articles in a publication called *Skeptical Inquirer*, a publication that I admit I had never heard of before. I took out a subscription and from then on would eagerly read the magazine from cover to cover as soon as it arrived.

The book also frequently referred to a certain James Randi, another name with which I was unfamiliar up to that point (and again someone I am now pleased to consider a personal friend). I bought copies of Randi's *Flim-Flam* (Randi, 1982a) and *The Truth about Uri Geller* (Randi, 1982b), and realized how naive I had been to take Uri Geller's claims at face value. Alcock also cited the work of Martin Gardner. I was already familiar with Gardner's books on recreational mathematics but had no idea that he also had an interest in pseudoscience. I read a copy of Gardner's classic *Fads and Fallacies in the Name of Science* (Gardner, 1957) and my conversion from believer to skeptic was complete.

I started working at Goldsmiths College, University of London, in 1985. At that point my interest in skepticism and pseudoscience was little more than a hobby. My academic research was limited to what I felt were more conventional and respectable academic topics, such as lateralization of cognitive functions between the cerebral hemispheres. But I did give a couple of very skeptical lectures on parapsychology to our second-year students as part of a Theoretical Issues module on our BSc (Hons) Psychology program.

Over the following years I met and got to know a number of individuals whose views confirmed me in my relatively newfound skepticism. These included Sue Blackmore and Richard Wiseman. Sue, like me, had once been a believer in the paranormal but had given up such beliefs after years of unsuccessfully trying to produce convincing evidence for the existence of paranormal powers (Blackmore, 1996). In contrast, Richard had always been skeptical regarding the paranormal (Wiseman, 2011) but had also been fascinated by the weird and wonderful since childhood. Sue and Richard had, at that point, much more direct contact with pro-paranormal parapsychologists than I had and yet they were totally unconvinced regarding the reality of paranormal forces. I also recall having the good fortune to meet Ray Hyman and the late Barry Beyerstein during this period. I suspect that many people would expect self-professed skeptics to be curmudgeonly naysayers, but that stereotype was the exact opposite of these two warm and friendly individuals. The writings of both Ray (Hyman, 1989) and Barry (Beyerstein, 1997, 2007) were very influential in informing my own views.

My first academic output in this area was a conference paper presented at the British Psychological Society's annual conference in Swansea in 1990 (French, 1990). It was an overview of cognitive biases that may underlie paranormal belief. In 1992, I published a version of this paper in *The Psychologist* (French, 1992a), the society's monthly magazine. This was the same year that I published my first empirical investigation into paranormal belief, in the *Australian Journal of Psychology* (French, 1992b).

In the mid-1990s I decided that I knew enough about parapsychology, pseudoscience, and the psychology of paranormal belief to provide an optional module on these topics for our final-year students as part of Goldsmiths' BSc program. The module proved to be popular and I must admit that I much preferred teaching it to some of the drier topics that I had taught up to that point. I also began gradually publishing a steady trickle of papers on the psychology of paranormal belief (e.g., French, 2001; Thalbourne and French, 1995, 1997), but I still concentrated my research efforts on more conventional areas. My feeling then was that my interest in the "weird stuff" would be tolerated, provided that I also produced enough papers in other "more respectable" areas, such as the relationship between cognition and emotion (e.g., Richards, French, and Dowd, 1995; Richards, French, and Randall, 1996).

Type I Skeptics

Let us pause to consider the views I then held about parapsychology and the paranormal. At the risk of oversimplification, my views of believers in the paranormal and of parapsychology in general were predominantly negative. With regard to believers in the paranormal, I viewed them as irrational (and probably less intelligent) compared to skeptics. I assumed that all forms of paranormal belief were dangerous and served no positive psychological functions. I viewed most psychic claimants as either being deliberate frauds, out to make as much profit as they could by exploiting the vulnerable, or perhaps suffering from some sort of serious psychopathology. I was very confident that there existed no good evidence in support of paranormal claims and that most parapsychologists were incompetent when it came to experimental design and data analysis. Finally, and most relevant to this chapter, I was convinced that parapsychology was nothing more and nothing less than a pseudoscience.

There is no doubt that the views described in the previous paragraph represented a strand of skeptical thought that was prevalent back then and still exists today. For convenience, I am referring to those who hold such views as type I skeptics. Such individuals see it as a priority to convince others that paranormal abilities do not exist and, as they see it, to turn back a rising tide of irrationalism that threatens to engulf us all. They tend to be particularly keen on military metaphors, viewing those who disagree with them as "the enemy" and using whatever tactics are necessary to gain "victory" including ridicule and ad hominem attacks.

Type II Skeptics

Rightly or wrongly, my views have mellowed. When I was a newly converted skeptic, the idea that we, the skeptics, were the "good guys" and in every way superior to the pro-paranormal "bad guys" was very appealing. It is very human to want simple answers, to see the world in black-and-white terms. The truth is, however, that real life is usually not so straightforward. As Oscar Wilde famously said, "The truth is rarely pure and never simple."

What were the factors that forced me to modify my views on parapsychology and paranormal belief? Essentially it was getting to know a number of parapsychologists personally and realizing that many of them did understand scientific methodology and knew how to analyze data properly. I found that I often felt I had more in common with moderate proponents of the paranormal than those with more extreme views on either side of the debate.

One person in particular who influenced my views at this time was the late Professor Bob Morris. Bob had been appointed as the Koestler Chair in Parapsychology at the University of Edinburgh in 1985 and from then, until his untimely death in 2004, he ran the Koestler Parapsychology Unit (KPU). His approach to the paranormal resulted in two distinct but related streams of research. On the one hand, much of the research carried out by the KPU focused on directly testing paranormal claims, such as investigating possible telepathy using the ganzfeld technique (e.g., Dalton et al., 1996).

The ganzfeld technique is based on the assumption that, if telepathy really does exist, telepathic signals might often be very weak in comparison to other signals, such as input from the various senses. Therefore, an attempt is made to dampen down these other input signals by putting participants into a state of mild perceptual deprivation. They lie on a comfortable couch with earphones on, listening to white noise with halves of ping-pong balls placed over their eyes. Most participants find this extremely relaxing and report that they quickly start to see vivid mental imagery. When they are in this state, a sender will wait for prearranged times and then attempt to telepathically send information about a randomly selected picture or film clip, and the participant will describe the mental images that they experience at those times. The participant (or independent judges) will subsequently attempt to match these descriptions to a set of four stimuli consisting of the actual target used plus three distractors. By chance alone, one would expect a direct hit rate of 25 percent, but some studies report significantly more hits than this.

In addition to directly testing for paranormal abilities, a second strand of research at the KPU focused on what Bob would have described as "what looks like it's psychic but isn't" (e.g., Wiseman and Morris, 1995). Thus members of the KPU were expected to be up to scratch not only with the current state of evidence relating to the possible existence of paranormal powers but also the myriad ways in which a person might come to think that he or she had experienced the paranormal, when in fact his or her experience could be accounted for perfectly well in non-paranormal terms, usually psychological.

This dual approach to the paranormal directly influenced the approach taken by the Anomalistic Psychology Research Unit at Goldsmiths, albeit the unit's primary focus is on investigating possible non-paranormal explanations for ostensibly paranormal experiences (e.g., Crawley, French, and Yesson, 2002; Richards, Hellgren, and French, 2014; Wilson and French, 2006, 2014) with relatively less emphasis on directly testing paranormal claims (e.g., Ritchie, Wiseman, and French, 2012a).

Bob Morris's influence is still felt in the the United Kingdom and internationally, thanks to the large number of PhD students that he supervised. Arguably the most skeptical of his academic offspring is Richard Wiseman; many others are very sympathetic toward the possible existence of genuinely paranormal phenomena. Caroline Watt, a founding member of the KPU and another of Bob's PhD students, is an excellent example of someone who has adopted Bob's open-minded but critical attitude toward paranormal claims (see, e.g., Irwin and Watt, 2007).

I believe I first met Bob at the Fifth European Skeptics Congress, held at Keele University in 1993. There had been a last-minute cancellation by one of the keynote speakers and Bob offered to step in. I thought this was a brave offer, given that many in the audience would hold extremely negative views regarding parapsychology. As it turned out, I need not have worried. Bob gave an excellent and measured presentation that made it abundantly clear that he was well aware of skeptical explanations for ostensibly paranormal experiences and, indeed, that he fully accepted that they must be taken seriously. As I got to know Bob better over subsequent years, I was impressed by his openness. He genuinely welcomed constructive criticism and firmly believed that the best chance of ever finally answering the question of whether or not paranormal abilities exist was through collaboration and discussion between moderates on both sides of the debate (see, e.g., Morris, 2000).

By this point, my conversion from a type I skeptic to a type II skeptic was almost complete. I no longer held that believers in the paranormal were irrational and that skeptics were rational. Instead, I believed that all human beings are prone to irrationality (albeit that the deliberate adoption of critical thinking can, to some extent, guard against such inherent irrationality). Furthermore, the evidence that intelligence was inversely related to paranormal belief was mixed at best (albeit that some specific reasoning biases do appear to show a more consistent relationship; French and Stone, 2014; French and Wilson, 2007).

Regarding the possible dangers of holding paranormal beliefs, I now believed that, while it was certainly true that such beliefs could have damaging consequences (e.g., belief in psychic healing preventing the proper treatment of treatable diseases until it was too late), it was also true that paranormal beliefs could sometimes bring psychological benefits (an obvious example being belief in life after death helping someone cope with a bereavement).

Were all who claimed psychic abilities either deliberate frauds or suffering from some sort of mental illness? This was certainly true of some claimants, but my personal view, having interacted with many self-professed "psychics" over the years, is that the vast majority are neither fraudulent nor crazy. Having said that, it is worth noting that it is very probable that the higher the profile of the "psychic," the more likely they are to resort to deliberate trickery, given the pressure on them to consistently demonstrate their alleged powers day after day on stage or on TV. (I half-jokingly refer to this as "French's Law.") For numerous examples of fraudulent psychics, see Randi (1982a, 1982b) and Kurtz (1985).

Was there any evidence to support paranormal claims? Yes, there was (e.g. Irwin and Watt, 2007). Much of this evidence came from what appeared to be, on initial inspection, reasonably well-controlled studies. Let me be absolutely clear at this point that, for reasons discussed in more detail elsewhere (French, 2010; French and Stone, 2014), I do not personally believe that paranormal abilities exist—but I am by no means certain that my view is correct. The evidence available has not been enough to justify the wholesale rejection of the conventional scientific view. But it is simply incorrect to assert, as some self-professed skeptics do, that there is no evidence in support of the existence of such abilities.

What about the competence (or otherwise) of parapsychologists? Although it was always possible, as it would be in all sciences, to find examples of shoddy methodology, poor reasoning, and inappropriate data analysis, it became apparent to me that many parapsychologists were actually more aware than their colleagues in more conventional disciplines of the possible artifacts and biases that can distort the scientific process. They had to be, given the intense critical scrutiny directed at any findings that might support, say, telepathy. As Bob Morris used to say, this was like chemists having to assure people that, "Yes, we did remember to wash the test tubes before we started."

I said that my conversion from type I to type II skeptic was "almost" complete at this stage. I was certainly beginning to question the overly negative impression of parapsychology that I had acquired from reading some skeptical literature but I still felt that, on balance, parapsychology was best thought of as being a pseudoscience rather than a true science, and this was the view I conveyed to my students. But I was beginning to doubt that too. It was time to reassess my position on the issue.

Science, Nonscience, and Pseudoscience

Philosophers of science have long struggled with the *demarcation problem*: how exactly can science be distinguished from nonscience (Chalmers, 1999)? Despite many proposed solutions, no consensus has emerged regarding a set of strict criteria that would allow one to unerringly categorize one discipline as a science and another as nonscience.

Instead, many commentators (e.g., Edge, Morris, Rush, and Palmer, 1986) would argue that disciplines can be characterized in terms of the degree to which they meet a number of different benchmarks such as falsifiability of hypotheses and theories, reproducibility of findings, generally accepted core knowledge, agreed-upon procedures, the employment of appropriate control conditions, and links with other branches of science. Some disciplines, such as chemistry and biology, are seen as scoring highly on most if not all benchmarks and can therefore confidently be classified as true sciences. Others, such as astrology and homeopathy, score very poorly on most if not all benchmarks and can therefore confidently be rejected as nonsciences. Although disciplines at these extremes are fairly easy to classify, those disciplines that lie between the two extremes on the continuum may be problematic in terms of classification (see Simonton, this volume).

Given the difficulty in producing a strict set of criteria to characterize science, it should come as no surprise that similar conceptual problems can arise when commentators attempt to produce lists of characteristics of pseudoscience. At a very general level, pseudoscience entails falsely claiming scientific status and adopting some of the superficial trappings of science but failing to meet the benchmarks of real science.

Several different sets of characteristics of pseudoscience have been proposed by a number of different commentators. For example, the following list of features of a pseudoscience was proposed by Mario Bunge (Alcock, 1981, p. 117):

- Its theory of knowledge is subjectivistic, containing aspects accessible only to the initiated;
- Its formal background is modest, with only rare involvement of mathematics or logic;

- Its fund of knowledge contains untestable or even false hypotheses that are in conflict with a larger body of knowledge;
- Its methods are neither checkable by alternative methods nor justifiable in terms of well-confirmed theories;
- It borrows nothing from neighboring fields; there is no overlap with another field of research;
- It has no specific background of relatively confirmed theories;
- It has an unchanging body of belief, whereas scientific enquiry teems with novelty;
- It has a worldview admitting elusive immaterial entities, such as disembodied minds, whereas science countenances only changing concrete things.

Wife and husband team, Daisie and Michael Radner, in their influential little volume *Science and Unreason* (Radner and Radner, 1982), listed nine "marks of pseudoscience" that, they claimed, were only ever found in "crackpot work and never in genuine scientific work." These were anachronistic thinking, the tendency to "look for mysteries," the "appeal to myths," a "grab-bag approach to evidence" (ignoring the actual quality of the evidence), irrefutable hypotheses, the use of the "argument from spurious similarity," "explanation by scenario," "research by exegesis," and a refusal to revise theories in the light of criticism. According to the Radners, if a discipline displayed even just one of these marks that was enough to condemn it as a pseudoscience.

Scott O. Lilienfeld (2005, 2018) argues that pseudosciences tend to have certain characteristics but, in contrast to the Radners, he argues that distinguishing between science and pseudoscience is not an all-or-none phenomenon, but instead that science and pseudoscience should be thought of as being at the opposite ends of a continuum. Each of the following characteristics are to be considered in terms of the degree to which they were met by the discipline in question:

- A tendency to invoke ad hoc hypotheses, which can be thought of as "escape hatches" or loopholes, as a means of immunizing claims from falsification;
- An absence of self-correction and an accompanying intellectual stagnation;
- An emphasis on confirmation rather than refutation;
- A tendency to place the burden of proof on skeptics, not proponents, of claims;
- Excessive reliance on anecdotal and testimonial evidence to substantiate claims;
- Evasion of the scrutiny afforded by peer review;
- Absence of "connectivity" [. . .], that is, a failure to build on existing scientific knowledge;
- Use of impressive-sounding jargon whose primary purpose is to lend claims a facade of scientific respectability;
- An absence of boundary conditions [. . .], that is, a failure to specify the settings under which claims do not hold.

Many other alternative lists of the features of pseudoscience could have been presented but these three will suffice to illustrate the fact that, although the lists certainly overlap to a considerable extent, they also differ in significant ways. For example, both the Radners (1982) and Lilienfeld (2005), following Popper (1963), emphasize the importance of falsifiability in assessing the scientific status of a hypothesis. "Irrefutable hypotheses" is one of the Radners's "marks of pseudoscience" and Lilienfeld refers to both "a tendency to invoke ad hoc hypotheses, which can be thought of as 'escape hatches' or loopholes, as a means of immunizing claims from falsification," and "an emphasis on confirmation rather than refutation." Bunge's list does not mention falsifiability.

This variation between different commentators is partly explained by the inherent slipperiness of the concept of pseudoscience, partly by the specific context in which the lists were put forward (e.g., Lilienfeld is particularly referring to pseudoscience in psychology), and partly by the general cultural context at the time. A nice example of the latter is the fact that the Radners (1982) spend a considerable amount of time discussing the "appeal to myths" that is apparent in the writings of some pseudoscientists. This refers to the tendency of some pseudoscientists at the time, including my beloved Erich von Däniken, to assume that myths and stories from ancient times were literally true and to then claim that they represented evidence in support of the pseudoscientific theory in question. This approach has waned in popularity since the 1980s and therefore "appeal to myths" is less likely to appear on more recent lists of the features of pseudoscience.

Some commentators (e.g., McNally, 2003; Truzzi, 1996) have argued that the inherent fuzziness of the concept of pseudoscience is so great that it would be preferable to simply stop using it. However, others have argued that just because it is sometimes a difficult concept to apply when considering particular disciplines in the gray area between the extremes of prototypical "real" science and prototypical pseudoscience, this does not mean it is of no use at all. In the words of Lilienfeld, Lynn, and Lohr (2015, p. 6), "From this perspective, pseudosciences can be conceptualized as possessing a fallible, but nevertheless useful, list of indicators or 'warning signs.' The more such warning signs a discipline exhibits, the more it begins to cross the murky dividing line separating science from pseudoscience . . ."

The Scientific Status of Parapsychology

We are now in a position to turn our attention specifically to the case of parapsychology. As discussed above, I had moved from being a supporter of parapsychology and a believer in a range of paranormal phenomena to being a harsh critic of the discipline and its practitioners and a complete skeptic regarding all paranormal claims. Then, when I actually met parapsychologists and got to know their views in more detail, I reevaluated my opinions. But I still believed that, on balance, parapsychology was better conceptualized as an example of pseudoscience than real science. Reading one particular paper finally changed my mind on that issue.

Marie-Catherine Mousseau (2003) adopted an empirical approach in her attempt to determine whether parapsychology was a science or a pseudoscience. She selected ten features that are commonly cited as distinguishing between science and pseudoscience and then performed content analyses on three mainstream journals (*British Journal of Psychology, Experimental Physiology,* and *Journal of Physics B: Atomic, Molecular and Optical Physics*) and four fringe journals (*Journal of Scientific Exploration, Journal of Parapsychology, Revue Française de Parapsychologie,* and *Journal of Psychical Research*[1]). Her results might come as something of a shock for some critics of parapsychology.

For example, it is sometimes argued that pseudoscience *suppresses or distorts unfavorable data* whereas real science seeks *empirical confirmations and disconfirmations*. But almost half of the reports published in the fringe journals were of negative outcomes (i.e., disconfirmations) in stark contrast to the complete lack of such reports in the mainstream journals. This bias toward statistically significant positive results on the part of mainstream journals has been acknowledged as a serious problem within psychology (see, e.g., Ritchie, Wiseman, and French, 2012b).

Pseudosciences are said to *use little mathematics or logic* but Mousseau (2003) found that all empirical articles in both the mainstream journals and the fringe journals used statistical tests. Pseudosciences are said to *rely too much on testimonials and anecdotal evidence* in contrast to science, which *proposes and tries out new hypotheses*. But in fact some 17 percent of fringe articles proposed new hypotheses. In the fringe journals, 43 percent of the articles were empirical in nature, with almost a quarter being lab-based experiments. In Mousseau's (2003, p. 274) words, "On all counts, this sample of fringe journals satisfies the methodological criteria for proper science."

Scientists, in contrast to pseudoscientists, are said to *admit their own ignorance and the need for more research and find their own field difficult and full of gaps*. However, 29 percent of the articles in fringe journals were found to be articles reflecting upon such epistemological issues, in contrast to a complete lack of such articles in the mainstream journals. Pseudosciences are said to have *no overlap with other fields of research*, whereas such connections are said to be a characteristic of real science. However, over a third of the citations in the fringe journals were to articles reporting mainstream science (such as physics, neuroscience, and psychology), although it is unclear whether these citations were generally positive or negative. By contrast, in the mainstream journals less than 10 percent of citations were to other fields (only 1 percent in physics journals).

It is apparent from the above that parapsychology appears, based on commonly asserted features said to distinguish pseudoscience from real science, to be better characterized as being a true science. Indeed, on several criteria it appears to outperform more mainstream sciences. It might be objected that Mousseau (2003) did not use the right list of features, but her chosen set of features certainly overlapped to a large extent with those included on many other such lists. As already discussed, there simply is no universally accepted set of criteria to distinguish between science and pseudoscience, and Mousseau's list appears to be reasonably representative.

Conclusion

It may seem odd to arrive at a position whereby one is far from convinced that paranormal abilities actually exist and yet to conclude that the discipline dedicated to testing the paranormal hypotheses is a true science. This is, to the best of my knowledge, very much a minority opinion among those on either side of the debate. But it seems to me the fairest conclusion to draw.

This does, of course, raise the question of how it is that some parapsychologists appear to produce significant positive findings in support of the existence of paranormal abilities in what appear to be reasonably well-controlled experiments. A detailed examination of this question is beyond the scope of the current chapter but I think that discussion regarding replication problems in psychology points toward an answer (for more details, see French and Stone, 2014; Ritchie, Wiseman, and French, 2012b).

There is no doubt that the biggest challenge that has always faced parapsychologists is their inability to reliably replicate the ostensibly paranormal effects they report in their journals. There is not a single paranormal effect that is robust enough to provide the basis of an undergraduate psychology practical class. Mainstream psychology also suffers from problems with respect to replication, but they are nowhere near as serious. Although it is now widely accepted that many effects reported in psychology journals are probably spurious, there are still hundreds of psychological effects that replicate reliably time after time.

It is almost certain that the spurious effects that are reported in psychology are largely caused by researchers adopting what have become known as "questionable research practices" (John, Loewenstein, and Prelec, 2012; Simmons, Nelson, and Simonsohn, 2011) rather than deliberate fraud (although such cases do arise, of course). Whether or not statistically significant results are obtained depends on numerous decisions regarding methodology and analysis. Researchers will often feel tempted to give themselves the benefit of any doubt in making such decisions, especially since careers are advanced by publication in high-impact journals, and these journals want to publish positive findings.

Questionable research practices will undoubtedly also be a feature of parapsychology, given that many parapsychologists were trained in experimental design and analysis as part of an undergraduate psychology degree. To give but one example, it is highly likely that Bem's (2011) controversial series of studies ostensibly supporting the existence of precognition involved a number of such practices (French and Stone, 2014). There is clear evidence of optional stopping—analyzing data as the experiment proceeds, and then stopping once a significant result has been obtained, as opposed to running a prespecified number of participants. Bem was also guilty of collecting data on a large number of variables that are not mentioned in the published report, thus allowing for the possibility of cherry-picking significant results from numerous analyses, most of which may go unreported.

All sciences aim to separate true signals from background noise. The more that the routine adoption of questionable research practices can be reduced, the clearer will be the separation between signal and noise. Is it possible that a science that was based on only noise with no true signals at all would look a lot like parapsychology?

There is one other argument that I feel supports the rather odd conclusion that parapsychology is best categorized as being a true science rather than a pseudoscience even if paranormal abilities do not really exist. Occasionally critics of parapsychology, such as James Randi, Susan Blackmore, Richard Wiseman, and even yours truly, will attempt to put the claims of self-proclaimed psychics to the test. When we do so, we attempt to apply well-controlled scientific methodologies that constitute fair tests of the claims in question. I do not see how it could really be denied that when we do so, we are engaged in parapsychological investigations—and we are doing so scientifically.

In the final analysis, science is a method for approaching the truth, not a body of established facts or a particular set of phenomena. In deciding whether parapsychology is better described as a science or a pseudoscience, we should look at the methods employed by parapsychologists. It would be unfair to judge any discipline only by considering the poorest work done within it (if we did so, psychology would certainly be dismissed as a pseudoscience). Parapsychology at its best—as exemplified by, for example, the articles in the *Journal of Parapsychology*—appears to meet most if not all of the benchmarks of true science as opposed to pseudoscience. That is so even if paranormal forces do not actually exist.

Note

1. This is probably meant to say either *Journal of the Society for Psychical Research* or *Journal of the American Society for Psychical Research*; to the best of my knowledge, there is no *Journal of Psychical Research*. The mistake does not undermine the basic thrust of the argument.

References

Alcock, J. E. (1981). *Parapsychology: Science or Magic?* Oxford: Pergamon Press.

Bem, D. J. (2011). Feeling the Future: Experimental Evidence for Anomalous Retroactive Influences on Cognition and Affect. *Journal of Personality and Social Psychology*, *100*(3): 407–425.

Beyerstein, B. L. (1997). Why Bogus Therapies Seem to Work. *Skeptical Inquirer*, *21*(5): 29–34.

Beyerstein, B. L. (2007). "The Neurology of the Weird: Brain States and Anomalous Experience." In S. Della Sala (Ed.), *Tall Tales about the Mind and Brain: Separating Fact from Fiction*, pp. 314–335. Oxford: Oxford University Press.

Blackmore, S. (1996). *In Search of the Light: The Adventures of a Parapsychologist.* Amherst, New York: Prometheus.

Chalmers, A. F. (1999). *What Is This Thing Called Science?* (3rd ed.). Queensland: University of Queensland Press.

Crawley, S. E., French, C. C., and Yesson, S. A. (2002). Evidence for Transliminality from a Subliminal Card Guessing Task. *Perception, 31*(7): 887–892.

Dalton, K. S., Morris, R. L., Delanoy, D. L., Radin, D.I., Taylor, R., and Wiseman, R. (1996). Security Measures in an Automated Ganzfeld System. *Journal of Parapsychology, 60*(2): 129–148.

Edge, H. L., Morris, R. L., Rush, J. H., and Palmer, J. (1986). *Foundations of Parapsychology: Exploring the Boundaries of Human Capability.* Boston, MA: Routledge and Kegan Paul.

Eysenck, H. J. (1957). *Sense and Nonsense in Psychology.* Harmondsworth, Middlesex: Penguin Books.

French, C. C. (1990). Belief in the Paranormal: Do Sheep and Goats Process Information Differently? British Psychological Society Annual Conference at University College of Swansea, April 5–8.

French, C. C. (1992a). Factors Underlying Belief in the Paranormal: Do Sheep and Goats Think Differently? *Psychologist, 5*(7): 295–299.

French, C. C. (1992b). Population Stereotypes and Belief in the Paranormal: Is There a Relationship? *Australian Psychologist, 27*(1): 57–58.

French, C. C. (2001). Dying to Know the Truth: Visions of a Dying Brain, or Just False Memories? *Lancet, 358*(9298): 2010–2011.

French, C. C. (2010). "Reflections of a (Relatively) Moderate Skeptic." In S. Krippner and H. L. Friedman (Eds)., *Debating Psychic Experience: Human Potential or Human Illusion?* pp. 53–64. Santa Barbara, CA: Praeger.

French, C. C., and Stone, A. (2014). *Anomalistic Psychology: Exploring Paranormal Belief and Experience.* Basingstoke: Palgrave Macmillan.

French, C. C., and Wilson, K. (2007). "Cognitive Factors Underlying Paranormal Beliefs and Experiences." In S. Della Sala (Ed.), *Tall Tales about the Mind and Brain: Separating Fact from Fiction,* pp. 3–22. Oxford: Oxford University Press.

Gardner, M. (1957). *Fads and Fallacies in the Name of Science.* New York: Dover.

Hyman, R. (1989). *The Elusive Quarry: A Scientific Appraisal of Psychical Research.* Buffalo, New York: Prometheus.

Irwin, H. J., and Watt, C. (2007). *An Introduction to Parapsychology* (5th ed.). Jefferson, NC: McFarland.

John, L. K., Loewenstein, G., and Prelec, D. (2012). Measuring the Prevalence of Questionable Research Practices with Incentives for Truth-Telling. *Psychological Science, 23*(5): 524–532.

Kurtz, P. (Ed.). (1985). *A Skeptic's Handbook of Parapsychology.* Buffalo, New York: Prometheus.

Lilienfeld, S. O. (2005). The 10 Commandments of Helping Students Distinguish Science from Pseudoscience in Psychology. *Observer, 18*(9): 39–40, 49–51.

Lilienfeld, S. O. (2018.) "Navigating a Post-truth World: Ten Enduring Lessons from the Study of Pseudoscience." In Allison B. Kaufman and James C. Kaufman (Eds.), *Pseudoscience: The Conspiracy against Science.* Cambridge, MA: MIT Press.

Lilienfeld, S. O., Lynn, S. J., and Lohr, J. M. (2015). "Science and Pseudoscience in Clinical Psychology: Initial Thoughts, Reflections, and Considerations." In S. O. Lilienfeld, S. J. Lynn, and J. M. Lohr (Eds.), *Science and Pseudoscience in Clinical Psychology.* (2nd ed.) pp. 1–16. New York: The Guilford Press.

McNally, R. J. (2003). Is the Pseudoscience Concept Useful for Clinical Psychology? The Demise of Pseudoscience. *Scientific Review of Mental Health Practice, 2*(2): 97–101.

Morris, R. L. (2000). Parapsychology in the 21st Century. *Journal of Parapsychology, 64*(2): 123–137.

Mousseau, M.-C. (2003). Parapsychology: Science or Pseudo-science? *Journal of Scientific Exploration, 17*(2): 271–282.

Popper, K. (1963). *Conjectures and Refutations: The Growth of Scientific Knowledge.* New York: Basic Books.

Radner, D., and Radner, M. (1982). *Science and Unreason.* Belmont, CA: Wadsworth.

Randi, J. (1982a). *Flim-Flam: Psychics, ESP, Unicorns, and Other Delusions.* Buffalo, New York: Prometheus.

Randi, J. (1982b). *The Truth about Uri Geller.* Buffalo, New York: Prometheus.

Richards, A., French, C. C., and Dowd, R. (1995). Hemisphere Asymmetry and the Processing of Emotional Words in Anxiety. *Neuropsychologia, 33*(7): 835–841.

Richards, A., French, C. C., and Randall, F. (1996). Anxiety and the Use of Strategies in the Processing of an Emotional Sentence-Picture Verification Task. *Journal of Abnormal Psychology, 105*(1): 132–136.

Richards, A., Hellgren, M. G., and French, C. C. (2014). Inattentional Blindness, Absorption, Working Memory Capacity, and Paranormal Belief. *Psychology of Consciousness: Theory, Research, and Practice, 1*(1): 60–69.

Ritchie, S. J., Wiseman, R., and French, C. C. (2012a). Failing the Future: Three Unsuccessful Attempts to Replicate Bem's "Retroactive Facilitation of Recall"' Effect. *Public Library of Science One, 7*(3): e33423. DOI: 10.1371/journal.pone.0033423

Ritchie, S. J., Wiseman, R., and French, C. C. (2012b). Replication, Replication, Replication. *Psychologist, 25*(5): 346–348.

Simmons, J. P., Nelson, L. D., and Simonsohn, U. (2011). False-Positive Psychology: Undisclosed Flexibility in Data Collection and Analysis Allows Presenting Anything as Significant. *Psychological Science, 22*(11): 1359–1366.

Story, R. (1976). *The Space-Gods Revealed: A Close Look at the Theories of Erich von Däniken.* New York: Harper & Row.

Taylor, J. (1975). *Superminds: A Scientist Looks at the Paranormal.* London: Viking Press.

Taylor, J. (1980). *Science and the Supernatural.* New York: E. P. Dutton.

Thalbourne, M. A., and French, C. C. (1995). Paranormal Belief, Manic-Depressiveness, and Magical Ideation: A Replication. *Personality and Individual Differences, 18*(2): 291–292.

Thalbourne, M. A., and French, C C. (1997). The Sheep-Goat Variable and Belief in Nonparanormal Anomalous Phenomena. *Journal of the Society for Psychical Research, 62*(1): 41–45.

Truzzi, M. (1996). Pseudoscience. In G. Stein (Ed.), *Encyclopedia of the Paranormal,* pp. 560–575. Amherst, New York: Prometheus.

von Däniken, E. (1971). *Chariots of the Gods? Was God an Astronaut?* London: Corgi Books.

Wilson, K., and French, C. C. (2006). The Relationship between Susceptibility to False Memories, Dissociativity, and Paranormal Belief and Experience. *Personality and Individual Differences, 41*(8): 1493–1502.

Wilson, K., and French, C. C. (2014). Magic and Memory: Using Conjuring to Explore the Effects of Suggestion, Social Influence and Paranormal Belief on Eyewitness Testimony for an Ostensibly Paranormal Event. *Frontiers in Psychology: Theoretical and Philosophical Psychology,* 5:1289. DOI: 10.3389/fpsyg.2014.01289

Wiseman, R. (2011). *Paranormality: Why We See What Isn't There.* London: Macmillan.

Wiseman, R., and Morris, R. L. (1995). Recalling Pseudo-psychic Demonstrations. *British Journal of Psychology, 86*(1): 113–125.

V Science Activism: How People Think about Pseudoscience

18 Using Case Studies to Combat a Pseudoscience Culture

Clyde Freeman Herreid

When you are a Bear of Very Little Brain, and you Think of Things, you find sometimes that a Thing which seemed very Thingish inside you is quite different when it gets out into the open and has other people looking at it.
—A. A. Milne, *Winnie-the-Pooh*

A. A. Milne, British novelist, poet, and playwright, chronicler of this winsome teddy bear living in the Hundred Acre Wood, reported that the honey-loving Pooh was fond of saying, "Think, think, think," and "Think it over, think it under," both delicious comments dear to the heart of teachers everywhere. Thus, he might be adopted as a mascot of STEM education. Further, the Pooh whose endearing ways eclipsed all of Milne's other accomplishments is also quoted as passing along these words of wisdom:

You can't stay in your corner of the Forest waiting for others to come to you. You have to go to them sometimes.

So we shall.

In the previous chapters we have read about pseudoscience and the dangers it presents. Here I wish to take up Pooh's challenge and consider how teachers can get out of the forest and deal with the hearts and minds in the classroom. Is there any way that we can cope with misinformation that issues forth from the media every day, thanks to self-proclaimed health experts or school board members wishing to dictate what happens in the science classroom? I argue the answer is yes, but that it cannot be done by using the same old approach that we use to teach STEM (science, technology, engineering, and mathematics) subjects in the majority of the world's classrooms. It is not enough to proclaim that pseudoscience is false or fake science. Saying something is not science will not do the trick. Art is not science either. Nor is sports or religion. But these are not pretenders. So how do we detect pseudoscience and grab it by the throat and put it to rest? First, we need to remind ourselves what constitutes science and who scientists are. This is what we need to instill in our students. The first part of this chapter outlines the key elements of the scientific enterprise that we need to deliver and how to do this. It is in the latter part of the chapter that I will deal with the pseudoscience culture and how to unmask its weaknesses.

Let's face reality. As STEM faculty, in the science classroom where most of us attend to our teaching responsibilities, we are in the business of delivering our subject matter to students; we are not typically warning students about bogus topics. In that setting our first obligation is to get a clear message across about what science is. It is when we are teaching special topics courses or nonmajors where we have the greatest opportunities to explore just what is and is not science. In neither of these settings is the lecture method the best way to deliver the goods.

First Things First: What Is Science?

Students need to understand that science is not simply a collection of facts that we teach in departments of chemistry, physics, geology, and biology. More importantly, it is the method of accumulating those facts and testing their veracity. This is the piece that is missing from many of our basic courses. Students need to deeply appreciate that it doesn't mean that scientific "facts" are irrevocable, but that their credibility has been tested repeatedly and found to have passed all reasonable tests. And it doesn't mean that there aren't facts that might be found in other fields, such as psychology or history. But if something is science, it needs to deal with facts that purport to explain the physical world.

School children are often taught a simplified version of what is called The Scientific Method. The steps are outlined as some variation of make Observations, ask Questions, formulate Hypothesizes, perform Experiments, and render Conclusions. As faculty we know, but students seldom do, that rarely are things so neat; many scientists go about their daily business without once referring to these steps. A marvelous website at the University of California at Berkeley explains how real science works (http://undsci.berkeley.edu). Teachers would do well to investigate all of the nuances of the site. The site makes the point that science:

- Aims to explain the natural world
- Uses testable ideas
- Relies on evidence
- Involves the scientific community
- Leads to ongoing research
- Leads to benefits for society

The important result that emerges from all of this guessing, testing, guessing, and retesting is that the "truth" finally begins to emerge. If you make a misstep along the way, for whatever reason, hopefully that mistake will be caught. The vital characteristic of science is that it is self-correcting. People change their minds when the evidence comes in. Not all at once. Not necessarily with a happy smile, especially when they have been shown to be wrong. And some scientific malcontents never change; they have to die before their wrongheaded view is abandoned. But truth will eventually come out. A well-known quote by Max Planck says it well: "A scientific truth does not triumph by convincing its opponents and making them see the light, but rather because its opponents eventually die and a new generation grows up that is familiar with it."

This is all well and good, but how do we teach all of this process? And moreover, how do we teach the limitations of science? This is important to keep in mind when we deal with controversies involving public policy such as abortion, stem cells, recreational use of marijuana, cloning, genetically engineered foods, or climate change. Again the Berkeley site comes to our aid:

- Science doesn't make moral judgments.
- Science doesn't make aesthetic judgments.
- Science doesn't tell you how to use scientific knowledge.
- Science doesn't draw conclusions about supernatural explanations.

To sum up the lessons: "The Berkeley site deals with how the average citizen should approach science stories when they don't know what to believe. The site doesn't give the reader a blueprint for evaluating sensational claims, but it raises fundamental questions that everyone should ask themselves when they hear a statement about the latest diet fad, cancer cure, or global warming. What is the evidence? Who says so? What does the scientific community at large say?" (Herreid, 2010)

And this is one of the key points about the scientific process: the role that the scientific community plays in all of this. This is another salient theme that is seldom appreciated by the nonscientist. Yet it should be obvious that scientists talk to one another: in elevators, in labs, in meetings, over the phone, by email, and via publications. They gossip, chat, speculate, and natter all of the time. They don't always agree: they argue. And they repeat experiments, their own and those of others—especially when disagreements occur. And this is where peer evaluation comes in. Scientists care what other scientists think of them and their ideas. After all, they are human. And the public only vaguely knows that scientists have developed a formal system of review, peer evaluation, that operates whenever they submit an article to a journal or a grant request to a government agency. Students need to know how this works.

They need to know that when a scientist (or more commonly a group of scientists today) thinks he or she has made a valuable discovery, the scientist will write up this discovery for scientific journal. The report is usually structured in a particular way with an introduction, a methods section, a results section, a discussion, and a list of references. The scientist sends this off to a journal for possible publication. The journal editors then forward the article to people who are experts in the field, and ask them to analyze the credibility of the work and its conclusions. Many articles are rejected if the work isn't up to snuff, and comments are sent back to the author. Rejection rates for high-profile journals are over 90 percent. Peer review sets a high standard of excellence.

Even after a scientific paper is published, the critical evaluation of the idea will continue. This is especially true if the ideas in the paper are controversial. If the finding is rather mundane, or had been expected, there will be only a passing interest. Discover a species of bat that hibernates in the winter; few eyebrows will be raised. But a bat that breathes underwater? Watch naturalists around the world perk up their ears. They'll demand to see the evidence and insist on testing the bats themselves. The National Geographic would certainly be at that scientist's doorstep. Marcello Truzzi, cofounder of the Society for the Scientific Investigation of Claims of the Paranormal, put it this way: "Extraordinary claims require extraordinary proof." The essence of peer review is captured in the Russian proverb that Ronald Reagan used to quote during the Cold War: "Trust, but verify." This is especially true if the claims are sensational, such as the assertion that an asteroid wiped out the dinosaurs.

If you want students to appreciate the story of dinosaur extinction, have them read how it all played out (Herreid, 2010; University of California, Berkeley, 2017). One of the valuable lessons is that one can see the key elements of a real scientific method and the role that the scientific society played in the process.

Who Are Scientists and Where Do We Find Them?

If we are looking for ways to clarify what science is, it is useful for students to see who the paragons in the field are, what kind of credentials they have, and what their identifying characteristics might be. Classroom exercises where students look up the background of distinguished scientists help establish a foundation for what makes a scientist.

Well-known figures include Edward Wilson at Harvard University, Albert Einstein at Princeton University, Carl Sagan of Cornell University, Steven Hawking of Cambridge University, Marie Curie of the University of Paris, Louis Pasteur at the University of Strasbourg, Richard Feynman of Cornell University and the California Institute of Technology, and Neil deGrasse Tyson, the director of the Hayden Planetarium of the American Museum of Natural History. These figures all became part of well-established educational institutions that vetted them and credentialed their expertise.

Scientists are not lonely hermits puttering along in the wilderness who suddenly burst forth with a cure for cancer or discovery of the Higgs boson particle. They usually work for conventional educational or commercial institutions; they have established their authority by their research and publication activity over years. In doing this, they have earned the title "scientist." It is true that some noted scientists became famous when they were not associated with an educational institution. Charles Darwin did it. So did Albert Einstein and Isaac Newton. But they too had to establish their qualifications by their outstanding work and by following the established canons of their disciplines. So what are these norms of scientific behavior?

The Canons of Science

Nations, societies, and even clubs have written laws of acceptable behavior. England has its Magna Carta forced upon King John by his rebellious barons; Christianity and Judaism have their commandments handed down by God to Moses; the Roman Code of Justinian and Salic law was imposed upon Germanic France; and there are surviving tablets from Mesopotamia with laws written in Sumerian dated 2100 BC. All lay out the basic rules of acceptable behavior. Scientists have no such laws, at least not in written form. But rules exist nevertheless, and anyone who breaks them risks paying a price.

The National Academy of Sciences (NAS) has produced three documents that deal with the topic of what science is and isn't (Committee on the Conduct of Science, NAS, 1989, 1995, 2009). Their first, *On Being a Scientist,* itemizes and discusses the principles that we associate with science. If we run into a person who claims to have found a major discovery, we not only should be skeptical of any unusual statement, we should see if they are following the traditional steps of science. One valuable exercise that I have had students perform is to have them read the first short pamphlet and ask them to itemize the rules of behavior that we expect from scientists; that is, I want them to develop scientific commandments. I follow this by asking them to read the historical account of the cold fusion debacle of 1989, have them identify where University of Utah researchers Martin Fleischmann and Stanley Pons might have erred (Rousseau, 1992). This can be done for any ignoble or noble study. See, for example, another unfortunate example in the Benveniste affair, which involved homeopathic claims that water can remember (Maddox, Randi, and Stewart, 1988). And now here are some of the student commandments.

Commandments for the Scientist in the Lab and Field

- Thou shalt acknowledge the limitations of your data.
- Thou shalt not select only the data that fit your prior expectations or hypothesis.
- Thou shalt repeat experiments.
- Thou shalt use double blind trials.
- Thou shalt use well-controlled experiments.
- Thou shalt pose hypotheses that are testable and falsifiable.
- Thou shalt be willing to change your mind when the evidence demonstrates that your favorite hypothesis is in error.

Commandments for the Scientist Interacting with Society

- Thou shalt recognize the contributions of others in your citations, authorship, and acknowledgments.

- Thou shalt not hold press conferences to announce discoveries before your work has been reviewed by peers in seminars, conferences, and published articles in scientific journals.
- Thou shalt not commit fraud; it is the gravest violation of scientific ethics. This means no selecting only the data that support your favorite hypothesis; no changing the readings to meet expectations; no fabricating data.
- Thou shalt not commit plagiarism, which passes someone else's ideas and verbiage as your own. This is the most blatant form of misappropriating credit.
- Thou shalt uphold the integrity of science by reporting serious transgressions of these basic canons to the appropriate granting agencies, academic institutions, and public at large.

The National Academy of Sciences has followed up with two more editions of *On Being a Scientist*. These publications (1995, 2009) can be especially helpful to the classroom teacher. They present case studies where students are asked to voice critical opinions about quandaries facing scientists dealing with data and ethical dilemmas such as conflict of interest, industrial sponsorship of academic research, sharing of research materials, assigning credit in publications, and various forms of misconduct, including plagiarism.

We need to teach students the accepted canons of scientific behavior and why they have been adopted. They need to absorb this knowledge into their souls, not just memorize some list. They should learn about physicist Richard Feynman's cautionary statement that before we scientists go public with new scientific results, we should consider every conceivable way we might be wrong (Lightman, 2015).

There are many examples of scientists who did not heed such advice: the much heralded 1970's miracle drug Laetrile, which supposedly destroyed tumors but poisoned people instead (Herbert, 1979; Lerner, 1981); the reported achievement of cold fusion (Taubes, 1993); the sensational account that scientists had found neutrinos that moved faster than the speed of light (Cho, 2011). All of these "discoveries" were soon disproved by other scientists. As MIT physicist Alan Lightman (2015) has commented, "The much-vaunted 'scientific method' and its objective pursuit of truth often cannot be found in the work habits of individual scientists. It's manifest only in the combined efforts of the scientific community with researchers constantly testing and criticizing one another's work. Individual scientists are driven by the same passions and biases and emotional attachments as non-scientists." In other words, beware of irrational exuberance.

How Scientists Can Go Wrong—"Methinks It Is Like a Weasel"

Researchers have examined how easily we humans fool ourselves (Nuzzo, 2015). People look for patterns even in a sea of randomness, and we invariably find it if we look hard enough. Stare at the sky for a moment and you are sure to see star constellations and cloud formations that emulate bears, camels, and other wondrous beasts. "Methinks it is like a weasel," says Hamlet, discussing the shape of a cloud with Polonius. We come by this talent naturally—it is in our genes. Searching for meaning may have served us well in the African veldt when our ancestors were scanning the horizon for predators. But it can lead unwary scientists astray. A recent study of a hundred psychology papers published by credible scholars revealed that only 39 percent had findings that could be replicated (Bohannon, 2015). This result could be due to many reasons, including unconscious bias. Scientists have a stake in the outcome of their research, as will be seen in the argument that immediately ensued after Bohannon's article appeared. (Anderson et al., 2016; Gilbert, King, Pettigrew, and Wilson, 2016). Researchers want to find what they are looking for. The temptation is great when the data are complex and voluminous. Then statistical methods are often repeated numerous times in countless ways, squeezed and wrung until they yield some positive results.

Nuzzo (2015) speaks of "hypothesis myopia." This is our tendency to collect evidence to support just one hypothesis and neglect other alternatives. And we are prone to neglect publishing negative results: we don't want to report failure, and journals and granting agencies are hardly interested in such papers either. Nuzzo tells of a 2012 study of two thousand psychologists, a study that found half of them selectively reported only positive results of experiments. Another pitfall is disconfirmation bias: when data can't be made to match the hypothesis, the researcher looks for ways to write off the data.

It is disappointing to admit but students need to learn that deliberate scientific misconduct happens, even from credible scientists. Recall the famous Piltdown man hoax (Russell, 2012), and Cyril Burt's apparently fabricated data about the heritability of intelligence (Tucker, 1997). Temptation has only grown, thanks to the fame and fortune that can be made off sensational claims that pan out. For example, universities have become intimately entangled with businesses, such as pharmaceutical companies, looking to make huge financial gains from patents. *Wired* magazine reported in 2004 that 3 percent of the 3,475 research institutions reporting to the US Department of Health and Human Services' Office of Research Integrity had indicated some form of scientific misconduct. (Of course, misconduct doesn't necessarily mean that a claim should be judged as pseudoscience.)

Discovery of misconduct can take years or be discovered in a trice. The consequences of discovery vary, but often enough a researcher's credibility and career are blighted, with attendant damage to the person's institution. A simple error may prompt a slap on the wrist. However, claim the existence of a perpetual motion machine or the like, and the researcher may be labeled a crackpot and ignored. These slighted individuals often respond by saying many of the world's most important discoveries have initially met with skepticism. Astronomer and scientific popularizer, Carl Sagan (1979; p. 64) reminds us, "The fact that some geniuses were laughed at does not imply that all who are laughed at are geniuses. They laughed at Columbus, they laughed at Fulton, and they laughed at the Wright Brothers. But they also laughed at Bozo the Clown." To decide a priori is not always easy, even for experts.

If these are ways that legitimate scientists can be led astray, it is not surprising that these pitfalls can play a role in pseudoscience claims where purposeful misbehavior may be in play.

My point in all of this discussion is that we must clearly establish the boundaries of good science and the shadow lands beyond, and we must do it by showing examples of first-rate science. One excellent way to do this is to present the details of our favorite experiments, but not the way we typically do in the classroom—in the lecture format. Instead try presenting the historical details, relating the personal and social details of the time. And look how engaging it can be if students are challenged to interact with data from the real experiments of classical molecular biology, identifying the molecules of heredity (http://sciencecases.lib.buffalo.edu/cs/files/mol_bio_classics.pdf) and the dramatic discovery of DNA's structure (http://sciencecases.lib.buffalo.edu/cs/files/dna_structure.pdf).

James Conant of Harvard University was a well-known advocate of this case history approach (Conant, 1949). He was Franklin Roosevelt's science adviser, and he returned to Harvard after World War II convinced that the American public did not understand how science was done. He created a course using historical case studies, lecturing on the great scientific discoveries such as the second law of thermodynamics with all the missteps along the way. He did not want science to be just revealed truths but wished to show the human enterprise in action.

I applaud his approach, but lecturing is not the ideal way to go about it. A massive amount of information has emerged that shows students learn best using active learning techniques, as in small group discussions like those promulgated by cooperative and collaborative learning. We now know the strengths of the case study approach (Herreid, 2007, 2012, 2014). And we have in repositories such as the National Center for Case Study Teaching in Science hundreds of cases studies that deal with good and bogus science.

Case Studies—A Rehearsal for Life

As I wrote once before, "Case studies are stories with an educational message" (Herreid, 2007). And people love stories. Stories have been the backbone of education in law schools since the early twentieth century, and in business schools for almost as long. Medicine has used them for years to train physicians and the rest of the health field has followed. Their use in STEM education is only recent (Herreid, 1994), but it has led to the publication of case studies in journals, books, and websites, such as the site maintained by the National Center for Case Study Teaching in Science at http://sciencecases.lib.buffalo.edu/cs/.

If we really want students to understand how science is done and to develop critical thinking skills, we want them to deal with real-life situations. We are not going to succeed if we teach science the way that it is normally done, with memorization given high priority. Science majors will tolerate this abuse, but nonmajors will hate it. We need to stop focusing on the lowest levels of Bloom's taxonomy (Bloom et al. 1956), which emphasize facts, and instead emphasize problem solving, application, and synthesis of information. Using stories does this brilliantly. Instantly the terms and principles become more tolerable and even interesting; we see that they have some use. This approach is particularly valuable in teaching students who are nonmajors; these are overwhelmingly the majority of our general science students. They are simply taking science because they have to; it is a general education requirement. These students will only take one or two semesters of science and that's it. They are off to become lawyers, business persons, merchants, clerks, policemen, firefighters, waiters, and the other folks who keep our world running. And they all vote and make decisions that affect everyone else.

Let's not bore them to tears trying to drum scientific factoids into their memory banks only to have the things leak out a few days later. Let's give them real stories and cautionary sagas. We can find case studies in many media sources every day. But they are easily found at the site maintained by the National Center for Case Study Teaching in Science (http://sciencecases.lib.buffalo.edu/cs/). This repository has more than 700 cases and teaching notes available in all STEM disciplines, and over 130 of these cases have the keyword "scientific method" attached. Here we will find stories about how scientists conduct their business, about missteps that have occurred in the scientific process and how they were uncovered and corrected, and about the community of scholars and how they criticize and check each other's work—not only out of the goodness of their hearts (the cooperative piece of the venture) but also because they are looking for flaws (the competitive nature of the enterprise). It is this community's self-criticism that generally exposes pseudoscience when it rears its head.

Using Case Studies to Teach about Science

What is it we want students to know about science: facts and process? Each of these can be taught using cases. And there are different methods to do the job, depending upon class size and the nature of the case (Herreid, 2007). If you are interested in how to design a research study, we can do this via a classroom exercise such as the Lady Tasting Coffee, in which the participant claimed that she could detect if milk was added before tea was poured into a cup or after (http://sciencecases.lib.buffalo.edu/cs/collection/detail.asp?case_id=414&id=414), or the Great Parking Debate, which addresses whether people leave their parking spaces faster if others are waiting (http://sciencecases.lib.buffalo.edu/cs/collection/detail.asp?case_id=578&id=578).

Are you interested in real studies where investigators follow the trail of an important discovery be it historical or contemporary—mysteries involving health or environment? How about the pioneering work of Ignaz Semmelweis in curing childbed fever (http://sciencecases.lib.buffalo.edu/cs/collection/detail.asp?case_id=429&id=429) or the revelation that bacteria cause ulcers (http://sciencecases.lib.buffalo.edu/cs/collection/detail.asp?case_id=571&id=571) or the data showing global climate change (http://sciencecases.lib.buffalo.edu/cs/collection/detail.asp?case_id=478&id=478) or the historical journey in the discovery of the atom (http://sciencecases.lib.buffalo.edu/cs/collection/detail.asp?case_id=667&id=667)?

Are you interested in ethical challenges in science? How about the ethics of physician-assisted suicide (http://sciencecases.lib.buffalo.edu/cs/collection/detail.asp?case_id=436&id=436)?

Or the ethics of human experimentation as seen in the infamous Tuskegee Institute syphilis project? Under the auspices of the US Public Health Service, 399 African American men with syphilis were recruited for a research study on the progression of the disease. It relates the story of physicians who wanted to find out if the progress of the venereal disease syphilis was the same in African Americans as it was in Norwegians. When the study began, in the 1930s, there was no adequate therapy for syphilis. Antibiotics were discovered in the 1940s, but the study remained observational. The men went untreated, with the scandal going unrevealed until 1972 (http://sciencecases.lib.buffalo.edu/cs/files/2-bad_blood.pdf).

Pseudoscience and How Do We Deal with It?

Pseudoscience is a belief, claim, or practice that is presented as science but fails to live up to the standards or canons of science. But let's not be too hasty.

Many weird ideas aren't necessarily "pseudoscience." They are hypotheses formed well in advance of the means to support or falsify them. When Alfred Wegner suggested the concept of continental drift in 1912, his colleagues thought him a little wacky and out of his field of expertise. But what he proposed wasn't pseudoscience, it was "unproven science." It could be falsified, just not with technology from 1912. In 1980 when Luis Alvarez and others proposed that the dinosaurs were wiped out by an asteroid sixty-five million years ago, there was little evidence and a lot of skepticism, but the idea wasn't pseudoscience. After thirty years, the scientific community did endorse the idea (Schulte et al., 2010). When Lynn Margolis tried to publish her endosymbiotic hypothesis of the origin of eukaryotic cells (Sagan, 1967), her theory was heretical and rejected by fifteen journals. But the idea certainly wasn't pseudoscience; she was a respected professor at Boston University. That is a real clue as what we can think of as science.

Science fiction author Arthur C. Clarke famously said, "Any sufficiently advanced technology is indistinguishable from magic" (Clarke, 1973). Should we call the idea of black holes, time travel, multiple universes, wormholes, and string theory pseudoscience? Even though these speculations are hardly accepted as facts, they have been proposed by reputable scientists and backed by substantial evidence. How about the claim that scientists experimentally determined that they had measured neutrino speeds faster than the speed of light? For a few days, in 2011, there was uproar in the media and physics community that one of Einstein's "laws" had been violated. Again, we wouldn't call that pronouncement pseudoscience—in fact it looked like real science. Perhaps a crazy measurement by wild and crazy guys—but science nevertheless. Why? Because the claim was being made by individuals who heretofore were seen as credible scientists with impeccable credentials.

Credentials and previous reputation play a role in determining just what we call pseudoscience and just whose pronouncements we should pay attention to. We should be wary of claims by respected individuals who say things that are out of their area of expertise. When a film celebrity endorses a brand of toothpaste or automobile, be a little skeptical. But when thousands of research scientists say that the Earth is warming and that humans are in large part responsible, this is not pseudoscience in action, whereas to dispute the evidence certainly verges on pseudoscience. This is especially so when the naysayers have strong economic ties to energy conglomerates that are among the largest contributors to CO_2 emissions. Here's a handy rule of thumb: when surprising claims are being made, ask who is financing the work.

Some claims move from the reasonable (possible) to pseudoscience because the nature of science moves on. Pre-Copernican astronomy was a perfectly acceptable scientific theory when it posited that the sun revolves around the earth, but with better telescopes the premise was thoroughly tested and found to be wrong. Now anyone arguing this position today would be flirting with the pseudoscientific culture.

If our purpose in the classroom is to alert students as to the dangers of nonscience, they and we will have to go over examples that display the earmarks of shoddy science. A good starting point is to have the students themselves list examples. Their list can be a surprise and allows a teacher an opportunity to discuss just what does or doesn't fit the definition. Another option is for the instructor to provide a list including examples that might make most people's top ten. Consider the following:

Astrology

Bermuda Triangle

Crop circles created by aliens.

Cryptozoology (Big Foot, Sasquatch, Loch Ness Monster)

Dowsing for water or gold or oil, and so on

Extrasensory perception (remote viewing, clairvoyance, telepathy, telekinesis, precognition)

Extraterrestrial UFOs and alien abductions.

Ghosts

Handwriting analysis

Intelligent design and creationism

Lie detectors

Perpetual motion machines

Phrenology (reading the character of a person by feeling the bumps on his head)

Psychic detectives

Tarot cards

Vaccination as a cause of autism

If these topics aren't science, students need to know why they don't fit the bill.

Primarily, they need to see why the data do not support the claims of their supporters; they need to see what the credentials are of the prominent sponsors. An excellent exercise is to have students try to first itemize the basic claims of each of the topics and then to apply the canons of science. Fundamentally we are asking them to identify the evidence for and against these claims—to consider how to apply Sagan's Baloney Detection Kit (1997) or Shermer and Linse's (2001) list of ten questions:

1. How reliable is the source of the claim?
2. Does the source make similar claims?
3. Have the claims been verified by somebody else?
4. Does this fit with the way the world works?
5. Has anyone tried to disprove the claim?
6. Where does the preponderance of evidence point?
7. Is the claimant playing by the rules of science?
8. Is the claimant providing positive evidence?
9. Does the new theory account for as many phenomena as the old theory?
10. Are personal beliefs driving the claim?

Or have students watch Joe Hanson of PBS Video on "How to Read Science News" (http://explore.brainpickings.org/post/70511854124/so-great-so-necessary-joe-hanson-of-its-okay). He has his own questions. Is the scientific method applied with peer review? Does the discovery sound too good to be true? Is the article trying to scare you? Does someone stand to gain financially from this? Hanson says we ought to approach every story with a balance of curiosity and skepticism. And finally, he reminds us that Carl Sagan (1997, p. 187) was fond of saying, "It pays to have an open mind, but not so open that your brains fall out."

But if the above list is obvious, let me touch upon some topics that have a large number of devotees in spite of the evidence against them. These are useful subjects for any science classroom discussion.

Astrology

In newspapers across the country there are regular columns that dole out advice on what the future will bring to you based on the positions of the planets and stars. Astrologers are ever present on radio, television, and social media. Sometimes free, sometimes with a price tag, they foresee your life's path. Astrology has been around for thousands of years. At one time it was hardly a pseudoscience, for the discipline we call science did not exist. But astrology has become a pseudoscience par excellence. It has been so repeatedly tested and found wanting that it is a poster child for the term pseudoscience (Pigliucci, 2010). This is one of the easiest pseudosciences to test, a process that can be done in a student lab. Yet thousands of people believe in its prophetic powers. Nancy Reagan, the wife of the late president, is said to have influenced her husband more than once to schedule events in accordance with astrological principles.

Acupuncture

This too has a long and noble history, one stemming from the ancient Asian medical tradition. Under acupuncture, the human body is claimed to be made of energy meridians. These energy lines can be manipulated by inserting needles into the skin at precise places to alter the health of an individual. This assertion has been repeatedly tested with mixed results. Some of the problems with evaluating the claims of acupuncture are revealed in this website on dealing with science-based medicine (https://www.sciencebasedmedicine.org/reference/acupuncture/) and further depicted in this student case study (http://sciencecases.lib.buffalo.edu/cs/collection/detail.asp?case_id=370&id=370). The largest clinical study of acupuncture to date was performed in Germany and the United States, and involved several thousand people; the conclusion was that "acupuncture and sham acupuncture treatments are no different in decreasing pain levels across multiple chronic pain disorders: migraine, tension headache, low back pain, and osteoarthritis of the knee" (Cherkin et al., 2009). Enthusiastic supporters persist. Negative data be damned, there is little doubt that the process will continue.

Chiropractic Medicine

This is another area where the psychological and financial investment is huge and the skepticism great. The original philosophical basis of chiropractic medicine has been repeatedly challenged: these are claims that spinal manipulation can cure asthma, diabetes, and assorted internal ailments, as well as muscular and spinal disorders. This declaration rankles the modern physician no end. It doesn't help the credibility of chiropractors that many members are opposed to vaccination. (A highly cynical critique of chiropractic can be found at https://www.sciencebasedmedicine.org/reference/chiropractic/).

Regardless of what the data say, it is clear that we are not going to see this field disappear any time soon. There are about two dozen chiropractic colleges in the United

States and another couple of dozen internationally. As a result, thousands of chiropractors hang out their shingles everywhere and are accepted by the lay public as a legitimate part of the medical profession. But there is hope on the horizon. In 2015 nine international chiropractic colleges, including those in France, Switzerland, and Australia, made a proclamation explicitly distancing themselves from the historical claim that the muscular skeletal system is the source of all ills: "The teaching of the vertebral subluxation complex as a vitalistic construct that claims that it is the cause of disease is unsupported by evidence. Its inclusion in a modern chiropractic curriculum in anything other than an historic context is therefore inappropriate and unnecessary" (Mirtz, 2009). The field may yet evolve into a legitimate clinical discipline whose principles place it within the STEM community.

Creation Science
Some Christians, including many Baptists, have as one of their main dogmas that the Bible is the literal word of God. By their interpretation, the book of Genesis states that the universe is only six thousand to ten thousand years old. This is in conflict with scientific discoveries demonstrating that the universe is 13.8 billion years old. Serious difficulties exist within the religious community as to how to reconcile these two different testimonies. One strategy is to impugn the science. This is easy to do if you ignore the fact that there is virtually no disagreement among scientists as to the age of the universe and that the organisms were not dumped here ten thousand years ago by a sudden miracle.

Eugenie Scott, former director of the National Center for Science Education, has called creationism her "My Favorite Pseudoscience," pointing out that it is a good foil for teaching students about the nature of science (Scott, 2001).

Scott points out that there are two basic principles of science that creationism violates: science explains the natural world in terms of natural processes, not supernatural ones, and the conclusions of science can change when new evidence is brought to bear. When a creationist simply says, "God did it," we can be confident we are not hearing science, says Scott. And when a creationist claims the universe is only six thousand years old, citing calculations based on Genesis, this flies in the face of all modern dating methods. And the claim that Noah's Ark carried pairs of each kind of animal on board for nearly a year to weather the great deluge is logically at odds with the logistic problems of space available on a boat whose dimensions are known—not to mention that only eight people were said to be on board to feed, water, and clean out the stalls.

Scott points out that showing creationism is factually wrong is insufficient. "People who support creation science do so for emotional reasons and are reluctant or unwilling to give it up unless those needs or concerns are otherwise assuaged," she argues. She believes much the same "can be said for believers in UFO's or out of body experiences, or paranormal phenomena in general: these beliefs are meeting some emotional needs, and consequently will be very difficult to abandon" (p. 254). In the case of creationism, the needs are being met by religion and the theology of biblical literalism, and this means that creation science will be especially difficult to give up.

Scott argues the public is divided into three parts: confirmed believers, confirmed skeptics, and a large middle group who doesn't know much science and doesn't have a major emotional investment in embracing a particular pseudoscientific view. She says that our focus of attention should be on the latter, that true believers are out of reach. Teachers will find this video about the subject useful because it shows how the intelligent design argument fails to meet several of the canons of science: https://www.youtube.com/watch?v=xO7IT81h200. In addition, there are student case studies devoted to this topic (http://sciencecases.lib.buffalo.edu/cs/collection/detail.asp?case_id=332&id=332c and http://sciencecases.lib.buffalo.edu/cs/collection/detail.asp?case_id=354&id=354).

I am afraid that Scott is right about not only this topic but others as well. Scientists may reject pseudoscience, astrology, chiropractic medicine, acupuncture, and creation science, but these beliefs are not going away. There is too much cultural and financial investment in these ideas for them to disappear soon.

What's the Harm of Pseudoscience?

It is important to deal with this question in the classroom. What's so terrible if someone believes in UFOs or astrology or Big Foot? Aside from the obvious, that they are based on erroneous facts or reasoning, there are indeed reasons to be troubled. It has to do with money, ethics, health, environmental safety, and public policy, just for starters. Is it benign for the United States to have a Congress that is unwilling to accept the verdict of the majority of the world's scientists that the Earth is warming and humans are playing a major role in bringing this about (Germain, 2015)? Is it a problem when a congressman shapes the federal research budget in favor of alternative medicine? Is it a problem when the president believes in astrology or listens to someone who does (Roberts, 1988)? Is it a problem when a homeopathic "doctor" prescribes drugs that have been diluted to a point where no molecule exists? Is it a problem when thousands of school children are taught that the world and all that is in it was created suddenly by a miracle ten thousand years ago? Is it a problem when people will consult "psychics" who claim to communicate with dead relatives and who need financial backing to continue? Is it a problem when a medical doctor falsifies evidence that vaccination causes autism, and when a celebrity pushes this view so persuasively that measles breaks out after frightened parents keep their children from being vaccinated (Editors of Lancet, 2010 and Wallace, 2009)? Is it a problem when a television health guru promotes products with doubtful merit? Or when the city of Berkeley, California, passes a measure requiring that cell phones be sold with a health warning about exposure to radio frequency radiation, even though scientists have repeatedly debunked the notion of ill effects (Storrs, 2015)?

Can We Train People (Students) to Detect Pseudoscience?

Yes and no. We certainly cannot train them to determine the validity of claims of a scientist who says the universe is 13.8 billion years old or that polio is caused by a virus. When the supporters of creationism say we should let the students decide on the validity of evolution, they are naive or disingenuous. How can we expect a fifteen-year-old student to evaluate evidence for evolution that few scientists outside of their specialty can? A molecular biologist is not trained to determine the age of a rock layer; a geologist does not determine the DNA sequence of a Neanderthal; a botanist does not calculate the likelihood of an earthquake in Indonesia. All of these scientists have their own areas of expertise and they do not know the details of other specialties firsthand. But they all have experience with the process of science, having taken many of the same STEM courses, and know they can depend on the integrity of its practitioners. Molecular biologists can appreciate the logic of radiometric dating. They believe the geologists who say the Earth is 4.6 billion years old, especially when there is almost universal agreement about this conclusion among geologists. And the geologists believe the molecular biologists who claim that the Neanderthals lived until about thirty thousand years ago. They believe it because they understand the general principles of the chemistry of DNA and how the number of mutations in a person can help determine how long ago humans split off from another group of animals. But most importantly they have faith in the system and principles of scientific discovery and testing. Will they be surprised if the numbers aren't exact? Hardly. They expect modifications of knowledge. And every so often they will have their fundamental beliefs thrown into a tizzy when a new big idea comes along. But when that happens, they expect that all of the relevant scientists will immediately check the data and challenge the idea in every conceivable way until they are satisfied beyond reasonable doubt that the new idea is here to stay. And only then will the idea be welcomed grudgingly into the scientific fold. One way to evaluate a debate is to look and see what the general consensus is on the topic by scientists. When virtually all scientists say that the evidence for evolution is overwhelming, you can be pretty sure that it is a fact; especially if the naysayers lack scientific credentials or have a vested interest in convincing you otherwise.

Here is the crux of the matter: neither students nor faculty can verify or test the conclusions of experts. Nor can experts do it outside of their own fields. But the key is to have faith in the integrity of the scientific process. Still the bottom line is we want the students to be cynical whenever an issue affects their well-being or wallet.

What Is a Teacher to Do?

To return to the theme of this chapter, if our first obligation is to show students how science works, we cannot do this solely by lecturing. Unfortunately, 86 percent of faculty in colleges and universities deliver their course material in this way (National Research Council, 2003). The trouble with this method of presentation is that students do not recall information well when it is delivered by the didactic method. There are dozens of studies that show this, and teachers and parents hardly need to be reminded; after all, we have thousands of our own exam scores to prove it, not to mention that on average 40 percent of the students in basic science courses fail, quit, or get Ds. Perhaps even worse, students who receive scientific information via lecture almost invariably think that information is revealed wisdom.

How to teach the process of science is the challenge. First, we recognize that we are already doing some of the right things—in some places, in some courses, and in some schools. When we get students into the laboratories, especially those that are not just teaching lab techniques, but are using inquiry labs, that is of help. Get students into research labs and that is even better—if we have seasoned researchers there as guides. Anytime we ask students to write grant proposals, or to write research reports and term papers about a classical experiment, that helps. But let's not kid ourselves. We are a long way away from producing a student who is educated in the wiles of science. We may have helped students understand the ethos of conducting individual science, but most often we do not open their eyes to what really goes on when scientists interact with society. To fill in this gap, they need to see scientists interacting, arguing, criticizing each other, going to seminars and meetings, giving talks, and presenting posters. They need to see how grants are evaluated and ranked and to read the summaries of papers by referees. They need to understand how misconduct has occurred and been uncovered, and they need to know the fates of the transgressors. They need not only to see how scientists got it right but how scientists got it wrong and why.

How do we share all of this wealth with our undergraduate students? The answer is we can't. There is too much to learn; that is why we have graduate schools and post-doctoral fellowships. It takes most PhDs years to come up to snuff. Let's accept that we can't produce mini-scientists in a couple of semesters of a gen ed class.

Is It Hopeless?

Faculty at Sam Houston State University didn't think so, and they created a course specifically focused on achieving these goals, with one of their techniques being to use active learning strategies (Rowe et al., 2015). Almost a century of research has shown that cooperative and collaborative learning strategies, typically using small groups of students, trumps the lecture style, which emphasizes the individualistic and competitive approach (Johnson, Johnson, and Smith, 2006; Smith, Sheppard, Johnson, and Johnson, 2005; Springer, Stanne, and Donovan, 1999). Case study teaching is one of the premier techniques for achieving such results and serves as one of the best ways to introduce the facts and methods of science.

"Foundations of Science," Sam Houston's interdisciplinary course, incorporates case studies (such as the vaccine-autism controversy). The class teaches "the basics of argumentation and logical fallacies; contrasts science with pseudoscience; and addresses psychological factors that might otherwise lead students to reject scientific ideas they find uncomfortable." The aim is to inculcate into the students the operational approach to critical thinking provided by Bernstein, Penner, Clarke-Stewart, and Roy (2006) via a set of questions: (1) What am I being asked to accept? (2) What evidence supports the claim? (3) Are there alternative explanations/hypotheses? And, finally, (4) what evidence supports the alternatives?

Sam Houston chose to focus on cases that we would judge pseudoscience because the topics are inherently interesting even to science-phobic students: astrology, homeopathy, Bigfoot, and even intelligent design. And they adopted a textbook that emphasizes the same approach: *How to Think about Weird Things: Critical Thinking for a New Age* (Schick and Vaughn, 2014). Key features in their curriculum are two case studies. The first deals with the claim that vaccination is a cause of autism (Rowe, 2010). The students work in small groups and analyze the data provided in Wakefield et al. (1998) study. Then they are asked to design a better study, and in the process they learn about experimental design and sample size, replication, double-blind studies, and scientific honesty (Editors of Lancet, 2010). In another case study, published recently in the *Journal of College Science Teaching* (Rowe, 2015), they apply ecology theory to evaluate the credibility of finding a plesiosaur moonlighting as the Loch Ness Monster. This case is especially interesting because it integrates traditional scientific facts and principles along with a skeptical approach to fantastic claims, illustrating how important it is to consider alternative hypotheses to unproven claims, especially ones that verge on the incredible. It uses a technique called the Interrupted Case Method. The students are given the story in pieces and, working in groups, they try to work their way through the questions at the end of each section. (The full version of the case, teaching notes, and answer key can be found at http://sciencecases.lib.buffalo.edu/cs/collection/detail.asp?case_id=779&id=779).

Final Thoughts

Learning to think like a scientist isn't easy. It doesn't happen overnight. What can we do when we only have a short time with students to help them develop critical thinking, especially when it concerns science? I have given a few suggestions that might help. But we must have forbearance and not scream and bay at the moon about those ignorant slobs who are the masses. We need to keep plugging away and remember Winnie-the-Pooh's advice:

"If the person you are talking to doesn't appear to be listening, be patient. It may simply be that he has a small piece of fluff in his ear."

We need to keep looking for ways to remove the fluff.

Acknowledgments

This chapter is based on work supported by the National Science Foundation (NSF) under Grant Nos. DUE-0341279, DUE-0618570, DUE- 0920264, DUE -1323355, and a Higher Education Reform Grant from the PEW Charitable Trusts. Any opinions, findings, conclusions, or recommendations expressed in this material are those of the author and do not necessarily reflect the views of the NSF or the PEW Charitable Trusts.

References

Anderson, C., Bahnik, Š., Barnett-Cowan, M., Bosco, F. A., Chandler, J., et al. (2016). Response to Comment on "Estimating the Reproducibility of Psychological Science." *Science*, 351: 1037–1039.

Bernstein, D., Penner, L., Clarke-Stewart, A., and Roy, E. (2006). *Psychology*. Boston: Houghton Mifflin.

Bloom, B. S. (Ed.), Engelhart, M. D., Furst, E. J., Hill, W. H., and Krathwohl, D. R. (1956). *Taxonomy of Educational Objectives, Handbook I: The Cognitive Domain*. New York: David McKay.

Bohannon, J. (2015). Many Psychological Papers Fail Replication Test. *Science*, *349* (6251): 910–911.

Cherkin, D., Sherman, K., Avins, A., Erro, J., Ichikawa, L., et al. (2009). A Randomized Trial Comparing Acupuncture, Simulated Acupuncture, and Usual Care for Chronic Low Back Pain. *Archives of Internal Medicine*, *169*(9): 858–866.

Cho, A. (2011). From Geneva to Italy Faster Than a Speeding Photon? *Science*, *333*(6051): 1809.

Clarke, A. (1973). "Hazards of Prophecy: The Failure of Imagination." In *Profiles of the Future: An Enquiry into the Limits of the Possible* (1962, rev. 1973), pp. 14, 21, 36. London: Macmillan.

Colyer, C. (1999). Childbed Fever: A Nineteenth-Century Mystery. Retrieved from http://sciencecases.lib.buffalo.edu/cs/collection/detail.asp?case_id=429&id=429 (accessed June 11, 2017).

Committee on the Conduct of Science, National Academy of Sciences. (1989). *On Being a Scientist*. Washington, DC: National Academy Press. Retrieved from http://www.pnas.org/content/86/23/9053.full.pdf

Committee on the Conduct of Science, National Academy of Sciences. (1995). *On Being a Scientist* (2nd ed.). Washington DC: National Academy Press. Retrieved from http://www.nap.edu/read/4917/chapter/1

Committee on the Conduct of Science, National Academy of Sciences. (2009). *On Being a Scientist* (3rd ed.). Washington, DC: National Academy Press. Retrieved from http://www.nap.edu/catalog/12192/on-being-a-scientist-a-guide-to-responsible-conduct-in#toc (Accessed December 29, 2015).

Conant, J. (1949). *The Growth of the Experimental Sciences: An Experiment in General Education.* New Haven, CT: Yale University Press.

Creation Science Made Easy. Retrieved from https://www.youtube.com/watch?v=xO7IT81h200 (Accessed June 11, 2017).

Editors of Lancet (2010). Retraction—Ileal-Lymphoid-Nodular Hyperplasia, Non-Specific Colitis, and Pervasive Developmental Disorder in Children. *Lancet,* 375: 445

Eichler, J. (2012). History of the Atom: From Atomism to the Nuclear Model. Retrieved from http://sciencecases.lib.buffalo.edu/cs/collection/detail.asp?case_id=667&id=667 (accessed June 11, 2017).

Explore: How to Read Science News. Retrieved from http://explore.brainpickings.org/post/70511854124/so-great-so-necessary-joe-hanson-of-its-okay (Accessed June 12, 2017).

Feenstra, J. (2011). The Great Parking Debate: A Research Methods Case Study. Retrieved from http://sciencecases.lib.buffalo.edu/cs/collection/detail.asp?case_id=578&id=578 (accessed June 11, 2017)

Fourtner, A., Fourtner, C., and Herreid, C. (2000). Bad Blood: A Case Study of the Tuskegee Syphilis Project. Retrieved from http://sciencecases.lib.buffalo.edu/cs/files/2-bad_blood.pdf (accessed June 11, 2017)

Germain, T. (2015). Here Are the 56 Percent of Congressional Republicans Who Deny Climate Change Moyers and Company. Retrieved from http://billmoyers.com/2015/02/03/congress-climate-deniers/

Gilbert, D., King, G., Pettigrew, S., and Wilson, T. (2016). Comment on "Estimating the Reproducibility of Psychological Science." *Science,* 351: 1037.

Herbert, V. (1979). Laetrile: The Cult of Cyanide Poisoning; Promoting Poison for Profit, *American Journal of Clinical Nutrition,* 32(5): 1121–1158. Retrieved from http://www.ajcn.org/content/32/5/1121.full.pdf

Herreid, C. (2004). And Now What, Ms. Ranger?: The Search for the Intelligent Designer. Retrieved from http://sciencecases.lib.buffalo.edu/cs/collection/detail.asp?case_id=332&id=332c (accessed June 11, 2017).

Herreid, C. F. (1994). Case Studies in Science—A Novel Method of Science Education. *Journal of College Science Teaching* 23(4): 221–229.

Herreid, C. F. (2007). *Start with a Story.* Arlington, VA: NSTA Press.

Herreid, C. F. (2010). The Scientific Method Ain't What It Used to Be. *Journal of College Science Teaching,* 39(6): 68–72.

Herreid, C. F., Schiller, N., and Herreid, K. (Eds.). (2012). *Science Stories: Using Case Studies to Teach Critical Thinking*. Arlington, VA: National Science Teachers Association Press.

Herreid, C. F., Schiller, N., and Herreid, K. F. (2014). *Science Stories You Can Count On: Case Studies with Quantitative Reasoning in Biology*. Arlington, VA: National Science Teachers Association Press.

Johnson, D., Johnson, R., and Smith, K. (2006). *Active Learning: Cooperation in the College Classroom*. Edina, MN: Interaction Book.

Knabb, M., Lutz, T., and Fairchild, G. (2010). Global Climate Change: Evidence and Causes. Retrieved from http://sciencecases.lib.buffalo.edu/cs/collection/detail.asp?case_id=478&id=478 (accessed June 11, 2017).

Lerner, I. (1981). Laetrile: A Lesson in Cancer Quackery. *CA: A Cancer Journal for Clinicians, 31*(2): 91–95. DOI: 10.3322/canjclin.31.2.91

Lightman, A. (2015). Nothing but the Truth. *Popular Science,* 50–51.

Maddox, J., Randi, J., and Stewart, W. (1988). "High-Dilution" Experiments a Delusion. *Nature, 334*(6180): 287–290.

Maynard, J., Mulcahy, M., and Kermick, D. (2009). Lady Tasting Coffee: A Case Study in Experimental Design. Retrieved from http://sciencecases.lib.buffalo.edu/cs/collection/detail.asp?case_id=414&id=414 (accessed June 11, 2017).

Meuler, D. A. (2011). Helicobacter pylori and the Bacterial Theory of Ulcers. Retrieved from http://sciencecases.lib.buffalo.edu/cs/collection/detail.asp?case_id=571&id=571 (accessed June 11, 2017).

Mirtz, T. A., Morgan, L., Wyatt, L. H., et al. (2009). An Epidemiological Examination of the Subluxation Construct Using Hill's Criteria of Causation. *Chiropractic & Osteopathy, 17*, 13.

Moitra, K. (2014). The Mona Lisa Molecule: Mysteries of DNA Unraveled. Retrieved from http://sciencecases.lib.buffalo.edu/cs/files/dna_structure.pdf (accessed June 11, 2017).

National Research Council, Committee on Undergraduate Biology Education to Prepare Research Scientists for the 21st Century, 2003. *BIO2010, Transforming Undergraduate Education for Future Research Biologists*. Washington, DC: National Academies Press. Retrieved from www.nap.edu/catalog.php?record_id=10497

Nuzzo, R. (2015). Fooling Ourselves. *Nature, 526* (7572): 182–185.

Pals-Rylaarsdam, R. (2010). The Evolution of Creationism: Critically Appraising Intelligent Design. Retrieved from http://sciencecases.lib.buffalo.edu/cs/collection/detail.asp?case_id=354&id=354

Pals-Rylaarsdam, R. (2012). Classic Experiments in Molecular Biology: The Transforming Principle: Identifying the Molecule of Inheritance. Retrieved from http://sciencecases.lib.buffalo.edu/cs/files/mol_bio_classics.pdf (accessed June 11, 2017).

Pigliucci, M. (2010). *Nonsense on Stilts: How to Tell Science from Bunk*. Chicago: University of Chicago Press.

Post, D., and Knutson, D. The Plan: Ethics and Physician Assisted Suicide. Retrieved from http://sciencecases.lib.buffalo.edu/cs/collection/detail.asp?case_id=436&id=436 (accessed June 11, 2017).

Roberts, S. (1988). White House Confirms Reagans Follow Astrology, Up to a Point. *New York Times*, May 4, 1988. Retrieved from http://www.nytimes.com/1988/05/04/us/white-house-confirms-reagans-follow-astrology-up-to-a-point.html

Rousseau, D. (1992). Case Studies in Pathological Science: How the Loss of Objectivity Led to False Conclusions in Studies of Polywater, Infinite Dilution and Cold Fusion. *American Scientist*, *80*(1): 54–63.

Rowe, M. P. (2010). Tragic Choices: Autism, Measles, and the MMR Vaccine. In C. F. Herreid, N. A. Schiller, and K. F. Herreid (Eds.), *Science Stories: Using Case Studies to Teach Critical Thinking*. Arlington, VA: NSTA Press. Retrieved from http://sciencecases.lib.buffalo.edu/cs/collection/detail.asp?case_id=576&id=576 (Accessed June 8, 2017).

Rowe, M. (2012). "Tragic Choices: Autism, Measles, and the MMR Vaccine." In C. Herreid, N. Schiller, and K. Herreid (Eds.), *Science Stories: Using Case Studies to Teach Critical Thinking*. Arlington, VA: NSTA Press. Retrieved from http://sciencecases.lib.buffalo.edu/cs/collection/detail.asp?case_id=576&id=576

Rowe, M. P. (2015). Crazy about Cryptids! An Ecological Hunt for Nessie and Other Legendary Creatures. National Center for Case Study Teaching in Science. Retrieved from http://sciencecases.lib.buffalo.edu/cs/ collection/detail.asp?case_id=779&id=779 (Accessed June 8, 2017).

Rowe, M., Gillespie, B., Harris, K., Koether, S., Shannon, L., and Rose, L. (2015). Redesigning a General Education Science Course to Promote Critical Thinking. *CBE—Life Sciences Education*, *14* (3): 1–12

Russell, Miles. (2012). *The Piltdown Man Hoax: Case Closed*. Stroud, UK: History Press.

Sagan, C. (1979). *Broca's Brain*. New York: Random House.

Sagan, C. (1997). *The Demon-Haunted World: Science as a Candle in the Dark*. New York: Ballantine Books.

Sagan, L. (1967). On the Origin of Mitosing Cells. *Journal of Theoretical Biology*, *14*(3): 225–274.

Science-Based Medicine. Acupuncture. Retrieved from https://sciencebasedmedicine.org/reference/acupuncture/ (Accessed June 11, 2017).

Science-Based Medicine. Chiropractic. Retrieved from https://www.sciencebasedmedicine.org/reference/chiropractic (Accessed June 11, 2017).

Schick, T., and Vaughn, L. (2014). *How to Think about Weird Things: Critical Thinking for a New Age*. New York: McGraw-Hill.

Schulte, P., Alegret, L., Arenillas, I., Arz, J., Barton, P., Brown, P., et al. (2010). The Chicxulub Asteroid Impact and Mass Extinction at the Cretaceous-Paleogene Boundary. *Science*, *327*(5970): 1214–1218.

Scott, E. (2001). "My Favorite Pseudoscience." In P. Kurtz, (Ed.), *Skeptical Odyssey*, pp. 245–256. Amherst, New York: Prometheus Books.

Shermer, M., and Linse, P. (2001). *Baloney Detection Kit*. Agawam, MA: Millennium Press.

Smith, K., Sheppard, S., Johnson, D., and Johnson, R. (2005). Pedagogies of Engagement: Classroom-Based Practices. *Journal of Engineering Education, 94*: 87–101.

Springer, L., Stanne, M. E., and Donovan, S. S. (1999). Effects of Small-Group Learning on Undergraduates in Science, Mathematics, Engineering, and Technology: A Meta-analysis. *Review Educational Research, 69*(1): 21–51.

Stonefoot, S., and Herreid, C. (2004). A Need for Needles—Acupuncture: Does It Really Work? http://sciencecases.lib.buffalo.edu/cs/collection/detail.asp?case_id=370&id=370 (accessed June 11, 2017).

Storrs, C. (2015). Cell Phones and Risk of Brain Tumors: What's the Real Science? Special to CNN. Retrieved from http://www.cnn.com/2015/07/28/health/cell-phones-brain-tumor-risk-berkeley/index.html

Taubes, G. (1993). *Bad Science: The Short Life and Weird Times of Cold Fusion*. New York: Random House.

Tucker, W. (1997). Re-reconsidering Burt: Beyond a Reasonable Doubt. *Journal of the History of the Behavioral Sciences, 33*(20): 145–162.

University of California, Berkeley. (2017). Understanding Science. Retrieved from http://undsci.berkeley.edu

Wakefield, A., Murch, S., Anthony, A., Linnell, J., Casson, D., et al. (1998). Ileal-Lymphoid-Nodular Hyperplasia, Non-specific Colitis, and Pervasive Developmental Disorder in Children. *Lancet, 351*(9103): 637–641.

Wallace, A. (2009, October 19). An Epidemic of Fear: How Panicked Parents Skipping Shots Endangers Us All. *Wired Magazine*. Retrieved from http://www.wired.com/magazine/2009/10/ff_waronscience/all/1

19 "HIV Does Not Cause AIDS": A Journey into AIDS Denialism

Seth C. Kalichman

AIDS, the end stage of HIV infection, is the most destructive epidemic in modern times. The first identified cases of AIDS occurred in the United States in 1981, and soon after that the cause was determined to be a newly identified virus, HIV. It is estimated that 1.2 million Americans are living with HIV, and that 40,000 new infections are diagnosed each year. There are nearly thirty-seven million people living with HIV/AIDS in the world and more than thirty-five million have died of its end-stage disease, AIDS. Medications that treat HIV infection have transformed it from a life-threatening diagnosis to a chronic disease. Individuals who are treated with anti-HIV (i.e., antiretroviral) medications can expect to have a nearly normal life expectancy, and these same medications are capable of preventing the transmission of HIV.

In parallel to these realities, there is an alternate universe that denies everything about HIV/AIDS, including its very existence. Led by a small group of vocal fringe scientists and pseudoscientists, there are numerous individuals who claim that HIV is harmless, or perhaps does not exist at all. Some of the most persistent HIV/AIDS skeptics have appointments at respected academic institutions, protected by the nearly sacred trust of academic tenure and freedom of expression. Nearly every facet of the reality we know to be HIV/AIDS has a mirror image. In response to AIDS scientists, there are AIDS pseudoscientists who have published their ideas in fake journals. AIDS pseudoscientists also write books and produce websites dedicated to discussing the hoax of HIV. They have held pseudoscientific conferences where AIDS nonbelievers convene.

Mirroring AIDS activism, there are counteractivists dedicated to exposing HIV conspiracies. There are holistic clinicians who have argued against HIV as the cause of AIDS, and politicians who have championed these ideas with destructive public policies. There are fund-raising schemes intended to fight the scientific establishment, and efforts to undermine legitimate AIDS charities. AIDS denialists have produced documentary films on the AIDS conspiracies and the scientific debate about whether HIV exists. Taken together, these activities have created the impression that there is social movement, mostly living on the Internet, that wants to counter the progress in combating AIDS. Those who deny the realities of HIV/AIDS identify themselves as AIDS dissidents or HIV skeptics. But they share the same characteristics and use the same strategies as other denialist groups, including climate change deniers (Lawler, 2002) and Holocaust deniers (Shermer and Grobman, 2000), and they are best described as AIDS denialists.

Discovering Denialism

My first exposure to AIDS denialism was in 2000, when denialists tried to hijack the thirteenth International AIDS Conference, held in Durban, South Africa. Like many others, I ignored them, thinking they merely sought attention and would go away in its absence. I was wrong; they did not go away. I again stumbled upon the world of AIDS denialism some seven years later. As the editor of a peer-reviewed journal, *AIDS and Behavior*, that is focused on the social and behavioral aspects of AIDS, I sent a request for a paper review to a social psychologist, Kelly Brennan at the College of Brockton, State University of New York. Kelly Brennan was at the time well known for her work in social relations and human attachment. I did not know her personally, but I was aware of the work she had done in graduate school. She declined the review and wrote me a note saying that she did not want anything to do with our journal. She also provided a link to the website Reappraising AIDS to see what she "thinks about AIDS." The website had posted her review of the 1996 book *Inventing the AIDS Virus* by University of California biologist Peter Duesberg, the leading voice in AIDS denialism. By this time Duesberg had been debunked countless times, and yet Kelly Brennan concluded:

Thanks to critical thinkers like Duesberg, the tide is turning. When the lawsuits are done, the CDC, NIH, and the FDA will have to be dismantled. If there is a just world, then these people, the pharmaceutical industry, and the greedy, craven scientists who propagated all these lies all this time will be tried for crimes against humanity. Duesberg is a hero and should be treated as such; I'm sure he'd be the last to want that though, as he seems rather modest, an accidental hero if ever there was one. If you've read this far, go read his book. You can't say now that you had no idea that HIV-AIDS was a myth.

It was indeed shocking to find someone I thought to be well-trained as a scientist, having made important contributions to the study of human relationships, would actually buy into AIDS denialism. I followed the links at the Reappraising AIDS website and found a world of AIDS denialism that seemed endless. As a psychologist, I found this fascinating. As someone who has dedicated his life to ending AIDS, I found it terrifying. AIDS denialism would come to occupy my attention for the next several years.

My journey into AIDS denialism would ultimately result in efforts aimed at understanding and exposing the AIDS deniers. My time in AIDS denialism was never intended to be an empirical study. Rather, I set out to document my personal account of this small but vocal group. The result was a book, *Denying AIDS: Conspiracy Theories, Pseudoscience, and Human Tragedy*. In *Denying AIDS* (Kalichman, 2009), I tried to synthesize the foundations of AIDS denialism and its impact on those most vulnerable to misinformation propagated by the denialist movement. Here I provide some of what I learned in the process of writing *Denying AIDS*. In this chapter, I draw from source material that I have not previously discussed, as well from experiences discussed in *Denying AIDS* and posted at the Denying AIDS blog (Kalichman, 2015).

Just the Fake Facts

It does not take long to learn about the alternative universe of AIDS denialism. Their self-published books, magazines, speeches, and blog posts are readily available on the Internet. A search for HIV and AIDS will result in a plethora of information. AIDS denialism coexists with the countless numbers of credible and factual-based websites. There are AIDS-denialist groups all over the world. The most vocal AIDS deniers come from the United States, Canada, United Kingdom, and Australia. It is readily apparent that AIDS denialism is built on a distorted view of reality. At its roots is a mistrust of science and government, a mistrust that has become enmeshed with politics. AIDS deniers exhibit a cognitive rigidity that is seen in what psychiatrists call encapsulated delusional disorders. The concept of an encapsulated delusion seems fitting because AIDS deniers can be functional in many respects, so they are not truly psychotic. And yet their beliefs about AIDS are divorced from reality, rigid, and impenetrable. Some AIDS denialists, however, represent more of a mixed bag of psychopathology with features of narcissism and paranoia. At the head of the movement are disgruntled "scientists" who believe they are right about AIDS being caused by toxins and poisons. These are rogue scientists who feel disrespected, ignored, and conspired against. They believe they have it right about AIDS despite all of the evidence to the contrary. They see a conspiracy fueled by government funding for AIDS research in collusion with corrupt researchers and a greedy pharmaceutical industry. They experience peer review as a corrupted censorship arm of the AIDS conspiracy.

Denialists transform basic scientific facts into disinformation using a common set of tactics or strategies. Denialists parse statements from legitimate sources and paste them into a text meant to sound scientific. The result is usually incomprehensible and devoid of logic. They morph science into pseudoscience by using jargon and senseless verbiage in torrents that can barely pass as streams of consciousness, much less science. Another common denialist tactic is cherry picking, selecting a lone scientific finding and presenting it out of context to suit their argument. Another is the single-study fallacy. While scientists are cautious when drawing conclusions from even a series of experiments, denialists demand to see a single study that proves HIV causes AIDS. There is, of course, no single scientific paper proving that HIV causes AIDS. Rather there are thousands of studies containing a wide range of evidence that, taken together, makes the overwhelming case that HIV causes immune system decline that ultimately results in AIDS. Is there the single study that definitively proves smoking causes cancer? How about a single piece of historical evidence that definitively proves millions of Jews were exterminated in Nazi Germany? Or a single physics experiment that proves it is possible to land a spacecraft on the moon?

Another common tactic of denialists is moving the goalpost: demand a specific piece of evidence, see it produced, and respond by saying that it is not enough, that some other piece of evidence is needed. Tara Smith and Steven Novella (Smith and Novella, 2007) offer this description:

> The strategy behind goalpost moving is simple: always demand more evidence than can currently be provided. If the evidence is then provided at a later date, simply change the demand to require even more evidence, or refuse to accept the kind of evidence that is being offered. In the 1980s, HIV deniers argued that drug therapy for AIDS was ineffective, did not significantly prolong survival, and in fact were toxic and damaged the immune system. However, after the introduction of a cocktail of newer and more effective agents in the 1990s, survival rates did impressively increase. HIV deniers no longer accept this criterion as evidence for drug effectiveness, and therefore the HIV theory of AIDS. Even stacks of papers and books published on the subject are not enough. (p. 1314)

At its core, AIDS denialism takes the opposite position of scientifically established facts. There are seven basic principles of AIDS denialism, summarized in table 19.1. First, HIV is dismissed as a harmless virus that cannot possibly cause immune system disease and could never result in AIDS. Second, AIDS deniers claim that HIV is merely present in people who develop AIDS, just as it is in many people who never develop AIDS. (AIDS denialists do not really deny the existence of AIDS; they deny that HIV causes it.) As a third principle, the cause of AIDS is explained by lifestyles and environmental toxins rather than HIV infection. Ultimately it is the use of drugs and impure foods, and the scourge of poverty, that degrades the immune system and causes AIDS. Fifth is to point to research that dates back to the 1980s, just as AIDS was identified and the cause was unknown. AIDS denialists claim that HIV has not been shown to fulfill the classic standard Koch's postulates for defining a disease, which involves isolating an agent and determining if it causes disease, even though it has many times over. The sixth principle is that drugs used to treat HIV are themselves toxic poisons, and that they are actually a cause of AIDS themselves. The illogic goes something like this: people carry the harmless virus, and when they test positive for its antibodies they are given antiretroviral medications, which cause AIDS. Or another avenue of thought is that the HIV test is invalid, and reacts to random biological events to test HIV positive, and then people are given antiretrovirals that cause AIDS. The final principle of AIDS denialism is that Africa's devastating AIDS problem is caused by poverty, just as AIDS in the United States and Europe is caused by illicit drug abuse. It is common for AIDS denialists to demand to see a single person with AIDS who has not been malnourished or ever used drugs. And when such individuals are produced, they are dismissed as

Table 19.1.

The Seven Principles of AIDS Denialism

Principle
HIV is a harmless retrovirus that does not and cannot cause AIDS.
HIV is present as a harmless passenger virus in some people who develop AIDS. Nevertheless, the virus is benign—an innocent bystander.
Lifestyles cause AIDS, in particular lifestyles that expose a person to environmental assaults on the immune system.
A person infected with HIV does not necessarily develop AIDS.
HIV does not fulfill the time-honored laws of biology, or Koch's postulates, that define infectious diseases.
AZT and other HIV treatments are toxic and can cause AIDS.
AIDS in Africa is an old condition caused by poverty, and only made worse by AZT and other drugs used to treat HIV, whereas AIDS in other regions stems from substance abuse.

lying. Infants and hemophiliacs with AIDS are considered the victims of an invalid HIV test followed by toxic treatments. Or they are said to have other illnesses, not HIV. AIDS denier Henry Bauer, for example, claims that people with hemophilia should have never been included in relation to AIDS. He confused hemophilia itself as immune-suppressing disease. Bauer stated that it is possible that Ryan White, diagnosed at age 13 in 1984 and dying of AIDS in 1990, could have contracted HIV at birth. Bauer concluded that:

Poor Ryan White had indeed been born in very ill health for a long time on account of his severe hemophilia. Testing for HIV just introduced in 1984—perhaps be might have been HIV positive at birth? It is possible that his death was hastened by HIV medications? (Bauer, 2007; p. 25)

AIDS denialism became increasingly active as the Internet flourished. As shown in table 19.2, the origins of AIDS denialism stem to the very first days of AIDS, when the cause was unknown and any theory was worthy of consideration. But the facts soon emerged, and HIV was discovered as a new virus that causes AIDS. Despite the advances of science and medicine, those old ideas persisted and emerged as AIDS denialism. By the year 2000, AIDS denialists were seeing "cracks in the HIV=AIDS orthodoxy" and used the Internet as a platform to "reveal the truth." AIDS pseudoscientists became increasingly frustrated because they felt censored by scientific journals. During this time, AIDS denialists and their presence on the Internet formed the basis for South Africa's president, Thabo Mbeki, to stall HIV testing, prevention, and treatment on the ground that HIV does not cause AIDS. South Africans with AIDS were denied access to antiretrovirals; instead the country's health minister told them to take vitamins and eat African potatoes and beetroot. The vitamins that were used to treat AIDS were promoted by German entrepreneur Mathias Rath and his assistant, David Rasnick, both

Table 19.2.
A Brief History of AIDS Denialism

Year	Milestone
1981	• The Perth Group (based in Perth, Australia) is established and formulates a theory that AIDS is caused by oxidation.
1982	• HEAL is established in New York as a support agency for people with AIDS. The group later becomes a network of support groups for people affected by AIDS who question the cause.
1984	• The first dissenting view on AIDS, "The Group-Fantasy Origins of AIDS" by Casper Schmidt, appears in *The Journal of Psychohistory*.
1985	• President Ronald Reagan mentions AIDS for the first time in a press conference. • Freelance journalist John Lauristen publishes his series of articles in *New York Native* critical of HIV research.
1986	• President Ronald Reagan urges the public not to panic.
1987–1988	• Peter Duesberg publishes a major scientific paper in *Cancer Research* that questions the role of HIV in causing AIDS. • *Science* publishes the Duesberg article's claims that HIV is not the cause of AIDS.
1990–1991	• Former President Ronald Reagan apologizes for neglecting AIDS. • John Lauristen publishes his book *Poison by Prescription: AZT*. • Robert Root-Bernstein questions whether enough evidence exists to conclude that HIV causes AIDS. • The Group for the Scientific Reappraisal of the HIV-AIDS Hypothesis is founded.
1992	• AIDS denialist Christine Maggiore tests HIV positive, only to later dispute the diagnosis. • Jody Wells creates London-based *Continuum* magazine, which later becomes a major outlet for AIDS-denialist pseudoscience and essays.
1993	• The Perth Group publishes an article in *Biotechnology* that questions the validity of HIV tests due to the "lack of a gold standard."
1995–1996	• Alive and Well AIDS Alternatives is formed by Christine Maggiore. • The Group for the Scientific Reappraisal of the HIV/AIDS Hypothesis publishes a letter in *Science* stating, "We propose that a thorough reappraisal of the existing evidence for and against this hypothesis be conducted by a suitable independent group. We further propose that critical epidemiological studies be devised and undertaken." • Peter Duesberg publishes his manifesto, *Inventing the AIDS Virus*.
1998–1999	• The Alberta Reappraising AIDS Society is founded and serves as a denialist clearinghouse on the Internet. • ACTUP San Francisco is formed as a vocal denialist activist group.
2000	• The South African Presidential AIDS Advisory Panel is formed, with half its members HIV/AIDS denialists. • President Thabo Mbeki of South Africa expresses doubt that HIV causes AIDS at the International AIDS Conference. • A rebuttal to the Durban Declaration is published in *Nature*.
2001	• ACTUP San Francisco's David Pasquarelli dies.

Table 19.2.

(continued)

Year	Milestone
2004–2005	• Vice President Dick Cheney of the United States is asked about the AIDS epidemic among African American women and replies, "I have not heard those numbers with respect to African American women. I was not aware that it was—that they're in epidemic." • Christine Maggiore's three-year-old daughter, Eliza Jane Scovill, dies.
2006–2007	• President Yahya Jammeh of Gambia announces he has discovered a cure for AIDS. His ancestors in a dream revealed the herbal treatment to him.
2008	• *Harper's Magazine* publishes Celia Farber's HIV/AIDS denialist story "Out of Control." • Henry Bauer publishes his book *Origins, Persistence, and Failings of the HIV/AIDS Theory*. • President Thabo Mbeki of South Africa fires his deputy minister of health, the progressive Nozizwe Madlala-Routledge, and expounds on his continued HIV/AIDS "dissidence." • South African courts rule against Matthias Rath and David Rasnick's clinical studies to prove vitamins can cure AIDS, finding them to be unlawful. • The Perth Group becomes involved in the criminal defense of Andre Chad Parenzee, convicted of endangering the life of three sex partners after having unprotected sex and not disclosing his HIV status. He had infected one of them. The Perth Group testifies for the defense that he cannot be guilty because HIV has not been proven to exist. The court throws out the testimony as baseless. • *Discover Magazine* publishes full feature article on Peter Duesberg. • Thabo Mbeki resigns as South Africa's president.
2009	• The popular television show *Law & Order: SVU* airs its "Retro" episode, which portrays an AIDS-denialist character whose child dies of neglected HIV infection. The story follows the events surrounding AIDS denialist Christine Maggiore. • AIDS denialist Clark Baker of the United States forms the Office of Medical and Scientific Justice to use AIDS denialists in criminal and civil cases involving HIV/AIDS. • Leading AIDS-denialist activist Christine Maggiore dies of AIDS. • Manto Tshabalala-Msimang, an AIDS denialist who served as South Africa's health minister, dies. • AIDS-denialist group Rethinking AIDS launches campaign against AIDS charities, targeting the Product (Red) campaign.
2010	• AIDS denialists hold an "alternative conference on AIDS" in Vienna during the International AIDS Conference. • Elsevier changes editors and editorial policy for its journal *Medical Hypotheses* after retracting two AIDS-denialist articles published without peer review. • AIDS denialists launch a widely distributed documentary film *House of Numbers*, which portrays AIDS science as confused, debated, and in disarray.
2011	• The Rethinking AIDS Society holds a convention in Washington, DC. • AIDS-denialist activists Emery Taylor and Karri Stokely die of AIDS.
2012	• Greek AIDS-denialist activist Maria Papagiannido dies of AIDS. • The Office of Medical and Scientific Justice, a US AIDS-denialist group, becomes involved in US military justice cases involving soldiers accused of infecting sex partners without prior disclosure of their HIV status.

of whom were found guilty of conducting illegal clinical studies in South Africa. As a direct result of Mbeki's AIDS-denialist policies, it is estimated that more than 330,000 lives were lost and over 35,000 babies were needlessly born with HIV infection (Chigwedere and Essex, 2010; Chigwedere, Seage, Gruskin, and Lee, 2008; Geffen, 2010; Nattrass, 2010). Other African countries followed Mbeki's example. In particular, the president of Gambia claimed that AIDS could be cured by a secret potion revealed in dreams from his ancestors (Associated Press, 2007). A group of AIDS deniers was formed as the Rethinking AIDS Society and began actively recruiting. In this context, I entered the world of AIDS denialism.

Meet Joe Newton

I knew that reading Internet posts and self-published books would only get me so far. I would have to enter the denialists' world to understand what motivated their propagating AIDS misinformation. I would have to directly interact with them. But I also knew that they would not be open and candid with me. I was, after all, part of the very establishment they did not trust. As a government-funded AIDS researcher, I would be the last person an AIDS denialist would open up to. I did not want to debate the cause of AIDS. What I did want to know was what might explain a belief system that touts the virtues of science while so blatantly defying objectivity.

I knew that I had to be deceptive to gain their trust. I would have to be curious and unthreatening. But I would not misrepresent myself as a person diagnosed with HIV and looking for answers. I was also unwilling to say I was one of them, a scientist turned rogue against the establishment. I decided to reach out to some of the leading voices of AIDS denialism as a student of public health, which of course I am. I would say that I am interested in AIDS, which is also true. I used the name Joe Newton, a nickname given to me by friends in South Africa.

The first thing I did as Joe Newton was to join the Internet list of "AIDS dissidents." At the time, there were 2,651 international signees, which included, among others, the scientists who dispute that HIV causes AIDS. It was apparent, however, that the vast majority of those who question that HIV causes AIDS were not in fact scientists. They were often nondescript clinicians, consultants, and alternative therapists. As Joe, I stated that I was a student of public health and signed the declaration questioning HIV. The keeper of the list at that time was a Canadian blogger named David Crowe. I emailed him and asked that I be included among the AIDS rethinkers. I quickly received a reply:

To Joe Newton

I will certainly put you on my society's email list. If you want to join Rethinking AIDS you need to sign the petition at: http://www.virusmyth.com/aids/statement/. If you want to be on the list http://rethinkingaids.com/quotes/rethinkers.htm you will need to supply us with some credentials or accomplishments as the purpose of this list is to address the complaint that people who don't believe HIV=AIDS are all ignorant (not that credentials are always that meaningful, but in our society they are taken as such).

And so our correspondence began. I replied,

To David Crowe

Thank you for replying. I have signed the petition at Rethinking AIDS. I am a student of public health so I do not yet have my MPH. I am very interested in learning and connecting with you and your group will help.

Thank you again and any further information will be very much appreciated.

As a student of public health questioning the facts of AIDS, and perhaps a budding denialist, I met the standard and Joe Newton was added to the list. I was now among the AIDS deniers.

To Joe Newton

Being a student of public health is a fine qualification. You are studying for your MPH I presume. Do you want us to list a location (Country and/or city) or institution?

There are a number of email discussion lists under "Links" at http://aras.ab.ca if you wish to get more involved.

Now in communication with David Crowe as Joe Newton, I had credibility to connect with others.

Entering the Dark Side

I interacted with only a few selected AIDS denialists. After being added to the list of AIDS rethinkers, I continued my correspondence with David Crowe. I had many questions about the motivations and beliefs behind the notion that HIV is not the cause of AIDS. He generously answered my questions and offered insights into why it is important to question science and not simply accept what emerges as a consensus. He clearly believed that the scientific establishment was corrupt and could not be trusted regarding many things, though AIDS was his most passionate cause. I started asking David Crowe about another AIDS skeptic, Henry Bauer. With my relationship to David Crowe, Bauer too trusted me and answered my many questions. David Crowe also connected me to the AIDS-denial activist Christine Maggiore, who in turn opened the door for me to meet Peter Duesberg. For more than a year I communicated with the insiders of HIV/AIDS denialism. They took me under their wings and mentored me, and these relationships filled in missing pieces of AIDS denialism.

In their minds the propagation of the HIV=AIDS "myth" is the product of a government conspiracy in cahoots with a multibillion-dollar pharmaceutical scam. AIDS denialists see themselves as the guardians of The Truth, one that in time will be revealed to all.I find it telling that the AIDS denialists were so welcoming to someone they thought to be a student in search of answers. It felt as though I was being groomed for initiation into a cult. And there also lies what is most concerning about AIDS denialism: that those most vulnerable to medical misinformation could easily be roped into denialism.

A Network of Denialism

David Crowe

A Canadian journalist based in Alberta, Crowe founded the Alberta Reappraising AIDS Society and is president of the group Rethinking AIDS. He manages the group's site, the most visible and up-to-date AIDS pseudoscience website on the Internet, posting numerous out-of-context clippings from scientific journals to support denialist claims. He had long been involved in environmental-naturalist movements and is the founder of the Green Party of Alberta. He is also a prominent agent of disinformation about cancer chemotherapy and other medical interventions. Crowe was on the advisory council of Another Look, an organization designed to distribute disinformation about breastfeeding-transmitted HIV.. He has signed on to numerous letters and documents challenging HIV/AIDS science. He himself is not, however, a scientist and describes himself as a telecommunications consultant, an environmentalist, and a critic of science and medicine. The aim of his efforts is to free science from the corrupting influence of money.

During my correspondences with David Crowe, I sought connection to another vocal member of Rethinking AIDS, Henry Bauer. I told Crowe that I had "heard that Dr. Henry Bauer at Virginia Tech University has demonstrated that HIV cannot cause AIDS." And I asked, "Are you familiar with his work and do you believe it?" His reply was an endorsement, but did not really answer my question:

To Joe Newton
Dr. Bauer has an excellent new book on HIV and AIDS ("The origin, persistence and failings of HIV/AIDS theory") and it is very much worth reading.

Crowe had posted a review of Bauer's book prior to publication. He was impressed with Bauer's innovative use of statistics and resolution of race differences in AIDS. The Alberta Reappraising AIDS Society displayed the cover of Henry Bauer's book on their home page, calling the book an "excellent summary of statistical anomalies in HIV/AIDS in the United States." When I asked David Crowe if Henry Bauer's belief that the Loch Ness Monster exists (Bauer, 1986, 2002) jeopardizes his credibility on AIDS, he responded by asking how would we know that the Loch Ness Monster was fiction if we didn't research it. What if we discovered that the monster didn't exist, but the phenomenon that convinced some people that it did was by itself interesting? He said that the problem is that if we rule out certain areas of inquiry, scientists will spend more time worrying about whether they should be asking certain questions than investigating nature. Crowe said that if you could have asked him in the nineteenth century "if gravity could bend light, if time slowed down the faster you traveled, if you got heavier if you moved faster . . . I would have said I'm 100% sure that the answer is 'No' to all . . . but Einstein proved all those amazing things true." Taking the stance of a truth seeker is common among denialists. Ironically, AIDS deniers state that all sides of the argument should be considered without censorship, while dismissing any and all evidence opposing their adamant belief that HIV does not cause AIDS.

Henry H. Bauer

A professor emeritus of chemistry and science studies, and dean emeritus of arts and sciences, at the Virginia Polytechnic Institute and State University, Henry Bauer has held academic positions at the Universities of Sydney, Michigan, Southampton, and Kentucky. He taught in humanities, science, and technology, but never did any notable research of his own. Based on the usual standards of peer-reviewed publications, Henry Bauer does not have what we would consider notable scientific accomplishments. I asked a friend on faculty at Virginia Tech about Henry Bauer, and he said that Bauer had been an able administrator and dean. My friend, an AIDS researcher, would not say anything about Bauer being an AIDS denialist. The sanctity of academic freedom creates silence about extreme views. The fear is that questioning any academic's expression of a view will become a slippery slope, opening the possibility of institutional censorship or worse.

Long before he got into AIDS, Henry Bauer went to the fringes of science. He has been editor-in-chief of the *Journal of Scientific Exploration*, which publishes pseudoscience on every far-out topic from alien abductions to telepathy. At the twenty-sixth annual convention for the Society of Scientific Exploration, Bauer presented his work on AIDS in sessions that included "Anomalous Energies and Balls of Light in Crop Circles," "UFO Research: Where Do We Go from Here?" and "Consciousness, Psi, and the Two Brains." He considers himself a leading authority on the Loch Ness Monster. In one of his writings on Nessies, Bauer said:

On my ninth visit to Loch Ness, I no longer expect to see the monster, even though I fully believe that it exists. Others may think that paradoxical, but I think it shows that I have learned a bit about the Monster, and about some other things as well. In 1958, on my first visit and as a casual tourist, I didn't expect to see the Monster because I didn't believe in it. In fact, I actively disbelieved in it. I knew the Monster to be a myth, a joke, or a hoax, good for the tourist trade but not for serious consideration. So I gave a mental sneer when, browsing in the library a few years later, I came upon a book entitled Loch Ness Monster. Superciliously I riffled the pages and found some glossy ones with photos, and those photos gave me pause. A long pause, for I took the book home and read it, and—for perhaps the first time in my life—I really didn't know what to believe. (Bauer, 1986; p. 238)

In 2007, Dr. Bauer published *The Origin, Persistence, and Failings of HIV/AIDS Theory* (Bauer, 2007), a book detailing his arguments against HIV as the cause of AIDS. On the cover is an image of HIV particles erupting from a human immune cell, with the cover caption stating, "Hypothetical depictions of HIV particles and HIV-infected cells." By examining HIV testing data and AIDS case rates, Bauer offers "proof" that HIV does not cause AIDS. This exercise defies any basic understanding of epidemiology, much less the differences among HIV, antibodies, and AIDS.

Bauer's logic concludes that because it would take ten years for people who test HIV positive to develop AIDS, past HIV testing data should predict future AIDS cases. Using data from a national study of US military recruits in the early 1980s, he claims that male-to-female ratios of HIV-positive tests do not correspond to the US national gender ratios of AIDS diagnoses ten years later. He also claims to prove that HIV is not transmitted sexually because women military recruits are just as likely as men to test HIV positive, and yet men in the United States are far more likely to develop AIDS. I debunked Bauer's flawed reasoning in *Denying AIDS* (Kalichman, 2009).

Our correspondences explored several dimensions of how the view that HIV causes AIDS allegedly misrepresents reality. Peer review in the scientific enterprise is, of course, one reason. Henry Bauer wrote in an email:

> I think that the system worked fairly well until the second half of the 20th century. Partly because if you wanted worldly success you didn't go into research, so there was a lot of idealism among those who did. Partly because there were sufficient numbers of competing journals, and granting sources, and one didn't need much in the way of research funds in most fields. So disputes were pretty much PUBLIC and that imposed some constraints. Nowadays with anonymous grant reviewers and manuscript reviewers and highly organized societies and journals, etc., there's really a governing Establishment that can effectively exclude what isn't on the current bandwagon in any given field.

Another problem lies in the media. In discussing how his book has not gotten the attention it deserved from media outlets, he explained:

> Maybe the most depressing aspect is what a firm grip the mainstream view of HIV/AIDS has. Anyone who questions it is automatically written off by almost every journalist, pundit, etc., as a crank. But I have several ideas for ways to attract attention to the book and its implications. It's early days yet, and the next year or so may tell whether one or other idea pays off.
>
> When I started to ask for advice, where to publish my HIV data analysis, many "dissidents" told me I had absolutely no chance of getting into a mainstream journal. I contacted quite a few journalists with whom I had had good interactions in the past, even about my unorthodox beliefs about Nessies and other things, but none of them were willing to look at my HIV/AIDS stuff. Most didn't even acknowledge my request. Have a look at what Gordon Stewart wrote in Index on Censorship. It is totally against everything that most people think goes on in science, but the fact is that people who question whether HIV causes AIDS are treated like pariahs by mainstream scientists, and the media take their cue from that.

Like David Crowe, Henry Bauer could seem harmless, a nice guy with offbeat ideas. He was generous with his time to a student of public health. He felt he had something to say about AIDS and was not being taken seriously. He felt he was simply dismissed on account of his "unorthodox" interests. At one point he even expressed concern for me, stating, "I really don't want you to get discouraged or cynical like some of us older folks, but I have to tell it like I see it. People like you will be the ones who make things better." But if there is harm in propagating the myth that HIV does not cause AIDS, that HIV tests are invalid, and that HIV treatments are toxic, Henry Bauer is without question causing harm. Someone who listens to Bauer and comes to believe that there are Nessies in Scottish waters may take a holiday to see for themselves. What harm could come of that? But someone who listens to Bauer and takes refuge in a university professor saying that HIV cannot cause AIDS may ignore their HIV diagnosis and, as has happened in well-known cases, die an early death. Delving in the fringes of sciences has consequences.

Meeting Peter Duesberg

Peter Duesberg is a professor of molecular and cell biology at the University of California, Berkeley. He received his PhD in chemistry from the University of Frankfurt in 1963. He is the son of two medical doctors; his father was a professor of internal medicine. He joined the faculty at Berkley in 1964 and isolated the first cancer gene through his work on retroviruses in 1970. His subsequent work on genetics in cancer resulted in his election to the prestigious National Academy of Sciences in 1986. But soon after, he abruptly abandoned his research on oncogenes and became a proponent of a cancer theory based on aneuploidy, the presence of an abnormal number of chromosomes. Duesberg claims, counter to mainstream cancer research, that aneuploidy is the sole cause of cancer. According to Duesberg, chemicals found in the environment, food and drugs cause aneuploidy. That is, all cancers are entirely caused by carcinogens. This notion is extended in Duesberg's thinking about AIDS—chemicals found in the environment, food, and drugs cause AIDS. Duesberg agrees that retroviruses, including HIV, exist, but he firmly holds that they are passenger viruses incapable of causing disease.

It became important to me, in my quest to understand AIDS denialism to meet Peter Duesberg. After asking David Crowe about whether Duesberg was accessible, I wrote the following email note:

Hello Dr. Duesberg

I hope this note finds you well.

Mr. Crowe suggested that you may respond to my email. I am a student of Public Health following the developments in AIDS. It would appear to me that with recent events such as the publication of the Rodriguez paper in *JAMA* and the continued failings of treatments as well as the pile of failed vaccines, the HIV tower may be ready to fall. I am curious if this is how you see it and whether you are working on any new papers or books? Also, will you be speaking publicly anytime soon? My dream is to see you talk on AIDS.

My best to you and thank you for your time!!

His reply was cordial and inviting, showed his sense of humor, and revealed his disdain for the government.:

Dear Joe Newton,

It's not very common that I get invited to give a talk on AIDS in this "freest of all countries" (president Bush). But, please let me know where you are, and then I contact you if I come close and we will have a "ball"—if just the two of us.

I learned from Christine Maggiore (see below) that Peter Duesberg was hosting a small conference on the chromosomal basis for cancer in Oakland, California, not far from the Berkeley campus. This would be my opportunity to meet Duesberg, and I did not let it pass me by.

The conference was more of a small gathering, around fifty attendees in a small hotel. Upon arrival, I needed to pay the modest registration fee. This posed a problem. I did not have a check or credit card in the name of Joe Newton. I offered to pay cash, and the conference secretary said they were not able to take cash; however, because I was a student, I could attend without paying. Another act of kindness, this experience confirmed my notion that AIDS denialism was not a moneymaking scheme, at least not for most of those involved.

At the two-day conference, I had the chance to watch Duesberg present his ideas on cancer. Some of those present shared his extreme views, but others did not. There was a vibrant exchange of ideas. Duesberg was engaging; actually, he was charismatic. He was without question one of the more intellectually stimulating people I have encountered. In a spirited and respectful professorial style, Duesberg challenged others who did not see things as he did. Presenters frequently made reference to Duesberg taking the opposite view of the mainstream, pushing back against the establishment to think out of the box. This small group who came to this self-organized conference clearly respected Duesberg and what he had to say.

My experience at Duesberg's conference demonstrated that he is no fraud. Duesberg believes what he says. He has taken a position, perhaps because no one else has, and argues for it. Still, if a core value of science is to accept the facts as they are revealed, to dispute evidence only with counterevidence that has been scientifically attained, then Duesberg is no longer acting as a scientist. Duesberg thrives on intellectual argument, taking a position opposed to the established and digging in deep. He is a true contrarian. He also commands considerable attention. During a break in the sessions, I asked him to sign my copy of his 1996 book *Reinventing the AIDS Virus*. He graciously did so, inscribing "It's been a while. To Newton, in my cordial regards, Peter Duesberg." He also looked me square in the eyes and said, "This is not an infectious disease. There is no vaccine for this, and there never will be." I was left with no doubt that he really believes this. He can ignore the science to protect his beliefs because he is convinced that the grant funding, peer review process, and medical establishment have censored any alternative explanations for the cause of AIDS. Duesberg has held fast to his beliefs, and he has done so to the detriment of his standing in the scientific community. There is no evidence that Duesberg has profited monetarily from his position on AIDS. In fact, the opposite seems to clearly be the case. What he has received is attention. By all accounts, Duesberg was a brilliant scientist. But something went astray. Some say that he was angered by others getting credit he felt he deserved. Others believe he has a character flaw, a narcissism that makes him crave attention above all else. I do not hold to either of these viewpoints. A contrarian who has abandoned science, Duesberg wants to argue. He wants to debate. He wants to be right. Duesberg has been unable to produce evidence that drugs and poverty cause AIDS, so he ignores the facts, discounts the science, and digs in to defend his beliefs.

It Gets Worse

It is not the AIDS denialists who are the real problem here. The likes of David Crowe, Henry Bauer, and Peter Duesberg would be of little importance if all they did was to spout ideas that run to the contrary of the mainstream. What harm is there in believing in UFOs, the Loch Ness Monster, or a clandestine government operation to fake the moon landing? But consider the denial of climate change, which can result in continued release of excess carbon in the atmosphere. Or denial of the Holocaust, which fuels anti-Semitism. The harm in AIDS denialism is its impact on those infected and affected by HIV/AIDS.

People who listen to AIDS denialists are people at high risk for HIV or who have been diagnosed with HIV, and their family and friends. AIDS deniers can appear credible when they say that HIV does not cause AIDS. With mistrust in science and medicine at an all-time high, there is a good chance that a person who tests HIV positive will buy into the idea that the test used to diagnose their infection is a fraud, and that the treatments are poison. Facing a life-threatening and socially stigmatized condition, many people would welcome this message. It is the vulnerability of those in denial that makes the denialists so dangerous.

No follower of AIDS denialism is better known than Christine Maggiore, the founder of the AIDS-denial activist group Alive & Well. Christine Maggiore is the mirror image of Elizabeth Glaser in the alternative universe of AIDS denialism. Elizabeth Glaser contracted HIV in a blood transfusion in 1981 while giving birth to her daughter, Ariel. She unknowingly passed the virus through breast milk before the cause of AIDS was even known. Tragically, her second child also contracted HIV. Elizabeth Glaser became a compelling voice for all those infected and affected by HIV, breaking stereotypes and dispelling myths.

Christine Maggiore's story is no less tragic, and in some ways even more so. Maggiore tested HIV positive after a negative HIV test and a test that seems to have been inconclusive. She became aware of Pete Duesberg and in 1995 founded Alive & Well. She also wrote a book, *What If Everything You Thought You Knew about AIDS Was Wrong?* (Maggiore, 1999). She refused HIV care and felt she was living in a perfectly healthy way while remaining completely untreated. But unlike Elizabeth Glaser, who had no way of knowing she could infect her children, Maggiore's denial resulted in her not taking steps that were then available to prevent the transmission of HIV. Indeed, her three-year-old daughter Eliza Jane died of AIDS-related pneumonia. But Christine Maggiore denied that too, claiming that it was the antibiotics used to treat the pneumonia that caused her baby's death.

In December 2008, at the age of fifty-two, Maggiore died at her home in Los Angeles. The world of AIDS denialism would not acknowledge that she died an early death caused by untreated HIV infection. Just as Christine Maggiore denied that her baby died of AIDS, the AIDS denialists claimed that she died from stress. AIDS deniers claimed that the cause of the stress could have been that her life, including the death of her child, was portrayed on the popular television show *Law & Order: SVU* (2008, Season 10, Episode 5 "Retro") that aired shortly before her death. This may seem unfathomable, but we are dealing, after all, with a movement of denialism.

Soon after Christine Maggiore died, there was a new public face of AIDS denialism, Karri Stokely. She was another HIV-positive mother of two who repeatedly shared her story of how HIV medications, not HIV, nearly killed her. After receiving an AIDS diagnosis in 1996, at the age of twenty-nine, she was treated, including with antiretrovirals for eleven years. But her disease advanced. After she saw an Internet video saying that HIV was a hoax, Stokely became aware of Rethinking AIDS. Influenced by Peter Duesberg (Deer, 2012), she stopped taking HIV treatments and believed that her health improved. She then stepped into the vacuum left by Christine Maggiore as a vocal AIDS denier. She proclaimed, "I'm not getting any answers from the mainstream as to why I'm healthy, and why my husband is negative, and why I can quit these drugs. I think it's a crime. It's crime against humanity." Soon after, in April 2011, she too died of complications from AIDS.

There is no way of knowing how many Christine Maggiore's and Karri Stokely's there have been and still are. Testing HIV positive used to mean a death sentence. HIV treatments have transformed HIV into a chronic disease that is managed with antiretroviral therapy. Nevertheless, the disease remains serious and stigmatized. People who test HIV positive should be expected to go through a brief period of denial as they adapt to their diagnosis. The same is true for those who care about them. Wanting to understand what it means to test HIV positive, such people are likely to turn to the Internet, where AIDS denialism remains readily accessible. AIDS denialism is not harmless when it falls on their ears.

AIDS Denialism in the Courts

AIDS denialists have found another forum in which they can be heard, courts of law. When an HIV-positive person knowingly exposes another person to HIV, it can result in an arrest for reckless endangerment or worse. In the absence of any credible basis for proclaiming their innocence, some of the accused have used the AIDS-denialist defense: no such thing as HIV, no harm done. The original denialist defense was used in 1999 by Nushawn Williams. He was convicted as a sexual predator in New York State, guilty of rape and reckless endangerment. Williams is living with HIV and he infected some of his victims, including a thirteen-year-old. He launched an appeal, claiming that he is not HIV infected, based on electron micrographs. With consultation from a denialist group in California, the Office of Medical and Scientific Justice, Williams claimed that the HIV-positive tests he received are invalid and the electron micrographs do not show the virus. While in prison, Williams had been receiving antiretrovirals, and successful treatment leads to viral suppression, making it hard to get an image of the virus in blood plasma. The defense did not work, and Williams remained incarcerated.

Nushawn Williams was just the beginning of what became a series of efforts. In Australia, Chad Paenzee was convicted of endangering life for knowingly exposing and infecting an uninformed woman. He appealed his case with the defense that HIV does not exist. The judges rejected the defense, which had involved a vocal AIDS-denialist group in Perth, Australia. The Los Angeles–based Office of Medical and Scientific Justice (OMSJ) has served a similar role. The group works with the Alabama lawyer G. Baron Coleman, the principal user of the denialist defense. In a 2014 case, a Georgia pastor named Craig Lamar Davis was found guilty of exposing women to HIV, which the defense claimed was not possible because HIV and AIDS do not exist. As reported by the *Atlanta Journal-Constitution*: "They [the jury] were able to weigh the validity of testimony of people who don't believe AIDS or HIV exist" (Joyner, 2014). The group's pseudoscientific expert in this case was Dr. Nancy Banks, a Harvard MD who practices as a gynecologist. She has never worked in research and has no peer-reviewed, or even non–peer reviewed, papers of any kind. Banks did write a book, *AIDS, Opium, Diamonds, and Empire: The Deadly Virus of International Greed*, in which she exposes the vast conspiracy of the Jewish-run African diamond trade, the pharmaceutical industry, the media, and the US government to kill Africans with toxic substances. Banks testified during the trial that the defendant's crack cocaine use at the time may have caused him to be misdiagnosed HIV positive, and that no test can definitively determine if someone has the virus. *"They don't know where those antibodies come from,"* she stated. *"They've never been able to isolate the virus from the protein."* Davis was found guilty on two counts of reckless endangerment.

The aim of the denialist defense is to create a sense of confusion about the validity of HIV testing that can cast reasonable doubt for the accused. The legal standard for what constitutes an expert is mostly based on credentials, not credibility. John Turner, one of Davis' attorneys, stated, "We clearly established reasons to question the results of [HIV testing]. We handed them reasonable doubt on a platter, but they chose to disregard it."

In its court cases with attorney Coleman, the OMSJ has also used David Rasnick as an expert. Rasnick is Peter Duesberg's closest collaborator, really his right hand when appearing in public together. He was a visiting scholar with the Department of Molecular and Cell Biology at the University of California at Berkeley (1996–2005), although the university retracted his appointment. Rasnick worked with Duesberg at Berkeley, and served with him on the now infamous panel of AIDS experts appointed by President Thabo Mbeki of South Africa in 2000. As noted earlier, Rasnick worked for the German vitamin salesman Mathias Rath in South Africa. Both were found guilty in 2008 of conducting illegal clinical trials in South Africa (Kapp, 2008). I met David Rasnick at Peter Duesberg's cancer conference, where he was about to market a service using technology that could detect cancer anywhere in the body before it developed (the technology sought out chromosomal anomalies). Like Banks, Rasnick's views on AIDS are conspiracy minded:

The HIV cult has transported AIDS beyond the domain of science and medicine, and into the realm of mythology. The discourse is controlled by powerful individuals and institutions with a professional or financial stake in HIV, who take it upon themselves to be the sole purveyors of "truth." Government institutions have compounded the difficulty of arriving at a true understanding of AIDS by doing everything in their power to suppress the views of scientists who disagree with established opinions. . . . President Clinton did his bit to thicken the protective fog encasing the AIDS Blunder. Last summer he declared AIDS to be a risk to the national security of the United States. That action allowed at least three additional federal institutions to play a direct role in maintaining and protecting the fiction of a global AIDS pandemic. These institutions are the Federal Bureau of Investigation (FBI), the Central Intelligence Agency (CIA), and the National Security Agency (NSA). The involvement of the FBI, CIA, and NSA in AIDS represents a far greater threat to our freedoms than to HIV. (Rasnick, 1997)

A third OMSJ consulting expert is Rodney Richards, who worked in the biotech industry until around 1992. He claims to have begun questioning HIV as the cause of AIDS after seeing Peter Duesberg give a talk. Like Banks and Rasnick, he has never published research as a lead author in a peer-reviewed journal. In court, Richards relies on package inserts and technical manuals from the testing kits to claim that the tests are invalid. For example, he points out that a positive HIV test result "presumes" the presence of HIV, using language intended to create doubt in the test.

Coleman and OMSJ became involved with the United States' military justice system, in cases where soldiers were prosecuted for exposing sex partners to HIV without disclosing their HIV status. These are serious cases where real people's lives are at stake. Coleman understandably told me that it does not matter if what Banks, Rasnick, and

Richards say is baseless or not if it can help him launch a zealous defense for his clients. But is such a defense in the accused's best interest? Unchallenged, the denialist defense may seem reasonable. But in a court of law, denialism is challenged by scientists for the prosecution, and here the credibility of the witness can weigh heavily.

In one case, military prosecutors called me as an expert on a case involving Coleman and the OMSJ. I was not called as an expert on AIDS, but as an expert in AIDS denialism. Richards and Banks were at first going to be called to testify on the questions surrounding HIV as the cause of AIDS. They would again question the reliability and validity of HIV testing. I testified at a preliminary hearing regarding their views and their association with Rethinking AIDS. Subsequently, Richard and Banks were not asked to testify at trial. But David Rasnick did testify on the invalidity of HIV testing. Rasnick sat in the court and slept through the afternoon on what was my first day, as well as the morning of the second day. When he testified, the judge limited his "expertise" to antibody testing. He could not discuss anything else, including genetic testing, which was part of the original defense plan. He was allowed to testify on HIV testing as an expert because he worked at Abbott Laboratories in 1978, a low bar for expertise on HIV testing.

As a sworn witness, Rasnick spoke of how his minority views on AIDS have been silenced by the research establishment. He was not sure what journals he had published in that were or were not peer reviewed. He claimed letters and commentaries as peer-reviewed articles. He claimed to be completely unaware of any charges being filed against him for conducting illegal clinical trials to test the effects of vitamins as a cure for AIDS in South Africa. He was surprised to find the verdict against him in the case documents—Rasnick said he had no idea about it. Rasnick went on to explain how HIV testing is completely unreliable. Again, the denialist defense failed. The defendant was convicted of aggravated assault and reckless endangerment.

Subsequently, the OMSJ was involved in another case of reckless endangerment, in the appeals courts of Military Justice. Here, the argument focused on the low risk of HIV transmission through vaginal intercourse. This is a variation of the denialist defense that can more easily be persuasive. AIDS deniers have repeatedly misrepresented research by one of the most respected researchers in HIV/AIDS, Nancy Padian, to claim that the chances of transmitting HIV are between 1 in 10,000 to 1 in 100,000. The actual research shows that under certain conditions, the risk can be as low as 1 in 500. But the conditions are complicated by multiple co-occurring factors that can substantially increase the risk. Obviously, tens of millions of people in the world have been infected this way. In this case, the denialist defense worked. The military judges ruled that the risk of HIV transmission was low and the victim was not infected. Thus, being HIV positive and exposing a sex partner to HIV does not meet the standard of aggravated assault. This ruling in effect overturned all previous such cases in the military, including the 2014 case in which I was involved.

Where We Stand

If AIDS deniers were confined to mimeographed newsletters, postal mailings, and the Annual Convention for the Society of Scientific Exploration, we would need not waste our time on their antics. But this is sadly not the case. Thanks to the permanence of Internet postings, AIDS denialism spreads a perpetual stream of medical misinformation. AIDS denialism directly harms individuals who avoid getting tested for HIV and are dissuaded from seeking treatment for their HIV infection. The adverse effects of AIDS denialism are indisputable. And yet they remain ignored.

People who are newly diagnosed with HIV, as well as those who become displeased with their care providers, are prone to seek answers online. Denial is normal in the process of coping with the trauma that comes with a serious medical diagnosis. It may be that a maladaptive form of denial characterizes those most vulnerable to AIDS-denialist rhetoric. Psychiatry has described the phenomenon of malignant denial, which occurs when the expected coping response of denial is protracted and interferes with medical care (Bahnson and Bahnson, 1966; Muskin, Feldhammer, Gelfand, and Strauss, 1998; Strauss, Spitzer, and Muskin, 1990). Christine Maggiore and Karri Stokely are two visible examples of an unknown number of people diagnosed with HIV who fell into AIDS denialisms only to refuse treatment and suffer the dire consequences. AIDS denialism will not end with the death of its loudest voices. In parallel to HIV infection itself, AIDS denialism cannot be entirely eliminated. However, also like HIV infection, denialism can be contained and its harms minimized. The impact of AIDS denialists diminishes when their lack of credibility is revealed and when they are drowned out by clearly communicated, interpretable, medically sound information. Continued medical progress toward ending AIDS is also progress toward ending its mirror image of AIDS denialism.

References

Associated Press. (2007). Gambia's President Claims He Has Cure for AIDS. Retrieved from http://www.msnbc.msn.com/id/17244005/

Bahnson, C. B., and Bahnson, M. B. (1966). Role of the Ego Defenses: Denial and Repression in the Etiology of Malignant Neoplasm. *Annals of the New York Academy of Sciences, 125*(3): 827–845.

Bauer, H. (1986). *The Enigma of Loch Ness: Making Sense of a Mystery*. Urbana: University of Illinois Press.

Bauer, H. (2002). The Case for the Loch Ness Monster: The Scientific Evidence. *Journal of Scientific Exploration, 16*: 225–246.

Bauer, H. (2007). *The Origin, Persistence, and Failings of HIV/AIDS Theory*. North Carolina: MacFarland.

Chigwedere, P., and Essex, M. (2010). AIDS Denialism and Public Health Practice. *AIDS and Behavior, 14*(2): 237–247. DOI: 10.1007/s10461-009-9654-7

Chigwedere, P., Seage, G. R., 3rd, Gruskin, S., and Lee, T. H. (2008). Estimating the Lost Benefits of Antiretroviral Drug Use in South Africa. *Journal of Acquired Immune Deficiency Syndrome, 49*(4): 410–415.

Deer, B. (2012). Death by Denial: The Campaigners Who Continue to Deny HIV Causes AIDS., *Guardian*. Retrieved from http://www.guardian.co.uk/science/blog/2012/feb/21/death-denial-hiv-aids

Geffen, N. (2010). *Debunking Delusions: The Inside Story of the Treatment Action Campaign*. Auckland Park, South Africa: Jacana Media.

Joyner, T. (2014). Defendant in HIV Trial Found Guilty. *Atlanta Journal Constitution*, Jan. 21. Retrieved from http://www.ajc.com/news/defendant-hiv-trial-found-guilty/6Or2oPFLfDX2pVjNhjclwJ/

Kalichman, S. C. (2009). *Denying AIDS: Conspiracy Theories, Pseudoscience, and Human Tragedy*. New York: Copernicus/Springer.

Kalichman, S. C. (2015). *Denying AIDS*. Retrieved from http://denyingaids.blogspot.com

Kapp, C. (2008). South African Court Bans Vitamins Trials for HIV/AIDS. *Lancet, 372*(July): 15.

Lawler, A. (2002). Climate Change: Battle Over IPCC Chair Renews Debate on U.S. Climate Policy. *Science, 296*: 232–233.

Leto, P. (Writer), Wolf, D. (Director). October 28, 2008, Law & Order: Special Victims Unit , Season 10, Episode 5 "Retro."

Maggiore, C., and American Foundation for AIDS Alternatives. (1999). *What If Everything You Thought You Knew About AIDS Was Wrong?* (4th ed.). Studio City, CA: American Foundation for AIDS Alternatives.

Muskin, P. R., Feldhammer, T., Gelfand, J. L., and Strauss, D. H. (1998). Maladaptive Denial of Physical Illness: A Useful New "Diagnosis." *Int J Psychiatry Med, 28*(4): 463–477.

Nattrass, N. (2010). Still Crazy After All These Years: The Challenge of AIDS Denialism for Science. *AIDS and Behavior, 14*(2): 248–251. DOI: 10.1007/s10461-009-9641-z

Rasnick, D., (1997). Blinded by Science. *Spin*, June. Retrieved from http://www.virusmyth.com/aids/hiv/drblinded.htm

Shermer, M., and Grobman, A. (2000). *Denying History: Who Says the Holocaust Never Happened and Why They Say It*. Los Angeles: University of California Press.

Smith, T. C., and Novella, S. P. (2007). HIV Denial in the Internet era. *Public Library of Science Medicine, 4*(8): e256. DOI: 06-PLME-PF-0343 [pii] 10.1371/journal.pmed.0040256

Strauss, D. H., Spitzer, R. L., and Muskin, P. R. (1990). Maladaptive Denial of Physical Illness: A Proposal for DSM-IV. *American Journal of Psychiatry, 147*(9): 1168–1172.

20 Swaying Pseudoscience: The Inoculation Effect

Kavin Senapathy

The good thing about science is that it's true whether or not you believe in it.
—Neil deGrasse Tyson

As science communicators go, Dr. Neil deGrasse Tyson is a paragon. Bringing science to the masses, he jumps from radio to TV to podcasts with the ease of an all-terrain vehicle, his coolness rivaling that of rock stars. Picking up where the late, great Carl Sagan left off, Tyson has brought the science "epic mic drop" to where mics have never been dropped before.

One of his most visible targets was rap star B.o.B., adored for popular hits like "Airplanes" and "Nothin' On You." B.o.B. had spent much of January 2016 tweeting about his belief in flat-earth conspiracy theories, insisting to 2.3 million followers that if the earth weren't flat, photos of the horizon would show curvature. He posted a flurry of photos purporting to prove that not only is our planet flat, but that NASA and the United Nations are in on a conspiracy to cover up the truth.

"[O]nce and for all," said Tyson, addressing the rapper from the stage of *The Nightly Show*. "The earth looks flat because, one, you're not far enough away at your size, and two, your size isn't large enough relative to Earth to notice any curvature at all." Then the mic drop: "And by the way, this is called gravity" (*Nightly Show*, 2016).

Media, social media, and science enthusiasts around literal and figurative watercoolers everywhere went wild. The way Tyson shut down the fact-denying chart topper was the perfect marriage of science and edge-of-your-seat drama. The elegance of his argument, the panache with which the scientist not only defended the truth but declared that the earth isn't flat, and the boldness of that mic hitting the stage sent a buzz across the science world. But the bravura element bothered some. Richard Grant, writing for *The Guardian*, argued that expert "smackdowns" against pseudoscience are a form of self-congratulatory tribalism, a self-affirmation for science insiders rather than an effective way to convince everyman of the truth (Grant, 2016).

So was this an example of successful persuasion, or a condescending flounce? Did Tyson shut down the famous rapper, and was changing B.o.B.'s mind even Tyson's intention?

Doubtful at best. Tyson knew what he was doing, and persuading a delusional rapper wasn't on the scientist's list of goals for the night. Swaying the public from science-scarce beliefs is a marathon, not a sprint, a reality that science communicators—from the celebrated Tyson to your everyday Internet arguer—encounter regularly.

Before I continue, a disclaimer: the following discusses what has worked for me—as a writer, avid social media participant, and science activist. I'm not a scientist, nor a doctor, not even an expert. Rather, I'm a science communicator and mom who knows how to read a scientific paper, and someone who asks experts when I don't understand something. My relevant credentials include having thrown myself heart and mind first into the pseudoscience fray and observing firsthand the bowels of science denial at the March Against Monsanto (MAM) annual protests accompanied by my grassroots organization, March Against Myths (MAMyths). MAM is pseudoscience manifest. Founded to oppose objectionable business practices, the group has since expanded to fight accepted science on vaccines, water fluoridation, and cancer treatments; it even promotes chemtrail and moon landing conspiracy theories. When staring MAM in the face, encountering the bowels of science denial and the tortured souls who inhabit it, it's clear that all brands of pseudoscience are represented in science-denial bastions.

In scientific research, pursuit of truth is the ultimate goal: to explain and to understand, using testable ideas and relying on evidence. Systematic and dynamic, science gleans and organizes information. It is a ceaseless process, a system for investigating phenomena, acquiring new knowledge, and correcting and integrating previous knowledge. For the scientific community to deem anything "science," it must be based on hard evidence subject to testing, rejection, and confirmation. If you're reading this book, whether as a bona fide scientist, science communicator, or science enthusiast, this is old news to you.

Fighting pseudoscience is a different beast and must rely on different techniques. Too often, fighting pseudoscience outside the realm of research devolves into the pursuit of being right; battles fought with citations, corpses of hurt pride left to decay. And while truth is a thing of beauty, fighting pseudoscience isn't necessarily about pedantry or self-righteousness, but progress and justice. To switch gears from science to confronting pseudoscience is a conscious back and forth, a tangible mind shift one must undertake consciously when called for. We must stay cognizant, and check ourselves and our colleagues when we slip into holier-than-thou fact flinging.

They say you can't see the forest for the trees. Trees are for science. Forests are for the front lines of fighting pseudoscience. From a mic drop on a mainstream TV stage to a social media argument, each discussion is only a mile in the marathon, a battle in a war. Here are just a few dos and don'ts.

Don't Forget the Middle Ground

Tyson didn't sway B.o.B. that night. The rap star was already too far gone. So why enter the ring with B.o.B., who predictably doubled down on his flat-earth hysteria post mic-drop despite mountains of contradictory evidence?

Because of the middle ground, the people who aren't part of the science-denial extreme but who might learn a little something about pursuit of the truth by watching a science hero tell is like it is. And though flat-earth theory doesn't appear to be an obvious pseudoscience—defined loosely as beliefs and/or practices mistakenly viewed as being based on valid science—it definitely is. Although we may think that flat-earthers believe our planet is flat because, well, *it looks that way*, they rely on plenty of seemingly scientific support. Take, for example, the Bedford Level experiment cited on the Flat Earth Society website (Flat Earth Society, 2006). When first carried out in 1838, it seemingly demonstrated a lack of curvature over a six-mile stretch of water. It was even subject to peer review, according to the society.

The flat-earth example is only illustrative though—while it's safe to say that people largely know and believe that the Earth is round and orbits the sun with the other familiar planets in our solar system, most issues of science aren't as clear in the public psyche. I'd bet good money that a randomly selected person knows that our planet isn't flat. I would not take the same bet on whether that same person knows that vaccines don't cause autism, that genetic engineering is one of many safe agricultural breeding (or pharmaceutical) techniques, or that humans actually use far more than 10 percent of their brains.

When the average person encounters these issues, the scientifically credible wheat becomes much trickier to separate from the bogus chaff, and we become mired in misinformation-driven fear.

Battle the Fundamentalist Bigwigs to Reach the Middle Ground

In order to prove itself superior to alternative modes of knowledge and be the only legitimate mode of knowing, reductionist science resorts to suppression and falsification of facts and thus commits violence against science itself, which ought to be a search for truth.
—Vandana Shiva (1987)

Think about strict morality, leader worship, sacred symbols, and unwavering, faith-like commitment to beliefs. Sounds like religious fundamentalism, right? These and other characteristics of fundamentalism evoke images of extremist religion, the furthest right or left in politics, and even Davidian compounds and gates to heaven.

Did marketing come to mind? In today's natural-is-better, gluten-free, paleo, amber-teething-necklace, and Autism-as-bogeyman world, fundamentalism is no longer confined to cults and the crazed political fringes. Fundamentalism lurks in grocery store aisles, mommy groups, and even the mainstream press, trickling down from leaders on high.

Science denial, especially with pseudoscience to legitimize it, is the twenty-first century fundamentalism, and we must accept this to even begin to change public perception. While the vast majority of the public is in the middle ground regarding issues like biotechnology, vaccines, genomics, evidence-based medicine, and more, the most vocal minority—and it's generally one with power and influence—is mired in ideology that is basically fundamentalist. This powerful vocal minority reaches the public via media and lobbying, and when done effectively, it doesn't appear extremist but objective, fact-based, and even benevolent. And pseudoscience mongers, especially well-funded ones, do it very well indeed. The GMO "right to know" labeling movement illustrates this.

Self-proclaimed consumer rights champions brandish a "right to know" placard when it comes to transgenic or gene-edited foods. The proclamation begins on high, broadcast by bigwigs, and trickles down to anti-GMO activists wielding figurative torches and pitchforks, demanding the right to know what's in their food. With visions of Frankenfoods and syringe-wielding scientists dancing menacingly in their heads, consumers wonder, "Why not just label it?" But does the "just label it" demand actually come from consumers? What is the "it" we want labeled, anyway? Is this a right we even need?

The "right to know" demand begins with covert profiteers who stand to gain from GMO labeling. They pit a fictional consumer David against a big-business Goliath, but the real Goliaths are those who have a vested interest in labeling. Consider anti-GMO leaders' reactions when President Barack Obama signed S. 764, which requires companies to inform consumers whether ingredients were derived from genetic engineering, but allows the information to be given by means of QR codes or toll-free numbers instead of explicit labels. After Obama signed the bill, opponents again admitted that labeling was no more than a tactic to eliminate GMOs from the market. Along with other leaders in the anti-GMO movement, Ronnie Cummins, director of the organic industry–funded Organic Consumers Association, has made it clear that the bill's requirements don't suffice (Cummins, 2016).

The labeling movement was never about our consumer rights, nor about a need to know. Other opponents leading the anti-GMO movement have explicitly said that the first step to eliminating these technologies is to label them. As alternative medicine champion and GMO opponent Dr. Joseph Mercola wrote in 2012, "Personally, I believe GM foods must be banned entirely, but labeling is the most efficient way to achieve this. Since 85 percent of the public will refuse to buy foods they know to be genetically modified, this will effectively eliminate them from the market just the way it was done in Europe" (Mercola, 2012).

In the developed world, our food supply is the safest, most abundant, and most varied it has ever been. It's a saturated market, with crowded shelves, overflowing counters, and colorful bins of fresh produce even as January snowdrifts pile up outside massive grocery stores in my Wisconsin city. Want crackers for the kids' lunchboxes? Some stores have an entire aisle dedicated to them. Need salad greens? The choices are opulent. Perhaps the only things that rival the variety of food choices are the range of prices for comparable products, and the labels—think non-GMO, gluten-free, "natural," made with "real" sugar, fair-trade, free-range, and more—that adorn food items.

The only way for organics and other so-called "natural" food products to grow their industries is to differentiate themselves in trivial ways disguised as meaningful benefits. We know that organic is no safer, healthier, or better for the environment, and we also know that the organic industry has intentionally misinformed the public to achieve growth (Chassy, Tribe, Brookes, and Kershen, 2014). Labeling is just another tactic in this game.

The consumer-focused "right to know" rhetoric is beguiling, and is pushed by myriad activists linked to organic industry-backed organizations like Just Label It and US Right to Know. The messaging is powerful and superficially persuasive, and it all starts with the fundamentalist linchpins.

There might be no better example of a fundamentalist pseudoscience leader than Vandana Shiva. Demanding $40,000 and round-trip business-class airfare from New Delhi for her promotional talks, the environmental activist may be the world's most visible anti-GMO leader. She's been called the "heir to Mahatma Gandhi's legacy" (Popham, 2014) and praised as a "rock star" of the anti-GMO movement. Consistently pushing evidence-scarce myths about genetic engineering, she exploits her culture and fellow countrymen, brandishing the Indian farmer–suicide bogeyman. And, like any fundamentalist leader in a position of power, Shiva plays the victim card, in her case assuming the role of the voice of the downtrodden Indian farmer, crying, *GMOs are an attack on seed sovereignty!* (Kent, 2016).

Though Shiva has repeatedly blamed Indian farmer suicide on genetically engineered crops, specifically Bt cotton (the "Bt" denotes an insecticidal protein from the *Bacillus thuringiensis* bacterium, which kills boring insect larvae but is harmless to humans and beneficial insects), the assertion simply isn't true. She also condemns companies like Monsanto for holding intellectual property rights to patented seeds, leading to a "seed monopoly" (Shiva, 2016). Either Shiva, who promotes organic agriculture, isn't aware that there are thousands of patented crop varieties, many of which can be grown and sold as organic, or she is cleverly disingenuous. The problem of Indian farmer suicide is very real, but Bt cotton is demonstrably not the reason for this complex and ongoing tragedy (Gilbert, 2013).

In *Seeds of Truth*, Shiva's response to a 2014 profile in *The New Yorker*, she complained about the article's description of a farmer with "skin the color of burnt molasses and the texture of a well-worn saddle" (Specter, 2014).

"In ways other than the obvious, Specter sounds like an *Angrez Sahib* (English Sahib) describing the 'natives' in 1943," Shiva wrote. "One can only hope that he may overcome his disdain of non-white, non-industrial populations, Indian farmers, and farmers in general" (Shiva, 2014).

Despite her endearing looks, Vandana Shiva's exotification of the country and India-as-victim messaging does her few favors in the science and wider agricultural communities, while her butchering of science has seen her earn monikers like "Luddite," "dangerous fabulist," and even "lunatic fringe" (Lynas, 2013).

With any form of fundamentalism, beliefs and feelings trump facts. Rather than relying on the broad weight of available evidence, and science-minded judgment of hazard versus risk, fundamentalist leaders convince the middle ground that feelings matter more than facts, using snippets of cherry-picked evidence to corroborate a worldview. Take the belief that vaccines and GMOs cause autism. The vast amount of evidence points to autism as a spectrum of disorders caused primarily by complex interactions between genetic variations across multiple loci, and by environmental influences that, beyond a shadow of a doubt, don't include vaccines or the food we eat. Yet far-reaching propaganda like the 2016 film *Vaxxed: From Cover-Up to Catastrophe* (intended to "reveal an alarming deception that has contributed to the skyrocketing increase of autism and potentially the most catastrophic epidemic of our lifetime") continuously bombards the public with doubt. Even mainstream and well-trusted public figures like Dr. Mehmet Oz provide platforms to claims that going GMO-free can cure "autism symptoms" (2014).

Just as other forms of fundamentalism have their sacred symbols and tenets, science deniers hold nature as sacred. You'll hear "the way Nature intended" or "x or y technology 'violates' what is Natural." Look no further than genetic engineering opponents and organic marketers, who insist that GMOs are a violation of nature and spin a natural versus unnatural false dichotomy. This is despite the fact that virtually all of the food we eat, whether labeled organic, natural, or even heirloom, has been genetically altered in the field or in a lab to be unrecognizable from wild ancestors.

Fundamentalism must pit a "them" against an "us." The most obvious example of the fabricated us versus them narrative is the anti-GMO camp casting themselves as David against Goliath Monsanto. Injecting big bad Monsanto fears into public perception of GMO has been a brilliant tactic, a way to automatically paint anyone speaking in favor of genetic engineering with a Monsanto apologist brush. With non-GMO and organic industry bigwigs pushing the myth that GMO is synonymous with big corporations and the worst of food system ills, these memes are popular, permeating the blogosphere, mommy groups, mainstream media, and street protests.

Fundamentalism always has leaders, and they must be adored blindly to keep a following. There are fundamentalist religious leaders, fundamentalist national leaders, and fundamentalist science-denier leaders. At the top of this camp we have alternative medicine proponents Drs. Joseph Mercola and Mark Hyman, seeds and oils tycoon John Roulac, anti-GMO activist Vandana Shiva, scientist Gilles Eric Serralini, disgraced former gastroenterologist Andrew Wakefield, consumer activist Vani Hari, and more.

They cleverly subvert their positions of power to appear like Davids in the face of Goliath. Take Wakefield, for example. He directed the film *Vaxxed: From Cover-Up to Catastrophe*. In the wake of backlash from the scientific, medical, and pro-vaccine communities, Tribeca Film Festival cofounder Robert De Niro announced that the film would no longer be screened at the event. Note that Wakefield is possibly the most prominent anti-vaccine leader of them all—one film reviewer wrote, "Wakefield doesn't just have a dog in this fight; he is the dog" (Kohn, 2016). Nevertheless, instead of chalking up Tribeca pulling the film to a calculated decision following backlash, Wakefield and his supporters cried conspiracy. *Natural News*, a website notorious for promoting evidence-scarce scare stories and hawking worthless alternative health products, claimed that Tribeca pulled the film because of "totalitarian censorship demands" and "pharma-funded media science trolls" (Adams, 2016). "We were denied due process," lamented an official statement from the Vaxxed team.

To change public perception and misconceptions based in pseudoscience, we must accept that we're dealing in the realm of fundamentalism, not of logic. We will never sway the Vandana Shivas of the world, nor the B.o.Bs, Food Babes, Drs. Oz, Jeffrey Smiths, Joseph Mercolas, or Deepak Chopras. So what's the point? Why try?

We will never sway them, but the public discussion has an impact—let's call it an inoculation effect, slowly but surely populating the masses with critical thinkers, and even providing herd immunity to quackery.

The Misinformation Hydra

My avatar is known in certain corners of the skeptic and science communicator community. A smiling, spectacled cartoon version of me, she wields a sword emblazoned with one word in all caps—SCIENCE. Slaying pseudoscience with the glimmering weapon of evidence, cutting through quacks with *POW, BANG, ZAP* citations like a science superhero, seems like a noble endeavor.

But in the real world of fighting pseudoscience, there's no heroism in flinging citations. They must be wielded carefully and with cultural context. In the real world, where our pens and voices are mightier than any avatar sword, slaying the pseudoscience mongers does not kill the beast. Cut off one head—as when the scientific community reduces the Food Babe to a media example of fact butchery, or when homeopathy or acupuncture is debunked—and another myth-spewing head will replace it. Don't get me wrong—slay them perpetually we must, as a necessary oil to maintain the wheels of progress. But the parallel objective, while less obvious, is perhaps the most important: arming enough of the public with misinformation radars to grow that herd immunity against quackery.

Well-meaning scientists and members of the science and science fan communities have cropped up Internet-wide. These advocates for good science and for GMOs are wonderful. They're passionate and tenacious, so many of them spend a lot of time and energy spreading good information. But a lot of them commit some common mistakes, mistakes that I've committed myself, especially early on.

Among these mistakes is making sweeping positive statements without any nuance. For example, touting genetic modification as a panacea for feeding the world and curing food-system ills without discussion of realities is a recipe to lose trust. But nuance is tricky, and using too much nuance and scientific detail can backfire. There's a delicate balance between not making sweeping positive statements and getting too technical or even pedantic. For example, there are certain times when we can say that genetic engineering is no more risky than traditional technologies. There are other situations when it makes a lot more sense to simply say, "GMOs are safe."

Another common mistake is throwing facts. It's almost instinctive to immediately respond to a concern with citations. *You're concerned about vaccines? Well here, have links to every single study that says they're safe.* Sometimes it's necessary to interpret a person's concerns, and have some dialogue. We know that presenting evidence without cultural context generally doesn't work. Along with facts, information, and basic understanding of a topic, context is important (Sturgis and Allum, 2004). We need to keep our facts straight, without treating them as ammunition.

Like vaccines, inoculation against bad information might benefit one's neighbors. Teach one person the truth and arm them with the means to detect misinformation, and they can spread these tools to those around them.

Remember, even scientists, so mired in the cultivation of their trees, must see the forest, the cultural realities, the well-placed concerns and fears of their audiences, to participate effectively in the noble fight against pseudoscience.

References

Adams, M. (2016). VAXXED Film Pulled from Robert De Niro's Tribeca Film Festival Following Totalitarian Censorship Demands from Pharma-Linked Vaccine Pushers and Media Science Trolls. Retrieved from http://www.naturalnews.com/053445_VAXXED_film_Robert_De_Niro_Tribeca_Festival.html

Chassy, B., Tribe, D., Brookes, G., and Kershen, D. (2014). Organic Marketing Report. *Academics Review*. Retrieved from http://academicsreview.org/wp-content/uploads/2014/04/Academics-Review_Organic-Marketing-Report1.pdf

Cummins, R. (2016, June 28). Organic Traitors Team Up with Monsanto and GMA on DARK act. Retrieved from https://www.organicconsumers.org/essays/organic-traitors-team-monsanto-and-gma-dark-act

Dr. Oz Show. (2014). New GMO Pesticide Doctors Are Warning Against. Retrieved from http://www.doctoroz.com/episode/new-gmo-pesticide-doctors-are-warning-against

Flat Earth Society. (2006, March 10). FAQ. Retrieved from http://www.theflatearthsociety.org/forum/index.php/topic,1324.msg1312141.html#msg1312141

Gilbert, N. (2013). Case Studies: A Hard Look at GM Crops. *News Feature, 497*(7447): 24. DOI: 10.1038/497024a

Grant, R. P. (2016). Why Scientists Are Losing the Fight to Communicate Science to the Public. *Guardian.* Retrieved from https://www.theguardian.com/science/occams-corner/2016/aug/23/scientists-losing-science-communication-skeptic-cox

Kent, J. (2016). 7 GMO Myths Debunked by Vandana Shiva. Retrieved from http://seedfreedom.info/7-gmo-myths-debunked-by-vandana-shiva/

Kohn, E. (2016). "Vaxxed: From Cover-Up to Catastrophe" Is Designed to Trick You. Retrieved from http://www.indiewire.com/2016/04/vaxxed-from-cover-up-to-catastrophe-is-designed-to-trick-you-review-21896/

Lynas, M. (2013). Time to Call Out the Anti-GMO Conspiracy Theory. Retrieved from http://www.marklynas.org/2013/04/time-to-call-out-the-anti-gmo-conspiracy-theory

Mercola, J. (2012). New Vermont GMO Labeling Policy Officially Introduced. Retrieved from http://articles.mercola.com/sites/articles/archive/2012/02/29/new-vermont-gmo-labeling-policy-officially-introduced.aspx

Nightly Show, The. (2016). Neil deGrasse Tyson Slams Flat-Earth Theorist B.o.B. Comedy Central. Retrieved from http://www.cc.com/video-clips/rca4i7/the-nightly-show-with-larry-wilmore-neil-degrasse-tyson-slams-flat-earth-theorist-b-o-b

Popham, P. (2014). Meet Vandana Shiva: The Deserving Heir to Mahatma Gandhi's Legacy. *Independent.* Retrieved from http://www.independent.co.uk/voices/comment/meet-vandana-shiva-the-deserving-heir-to-mahatma-gandhis-legacy-9681770.html

Shiva, V. (1987). The Violence of Reductionist Science. *Alternatives: Global, Local, Political, 12*(2): 243–261. DOI: 10.1177/030437548701200205

Shiva, V. (2014). Seeds of Truth. Retrieved from http://vandanashiva.com/?p=105

Shiva, V. (2016). Dr Vandana Shiva. Retrieved from http://vandanashiva.com/?p=402

Specter, M. (2014). Seeds of Doubt. *New Yorker.* Retrieved from http://www.newyorker.com/magazine/2014/08/25/seeds-of-doubt

Sturgis, P., and Allum, N. (2004). Science in Society: Re-Evaluating the Deficit Model of Public Attitudes. *Public Understanding of Science, 13*(1): 55–74. DOI: 10.1177/0963662504042690

Vaxxed: From Cover-Up to Catastrophe. (2016). Retrieved from http://vaxxedthemovie.com/

21 The Challenges of Changing Minds: How Confirmation Bias and Pattern Recognition Affect Our Search for Meaning

Indre Viskontas

There is a certain formula that some Hollywood executives use to decide whether an idea for a reality television show has legs: Are the characters unique and does the show take the audience to places where they do not usually have access? When these questions were asked of the producers of *Miracle Detectives*, a six-episode docuseries that aired on the Oprah Winfrey Network in 2011, the answer to both was a resounding yes. After all, the show would bring audiences into the living rooms of regular Americans who reported experiencing anything but regular events: their experiences were so unexpected that they called them miracles, and felt their lives were changed as a result.

I knew nothing about how deals got struck in the backrooms of the industry's elite production companies, but it was these features of the show's pitch that caught my interest as well. Before I heard from the production company, I had not given much thought to whether miracles, real miracles (stigmata, stones exuding milk, the dead brought back to life) truly happen. But I was pretty confident that, with the right expertise and experimentation, science could weed out spooky coincidences from events that actually defied explanation. I was raised Roman Catholic and considered myself to be a spiritual person, though I tended toward skepticism when it came to extraordinary claims.

I had also completed a PhD in cognitive neuroscience, specializing in autobiographical memory. I was well aware that remembering is subject to bias, suggestion, forgetting, and reconstruction. It would take more than a personal testimony to convince me that the laws of nature had been broken or bent under the hands of a supernatural power. But at the same time, it seemed like the perfect subject pool to study for someone interested in how memories of life events are built, and how dramatically a memory can be affected by what happens after an event.

With this interest in human memory, I was excited to meet the people who had witnessed life-changing occurrences and hear their stories. Emotional experiences can feel especially memorable, though they are also known to be more variable (and therefore perhaps more suspect) than memories of emotionally neutral events (Talarico, Labar, and Rubin, 2004). In the laboratory, most studies of autobiographical memories are carefully crafted to give rememberers the best chance of recalling events as accurately and objectively as possible.

But remembering in conversation—in the field, as it were—involves telling stories, and these stories change depending on the speaker's goals, the audience, and the social context (Marsh, 2007). The accuracy of the information matters less than the performance of the speaker. From the beginnings of experimental psychology, we have known that stories change when they are retold and that remembering is a constructive process (Bartlett, 1932; replicated by Bergman and Roediger, 1999). Yet most laypeople will refer to their memories as definitive evidence of what happened in the past.

When participants are explicitly told to avoid errors, they retell prose passages more accurately (Gauld and Stephenson, 1967). They are also quite good at finding their own errors; if they are conscientious, they make even fewer mistakes (Gauld and Stephenson, 1967). But when describing an event as a miracle, the speaker's goals have less to do with accuracy than with impressing their listeners with the uniqueness of what they experienced. In short, participants are highly motivated to construct a memory that convinces others of the validity of their experience, rather than to stick only to the facts. But perhaps with me, a skeptical scientist, as part of their audience, they would be more motivated to fact-check their remembering, since I would be looking for corroborating evidence as well. Because after all, Dudukovic, Marsh, and Tversky (2004) have shown that when undergraduates report life experiences with the goal of informing rather than entertaining their audience, they include fewer exaggerations.

But we also know that retelling stories and recalling events changes the memory for that information in ways that are hard to undo (Anderson, Bjork, and Bjork, 1994; Dudukovic et al. 2004). Telling memories as stories involves selective remembering and often relies on existing schemas for construction, which can be the source of errors (Marsh, 2007). For example, if I'm asked to describe an event in which I witnessed a robbery of a convenience store, I might use my schema of robberies to help scaffold the memory. That is, I might falsely remember that the robber was wearing a mask, pointed a gun, and had features that were masculine because that fits my schema, rather than realize that I do not recall any of these details. In the case of remembering what one now believes was a miracle, schemas may play a big role in how the story is told.

My goal in taking on the job of the scientific foil to a journalist who himself was a believer in miracles was not to simply poke holes in people's memories, but to use the available evidence to find the simplest explanation for the odd occurrence. Most of all, I wanted to apply scientific tools to these curious phenomena and I hoped I would be surprised. The best outcome, I thought, would be the one in which an event so grossly defied our expectations of how the world works, that we would have to rethink some of our fundamental assumptions. If that failed, at least I would have brought scientific rigor to some of the most interesting questions we can ask, such as why and when does nature (including humanity) behave in ways that we do not expect. If the experience was not quite as out of the ordinary as we might reasonably expect, could my scientifically informed explanations change the beliefs of others?

In all, we investigated twelve cases, ranging from a woman who seemed to be bleeding the wounds of Christ to a man who claimed he had x-ray vision. Each case was fascinating, and I enjoyed trying to puzzle out how we might test the claims to see whether there was a satisfactory explanation that did not require invoking the supernatural. From a scientific perspective, what was the most parsimonious explanation, the one requiring the fewest assumptions and accounting for the largest amount of data?

Producing television shows is expensive. There are a lot of people and equipment involved, and all kinds of other costs. Time, therefore, is very limited, both on and off camera. Typically, I would hear about the pitch for a particular story a few weeks in advance of the shoot. Then I would tell the producers what type of investigation I would want to do, which experts I wanted to interview, and what I would hope to learn from the witnesses.

But most importantly, in each case, I had to consider the base rates of the events under consideration, which can be easily ignored or miscalculated (Bar-Hillel, 1980). Simply defined, a base rate is the raw likelihood of an event happening without intervention: it's prior probability, before the influencing factor might have had a chance to operate. In other words, for each potential miracle, I had to ask the same question: Given the same circumstances, and no supernatural intervention, what are the chances that we would still observe this strange phenomenon?

For example, we often feel as though there is something strange afoot when we are in the midst of thinking about someone and suddenly the phone rings and it is them! In those situations, it's common for people to wonder whether some extrasensory perception is at work, without considering the average number of times that the same person might call regardless of what one had been thinking about, or even how many times one thinks about that person and he or she does not call. An analogous example is when someone remembers having had a dream, only to wake up to find that some of what was dreamt was coming true.

Quantifying random noise—calculating the odds of the pattern occurring just by chance, especially when the pattern involves multiple factors—is difficult. What's more, we are predisposed to evaluate our beliefs by looking for evidence that confirms, rather than disconfirms them: this is called the confirmation bias (for a review, see Nickerson, 1998). For example, believers in the paranormal are more likely to invoke telepathy when shown a magic trick than skeptics, even when given the same information (Hergovich, 2004). We might think that we weigh all evidence equally and build personal theories only after considering the data. The data have shown time and again that we do not. Psychics and mentalists depend on the combination of the base-rate fallacy and the confirmation bias to make a living.

Take the following fictional problem. Let's say that among men over sixty, 2 percent have prostate cancer and do not know it. On a routine exam, 90 percent of men with cancer will show a high prostate-specific antigen (PSA) count. Finally, let's suppose that 20 percent of all men who do not have cancer will show a high PSA count on the routine exam. Now let's say that Patient X is over sixty and just got the result that his PSA count is high. What are the odds that he has cancer?

A first impulse might be to say somewhere around 90 percent, since that's the number of men with cancer who will show a high PSA count. But that's incorrect. If you've made this error, however, you are in good company: many physicians make similar mistakes, and are just as prone to using the availability heuristic. They give more weight to things that come to mind quickly, and not enough weight to the base rate (Bornstein and Emler, 2001).

The critical piece of information is that the likelihood of cancer is low, only 2 percent. At the same time, the false positive rate of this test is fairly high, 20 percent. Let's look at this problem again, but from a slightly different perspective. Let's put the percentages into actual numbers. Let's say that we have a population of 10,000 men over sixty. Two percent have prostate cancer, which means that in our population, 200 of them have the disease.

Now let's say that all 10,000 men get tested. How many will show a positive result? Well, of the 200 that actually have the disease, 90 percent will show a high PSA count—so that's 180 men. Of the remaining 9,800 who do not have the disease, 20 percent will show a positive result—that's 1,960.

So now, out of the 10,000 men, 2,140 of them will show a high PSA count—1,960 plus 180. But of those 2,140 men, only 180 actually have cancer. So the odds of having cancer, given a high PSA count in this fictional example, are only 8 in 100 minus 8 percent. In other words, the overwhelming majority of the men with a high PSA count—92 percent—do *not* have cancer.

This example underscores the importance of considering several important factors when we're evaluating the probability of some set of circumstances. We need to know how likely the circumstances are—in our example, the rate of cancer in men over sixty. We also need to know how good the test is in detecting cancer—90 percent of men with cancer will test positive. And critically we need to know how selective the test is—what is the false positive rate? If 20 percent of men without cancer will show a positive test, and the rate of cancer is low, but the frequency of the test is high, then lots of men with a positive test will be walking around thinking they have cancer when they really do not.

The controversy surrounding PSA screenings and false positives in prostate cancer diagnoses is beyond the scope of this chapter (see Etzioni et al., 2002). But the minute we get the result, we cannot help but jump to the conclusion that we have the disease. If we were fully rational about the beliefs we hold and the conclusions we reach, we would always follow the insight of Thomas Bayes (Barnard and Bayes, 1958). In order to understand the likelihood of a given situation, we need to take into account the base rates, not just the confirmatory evidence. But it seems more instinctive for us to reach for evidence that readily confirms beliefs we have already formed or that supports whatever conclusions we find most "obvious."

The truth is that we do not entirely ignore base rates—we are not quite as irrational as that—but we do underestimate them, and fail to see their significance. As Koehler (1996) points out, the base-rate fallacy (that is, the idea that people ignore base rates entirely) is often overstated. Instead, in cases in which the motivation of the subject strongly favors one interpretation, base rates are given less weight than other criteria. Thus, in many of the exceptional cases that we examined on the show, the base rate was underestimated, perhaps largely because the individuals were motivated to find meaning in a challenging experience. After all, individuals in these kinds of reality shows are handpicked from a large population. Their experiences are rare, but given the number of people in the United States alone, the chances that some of them would have encountered strange coincidences are high. The chances that any of these events were caused by a supernatural force that bends the laws of nature but has yet to be proven to exist are much smaller.

In each of the episodes, we can see evidence of one of the most powerful aspects of human cognition: we look for, and often find, patterns, even when they are coincidental. We are conscious beings in a chaotic world. We want to find regularities in our environment, perhaps so that we can predict the consequences of our actions and plan accordingly—to know what or whom to avoid, what or whom to approach. Our brains are so efficient at detecting these patterns that we often err on the side of seeing meaning in something that, in fact, is just random noise. This efficiency might be especially persuasive in situations that are highly emotional, as is the case in every one of the experiences that we investigated.

Now it makes sense that people are selective with what type of evidence they consider when evaluating beloved beliefs that they are motivated to maintain. People who want to believe in the existence of a higher power, for example, and who see miracles as evidence for such an existence might be motivated to look for evidence that a miracle happened rather than considering evidence that supports a simpler explanation.

I came across this type of bias often during my time as host of *Miracle Detectives*.

One story in particular has stuck with me. It's a story of a child, born with severe developmental disabilities, who was in and out of the hospital often. It's a rather heartbreaking story, albeit one with a happy ending. I vividly remember meeting the young girl in her home. Albeit showing some apprehension at first, once she became comfortable with us, her face lit up with pleasure at seeing the video cameras and the lights and the other TV paraphernalia that crowded her living room.

We communicated for a while, gesturing and exchanging smiles, and I noticed that she seemed to check in with her mother frequently, as if taking emotional cues from her. Then she was wheeled into another room while her mother tearfully told us her story.

One particularly bad illness landed the girl in the hospital once again. She was severely ill, with some kind of infection, though the doctors did not seem to know what it was. In the pediatric intensive care unit (ICU), she was breathing only with the help of ventilators, staying alive attached to several different machines. The mother described the agony of seeing her child suffering, waiting for the medication and interventions that the doctors were prescribing to kick in.

Finally, she was asked by the medical staff if she wanted her child to continue treatment or if she was ready to say goodbye and remove the machines. A horrible choice for anyone, especially a parent. She turned to God for guidance and, after praying for a while, one of the nurses called her to the nursing station to show her something odd. It seemed as if an angel was appearing on the security monitors in the ICU.

In this sign, the mother found comfort. She believed that an angel was sent by God to reassure her. Whatever happened, her child was in his hands—whether she recovered and lived or succumbed and found her place in heaven.

She then decided to remove the ventilators and give her daughter peace. And when the staff removed all the machines, the child began to breathe on her own. She recovered.

This story was a difficult one to investigate, given the emotional nature of the outcome and the fact that a vulnerable child was involved. But I believe in the power of science to help humanity. If we could find out what saved the child, maybe we could use that knowledge to reduce suffering in others, whether the answer was in prayers or something else.

Bizarrely, the image of the angel only appeared when the doors to the ICU were opened. And there did not seem to be any windows or other unexpected sources of light in the corridor where the cameras were pointed at the doors. Once the child recovered and was released from the hospital the angel image stopped showing up on the monitors.

Back in Hollywood, we reconstructed the hallway in the ICU, including the position of the camera as it projected the video onto the monitor in the nurses' station. We checked the weather and charts indicating the position of the sun at the time of the incident, along with other sources of light. It turned out that when the doors were opened, there was a secondary hallway down which there was a room with a window. If the door to that room was left open, the sunlight from the window could hit the doors to the ICU but only when they were open. And ultimately, by positioning a light source emulating the setting sun's position on that fall day, we were able to recreate the angel on the monitor in our Hollywood studio.

What's more, after talking to several experts, I learned that the antibiotics that were administered to the child can begin working abruptly and can show significant changes very quickly. Furthermore, breathing on one's own as the tubes are removed is difficult

The Challenges of Changing Minds

and can induce anxiety. But perhaps the calm that the child sensed in her mother, after she had taken comfort in the idea that an angel was looking after her child, helped her child take those first breaths without as much fear.

So here were the simple explanations: the setting sun caused a flare on a video monitor and the child coincidentally recovered because the antibiotics finally kicked in, and she was able to breathe on her own without anxiety.

But these simple explanations were set aside in favor of explanations that were more comforting—that God sent an angel to comfort the parents and that their child had been miraculously healed.

It's not for me to evaluate a person's spiritual beliefs, and certainly there is room in a scientific explanation for the hand of a loving and all-powerful being. But the scientific explanation does not require that assumption. The setting sun being caught on the cameras on that particularly sunny fall day, at just the right angle, can produce the effect.

But what about the sense of calm that the mother got by seeing an angel, where all I saw was a flare? For that, science has less to say, except to point out that our brains have evolved to find meaning in our world. We search for evidence that confirms our beliefs. As we retell the experience, with the goal of convincing other that we have experienced something extraordinary, we pull it into a story that helps us make sense of chaos and the frightening thought that the universe is random.

The confirmation bias captures the fact that beliefs and opinions are based on cherry-picked data: years or even decades of situations in which evidence that confirms them grabs attention, while disconfirming evidence is ignored. It's why sometimes when a person learns a new fact or becomes engrossed in some topic, elements of that theme seem to follow him or her everywhere. They get excited about electric cars, for instance, and only then do they notice all the charging stations on their way to work that they had not seen before. Or a woman trying to get pregnant might notice babies and other pregnant women everywhere and wonder if the world really is becoming overpopulated.

The confirmation bias is especially rampant in political arguments, where one side or more will cherry-pick data points to support policies, as opposed to choosing policies based on consideration of all the data. The climate science debate will go down in history as a prime example (Farmer and Cook, 2013). For every new study demonstrating the link between human activity and climate change, deniers point to a data point or two showing that parts of the earth are cooler now than in the past.

For example, more snow in winter on the United States' East Coast is used by climate change deniers as proof that the Earth is cooling, while, in fact, scientists know that the increased precipitation is a predicted effect of warming ocean currents. Rather than looking at all the data and trying to understand how counterintuitive effects can result from global warming, deniers grab hold of bits of information that, on the surface, seem to confirm their opinions. This is a very human thing to do. In fact, even when the hypothesis to be tested is not a pet project, people still tend to search for confirming rather than disconfirming evidence. Why?

Part of the answer comes from work showing that we are just not very good at calculating or considering probabilities, as we considered earlier in the chapter. The world is vast and there is a lot going on at any given moment; coincidences are common. Yet we have a tendency to correlate or even infer causality when two things happen close in time, or when we hear a story about something that we think is pretty rare, but forget that the world is vast and that the number of people on it is staggering. We attribute a special aura to something that is much more ordinary that it might seem. While we might understand that coincidences happen, we tend to overestimate the potential influence of novel causal factors (Griffiths and Tenenbaum, 2007).

This type of inferred causality and confirmation bias can even be readily apparent in certain disciplines that *sound* scientific, but that really are pseudoscience: claims that seem to be grounded in the scientific method, but that actually are not.

Pyramidology is a quintessential example. For many people, the pyramids in Egypt have taken on a great significance and they are hailed as more than just great feats of human engineering (Derricourt, 2012). Some look at the pyramids and discover astronomical alignments that predict future events (like World War I, the crucifixion of Jesus, even the apocalypse). Some people believe that they contain supernatural powers, others that they were built by aliens. To back up their claims, pyramidologists often bombard us with data points, involving measurements taken of various aspects of the pyramids. The more measurements they make, the more significant numerical patterns they find.

For example, pyramidologists have noted that the height of the Great Pyramid, if multiplied by a thousand million, equals the distance between the Earth and the sun, give or take a bit, which they then claim is evidence that the construction of the pyramid must have been directed by extraterrestrials because surely the ancient Egyptians would not have known this number.

What this observation shows, however, is evidence that if you take enough measurements and massage them in just the right way, you can find numbers that correspond to whatever it is that you want to believe. The different ways that we can interpret and play with numbers is, of course, infinite.

This multiple comparisons issue is not just a problem in pyramidology: it's even a problem in neuroscience. It's especially prevalent in neuroimaging, where statistics play a central role.

By setting a statistical threshold fairly low and not correcting for multiple comparisons, one can find activation in virtually any part of the brain during any task. If you are looking to confirm a hypothesis, there is a gold mine of data to pick from. But that is not how science is supposed to work, of course. Even scientists, who know better, can get sucked in by the confirmation bias (Kriegeskorte et al., 2009).

So is the confirmation bias a bug in our brain that we could do without? Perhaps not. After all, it's part of the pattern-detection process that also gives us some truly sublime experiences like appreciating music, for example.

Diana Deutsch at the University of California, San Diego, has discovered a series of musical illusions that defy expectation. One of her most famous illusions is called "sometimes behave so strangely." First, Deutsch reads a sentence that contains the phrase "sometimes behave so strangely." Naturally enough the phrase just sounds like words in a sentence. But then, by repeating that one phrase over and over again, she succeeds in transforming our perception of the phrase.

No longer does it sound just like words being spoken; instead it sounds like music being sung! After listening to the repetitions, if you once more listen to the original sentence, it seems as though she bursts into song when she gets to that phrase (Deutsch, Lapidis, and Henthorn, 2008).

The mere act of repetition turns speech into song. This wonderful illusion at first seems to depreciate music, though a cynic might argue that it explains the popularity of many repetitive pop songs. But Deutsch has put a finger on a simple but key insight: repetition is found in music across cultures and genres (Nettl, 1983). In fact, it is a good candidate for the one feature that almost all music has in common. There are far more repetitions in music than in regular speech, and as Elizabeth Margulis has pointed out, functional-MRI studies have shown that emotional engagement with music depends at least in part on familiarity, which is achieved through repetition (Margulis, 2013).

It almost seems too simple to explain the power of music. Yet those of us who perform music by living composers—whose music is often unpredictable and difficult to understand in large part because it's devoid of repetitions—reserve judgment of a given piece until we have heard or played it more than once. It's often only after many listens that some pieces begin to finally sound musical.

In 2013, Margulis wondered whether repetition, when artificially inserted into a piece, can actually make it more enjoyable. She took music by renowned modernist composers Elliott Carter and Luciano Berio, and added in repetitions. She then asked listeners without specialized knowledge in contemporary art music to rate the pieces in three categories: (1) how much they liked them, (2) how interesting they found them, and (3) how artistic they were. Her listeners rated the repetition-hacked examples as more likeable, more interesting, and more artistic.

Margulis's (2013) finding fits perfectly with what I observed during the shooting of *Miracle Detectives*. I remember the details of another case, that of a young boy who was thought to be the reincarnated soul of a World War II pilot, in which repetitions played a role in setting up what became a meaningful pattern to those observing him. He was said to be obsessed with flying and airplanes from a very early age, watching a DVD of the Blue Angels over and over again. He would say all kinds of things that baffled his parents, and ultimately made them convinced that he was a pilot in a former life.

Our brains tend to look for patterns, whether they are faces in clouds and cliffs, or signs from a supernatural power. Our search for patterns, for *meaning* in images, music, and events is both automatic and often intensely pleasurable. Some people argue that meaning, or finding significance, in what is more likely to be arbitrary (such as life span) is what makes life worth living. We hold tightly to the need to connect with our world and each other. When children say things that we do not understand fully, we look for meaning in their words, often giving them more credit than is due.

Thus, we tend to look for evidence that confirms our existing beliefs, thoughts, and feelings. Even when we are testing an idea, we often succumb to the temptation to look for confirmation rather than evidence that would demonstrate that we are wrong. Confirmation bias crops up in many different fields and domains, and it can have ill effects. It can lead to superstitious beliefs, some of which can be harmful, if they cause anxiety or lead a person to make poor decisions (Irwin, Dagnall, and Drinkwater, 2012). It can cause paranoia and prolong depression, as a person ruminates and elaborates on negative thoughts. Snake-oil salesmen and other peddlers of misinformation can exploit our confirmation bias and use it to deceive us, selling us treatments or other products that do not actually work as they are purported to, relying on the placebo effect to keep us coming back. Confirmation bias can perpetuate stereotypes and hostility between different groups of people, as many counterexamples of a false characterization cannot override a positive one.

This last ill effect of the confirmation bias, that it can drive people apart into camps of us versus them, has been illustrated in a number of studies. One of these was famously conducted by Charles Lord, Lee Ross, and Mark Lepper in 1979. These authors were interested in understanding how the confirmation bias might contribute to attitude polarization, an increase in disagreement between two groups of people when presented with more information.

We see this effect when it comes to emotionally powerful issues. Often these issues tend to be political, like gun control, gay rights, and capital punishment. That last issue was the one used by Lord, Ross, and Lepper, who brought into the lab a collection of people with different views on capital punishment and had them sit around a table. A researcher (who did not know the various participants' views) told the participants that they were going to read two out of twenty randomly selected studies on how effective capital punishment is as a deterrent. The participants' job would be to evaluate the studies with respect to whether the research helped or hurt the case for using the death penalty.

But first they were asked to read a statement. Some were given this statement: "Kroner and Phillips (1977) compared murder rates for the year before and the year after adoption of capital punishment in 14 states. In 11 of the 14 states, murder rates were lower after adoption of the death penalty. This research supports the deterrent effect of the death penalty." Others were given this one: "Palmer and Crandall (1977) compared murder rates in 10 pairs of neighboring states with different capital punishment laws. In 8 of the 10 pairs, murder rates were higher in the state with capital punishment. This research opposes the deterrent effect of the death penalty."

Participants then read the first of the two research studies. After that, the experimenter asked the participants how strongly they believed the death penalty was or was not a deterrent, and what effect the research had on their beliefs. Next, they were given detailed research descriptions including critical comments on the methods of the study, the author's rebuttals to those criticisms, and so on. Finally, they were asked to write whether the study was pro- or anti-deterrent, and to evaluate their own beliefs again. The procedure was then repeated with the second of the two studies, this time with the opposite conclusion.

In line with the confirmation bias, Lord, Ross, and Lepper (1979) found that people reported that the study that was in line with their original opinions on capital punishment was more convincing. It had fewer flaws, better methods—in short, the science was more sound. What's more, they held their positions even more strongly at the end of the experiment. No matter which view a person held, he or she found support in the materials provided. The more a person feels he or she belongs to a group of like-minded individuals, the more easily we see attitude polarization (Mackie and Cooper, 1984).

What then might be the benefits of an instinct to confirm our beliefs and hold on to them in the face of contradictory evidence? Perhaps one answer to that question lies in the fact that beliefs largely bring people together. They can be a powerful glue, and that makes them important to society (Bar-Tal, 2000). Thankfully, in addition to cognitive factors that were shaped by social interactions during our evolutionary history, we are also in possession of well-developed frontal cortices that allow us to check our impulses and ultimately change our minds. One of the most rewarding aspects of my time as cohost of *Miracle Detectives* was the surprisingly large number of emails and messages that I received from people who found hope and inspiration from my scientific explanations. Admittedly, I might be falling prey to the availability heuristic here, but there were certainly a number of people whose beliefs were transformed and who found meaning in the more parsimonious explanations.

References

Anderson, M. C., Bjork, R. A., and Bjork, E. L. (1994). Remembering Can Cause Forgetting: Retrieval Dynamics in Long-Term Memory. *Journal of Experimental Psychology: Learning, Memory, and Cognition*, *20*(5): 1063.

Bar-Hillel, M. (1980). The Base-Rate Fallacy in Probability Judgments. *Acta Psychologica*, *44*(3): 211–233.

Barnard, G. A., and Bayes, T. (1958). Studies in the History of Probability and Statistics: IX. Thomas Bayes's Essay towards Solving a Problem in the Doctrine of Chances. *Biometrika*, *45*(3/4): 293–315.

Bar-Tal, D. (2000). *Shared Beliefs in a Society: Social Psychological Analysis*. London: Sage Publications.

Bartlett, F. C. (1932*). Remembering: An Experimental and Social Study*. Cambridge: Cambridge University Press.

Bergman, E. T., and Roediger, H. L. (1999). Can Bartlett's Repeated Reproduction Experiments Be Replicated? *Memory & Cognition*, *27*(6): 937–947.

Bornstein, B. H., and Emler, A. C. (2001). Rationality in Medical Decision Making: A Review of the Literature on Doctors' Decision-Making Biases. *Journal of Evaluation in Clinical Practice*, *7*(2): 97–107.

Derricourt, R. (2012). Pyramidologies of Egypt: A Typological Review. *Cambridge Archaeological Journal*, *22*(3): 353.

Deutsch, D., Lapidis, R., and Henthorn, T. (2008). The Speech-to-Song Illusion. *Journal of the Acoustical Society of America*, *124*(4): 2471.

Dudukovic, N. M., Marsh, E. J., and Tversky, B. (2004). Telling a Story or Telling It Straight: The Effects of Entertaining versus Accurate Retellings on Memory. *Applied Cognitive Psychology*, *18*(2): 125–143.

Etzioni, R., Penson, D. F., Legler, J. M., Di Tommaso, D., Boer, R., Gann, P. H., and Feuer, E. J. (2002). Overdiagnosis due to Prostate-Specific Antigen Screening: Lessons from US Prostate Cancer Incidence Trends. *Journal of the National Cancer Institute*, *94*(13): 981–990.

Farmer, G. T., and Cook, J. (2013). "Understanding Climate Change Denial." In *Climate Change Science: A Modern Synthesis*, pp. 445–466. Dordrecht, Netherlands: Springer Science+Business Media.

Gauld, A., and Stephenson, G. M. (1967). Some Experiments Relating to Bartlett's Theory of Remembering. *British Journal of Psychology*, *58*(1–2): 39–49.

Griffiths, T. L., and Tenenbaum, J. B. (2007). From Mere Coincidences to Meaningful Discoveries. *Cognition*, *103*(2): 180–226.

Hergovich, A. (2004). The Effect of Pseudo-psychic Demonstrations as Dependent on Belief in Paranormal Phenomena and Suggestibility. *Personality and Individual Differences*, *36*(2): 365–380.

Irwin, H. J., Dagnall, N., and Drinkwater, K. (2012). Paranormal Beliefs and Cognitive Processes Underlying the Formation of Delusions. *Australian Journal of Parapsychology*, *12*(2): 107.

Koehler, J. J. (1996). The Base Rate Fallacy Reconsidered: Descriptive, Normative, and Methodological Challenges. *Behavioral and Brain Sciences*, *19*(1): 1–17.

Kriegeskorte, N., Simmons, W. K., Bellgowan, P. S., and Baker, C. I. (2009). Circular Analysis in Systems Neuroscience: The Dangers of Double Dipping. *Nature Neuroscience*, *12*(5): 535–540.

Lord, C. G., Ross, L., and Lepper, M. R. (1979). Biased Assimilation and Attitude Polarization: The Effects of Prior Theories on Subsequently Considered Evidence. *Journal of Personality and Social Psychology*, *37*(11): 2098.

Mackie, D., and Cooper, J. (1984). Attitude Polarization: Effects of Group Membership. *Journal of Personality and Social Psychology*, *46*(3): 575.

Margulis, E. H. (2013). Repetition and Emotive Communication in Music versus Speech. *Frontiers in Psychology*, *4*: 167–168.

Margulis, E. H. (2013). Aesthetic Responses to Repetition in Unfamiliar Music. *Empirical Studies of the Arts*, *31*(1): 45–57.

Marsh, E. J. (2007). Retelling Is Not the Same as Recalling: Implications for Memory. *Current Directions Psychological Science*, *16*(1): 16–20.

Nettl, B. (1983). *The Study of Ethnomusicology: Twenty-Nine Issues and Concepts* (No. 39). Chicago: University of Illinois Press.

Nickerson, R. S. (1998). Confirmation Bias: A Ubiquitous Phenomenon in Many Guises. *Review of General Psychology*, *2*(2): 175.

Talarico, J. M., LaBar, K. S., and Rubin, D. C. (2004). Emotional Intensity Predicts Autobiographical Memory Experience. *Memory & Cognition*, *32*(7): 1118–1132.

VI   Conclusion

22 Truth Shall Prevail

Paul Joseph Barnett and James C. Kaufman

Truth shall prevail
For want of me the world's course will not fail:
When all its work is done, the lie shall rot;
The truth is great, and shall prevail,
When none cares whether it prevail or not.
Coventry Patmore
From *Magna est Veritas* (1890, I:XII)

Truth

Truth is the ultimate goal of every scientist. Discovering truth is not an easy process, and accepting the truth can be even more difficult. Scientific discoveries that disrupt the status quo or challenge those in power have traditionally met with extreme resistance, ire, or violence, whether in the form of the church denying Galileo's claim that the Earth orbits the sun, or the president of South Africa denying that HIV causes AIDS (Kalichman, this volume). Some truths are hard to accept because the implications are too much to bear. If a young child is diagnosed with autism and there is no identifiable cause, a parent has to consider the possibility that some aspects of life are up to chance with no one to blame. Many of the issues in pseudoscience are very personal and hard to discuss, particularly in a public forum. Scientists should be sensitive to that fact, but not at the expense of truth and progress. The United States in particular has become flooded with the false notion that everybody has the right not to be offended. Large sections of pseudoscience, such as creationism/intelligent design and the anti-vaccine movements, have been allowed to flourish largely because it is politically incorrect to question or criticize the beliefs of others.

Fighting for truth is a battle against an amaranthine flow of true believers armed with ignorance and misinformation. Becoming a proponent of truth can have real costs both personally and professionally (French, this volume; Hermes, this volume). Those who would take up the mantle of science in the search for knowledge and truth require a stoic fortitude. As Nietzsche said, "He must have become indifferent; he must never inquire whether truth is profitable or whether it may prove fatal" (1911, p.1). But science cannot become insulting, dismissive, or combative. Humility, or a reticence to speak in absolutes, is the single greatest distinction between the sides of pseudoscience and evidence-based science (also facts, but mostly humility). Scientists should always be willing to entertain the possibility that they may be incorrect (Orzel, this volume).

Historically, science is riddled with inaccurate ideas that at one time were believed to be true, but that now seem ridiculous. Consider the once-respected theories that humans have four humors (blood, yellow bile, black bile, and phlegm) that affect health and temperament (Harvard University Library Open Collections Program, n.d.), that geese grow on trees (Mandeville, 1900), or that the earth is flat (Fleming, 2016). As humanity and our understanding of the universe has evolved, we can recognize these ideas as misunderstandings or mistakes. Each correction marks an advance, but also an opportunity to sprout another vestigial growth of half-truths, misinformation, and scientific illiteracy that is pseudoscience (Hecht, this volume).

Science is not dogma. It is a process of study, learning, and growth. Empirical observations are recorded and compiled to establish patterns that can be used to predict the most likely outcome of a given event. As we develop, so too do our tools and methods of observation. These advances can confirm an existing theory, reveal that the current understanding is only a partial truth, or completely negate all previous ideas and supplant them with a new understanding.

This self-correcting nature is not simply the principal strength of science; it is the fundamental core upon which all science is built. The goal of all scientific exploration is not to support a pet theory or confirm a hypothesis. It is to discover the truth. Nothing in science is above being revisited if our understanding of the universe changes. It is this ability to respond to new information that is cited by many pseudoscience believers as evidence of weakness in the scientific community. They misunderstand the scientific process. Science does not dictate absolute truths; it offers theories and predictive models based on empirical evidence. When that evidence grows or changes, explanations and predictions are adjusted accordingly. The facts did not change. Our understanding of them did. Such developments are not evidence of flaws in established science, but merely examples of unavoidable human error. The entire structure of the scientific process is built to combat inaccurate information. Credible peer-reviewed journals and replication studies serve to filter out as much human error, bias, or deception as possible (Beall, this volume; Marcus & Oransky, this volume). When that system is compromised, mistakes can be accepted as fact—but the truth endures. In a world that is focused on short-term outcomes, science takes the long-term view; to quote Shakespeare, "at the length truth will out" (Shakespeare, 1596–1599; *Merchant of Venice* 2.2.645).

Perspective

A mountain looks very different to people who live at its base and people who live miles away. Perspective can have dramatic effects. Archimedes and Eratosthenes not only theorized that the Earth was spherical, they calculated its circumference with remarkable accuracy (Upton, 2014). Even with the knowledge of a spherical Earth, human perspective changed again in 1947, when NASA used V2 rockets fitted with cameras to take the first pictures of the earth from over 100 miles in space (NASA, 2009). Each time an advance in technology allows humans to travel deeper into space, this "mountain" on which we all live becomes a much smaller piece of the overall panorama. The more we can observe, the better we can understand and predict—and the less room there is for pseudoscience (Simonton, this volume).

What Have We Learned?

This volume has addressed numerous areas of pseudoscience, yet has barely grazed the surface. With the continued growth of technology, new methods of communication are developed on a daily basis. The same resources that help share the benefits of advances such as stem cell research also provide a platform for those bearing well-meaning misinformation, conspiracy theories, simplistic reasoning, cynical cash grabs, and misplaced paranoia to connect and communicate. These nebulous groups that might once have been solitary voices can now unite into a larger and more powerful presence.

We are aware that we are likely preaching to the choir; readers who have made it this far are most likely not ardent pseudoscience supporters. It is unfortunate that most people (on both ends of the science-pseudoscience spectrum) gravitate toward sources and people who share their beliefs. At first glance it may seem similar to people blocking Facebook friends with opposite political opinions—a confirmation bias in which one selectively attends to people or information that supports a predetermined opinion. The difference in the case of science versus pseudoscience is that science is not ignoring any contrary arguments. Most pseudoscientific claims have been studied and falsified; the judicial system calls this "asked and answered," with the goal of discouraging wasted time, excessive irrelevant tangents, and even witness harassment (Fed. R. Evid. 611[a]). But, as many of the chapters in this book detail, each new study that responds to pseudoscience is either dismissed or subjected to a new pseudoscientific claim (or, even worse, ad hominem attacks; see Folta, this volume). In evolutionary science, a common bon mot is that scientists see the discovery of new fossils as filling in gaps, while creationists see it as creating two new ones.

Truth is not like a political debate; it cannot be changed by bullying, propaganda, or talking over the other side. There is no mercury in modern vaccines (CDC n.d., a). Thimerosal, the mercury-based preservative once used in multidose vials, was phased out between 1999 and 2001 (with the exception of multidose flu vials) and is not the same as methylmercury which the anti-vaccine movement claims, incorrectly, causes autism (CDC, n.d., b). The physicist and Nobel laureate Wolfgang Pauli, upon reading a paper with overwhelmingly illogical and unscientific claims, commented that beyond being not right, "It is not even wrong" (Peierls, 1960, p.186). The same thought applies here.

Responding to pseudoscience takes its toll. Scientists get frustrated when they have to repeat facts, refute the same fallacious arguments, and end up in defensive positions because their detractors are not bound to the same rules. Like nuisance lawsuits, pseudoscience rarely plays to win; the goal is simply not to lose. Getting people to doubt science counts as a victory. Even having the truth as a weapon does not fend off attacks by large organizations that can boast celebrity endorsements, articles in impressive-sounding journals, and scores of members with Internet connections and too much time on their hands. (Beall, this volume; Marcus & Oransky, this volume).

Mind the Media

The human mind is complex and not inherently wired for objective logical evaluation (Blanco & Matute, this volume; Lobato & Zimmerman, this volume; Viskontas, this volume). It is susceptible to multiple forms of bias and flawed reasoning and is terrible at objective and accurate risk assessment (Kennair, Sandseter, & Ball, this volume). The media provides repeated exposure to anomalies that can create a perception of risk disproportional to reality. There are reasons most people are more afraid of foreign terrorist attacks than handguns or driving. The challenge of accurate assessment is further complicated when individuals intentionally try to deceive others using techniques ranging from spin to outright lies, as in the case of homeopaths, creationists, and virtually all other proponents of pseudoscience.

One of the most powerful ways to refute pseudoscience is to teach and share an accurate understanding of science with the public. Unfortunately, learning is not a passive activity. Within a single broadcast or article, the media can discuss two contradictory studies. Most laypeople would have no reasonable way of discerning the quality of the work. Even academics working outside of their specific discipline can be fooled. Without investigating it, how many researchers outside of medicine would know that *Cancer Research* is an established and prestigious journal, whereas *Cancer Research Journal* is a predatory scam (e.g., Beall, this volume)? Even if consumers are not being actively misled, wrong information still permeates through. People can legitimately misunderstand work outside of their expertise; others can cherry-pick findings or vastly simplify material to better serve a compelling narrative. Consider how many bits of information the average person is exposed to daily. It would be impossible to devote the time and effort needed to check each one (or even a small fraction).

Pseudoscience is different from junk science, but still related. Junk science can be the result of careless test design and execution, misinterpretation of data, cognitive bias on the part of the experimenter, or even manipulated or fudged results. Whatever the underlying cause, junk science damages genuine science. For example, junk science is responsible for creating the false and deadly idea that vaccines cause autism (Deer, 2004), whereas pseudoscience is responsible for spreading that idea into the larger anti-vaccine movement. The technique of proposing baseless loaded questions is extremely effective in generating doubt and requires no knowledge or understanding to have a harmful impact. When facts are intentionally misrepresented, this technique is even more powerful.

Trust Issues

The default approach for most people is neutrality, if not a slight degree of trust for others. If we are told something, we will often assume it is true unless we have reason to believe otherwise. Many children are told that when they lose a tooth they should place it under their pillow and a magical fairy creature will enter in the night and exchange their tooth for money. (The exact amount of money varies for each family, depending on the parents' socioeconomic status and the number of children in the household who have not yet received all of their permanent teeth.) Children believe this obviously fictional (and slightly disturbing) story because they have no reason to doubt their parents' honesty. It's not until adolescence that we start to realize the extent and frequency of familial lies, lies about everything from supernatural beings that bring gifts and treats on holidays to boogeymen who appear in the event of uncleaned rooms, uneaten vegetables, or unfinished homework. The tendency of most people to give others the benefit of the doubt has been exploited for generations by those who aim to deceive. Deception significantly complicates the already difficult process of changing a long-held belief (Viskontas, this volume).

Apostasy

A true believer, upon realizing that a long-held belief is wrong, may be upset but will recover. But if it is discovered the belief was caused by intentional deception, the believer may be devastated.

One of the saddest lessons of history is this: If we've been bamboozled long enough, we tend to reject any evidence of the bamboozle. We're no longer interested in finding out the truth. The bamboozle has captured us. It's simply too painful to acknowledge, even to ourselves, that we've been taken. Once you give a charlatan power over you, you almost never get it back. So the old bamboozles tend to persist as the new ones rise. (Sagan, 1997; p. 241)

Harry Houdini was very close to his mother. When she died, he was out of the country. Devastated by the loss and his own guilt, he tried repeatedly to contact her through a psychic or spirit medium. Houdini was friends with Arthur Conan Doyle, the creator of Sherlock Holmes. Doyle's wife, Jean, considered herself a spiritualist capable of communicating with the dead. She offered to contact Houdini's mother in a private séance. Using a practice called automatic writing, Jean produced a fifteen-page document supposedly from his mother. Two things that made Houdini skeptical: the letter was written in English, which his mother barely spoke, and it started with a Christian cross, a symbol his Jewish mother would never have used (Gardner, 2015). From that point forward, Houdini did everything in his power to expose the charlatans of pseudoscience who preyed on the vulnerable.

Houdini started a tradition that continues to this day. Exposing frauds and con artists is an institution in the magic community, one that has been adopted by Penn and Teller, Banachek, and (perhaps most famously) James Randi, known as the Amazing Randi. He uses his knowledge of illusion and deception to uncover frauds and hucksters and then duplicate their supposed supernatural powers. Notable targets have included self-proclaimed psychic spoon bender Uri Geller and the faith healer Peter Popoff. In 1979, Randi executed his most ambitious experiment, one that demonstrated how easy it is to deceive people when proper scientific controls are not used.

It began when Peter Phillips at Washington University received a $500,000 grant to perform psychic research. Randi offered him advice on how to avoid participant deception and experimenter bias, but he was ignored. Randi then trained two teenage aspiring magicians (Mike Edwards and Steve Shaw, who is now known as Banachek) on the methodologies of mentalism and magic. The two teenagers consistently fooled the researchers for nearly two years until Randi ended the hoax (Randi, 1986). The Washington University team continued to defend their methods for the next thirty-five years (e.g., Phillips, 2015; Thalbourne, 1995). Some of their points are valid: Randi would call his intervention an experiment, yet no ethics board would have ever approved his plan. In addition, one of the main reasons why Randi's initial advice went unheeded was that he was perceived as more of a showman than a scholar. Randi's end game, in which he gave *Discover* magazine the scoop and sent out press releases, did not negate this impression. That said, much of the lab's reason for not suspecting a deception even when rumors began spreading can be summarized as: "Why would anyone go to so much trouble?" As reasons go, that's not especially inspiring.

Randi made a career of exposing frauds and pseudoscience claims. His nonprofit organization offered a cash prize of $1 million to anyone who could (under controlled conditions) prove paranormal, supernatural, or otherwise unworldly claims. After nearly twenty years of the contest, Randi retired in 2015 with no one earning the prize money (Denman and Adams, 2015). Randi's million dollar challenge, like his debunking of charlatans, has historical roots. Ralph Waldo Emerson, in his essay "Truth," recounts the story of a similar challenge from the 1850s, one inspired by the infamous Rochester rappings, a fraud perpetrated by the Fox sisters (Abbott, 2012).

> Thus when the Rochester rappings began to be heard of in England, a man deposited £100 in a sealed box in the Dublin bank, and then advertised in the newspapers to all somnambulists, mesmerizers and others, that whoever could tell him the number of his note should have the money. He let it lie there six months, the newspapers now and then, at his insistence, stimulating the attention of the adepts; but none could ever tell him; and he said, "Now let me never be bothered more with this proven lie." (Emerson, 1856, p. 527)

The Cost

Despite all of these reasons to be wary of pseudoscience, a common counterargument from true believers is to ask, "What's the harm?" We would argue that pseudoscience has a wide array of costs that assume many forms. The most apparent cost is the astounding amount of money flushed down the drain by suckers. According to the National Center for Health Statistics, Americans spent over $30 billion out-of-pocket in 2012 on "complementary health" (Nahin, Barnes, and Stussman, 2016). Unfortunately, the costs of pseudoscience can be far more tragic than wasted money.

Noah Maxin, eleven, was three months into his chemotherapy treatment and his leukemia was in remission. His parents were concerned about the long-term effect of chemotherapy so they discontinued the treatment and opted for a holistic approach. Noah's cancer returned and he died a few years later (Gorski, 2007). Ryan Lovett, seven, died from a blood infection after battling strep throat for over two weeks. Not only had his mother been treating him with dandelion tea and oil of oregano, the autopsy showed no evidence of Tylenol or Advil in his body (Grant, 2016). The victims are not just children of parents who favor "alternative" medicine. Katie May, thirty-four, died from a stroke caused by a torn artery in her neck, an injury she received from a chiropractic manipulation (Lee, 2016). It is true that science-based medicine has risks as well, but there are also demonstrated potential gains. None of the treatments that contributed to the deaths of these three individuals have any demonstrable, independently verifiable benefit. These are just three casualties of alternative medicine, but the tragic consequences of pseudoscience extend farther than individual health issues.

Advanced Tactical Security & Communications Ltd. sold the ADE 651, a device purporting to use "electrostatic ion attraction" to detect multiple types of explosives and drugs at below trace levels. It also promised to detect currency, bodies, and even guns, regardless of the method of concealment. The device, principally implemented to detect bombs in countries such as Iraq, Afghanistan, and Pakistan, consisted of two handles with antenna-like extensions that swiveled back and forth. If the arrangement sounds like a pair of dowsing rods (a centuries-old method, long proven false, that was used to "find" hidden water, gold, or oil), it's because that's exactly what they were (Hambling, 2016; Plait, 2013). The devices continually failed to find explosives, thereby directly resulting in the injury or death of thousands of people (Nordland, 2009). James McCormick, the conman responsible for selling this device, was found guilty of fraud and sentenced to ten years in prison (Hawley, 2014). Throughout the course of the entire investigation, McCormick maintained that he believed the devices worked, until police discovered that he had been ordering novelty golf-ball finders, relabeling them, and then selling them as bomb detectors (Martin and Gye, 2013; Pasternack, 2013). McCormick convinced dozens of government officials in multiple countries by exploiting scientific ignorance and a lack of critical thinking (Kozak, this volume). McCormick's claims had no scientific basis and should have been immediately exposed as pseudoscience. But his deception is not an isolated occurrence. In 2002, the US Department of Energy spent over $400,000 testing a dowsing process referred to as Passive Magnetic Resonance Anomaly Mapping (PMRAM), for which there was only one qualified operator in the world who could "sense changes in magnetic fields" (Friedman, 2002, p.1). The PMRAM was at least the third dowsing device tested by the DOE (Jaroff, 2002).

The most significant cost of pseudoscience is progress. Disagreement among scientists can be productive and lead to a more accurate understanding of the universe and everything in it (Orzel, this volume). Sharing such knowledge and explaining it to the younger generation allow the exponential growth of humankind. Space travel would be impossible without the work of ancient Greek mathematicians. The universe is beautiful and complex. That humans can study and understand it is a testament to generations of collective mental efforts. Yet material and intellectual resources are limited. Time, money, and energy spent refuting pseudoscientific claims (over and over again) are not spent on curing diseases, discovering renewable energy, or exploring the galaxy.

What Next?

Making mistakes is not simply part of the human experience; it is a core fiber of our existence. Our brains are susceptible to bias, misinterpretation, and fallacious logic. It is not only possible but highly probable that every reader of this chapter has at least one long-held belief about the world that is incorrect. Society, the media, and cultural influences create a bastardized representation of the truth that establishes a network of wrong information permeating our popular culture (Benisz, Willis, & Dumont, this volume; Gorman, this volume; Kozak this volume; Lynn, Gautam, Ellenberg, & Lilienfeld, this volume). The first step for people who want to be less susceptible to pseudoscience is to genuinely consider the possibility that their current beliefs about the world may be wrong (Viskontas, this volume). Allowing for the possibility of being wrong introduces a healthy bit of humility and creates a loop of reevaluation for every new discovery.

Reevaluating the validity of a long-held belief can have two possible outcomes, either of which is arguably beneficial. The first is that the belief is confirmed and fortified through critical and objective review of the currently available information. The second is that an incorrect belief is corrected through critical and objective review of the currently available information. Either way, accurate knowledge is attained.

It would be easy to vilify the true believers of pseudoscience or dismiss them as ignorant or stupid. Yet doing so is foolish and potentially dangerous. It is also a mistake to consider them unintelligent villains. Humans want what they believe to be correct. We want to protect our families and the people we love. Many of the chapters herein addressed issues of health and medicine, which are naturally of utmost importance to everyone. The food we eat (Folta, this volume), the treatment and medicine we seek (Gorski, this volume; Hermes, this volume), and the way we respond to perceived risk (Kennair, Sandseter, & Ball, this volume), infection, and disease (Howard & Reiss, this volume; Kalichman, this volume) determine the quality and length of our lives. Most people who believe in pseudoscience are trying to follow the path that they believe is best. People persist with wrong beliefs for many reasons: how our brains function (Blanco & Matute, this volume; Lobato & Zimmerman, this volume; Viskontas, this volume), how our society misunderstands and miscommunicates science (Benisz, Willis, & Dumont, this volume; Gorman, this volume; Kozak, this volume; Lynn, et al., this volume), or because of outright deception (Blanco & Matute, this volume; Lobato & Zimmerman, this volume).

How to Promote Change

It doesn't take an expert to combat pseudoscience. In 1996, eight-year-old Emily Rosa saw a video about therapeutic touch, a pseudoscience that claims practitioners can heal without touching their patients, via the manipulation of the "human energy field." Emily thought the idea was interesting, so she designed an experiment. With the help of her parents, she tested twenty-one practitioners and found that their ability to sense the presence of her hand suspended slightly above their own was only 41 percent (or less than chance) when their view was blocked. Nobody tested could consistently identify the "energy field" that is the foundation of the therapeutic touch approach (Saad, 2011). At the age of 11, Emily published her results in the *Journal of the American Medical Association* (Rosa, Sarner, Barrett, and Rosa, 1998).

Directly challenging followers of pseudoscience is not simply ineffective. It focuses on the symptoms instead of the disease. Scientific literacy and critical thinking skills serve as a strong defense against the missteps of pseudoscience (Herreid, this volume). True believers have the challenge of persuading others to believe in a specific idea. One of the strengths of science is that it does not promote a specific ideology or belief. It is the endeavor to seek truth.

Conclusion

Nietzsche's Zarathustra descended from the mountain to share his knowledge. Socrates contended that the prisoner, having escaped the cave and seen reality, should return to its depths to help free those living in the world of shadows.

You must go down, then, each in his turn, to live with the rest and let your eyes grow accustomed to the darkness. You will then see a thousand times better than those who live there always; you will recognize every image for what it is and know what it represents . . . (Plato, *Republic* 520[c])

Openly discussing issues of science and pseudoscience in everyday life can help relieve some of the tension and produce a beneficial trickle-down effect on others (Senapathy, this volume). Individual opinions and beliefs should be respected. But if public proponents of pseudoscience present their beliefs as science, they are then open to debate. The fight against pseudoscience can feel like Sisyphus leading the Light Brigade into battle against the Hydra, but giving up is not an option. If we hear a coworker talk about vaccines causing autism or homeopathy curing cancer, our first impulse may be to ignore them. The better (and harder) path is to engage. It is possible to have a dialogue without being confrontational. We can respond without reacting. We can discuss the underlying reasons for their beliefs. The coworker may not know homeopathic "medicine" is just water and alcohol with no active ingredients, or that the study that created the autism myth was falsified and retracted (Harris, 2010), leading to the author losing his medical license (Park, 2010). Having a conversation may be the most productive thing any of us can do to promote critical thinking, both in those around us and in ourselves. Knowledge is growth, and growth can be painful. But without growth, there are no prospects for a better future.

References

Abbott, K. (2012, October 30). The Fox Sisters and the Rap on Spiritualism. Smithsonian.com. Retrieved from http://www.smithsonianmag.com/history/the-fox-sisters-and-the-rap-on-spiritualism-99663697/

Centers for Disease Control and Prevention (US). (n.d., a). Thimerosal in Vaccines. Washington, DC: CDC. Retrieved from https://www.cdc.gov/vaccinesafety/concerns/thimerosal/

Centers for Disease Control and Prevention (US). (n.d., b). Vaccines Do Not Cause Autism. Washington, DC: CDC. Retrieved from https://www.cdc.gov/vaccinesafety/concerns/autism.html

Deer, B. (2004, February 22). Revealed: MMR Research Scandal. *The Sunday Times*. Retrieved from http://www.thetimes.co.uk/tto/health/article1879347.ece (Accessed January 31, 2017).

Denman, C., and Adams, R. (2015). JREF Status. James Randi Educational Foundation. Retrieved from http://web.randi.org/home/jref-status

Emerson, R. W. (1856). "English Traits: VII Truth." In B. Atkinson (Ed.), *Selected Writings of Ralph Waldo Emerson*, pp. 523–528. New York: Modern Library.

Encyclopædia Britannica. (n.d.). Eratosthenes. Retrieved from https://www.britannica.com/biography/Eratosthenes

Federal Rules of Evidence. 611(a). Mode and Order of Examining Witnesses and Presenting Evidence.

Fleming, C. (2016, January 29). Flat Wrong: The Misunderstood History of Flat Earth Theories [Blog post]. Phys.org. Retrieved from https://phys.org/news/2016-01-flat-wrong-misunderstood-history-earth.html

Friedman, G. (2002). Audit Report on "Passive Magnetic Resonance Anomaly Mapping at Environmental Management Sites." US Department of Energy, DOE/IG-0539. Retrieved from https://energy.gov/sites/prod/files/igprod/documents/CalendarYear2002/ig-0539.pdf

Gardner, L. (2015, August 10). Harry Houdini and Arthur Conan Doyle: A Friendship Split by Spiritualism. *Guardian*. Retrieved from https://www.theguardian.com/stage/2015/aug/10/houdini-and-conan-doyle-impossible-edinburgh-festival

Gorski, D. (2007). Another Young Life Claimed by a Misguided Faith in Alternative Medicine. Science Blogs. Retrieved from http://scienceblogs.com/insolence/2007/05/24/another-young-life-claimed-by-a-misguide-1/

Grant, M. (2016). Ryan Lovett Was Dying for Days, Pathologist Testifies at Mother's Trial. CBC/Radio-Canada. Retrieved from http://www.cbc.ca/news/canada/calgary/tamara-lovett-ryan-failing-necessaries-trial-doctors-medical-examiner-strep-1.3873049

Hambling, D. (2016). The Military Pseudoscience That Just Won't Die. Popular Mechanics. Retrieved from http://www.popularmechanics.com/military/research/a21678/dowsing-iraq-bomb-detectors/

Harris, G. (2010, February 2). Journal Retracts 1998 Paper Linking Autism to Vaccines. *The New York Times*. Retrieved from http://www.nytimes.com/2010/02/03/health/research/03lancet.html

Harvard University Library Open Collections Program. (n.d.). Contagion: Historical Views of Disease and Epidemics: Humoral Theory. Retrieved from http://ocp.hul.harvard.edu/contagion/humoraltheory.html

Hawley, C. (2014, October 3). The Story of the Fake Bomb Detectors. BBC. Retrieved from http://www.bbc.com/news/uk-29459896

Jaroff, L. (2002). At the DOE, Dowsing for Dollars. *Time*. Retrieved from http://content.time.com/time/health/article/0,8599,231110,00.html

Lee, B. (2016, October 10). Model Katie May's Death Raises More Questions about Chiropractors. *Forbes*. Retrieved from http://www.forbes.com/sites/brucelee/2016/10/23/model-katie-mays-death-raises-more-questions-about-chiropractors/#3276c87212aa

Mandeville, J., (1900). *The Travels of Sir John Mandeville*. (Posted by Project Gutenberg, 1997; retrieved from http://www.gutenberg.org/ebooks/782)

Martin, A. and Gye, H. (2013, April 24). British Government Helped £50million Fraudster Market Fake Bomb Detectors Based on Novelty Golf Ball Finders to UN Agencies. *Daily Mail*. Retrieved from http://www.dailymail.co.uk/news/article-2313508/James-McCormick-50m-selling-fake-bomb-detectors-bribed-Iraqi-officials-win-huge-contract.html

Nahin, R., Barnes, P., and Stussman, B. (2016). *Expenditures on Complementary Health Approaches: United States, 2012*. (National Health Statistics Reports No. 95). Hyattsville, MD: National Center for Health Statistics. Retrieved from https://www.cdc.gov/nchs/data/nhsr/nhsr095.pdf

NASA. (2009). First Pictures of Earth from 100 Miles in Space, 1947. Retrieved from https://www.nasa.gov/multimedia/imagegallery/image_feature_1298.html

Nietzsche, F. (1911). *The Antichrist* (A. Ludovici, trans.). (Reprinted 2000; Amherst, New York: Prometheus Books.)

Nordland, R. (2009, November 3). Iraq Swears by Bomb Detector U.S. Sees as Useless. *The New York Times*. Retrieved from http://www.nytimes.com/2009/11/04/world/middleeast/04sensors.html?_r=0

Park, A. (2010). Doctor behind Vaccine-Autism Link Loses License. *Time*, May 24. Retrieved from http://healthland.time.com/2010/05/24/doctor-behind-vaccine-autism-link-loses-license/

Pasternack, A. (2013, March 21). The Worst Gadget: Iraq's Most Popular Bomb Detector Is Actually Just a Toy. Motherboard. Retrieved from https://motherboard.vice.com/en_us/article/iraqs-most-popular-bomb-detection-device-is-useless-video

Patmore, C. (1890). *The Unknown Eros* (3rd ed.), Book I, XII. (Posted by Project Gutenberg, 2004; retrieved from https://www.gutenberg.org/ebooks/13672)

Peierls, R. (1960). *Biographical Memoirs of Fellows of the Royal Society*, vol. 5, pp.174–192. London: Royal Society Publishing. Retrieved from http://www.jstor.org/stable/769285

Phillips, P. R. (2015). *Companion to the Project Alpha Papers*. Pari, Italy: Pari Publishing.

Plait, P. (2013, April 29). Maker of Useless Dowsing Rod for Bombs Convicted for Fraud. Slate: Bad Astronomy. Retrieved from http://www.slate.com/blogs/bad_astronomy/2013/04/29/dowsing_for_bombs_maker_of_useless_bomb_detectors_convicted_of_fraud.html

Randi, J. (1986). "The Project Alpha Experiment." In K. Frazier (Ed.), *Science Confronts the Paranormal*, pp. 158–165. Amherst, New York: Prometheus Books.

Rosa, L, Sarner, L., Barrett, S., and Rosa, E. (1998). A Close Look at Therapeutic Touch. *Journal of the American Medical Association*, *279*(13): 1005–1010. DOI: 10.1001/jama.279.13.1005

Saad, G. (2011). Eleven-Year-Old Debunks Therapeutic Touch: The Case of Emily Rosa. *Psychology Today*. Retrieved from https://www.psychologytoday.com/blog/homo-consumericus/201109/eleven-year-old-debunks-therapeutic-touch-the-case-emily-rosa

Sagan, C. (1997). *The Demon-Haunted World: Science as a Candle in the Dark*. New York: Random House.

Shakespeare, W. (1596–1599). *The Merchant of Venice*. In D. Bevington (Ed.), *Complete Works of Shakespeare* (4th ed.), pp. 182–215. New York: Addison-Wesley.

Thalbourne, M. A. (1995). Science versus Showmanship: A History of the Randi Hoax. *Journal of American Society for Psychical Research*, *89*: 344–366.

Upton, E. (2014). The Man Who Accurately Estimated the Circumference of the Earth over 2,000 Years Ago. TodayIFoundOut.com. Retrieved from http://www.todayifoundout.com/index.php/2014/01/amazing-eratosthenes/

Contributor List

David Ball   Middlesex University, London

Paul Joseph Barnett   University of Connecticut

Jeffery Beall   University of Colorado, Denver

Mark Benisz   School Psychologist, Kiryas Joel School District (New York State)

Fernando Blanco   Deusto University; Bilbao, Spain

Ron Dumont   Fairleigh Dickenson University

Stacy Ellenberg   Binghamton University

Kevin M. Folta   University of Florida

Christopher C. French   Goldsmiths, University of London

Ashwin Gautam   Binghamton University

Dennis M. Gorman   Texas A&M University

David H. Gorski   Wayne State University School of Medicine, Barbara Ann Karmanos Cancer Institute

David K. Hecht   Bowdoin College

Britt Marie Hermes   University of Kiel; Kiel, Germany

Clyde Freeman Herreid   University at Buffalo, State University of New York

Jonathan Howard   New York University Medical Center

Seth C. Kalichman   University of Connecticut

Allison B. Kaufman   University of Connecticut, Avery Point

James C. Kaufman   University of Connecticut

Leif Edward Ottesen Kennair   Norwegian University of Science and Technology; Trondheim, Norway

Arnold Kozak   University of Vermont School of Medicine

Scott O. Lilienfeld   Emory University

Emilio J. C. Lobato   Illinois State University

Steven Jay Lynn   Binghamton University

Adam Marcus   Retraction Watch

Helena Matute   Deusto University; Bilbao, Spain

Ivan Oransky   New York University

Chad Orzel   Union College

Dorit Rubinstein Reiss   University of California, Hastings

Ellen Beate Hansen Sandseter   Queen Maud University of Early Childhood Education; Trondheim, Norway

Kavin Senapathy   March Against Myths About Modification

Dean Keith Simonton   University of California, Davis

Indre Viskontas   University of San Francisco

John O. Willis   Rivier University

Corrine Zimmerman   Illinois State University

Index

Abbott Laboratories, 437
Ability-achievement calculations, IQ tests and, 363–364
Abramson, L. Y., 66
Abraxis, 129–130
Absolute knowledge, 226
Academic and Scientific Publishing, 285
Academic Consortium for Integrative Medicine and Health (ACIMH), 314
Academic freedom, 184
Academic medical centers, integrative medicine programs in, 310–314
Accelerated Resolution Therapy (ART), 272–273
Achievement, IQ test scores and, 363–364
ACT (Acceptance and Commitment Therapy), 229
ACT group, 272–273
Action, causal illusion and, 65
Active learning
 case studies and, 402, 412
 strategies for, 411–412
ACT test, 362
ACTUP, 424
Acupuncture, 139, 310, 316, 319, 407
Adams, J., 178
Adams, Mike, 212
Adaptive bias in pattern detection, 47–50
ADE 651, 473–474
Advanced Tactical Security & Communications Ltd., 473–474
Affect heuristic, xi

Agency of Healthcare Research and Quality (AHRQ), 229
Agriculture. *See also* Crop biotechnology
 history of, 107–108
AIDS, Opium, Diamonds, and Empire (Banks), 435
AIDS, worldwide incidence of, 419
AIDS and Behavior (journal), 420
The AIDS Conspiracy (Nattrass), 289–290
AIDS denialism, 6, 289–290, 301–302, 419–439
 author's pseudonym and contact with deniers, 426–428
 contemporary status of, 438
 in the courts, 435–437
 harm from, 433–434, 438
 history of, 423–426
 overview, 419–421
 prominent AIDS deniers, 428–433
 seven principles of, 422–423
 tactics, 421–422
AIDS dissidents, 419–420
Alberta Reappraising AIDS Society, 424, 428
Albuterol, 157
Alchemy, 8–9, 12
Alcock, James, 378
Aldis, O., 173
Alert hypnosis, 336
Alien visitations to earth, xii
Alive & Well AIDS Alternatives, 424, 434
Al-Khalili, Jim, 245, 248
Allin, L., 178

Allopathic medicine, 12–14, 142, 163n3
Alloy, L. B., 66
Alternative medicine, 318. *See also* Complementary and alternative medicine (CAM); Integrative medicine
  harm from, 45–46
  history of, 13–14
Alvarez, Luis, 404
American Association for the Advancement of Science, 225
American Association of Naturopathic Physicians, 146, 152
American Board of Medical Specialties, 325
"The American Century," science and, 5
American Consumer Product Safety Commission, 176
American Diabetes Association, 315
American Journal of Immunology, 288
American Journal of Modern Physics, 286
American Medical Association, 13, 14
American Psychologist (journal), 232
Americans, belief in pseudoscience, 46
American Society of Clinical Oncology, 325
Americans with Disabilities Act Amendments Act (2008), 362
AMIkids Personal Growth Model, 272
Ammirati, R., 68
Analytic thinking styles, 26
An Apple a Day program, 272
Anecdotal evidence
  in drug prevention research, 264
  pseudoscience and, xiv, 386
Anecdotal fallacy, anti-vaccine movement and, 200–203
Aneuploidy, cancer theory based on, 431, 432
Annual Convention for the Society of Scientific Exploration, 438
Another Look (organization), 428
Anthroposophic medicine, 313
Anthroposophy, 313–314
Anticipation frequency, hierarchy of the sciences and, 85–86
Anti-evolution pseudoscience, 25
Anti-HIV medications, 419
Antinuclear activists, use of pseudoscience, 288
Antiphobic effect, of risky play, 175
Anti-science beliefs, 22
Anti-vaccine movement, 45, 195–219, 239
  anecdotal fallacy and, 200–203
  appeal to motive, 209
  bad science or math and, 210–211
  cherry picking the data, 209–210, 446
  chiropractic medicine and, 407
  conspiracy theories and, 212–213
  deception and, 212–214
  demographics of adherents, 195
  emerging evidence and, 211–212
  facts about vaccines, 195
  false authority and, 203–206
  genetic fallacy and, 208–209
  middle ground and, 211
  moving the goalposts and, 206–208
  naturalistic fallacy and, 198–199
  naturopathic medicine and, 159–161
  nirvana fallacy and, 199–200
  overview, 195–199
  shifting the burden of proof, 200
  thimerosal and, 206, 312, 469–470
  Wakefield and, 11–12, 198, 202, 239, 242, 243, 412, 447
Anxiety
  generalized anxiety disorder, 181–182
  pathological, 179
  treatment for, 184–185
Apostasy, pseudoscience and, 471–473
Appeal to motive, anti-vaccine movement and, 209
Applied kinesiology, 154
Applied mathematics, pseudoscience and, 88–89
Applied Mathematics (journal), 303
Applied science, 89
Archimedes, 469
Arkani-Hamed, Nima, 246
Aronson, Elliott, xiii

Arxiv preprint server, 241
  paper on hydrino physics, 252–253
  papers on BICEP2 experiment, 249
  papers on OPERA, 244, 245, 247
Associative learning, causal illusions and, 59–63
Astrology, xii, 11, 89, 224–225, 407
Astronomy, 89
Atkins v. Virginia, 364
Atlanta Journal-Constitution (newspaper), 435
ATLAS Program, evaluation of, 267
Atomic bomb, 6
Attitude polarization, 28
  conformation bias and, 460–461
Australian Journal of Psychology, 379
Auterio, Dario, 244, 246
Authority, question of, 16
Author-pays model, 284, 291–292, 296
Autism
  anti-vaccine movement and, 446
  GM foods and, 446
  MMR vaccine and, 198, 202, 212
  thimerosol and, 312
Autobiographical memory, 451–452
Avoidance behaviors, 179
Avoidance strategy, worry and, 181
Ayurveda, 293
Ayyadurai, Shiva, 124

Background Imaging of Cosmic Extragalactic Polarization. *See* BICEP2 experiment
Backward conditioning, 61
Baker, Clark, 425
Ball, D. J., 173
Baloney Detection Kit (Sagan), 406
Banachek, 472
Bananas, antecedents of modern, 107
Banks, Nancy, 435, 437
Barberia, I., 56, 57, 59, 68
Bark, Toni, 203
Bartelme, Ricardo R., 313
Base-rate fallacy, 453–455
Base rate of event, 453

Bastyr University, 148, 151, 152, 154, 156, 158–159
Bauer, Henry H., 423, 425, 427, 428, 429–431
Bayes, Thomas, 454
Beall, Jeffrey, 301
Beard fallacy, xii
Beaumont Health System, 320
Bedford Level experiment, 443
Behavior
  causal-density bias and animal's, 55–56
  fear's impact on, 179
Behavioral norms, of "real" science, 240–241
Behavioral persistence, causal illusion and, 65
Bekoff, M., 174
Belief in fair world, causal illusions and, 66–67
Belief revision, 28–29
Beliefs
  competing mental models and, 27–28
  reevaluating long-held, 474–475
  search for evidence confirming (*see* Confirmation bias)
Belladonna, 157
The Bell Curve Debate (Russell & Glauberman), 362
The Bell Curve (Herrnstein & Murray), 361
Bem, Daryl, 92, 387–388
Berio, Luciano, 459
Bertolucci, Sergio, 247
Beth Israel Deaconness Medical Center, 320
Beyerstein, Barry, 332, 379
Bias, xiii. *See also* Confirmation bias
  cause-density, 55
  disconfirmation, xv, 401
  outcome-density, 55
  science and, xiii
Bias blind spot, xiii, 35–36
Biased assimilation, 28
Biased estimation of causality, 53–55
BICEP2 experiment, 240, 247–251
  commentary, 250–251
  OPERA experiment *vs.*, 249, 250
  overview, 256
  timeline, 248–249

Bigfoot, xii
Binet, Alfred, 353, 356
Binet-Simon test, 355–356
Biodynamic farming, 313
Biofield therapies, 318
Biology, teleological thinking and, 24
Bjorklund, D. F., 173–174
BlackLight Power, 252
Blackmore, Susan, 379, 388
Blanco, F., 56, 57, 59
Blinn, 175
Blogging, enforcing demarcation via, 296
Blueprints for Violence Prevention (University of Colorado), 264
Board certification for integrative medicine, 325
B.o.B., 441–442, 443
"Bodywork," 313
Bohannon, John, 223, 289, 294–295, 401
Bohr, Niels, 251
Bomb detection, 473–474
Boring, Edwin, 355
"Both sides" science journalism, 30
Bothwell, R. K., 340
Bought (film), 203
Boundaries of science, 402
Boundary work, 12–16, 289–290, 295–296
Bowlby, John, 174
Boyesen, M., 173
Boyle, Robert, 9
Braid, J., 336
Brain, learning and, 47–48
Brain training, 364–366
Breast milk, claim of finding glyphosate in, 127–128
Brennan, Kelly, 420
Brewer, W. F., 28
Brilliant Light Power, 252, 254
British Journal of Psychology, 386
British Psychological Society, 379
Britton, Willoughby, 232
Brogan, Kelly, 198, 203–204, 208, 209–210
Bruner, J. S., 173

Bt cotton, Indian farmer suicide and, 445–446
Buchanan, T., 186
Buddhism, mindfulness and, 234
Bunge, Mario, 383–384, 385
Burden of proof, anti-vaccine movement and shifting, 200
Burnell, Jocelyn Bell, 248–249
Burnham, John C., 7
Buros Mental Measurement Yearbook, 352
Burt, Cyril, 401
Burzynski, Stanislaw, 145
Byers, J. A., 174

"Call for papers," 284–285
Cameron, Vinoo, 287
Cancer
  Gonzalez protocol, 324
  "integrative" treatment of, 310, 324–325
  naturopathic treatment of, 143–147
  PSA testing, 454–455
Cancer Research Journal, 287, 470
Cancer Research (journal), 470
Cancer Treatment Centers of America, 310
Canola, genetic engineering of, 103
Capital punishment. *See* Death penalty
Carlstrom, Jon, 249
Carnegie Institute First-Professional Degree, 152
Carrey, Jim, 242
Carroll, John, 357
Carson, Rachel, 5
Carter, Elliott, 459
Cartwright, Susan, 245
Case studies, 402–403
  for science education, 403–404, 412
Cattell, Raymond, 357
Cattell-Horn-Carroll (CHC) theory, 357–359
Causality, 89
  biased estimation of, 53–55
  correct estimation of, 50–53
  illusion of (*see* Illusion of causality)
  inferred, 458

Causal loop models in drug prevention research, 267–271, 275
Causal relationships, principles of, 50–53
Causation, confusion with correlation, 111–112
Cause-density bias, 55
　animal's behavior and, 55–56
Cause-effect relationship. *See* Causality
Center for Integrative Medicine at the George Washington University Medical Center, 312
Center for Substance Abuse Prevention (CSAP), 264
Centers for Disease Control and Prevention, 176, 212
CERN (European Organization for Nuclear Research), 244, 245, 246, 247
Chan, Moses, 242
Chariots of the Gods (von Däniken), 376
Chelation therapy, 323–324
Chelidonium, 147
Chemistry, teleological thinking and, 24
Chemtrail conspiracy theory, 288, 302–303
Cheney, Dick, 425
Cherry picking data, 470
　AIDS denialism and, 421
　anti-vaccine movement and, 209–210, 446
　confirmation bias and, 457
　in drug prevention research, 267, 274
　pseudoscience and, 386, 388
Chi, 138, 311
Chicken pox, 198–199
Chinese medicine. *See* Traditional Chinese medicine (TCM)
Chinn, C. A., 28
Chiropractic medicine, 407–408
Chronic Lyme disease, 139, 162–163n2
Citation concentration, hierarchy of the sciences and, 84–85
Citation immediacy, hierarchy of the sciences and, 86–87
Clarke, Arthur C., 404

Cleveland Clinic Foundation, integrative medicine and, 311, 312, 320
Climate change, group membership and opinion on, 34
Climate change denialism, 25–26, 239, 243
　confirmation bias and, 457–458
Climbing, benefits of, 172
Clinical training, in mindfulness, 233
Clinical trials, integrative medicine and, 322–324
ClinicalTrials.gov, 276
Coal fly ash conspiracy, 302–303
Coffee enemas, 324
Cognition
　cultural, 33–34
　motivated, 33
Cognitive Assessment System—2nd edition, 358, 359
Cognitive behavior manuals, to treat anxiety, 184–185
Cognitive bias. *See also* Bias; Confirmation bias
　paranormal beliefs and, 379
　reduction of, 68
Cognitive components, of scientific thinking, 23
Cognitive conflict, 27
Cognitive dissonance, 351
Cohen, Andrew, 245, 247
Coincidence, contingency and, 53–54
Cold fusion, 11, 95, 252, 257–258n1, 400
Cole, S., 84
Coleman, G. Baron, 435, 436–437
Colleges and universities
　academic and intellectual freedom at, 184
　IQ tests and admission to, 362
　mental health counseling at, 183–184
　need for conceptual challenges at, 185–186
Communication of scientific results, 241. *See also* Science publishing
Competitive ELISA (enzyme-linked, immunosorbent assay), 129–130

Complementary and alternative medicine (CAM), 45. *See also* Alternative medicine; Integrative medicine
  clinical trials, 322–324
  definition of, 315–316
  harms *vs.* benefits, 322–325
  money spent on, 473
  pseudoscientific articles on, 292–293
  subtypes, 317–319
Composite Intelligence Index (CIX), 355
Comte, August, hierarchy of the sciences, 79–81, 89
Conant, James, 402
Conditioning models, causal illusions and, 59–63
Confirmation bias, 453–460. *See also* Bias
  climate science debate and, 457–458
  death penalty opinion and, 460–461
  hierarchy of the sciences and, 87–88, 95
  pattern recognition and, 455–457, 459–460
  PSA testing and, 454–455
  pyramidology and, 458
  repetition and, 459–460
  us *vs.* them and, 460–461
Conspiracy theories
  AIDS denialism and, 421, 427, 436
  anti-vaccine movement and, 212–213
  coal fly ash, 302–303
  flat-earth, 441–442, 443
  pseudoscience and, 16–17
  publication of, 288
Consultation rate, hierarchy of the sciences and, 82–83
Contiguity
  causality and, 50–51
  Rescorla-Wagner model and, 61
Contingency
  causality and, 50–51
  computing, 51–53
  Rescorla-Wagner model and, 61
  sensitivity to manipulations of, 53
Contingency matrix, 51–53, 54
Continuum magazine, 424
Control, illusion of, 65–66
Conway, Erik M., 16
Core of a discipline, 83
Corn
  antecedents of modern varieties of, 107
  genetic engineering of field, 103
  hybrid, 108–109
  "Stunning Corn Comparison," 123–124
Correlation, confusion with causation, 111–112
Cosmology, 289
Cost, of pseudoscience, 473–474. *See also* Harm from pseudoscience
Coster, D., 173
Council on Naturopathic Medical Education (CNME), 151
Courts
  AIDS denialism in, 435–437
  hypnosis and courtroom testimony, 339–340
Covariation, causality and, 51
Craniosacral therapy, 316, 320
"Crank magnetism," 204
Creation science/creationism, 16, 89–90, 408–409
Creativity in Science (Simonton), 77–78
Creativity research, parsimony and, 94
Credentials, evaluating scientific claims and, 405, 406
Credibility contests, 15
Crick, Francis, 86
Critical thinking
  as defense against pseudoscience, 225–226, 475
  defined, 225
  reducing illusions of causality and, 69
  science education and, 412
Crop biotechnology, 103–104
  agronomic hybrid plants, 108–109
  formaldehyde claims and, 122–123, 124–125
  glyphosate claims and, 112–113, 117, 118–122, 123, 125–130
  harm of pseudoscience claims against, 106–107

Index

history of agriculture and, 107–108
labeling GM foods, 444–445
new technologies, 131
polyploidization, 109–110
pseudoscience examples, 118–130
pseudoscientific arguments against, 104–106, 110–118
radiation and, 109
safety of, 104
Crowe, David, 426–427, 428–429, 431
Cultural cognition, scientific thinking and, 33–34
Cultural factors, in scientific thinking, 32–35
Cultural groups, IQ test results and, 360–361
Cultural identity, scientific thinking and, 32–33
Cummins, Ronnie, 444
Curcumin, 161
Curie, Marie, 398
Curriculum, naturopathic medicine, 149–151
Cynicism, skepticism *vs.*, xv

DARE program, 264, 276n1
Darwin, Charles, 10, 89, 107, 398
Das, J. P., 359
Databases, predatory journals listed in, 293–295, 297
Davidson, R. J., 231
Davis, Craig Lamar, 435–436
Dawkins, Richard, 321
DBT (Dialectical Behavior Therapy), 229
Death penalty
  confirmation bias and opinion on, 460–461
  IQ test scores and, 360, 364
Deception, anti-vaccine movement and, 212–214
Decision making, causal illusion and, 63–64
Declaration of Helsinki, 321–322
"Deep time," 26
Deferential belief, 27
Demarcation, 3, 289, 383
  boundary work, 295–296
  payments from authors and, 291–292
Demarcation fallacy, 7–12
DeNiro, Robert, 447

Dent, C. W., 269, 270
Denying AIDS: Conspiracy Theories, Pseudoscience, and Human Tragedy (Kalichman), 420–421, 430
Department of Education, Exemplary and Promising programs, 264
Department of Health and Human Services, Office of Research Integrity, 401
Depopulation agenda, anti-vaccine movement claim of, 212
Depression and Anxiety (journal), 291
Depressive realism effect, 66
DesJardins, Julie, 14
Detoxification programs, naturopathic, 153, 154–155
Deutsch, Diana, 459
Developing world, genetically engineered crops in, 106
Development, risky play and, 171, 187
Dewhirst, B., 342
Diagnostic and Statistical Manual, Fifth Edition, 364
Digitalis, 157
Digital object identifiers (DOIs), 296
Digital preservation, 285
Digoxin, 157
Dimidjian, Sona, 228, 232
Dirac equation, 257
Discipline
  core of, 83
  research frontier of, 83
Disconfirmation bias, xv, 401
Discover magazine, 425, 472
Disease, naturopathic medicine and root cause of, 148–153
Dissemination of scientific results, 241
Divination, 7
Dogmatism, 234
"Do GMOs Accumulate Formaldehyde and Disrupt Molecular Systems Equilibria?" (Ayyadurai), 124
Domain specificity of scientific thinking, xiv
Donnell, Robert W., 310

Douglas, Karen, 212
Doyle, Arthur Conan, 472
Doyle, Jean, 472
Drisko, Jeanne A., 312
Drug abuse, AIDS and, 422, 423
Drug Abuse Resistance Education (DARE) program, 264, 276n1
Drug prevention research, 263–281
   flexible data analysis and selective reporting in, 267–271
   minimal adherence to study design criteria in, 271–274
   National Registry of Effective Prevention Programs, 264–266, 274–276
   overview, 263–264
   Project Towards No Drug Abuse (Project TND), 267–271
   pseudoscience in, 266–274
DuBreuil, S. C., 342
Dudukovic, N. M., 452
Duesberg, Peter, 420, 424, 425, 427, 431–433, 434, 436
du Maurier, George, 332
Dunbar, K., 23
Duncan, David Ewen, 224
Durban Declaration, 424

EBSCO, 293
Echinacea, 161
Economic analysis of vaccines, 195
Edison, Thomas, 5, 159
Education, to reduce illusions of causality, 68–69. *See also* Science education
Edwards, Mike, 472
Efficacy, mindfulness and, 232–233
Efficiency, heuristics and, 63–64
Einstein, Albert, 241, 398
   attack on, 289
   dissertation, 81
   parsimony and, 90
   Piaget and, 22
   pseudoscience attacks on theory of, 77, 78, 95, 286–287

Elderberry plant, 157
Electromagnetic-based therapies, 318
Elements, play near dangerous, 172
El Naschie, Mohamed S., 286–287
Elsevier, 283–284, 293, 303, 425
Emerson, Ralph Waldo, 472–473
Emotion, anti-vaccine movement and appeals to, 201
Emotional avoidance, fear and, 179
Emotional states, need for control and, 66
"Empirical Data Confirm Autism Symptoms Related to Aluminum and Acetaminophen Exposure" (Seneff), 296
Employment, IQ tests and, 362
Encapsulated delusional disorders, 421
Endocrine disrupters, glyphosate and, 112–113
Endocrine Disruptions and Cytotoxicity of Glyphosate and Roundup in Human JAr Cells In Vitro (Young), 113
"Energy medicine," 310, 311, 316, 318, 475
Ennis, Robert, 225
Entropy (journal), 296
Environmental Toxicology and Chemistry (journal), 126
Environment of evolutionary adaptedness (FEA), 174
Epistemological development
   critical thinking and, 225–226
   four-stage model, 226–227
   pseudoscience vulnerability and, 226–227, 233–234
Equal Employment Opportunity Commission, 362
Eratosthenes, 469
Ereditato, Antonio, 244, 246
Erickson, Milton, 336
Ernst, Ezard, 313
Essentialist thinking, 25–26
Eugenics, 361–362
Europeans' belief in pseudoscience, 46
Evaluative stage of epistemological development, 227

Evidence-based medicine, blind spot of, 321–322
Evidence-based practice. *See* Drug prevention research
Evolution
  causal illusion and, 67
  history of evolutionary science, 10–11
  naturalism and, 89–90
  parsimony and, 93–94
Evolutionary function of play and risky play, 173–176, 187
Evolutionary psychology, 11
Exercise therapies, 318, 323
Expectancies, hypnosis and, 334–335
Experimental Physiology, 386
Expertise, 35
  evaluating, 30–32
Experts
  anti-vaccine movement and dubious, 203–206
  pseudoscience and dismissing, 117–118
Exposure therapy, 180, 185
Extrasensory perception (ESP), 376
Eye Movement Desensitization and Reprocessing, 267
Eysenck, Hans J., 366, 377

Fads and Fallacies in the Name of Science (Gardner), 378
Fair world, causal illusions and belief in, 66–67
False authority, anti-vaccine movement and, 203–206
False-positive errors, causal illusions and, 58–59, 64, 67
Falsifiability, xii
  assessing status of hypothesis and, 385
  as attribute of genuine science, 90–91
  pseudoscience and, 92
Falsifiability principle, 12
Fan, J., 338
Fanelli, D., 87–88
Farber, Celia, 425

Farm and Ranch Freedom website, 118
Fear
  function of, 179
  normal, 179
  phobias, 180
Federal Trade Commission, 366
"Feeling the Future: Experimental Evidence for Anomalous Retroactive Influences on Cognition and Affect" (Bem), 92–93
Fellow of American Board of Naturopathic Oncology (FABNO), 144–145
Fermilab, 246
Feynman, Richard, 398, 400
"First, do no harm," naturopathic medicine and, 139, 141–143
First-professional degrees, 152
Flat-earth conspiracy theories, 441–442, 443
Flat Earth Society website, 443
Flauger, Raphael, 249
Fleischmann, Martin, 11, 257n1, 399
Flim-Flam (Randi), 378
Florida Organic Growers, 120
Fluid-Crystallized Index (FCI), 355
Flynn effect, 360
Food and Drug Administration, 227
  homeopathic medication and, 157, 321
  naturopathic treatments and, 145–146, 157
Food intolerance, 155–156
Formaldehyde
  fear of, 130
  GE crops and, 123, 124–125
  pseudoscientific claims about, 122–123
"Foundations of Science" course (Sam Houston State University), 412
Fox sisters, 473
Framing effects, 32–33
Frankenfood Paradox, 111
Franklin, Benjamin, 331–332, 334–335
Franklin Commission, on hypnosis, 331–332, 334–335
French, Elijah Daniel, 214
French, Rachel, 214
Frequency of cause, causal illusion and, 55

Frequency of effect, causal illusion and, 55
F1000Research (journal), 295
Freud, Sigmund, 86
Frontiers in Public Health (journal), 211, 302–303
Frontiers (publisher), 292, 302–303
"Full Scale" Composite Index (FS), 355
Full Scale IQ, 354, 355
Functional medicine, 311–312, 313
Fundamentalism, pseudoscience, 443–447

g (general factor of intelligence), 353–355
Gabora, N. J., 342
Galak, J., 92
Gale, 293
Galileo, 467
Gambia, AIDS denialism in, 425, 426
Ganzfeld techniques, 380–381
Gardner, Howard, 359
Gardner, Martin, 378
Gc protein-derived macrophage activating factor (GcMAF), 287–288
Geller, Uri, 376–377, 378, 472
Gelman, S. A., 27
General Conceptual Ability (GCA), 355
General Intellectual Ability (GIA), 355
Generalized anxiety disorder, 181–182
  treatment of, 184–185
Genetic engineering (GE). *See* Crop biotechnology
Genetic fallacy, anti-vaccine movement and, 208–209
Genetics, intelligence and, 360–362
Getting lost, risky play and, 172
Gieryn, Thomas, 14, 15, 289, 295
Gilbreth, Frank and Lillian, 14
Gillespie, Graeme, 377
Giudice, Gian, 245
Glaser, Elizabeth, 433
Glashow, Sheldon, 245, 247
Glauberman, Naomi, 361–362
Gleave, J., 173
Gleiser, Marcelo, 251

Global Institute for Research & Education, 285
Glymour, Clark, 310, 325
Glyphophobia, 130
Glyphosate, 296
  in breast milk, 127–128
  claim causing microfungus, 118–122
  competitive ELISA of, 129–130
  detection of, 125–127, 129–130
  estrogen production and, 112–113
  GE corn and, 123
  pseudoscientific claims about, 112–113, 117, 118–122, 123, 125–130
  in urine, 128
  in wine, 129
GMO. *See* Crop biotechnology
Gold open-access model, 284
Goldsmiths College, Anomalistic Psychology Research Unit, 381
Gonzalez, Nicholas, 324
Gonzalez protocol for pancreatic cancer, 324
Goodson, Patricia, 302
Google Scholar, 287, 289, 293
Gordin, Michael, 17
Gorski, David, 199, 296
Government
  AIDS denialists' distain for, 421, 432
  approval of pseudoscience by indexing it in database, 297
  funding research in integrative medicine, 314
Goyal, M., 229
"Grand Unified Theory of Classical Quantum Physics" (Mills), 252
Grant, Richard, 441
Graph prominence, hierarchy of the sciences and, 82–83
Gray, P., 183
The Greater Good (film), 203
Greatorex, P., 174
Great Parking Debate exercise, 403
GreenMedInfo, 126, 127, 199, 207
Gromet, D. M., 32–33

Group for the Scientific Reappraisal of the HIV-AIDS Hypothesis, 424
Group membership, scientific thinking and, 33–34
Gruzelier, John, 332
The Guardian (newspaper), 441
Guillotin, Joseph-Ignace, 331–332

Haidt, Jonathan, 183, 184–185
Hallucination,, hypnosis and, 337
Hammill, D. D., 363
Hanaway, Patrick, 312
Hanson, Joe, 406
Hard sciences, objectivity in, 95
Harm from pseudoscience, 473–474
  of AIDS denialism, 433–434, 438
  of CAM, 45–46, 322–325
  on crop biotechnology, 106–107, 131
  explaining in science education, 409–410
  from homeopathy, 45, 46
Harris, Dan, 230–231
Harris, Sam, 208
Hart, Vani, 447
Harvard Educational Review, 361
Harvard Smithsonian Astrophysical Data Service, 248, 253
Hawking, Steven, 398
Hawthorn, 138
Healing power of nature, naturopathic medicine and, 143–147
Healing principles, naturopathic, 138–141
Healing touch, 311
Health, pseudoscience and decisions related to, 45–46
Health (journal), 301
Health of Business Business of Health (HBBH) program, 154
Heights, risky play and, 172, 174, 175
Helicopter parenting, 186
Helium, supersolid, 242
Hemophilia, AIDS and, 423
Henry, D. C., 335
Herbalism/herbal medicine, 139, 157
Herd immunity, 197
Herndon, J. Marvin, 303
Herreid, Clyde, 233
Herrington, S., 176
Herrnstein, R. J., 361
Heuristics, xv, xi
  causality and, 54
  efficient decision making and, 63–64
  usefulness of, 46–47
Hewitt, E. C., 338
Hidden observer, hypnosis and, 337–338
Hierarchy of the sciences, 79–81
  confirmation bias, 87–88
  consultation rate and graph prominence, 82–83
  empirical indicators of placement, 81–88
  falsifiability and, 92
  multiples probability and anticipation frequency, 85–86
  naturalism and, 92–93
  objectivity and, 94–95
  obsolescence rate and citation immediacy, 86–87
  parsimony and, 93–94
  peer-evaluation consensus, early impact rate, citation concentration, a *h* index, 84–85
  positivism and, 96
  pseudoscientific beliefs within, 91–92
  pure *vs.* applied sciences and, 89
  related hardness and paradigm development, 81–82, 85
  superstring theory and, 95–96
  theories-to-laws ratio and lecture fluency, 83–84
Hilgard, Ernest, 337–338
Hill, Colin, 249
h index, hierarchy of the sciences and, 84–85
Hippocrates, 309
HIV conspiracies, 419
HIV skeptics, 420. *See also* AIDS denialism
HIV testing, questioning validity of, 435–437
HIV vaccine, 213
Hobbes, Thomas, 78

Holistic medicine, 13, 310, 325
Holmes, Oliver Wendell, 13
Homeopathic Pharmacopeia, 321
Homeopathic teething tablets, 157
Homeopathy, 13, 314
  in academic medical centers, 310
  harm from, 45, 46
  integrative medicine and, 317–318, 319, 320–321, 322
  naturopathic medicine and, 139, 161, 320–321 (*see also* Naturopathic medicine)
  perceived as harmless, 57, 58–59
  perception of effectiveness of, 56–59
Homeostasis, 47–48
Homoeopathy (journal), 283–284
Honeycutt, Zen, 130
Hooker, Brian, 202
Horn, John, 357
Houdini, Harry, 472
House of Numbers (film), 425
"How Much Can We Boost IQ and Scholastic Achievement?" (Jensen), 361
How Superstition Won and Science Lost (Burnham), 7
"How to Read Science News" (Hanson), 406
How to Think about Weird Things (Schick & Vaughn), 412
HPV vaccine, 201, 207
Huber, Don M., 118–122, 128
Huffington Post, 231
Hull, Clark Henry, 333
Humility, evidence-based science, pseudoscience and, 468
Humphries, Suzanne, 203
Hutter, M., 355
Hybrid plants, 108–109
Hydrino physics, 251–254
  history of, 251–252
  overview, 257
  as pseudoscience, 252–254
HydroCatalysis, 252
Hydrogen. *See also* Hydrino physics
  Bohr's model, 251

Hyman, Mark, 311–312, 447
Hyman, Ray, 379
Hypnosis, 331–349
  defined, 333
  hypnotic phenomena, 335–342
  is hypnosis a trance state?, 333–335, 343–344
  myth of, 331
  popular media and, 332–333
  psychotherapy and, 343
  stage, 332–333
Hypnosis Black Secrets website, 332, 334
Hypnotic age regression, 341
Hypnotic inductions, 333–335
Hypnotic phenomena, 335–342
  distant memories and, 341
  hidden observer and, 337–338
  literalism and, 336–337
  optokinetic reflex and, 339–340
  past-life memories and, 341–342
  recent memory and, 340
  sleep-like state and, 336
  Stroop effect and, 338–339
  trance logic and, 337
Hypnotic stare, 339
Hypotheses, falsifiability of, 385
Hypothesis myopia, 401
Hypothesis testing, 87–88

Iatrogenic effects, system research and, 266–267
ICARUS collaboration, 245–246
"If Google Was a Guy" (collegehumor.com), 209
IgG food allergy panel, 155–156
Illusion of causality, 54–55
  animal's behavior and, 55–56
  attempts to reduce, 68–69
  belief in pseudoscience and, 46, 47
  benefits of, 63–67
  conditioning models and, 59–63
  factors that produce, 55–56
  machines and, 61–63
  survival benefits, 67

Index

Illusion of control, 65–66
Image transparency response, hypnosis and, 337
Imaginative suggestibility, hypnosis and, 334
Imaginative suggestions, hypnosis and, 333
iMedPub Journals, 294
Impact rate, hierarchy of the sciences and early, 84–85
Inbreeding depression, 108
Indexing, of predatory journals, 293–295, 297
Individual factors in scientific thinking, 24–29
Inferred causality, 458
Information-relating, critical thinking and, 225
Information seeking, critical thinking and, 225
Injuries, from risky play, 176–177, 187
Inoculation effect, pseudoscience and, 441–449
Inquiry labs, for students, 411
Institute of Electrical and Electronic Engineers, 303
Integrative Cancer Therapies (journal), 144
Integrative medicine, 309–329. *See also* Complementary and alternative medicine (CAM)
 blind spot of evidence-based medicine, 321–322
 future of medicine, 325
 harms *vs.* questionable benefits of, 322–325
 overview, 315–319
 problem with, 319–321
 pseudoscience in medical academia and medicine, 310–314
Integrative oncology, 310, 324–325
Intellectual freedom, 184
Intelligence, defined, 355
Intelligence tests, 351–373
 alternative theories of intelligence, 359–360
 definition of intelligence, 355
 Flynn effect, 360
 genetics and intelligence, 360–362
 intelligence as fixed construct, 356–357

intelligence as quotient, 355–356
intelligence defined as singular construct, 353–355
overview, 351–353
pseudoscience and, 362–366
real-life outcomes and, 362
structures of intelligence, 357–359
theories of intelligence, 357–362
widespread use of, 352–353
Intelligent Design, 16, 89–90, 409
Intentionality, teleological thinking and, 24–25
Intergovernmental Panel on Climate Change (IPCC), 34
Internal Family Systems (IFS), 272, 273–274
International AIDS Conference (2000), 420, 424
International AIDS Conference (2008), 425
The International Archives of Medicine, 223, 294
The International Ayurvedic Medical Journal, 293
International Journal of Environmental Research and Public Health, 288
An International Quarterly Journal of Research in Ayurveda, 293
Internet
 AIDS denialism on, 423
 anti-vaccine movement use of Internet surveys, 210–211
 attack on crop biotechnology and, 105
 claims about hypnosis on, 332
 IQ tests on, 366
 online science journals, 284
 past-life therapy, 342
 as source of scientific information, 31–32
Interrupted Case Method, 412
Intuition, xi
Intuitive thinking styles, 26
Inventing the AIDS Virus (Duesberg), 420, 424
In vitro data, overstepping and, 116–117
Involuntariness, hypnosis and, 334
IQ Festishism, 363
IQ Testing 101 (Kaufman), 352
IQ tests. *See* Intelligence tests

Jacoby, Russell, 361–362
Jammeh, Yahya, 425
Jenner, Edward, 195, 202
Jensen, Arthur, 361
Ji, Sayer, 199, 210
  published papers, 213–214
Jobs, Steve, 46
Journal of AYUSH: Ayurveda, Yoga, Unani, Siddha and Homeopathy, 293
Journal of College Science Teaching, 412
Journal of Depression & Anxiety, 291
The Journal of Infectious Agents and Cancer, 203
Journal of Parapsychology, 386, 388
Journal of Personality and Social Psychology, 92
Journal of Physics B: Atomic, Molecular and Optical Physics, 386
Journal of Psychical Research, 386
Journal of Scientific Exploration, 386, 429
Journal of the American Medical Association, 475
Journal of the American Society for Psychical Research, 388n1
Journal of the Society for Psychical Research, 388n1
The Journal of Translational Science, 211
Journals. *See also* Science publishing
  open-access, 78, 285, 286, 289, 293, 296
  peer-reviewed, 241, 301–305
  predatory, 284–290, 301–302
Junk science
  in predatory journals, 286
  pseudoscience *vs.*, 471
Just Label It, 445

Kabat-Zinn, John, 229, 230
KABC-II, 362
Kahan, D. M., 33
Kahneman, Daniel, 63
Kallio, S., 339
Kamin, L. J., 361
Kaufman, Alan, 352
Kaufman Assessment Battery for Children—2nd edition, 358, 359

Keely, Brian, 16, 17
Kelman, M., 363
Kennair, L. E. O., 175–176
Kennedy, Robert F., Jr., 312
Kepler, Johannes, 9, 89
Kerr, Catherine, 231, 232, 234
Kip, K. E., 272, 277n3
Klahr, D., 22, 23
Klein-Gordon equation, 257
Knowledge, social transmission of, 29
Koehler, J. J., 455
Koestler Parapsychology Unit (KPU), 380–381
Korpan, C. A., 31
Kovac, John, 248
Krauss, Lawrence, 245, 246, 248
Kuhn, D., 22, 23, 29
Kuo, Chao-Lin, 250

Labeling, of GMO foods, 444–445
Lady Tasting Coffee exercise, 403
Laetrile, 400
The Lancet (journal), 242
Landfield, K., 68
Langer, E. J., 65
Larry P. v. Riles, 352
Lasane, T. P., 28
Laudan, Larry, xiv
Lauristen, John, 424
Lavoisier, Antoine, 331–332
Law & Order: SUV (television program), 425, 434
Learned, John, 245
Learning, 48
  visual illusion and, 50
Learning disabilities, IQ scores and, 362–364
Lears, Jackson, 7
Leary, S. P., 28
Least-costly mistake, causal illusion and, 64–65
LeBoeuf, R. A., 92
Lecture fluency, hierarchy of the sciences and, 83–84

Lee, Adrian, 249
Legare, C. H., 27
Legg, S., 355
Leibnitz, Gottfried Wilhelm, 86
LeMoyne, T., 186
Lepper, Mark, 460, 461
Lester, G., 363
Lewontin, R. C., 8
Life Skills Training program, 267
Lightman, Alan, 400
Lilienfeld, S. L., 332
Lilienfeld, Scott O., 68, 228, 384, 385
Lind, James, 309
Linde, Andrei, 250
Linnaeus, Carl, 90
Linse, P., 406
Literalism, as marker of hypnosis, 336–337
Lobelia, 157
Loch Ness Monster, 412
  Bauer and, 428, 429, 431
Logical fallacy, pseudoscientific attacks on crop biotechnology and, 110–112
Lohr, J. M., 228, 385
Lord, Charles, 460, 461
Lovett, Ryan, 473
Luce, Henry, 5
Lukianoff, G., 183, 185
Lumosity 'brain training' program, 366
Lumos Labs, 366
Lumpy Rat Paper, 114–115
Luria, Alexander, 359
Lyme disease
  alleged sexual transmission of, 295
  chronic, 139, 162–163n2
Lynas, Mark, 122
Lynn, S. J., 228, 332, 385

Machines, causal illusions and, 61–63
MacLeod, C. M., 338–339
Madlala-Routledge, Nozizwe, 425
Magazines, IQ tests in, 366
Maggiore, Christine, 424, 425, 427, 432, 433–434, 438

Magic community, exposing fraud and, 377, 378, 388, 472–473
Majewski, M. S., 126–127
Malignant denial, 438
Manicavasagar, V., 174
Manipulative and body-based practices, 317, 318
Manufactured risk, crop biotechnology and, 105
Marano, Hara Estroff, 183, 184–185
March Against Monsanto (MAM), 442
March Against Myths (MAMyths), 442
Marek-Crnjac, Leila, 287
Margolis, Lynn, 404
Margulis, Elizabeth, 459
Margulis, Jennifer, 198–199
Marketing, of naturopathic medicine, 153–154
Mars, microbial life on, 242
Marsh, E. J., 452
Maskell, James, 211–212
Massage therapy, 315
Mathematics, pseudoscience and, 78–79, 88–89, 386
Matute, H., 56, 57, 59
Maxin, Noah, 473
May, Katie, 473
Mbeki, Thabo, AIDS denialism and, 423–426, 436
McCarthy, Jenny, 211, 242
McCormick, James, 474
McGuire, Shelley, 128
McNutt, G., 363
MDPI (publisher), 288, 296
Measles outbreaks, 197–199, 210, 213, 239
Media
  AIDS denialism and, 430
  coverage of mindfulness movement, 230–231
  distortions of science in, 227
  hypnosis and popular, 332–333
  predatory journals as information source for, 294
  pseudoscience and, 470–471

Medical academia, integrative medicine programs in, 310–314
Medical schools, integrative medicine and, 314
Medicine. *See also* Complementary and alternative medicine (CAM); Integrative medicine
  anthroposophic, 313
  evidence-based, 321–322
  functional, 311–312, 313
  history of, xi, 12–14, 309
  holistic, 325
  mind-body, 318
  orthomolecular, 312–313
  science-based, 309
Medimorec, S., 34
MedKnow (publisher), 293
Megavitamin therapy, 313
Memorial Sloan Kettering Cancer Center, integrative medicine at, 311
Memory
  autobiographical, 451–452
  hypnosis and distant, 341
  hypnosis and past-life, 341–342
  hypnosis and recent, 340
Memory hardening, 340
Menary, E., 342
Mental age, 355–356
Mental health counseling, at US colleges, 183–184
Mentalists, confirmation bias and base-rate fallacy and, 453
Mental models or schemas, competing, 26–28
Mental Processing Index (MPI), 355
Mental quotient, 356
Menzies, R. G., 175
Merchants of Doubt (Oreskes & Conway), 16
Merck, 209
Mercola, Joseph, 203, 204, 444, 447
Mercola.com website, 123
Merrill, Maud, 353, 356
Merton, Robert K., 81–82, 86
Mesh, Marty, 120

Mesmer, Franz Anton, 331
Mesmerism, 9, 10, 77
Metacognition, scientific thinking and, 23, 29, 35
Metacognitive strategies, critical thinking and, 225
Metacognitive therapy, 181
Metaworry, 181
Methylmercury, 470
Microbiome, 212
Microfungus, glyphosate and, 118–122
Military justice system, AIDS denialism and, 436–437
Mills, Randell, 252–254, 257
Milne, A. A., 395, 413
Mind and body practices, 317
Mind-body medicine, 318
Mindful America (Wilson), 228
Mindfulness-Based Cognitive Therapy (MBCT), 232
Mindfulness-Based Stress Reduction (MBSR), 229
Mindfulness Magazine, 230–231
Mindfulness movement, 227–233
  beyond the hype, 231–232
  media claims by high-profile mindfulness figures, 230–231
  realizing potential of, 234–235
  recommendations for field, 232–233
  state of the research, 229–230
Minorities, IQ tests and, 352, 360–361
MINOS experiment, 246
Miracle Detectives (television program), 451–453, 455–457, 459–460, 461
Miracles, investigating, 451–461
Misinformation age, xi
"Misleading chocolate study" (Bohannon), 294–295
Mistletoe extract, naturopathic cancer treatment, 145
MIT Technology Review, 224
MMR vaccine, 198, 202, 212
Moment Program, 271

Moms Across America, 123, 127–128, 129, 130
Monetary gain, from publishing pseudoscience, 287–288
Monsanto, 119, 445, 446
Morris, Bob, 380–383
Morris-Yates, A., 174
Motivated cognition, 33
Motivation
  authoring pseudoscience and, 287–289
  critical thinking and, 225
  hypnotic responsiveness and, 334, 335
Mousseau, Marie-Catherine, 386–387
Movies
  anti-vaccine movement, 203, 446, 447
  hypnotism in, 332
Moving the goalposts
  AIDS denialism and, 422
  anti-vaccine movement and, 206–208
Müller-Lyer illusion, 48–50, 54
Multiple comparisons issue, 458
Multiple intelligences, 359
Multiples probability, hierarchy of the sciences and, 85–86
Multisystemic Therapy, 267
Munro, G. D., 28
Murray, C., 361
Muscle testing, 154
Musical illusions, 459
Mutations, direction of, 93
Myths, pseudoscience and, 385

Naglieri, Jack, 359
Naming system, 90
NASA, xii, 242, 469
Nash, M., 341
National Academy of Sciences (NAS), 431
  on norms of scientific behavior, 399–400
National Cancer Institute (NCI), 314
National Center for Biotechnology Information, 293
National Center for Case Study Teaching in Science, 402, 403
National Center for Complementary and Integrative Health (NCCIH), 314, 321
  definition of alternative medicine, 315
  definition of integrative medicine, 316
National Center for Health Statistics, 473
National Center for Science Education, 408
National College of Natural Medicine, 148
National Institute on Drug Abuse (NIDA), 264
National Institutes of Health (NIH), 293
  integrative medicine research funding, 323–324
National Registry of Effective Prevention Programs (NREPP), 264–266, 274–276
  failure to ensure review quality, 274–276
  review of Accelerated Resolution Therapy, 272–273
  review of Internal Family Systems, 273–274
  review of Moment Program, 272–274
  review of Project TND, 267–271, 276n2
National Research Council, 21
National Science Foundation, Science and Engineering Indicators report, 224–225
National University for Natural Medicine, 154
National Vaccine Injury Compensation program, 202
Nattrass, Nicoli, 289–290
Natural, meaning of term, 107
Naturalism
  as attribute of genuine science, 89–90
  parsimony and, 90
  pseudoscience and, 92–93
Naturalistic fallacy, 25
  anti-vaccine movement and, 198–199
Natural laws, 89
Natural News website, 123, 212, 447
Natural philosophy, 9, 12
Natural products, 317
  complex, 319
Natural sciences
  in hierarchy of the sciences, 79–81
  naturalism and, 93
Nature (journal), 77, 250

Naturopathic education. *See also* Naturopathic schools
  content of, 156–157
  standards for, 157–159
Naturopathic medical doctor (NMD), 137
Naturopathic medicine, 137–169
  academic medical centers, 311
  cancer treatment, 143–147
  doctor as teacher, 153–156
  education for, 148–151
  "first, do no harm" and, 141–143
  healing power of nature and, 143–147
  history of naturopathic theory, 137–138
  homeopathy and, 139, 161, 320–321 (*see also* Homeopathy)
  licensing of practitioners, 152–153
  prevention and, 159–161
  root cause of disease and, 148–153
  six principles of, 138–141
  vaccines and, 140, 159–161
  whole-person treatment, 156–159
Naturopathic medicine doctorate degree (ND), 137, 151–152
Naturopathic oncology, 143–147
Naturopathic Physicians Licensing Examination (NPLEX), 152
Naturopathic schools, 148
  accreditation of, 151–152
  curriculum, 149–151
Naturopathy. *See* Naturopathic medicine
Naudts, Jan, 252–253
NBC News, 311
Neides, Daniel, 312
Nelson, L. D., 92
Neuroimaging, multiple comparisons issue in, 458
Neurolaw, 32–33
Neutrality, as default position, 471
Neutrinos, faster-than-light, 243–247, 255, 400, 405
New England Journal of Medicine, 310
New Scientist (magazine), 377
Newton, Isaac, 9, 86, 398

Newton, Joe (Seth Kalichman), 426–428
New Yorker (magazine), 445
New York Times Magazine, 351
Nicholls, J., 176
Niebuhr, Reinhold, 6
Nietzsche, Friedrich, 468, 476
Nightline (ABC), 230
The Nightly Show (television program), 441
Nirvana fallacy, anti-vaccine movement and, 199–200
Nisbet, E. C., 33
Nongovernmental International Panel on Climate Change (NIPCC), 34
Norris, S. P., 30, 31
Novella, Steven, 422
Nuance, scientific detail and, 448
Nuclear fusion, 89
Nutrition, naturopathic medicine and, 154–155
Nutritional therapeutics, 318
Nuzzo, R., 401
Nye, Bill, 239

Obama, Barack, 444
Objectivity
  as attribute of genuine science, 91
  pseudoscience and, 94–95
Obsolescence rate, hierarchy of the sciences and, 86–87
Occam's razor, 90, 93, 342
O'Connell, D., 174
Office of Cancer Complementary and Alternative Medicine (ACIMH), 314
Office of Medical and Scientific Justice (OMSJ), 425, 435, 436–437
On Being a Scientist (NAS), 399–400
One-off reports, pseudoscience and, 113–115
One-tailed tests, 263, 276n2
Online science journals, 294–295
Open-access journals, pseudoscience and, 78, 285, 286, 289, 293, 296
Open-access publishing model, 284, 285, 291–295, 296

Open Journal of Pediatrics, 288
OPERA experiment, 240, 243–247
  BICEP2 experiment *vs.*, 249, 250
  commentary, 246–247
  overview, 255
  reactions to, 244–245
  timeline, 244–246
Operationalization, 355
Oppenheimer, Robert, 6–7
Oprah Winfrey Show (television program), 341, 342
Optokinetic reflex, hypnosis and, 339–340
Orac (David Gorski), 296
Oreskes, Naomi, 16
Organic Consumers Association, 444
Organic food producers, campaign against GM foods, 444–445
The Origin, Persistence, and Failings of HIV/AIDS Theory (Bauer), 425, 429–430
Orne, M. T., 337
Orthomolecular medicine, 312–313
Orthorexia, 227
Oscillation Project with Emulsion-tRacking Apparatus. *See* OPERA experiment
Outcome-density bias, 55
Overbye, Dennis, 247
Overparenting, 186
Overstepping results, pseudoscience and, 115–117
Oz, Mehmet, 341, 342, 446

Padian, Nancy, 437
Paenzee, Chad, 435
Pakistan, predatory journals and, 285
Palevsky, Lawrence "Larry," 203, 204, 206
Palm reading, 7
Pancreatic cancer, Gonzalez protocol for, 324
Papagiannido, Maria, 425
Papaya, genetically engineered, 104, 106
Paradigm development, hierarchy of the sciences and, 81–82, 85
Paranormal, 376

Parapsychology, 375–391
  author's early interest in, 375–378
  questionable research practices, 387–388
  science, nonscience, and pseudoscience, 383–385
  scientific status of, 385–388
  skepticism about, 378–379
  Type II skeptics, 380–383
  Type I skeptics, 379–380
Parapsychology: Science or Magic? (Alcock), 378
Parental worry, risky play and, 180–181
Parenting, helicopter, 186
Parenzee, Andre Chad, 425
Parsimony, 22, 35
  as attribute of genuine science, 90
  past-life experiences and, 342
  pseudoscience and, 93–94
  superstring theory and, 95–96
Pasquarelli, David, 424
Passive Magnetic Resonance Anomaly Mapping (PMRAM), 474
Pasteur, Louis, 398
Past-life therapy, 342
Pattern detection, adaptive bias in, 47–50
Pattern recognition, meaning and, 455–457, 459–460
Pauli, Wolfgang, 470
Pauling, Linus, xiii, 86, 312
Pavlov, Ivan, 59
Pay-to-publish journals, 285–286, 296
pdf versions of unpublished manuscripts, pseudoscience, 78
Pearl, Sharrona, 10
Peer-evaluation consensus, hierarchy of the sciences and, 84–85
Peer review, 397–398
  AIDS denialism and, 430
  post-publication, 295
  predatory journals and, 285–290
  pseudoscience and avoidance of, xiv
Peer-reviewed journals, 241
  pseudoscience in, 301–305
Peer-reviewed outlets, 31

Peer review report, 290–291
Pellegrini, A. D., 171, 173–174
Penn and Teller, 472
Pennycook, G., 34
Perfect solution fallacy, anti-vaccine movement and, 199–200
Performance hypnosis, 332–333
Perry, W. G., 226
Personality theories, 94
Personal level of thinking, 226
Perspective, 469
Perth Group, 424, 425
Pertussis vaccine, 209–210, 213
Pew Research Center, American Trends Panel, 224
Pharmaceutical companies, vaccines and, 208–209
Pharmacological and biologic treatments, 319
PharmaCompass.com, 208
Phillips, Jonathan, 252
Phillips, L. M., 31
Phillips, Peter, 472
Phipps, James, 195
Phobias, 180
    treatment of, 184–185
Phrenology, 9, 77, 352
Physical sciences in hierarchy of the sciences, 79–81
Physician's Association for Anthroposophic Medicine, 313–314
Physiognomy, 9, 10
pi, 287
Piaget, Jean, 22, 86, 226
Pigliucci, Massimo, 6
Piltdown man, 88, 401
Placebo effect, 45, 56
Plain Dealer (newspaper), 311
Planck, Max, 397
Planck satellite experiment, 248, 249, 250–251, 256
Planning, Attention, Simultaneous, and Successive (PASS) theory, 358, 359
Plant/Animal Genome meeting, 121–122

Play. *See* Risky play
Playground safety, 171, 176–177
Poison by Prescription: AZT (Lauristen), 424
Poisson distribution, 86
PolarBear, 249
Political affiliation, framing and, 33
Polyploidization, 109–110
Pons, Stanley, 11, 257n1, 399
Pop-culture science myths, 22
Popoff, Peter, 472
Popper, Karl, xii, 12, 90, 385
Positivism, hierarchy of the sciences and, 96
Posner, M. I., 338
Post-publication peer review (PPPR), 295
Poulton, R., 175
Poverty, AIDS and, 422, 423
Prana, 138
Preclinical studies, 322
Precognition, 387–388
Predatory journals, 284–290, 301–302. *See also* Science publishing
"Pregnancy-Friendly Protection? The Truth About Whooping Cough Vaccine" (Brogan), 209
Primordial gravitational waves, 247–251, 256
Principe, Lawrence, 9
Priority
    causality and, 50–51
    Rescorla-Wagner model and, 61
Progress in Physics (journal), 289
Prohaska, Meredith, 201–202
Project ALERT, 267
Project Towards No Drug Abuse (Project TND), 267–271, 276n2
Promiscuous teleology, 24
Prostate cancer screening, 454–455
Pryke, Clem, 249
Pseudo-medicines, 45
    causal illusion and, 64–65
    perception of effectiveness of, 56–59
Pseudoscience. *See also individual topics in pseudoscience*
    alternative publication venues, 78

apostasy and, 471–473
attacks on glyphosate and Roundup, 112–113, 117, 118–122, 123, 125–130
boundary work and, 12–16, 289–290, 295–296
combating, 447–448
confusing correlation and causation, 111–112
on continuum with science, xiv, 384
crop biotechnology and harm caused by, 106–107, 131 (*see also* Crop biotechnology)
culture of change and, 7
defenses against, 475
defined, 22
demarcation fallacy and, 3, 7–12
detecting, 410
dismissing experts and, 117–118
in drug prevention research (*see* Drug prevention research)
erroneous claims distinguished from, xv
etymology of, 17
exposing fraud, 472–473
features of, 228
fundamentalist, 443–447
harm from (*see* Harm from pseudoscience)
health-related decisions and, 45–46
within the hierarchy of the sciences, 91–92
as historical development, 3–4, 8–10
hydrino physics and structure of, 251–254, 257
illusion of causality and, 46, 47
increase in belief in, xii
inoculation effect and, 441–449
intelligence tests and, 362–366
junk science *vs.*, 471
list of features, 383–384
list of topics, 405–406
media and, 470–471
misappropriation of risk and, 125–127
motivations for authoring, 287–289
one-off reports and DOA science, 113–115
overstepping results and, 115–117
peer-reviewed journals and, 301–305
predatory journals and, 286–293, 296–297
reasons flourishing in United States, 467
reevaluating long-held beliefs and, 474–475
reinterpreting scientific results to fit agenda of, 112–113
science education and, 404–409
science *vs.*, xii, 88–91
threat from, 16–17
vulnerability to (*see* Pseudoscience vulnerability)
warning signs of, xiv
Pseudoscience vulnerability, 223–238
addressing, 233–325
critical thinking and, 225–226
epistomological development and, 226–227, 233–234
media distortions of science and, 227
mindfulness movement and, 227–233, 234–235
recommendations to combat, 232–233
science literacy and, 224–225
Pseudoscientific thinking, scientific thinking and, xv
Psi phenomenon, 92–93
Psychics, xii, 382, 453
Psychokinesis, xii, 376. *See also* Geller, Uri
Psychological essentialism, 25
The Psychologist (journal), 379
Psychology
drug prevention research and, 264
evolutionary, 11
in hierarchy of the sciences, 79–81
parsimony and, 94
social, 92
subjectivity and, 94–95
Psychology of science, 22
Psychotherapy, hypnosis and, 343
Public, scientists' relationship with, 34–35
PubMed, 209, 293
PubPeer, 296
Purdue University, Huber and, 119, 120–121
Pure and Applied Mathematics Journal, 287
Pure science, 89
Pyramidology, 458

"Quackademic medicine," 310
Quote mining, 209–210

Race, IQ test results and, 360–361
Radner, Daisie and Michael, 384, 385
Rain dances, 56
Randall, Lisa, 245
Randi, James (The Amazing Randi), 377, 378, 388, 472–473
Randomized controlled trials (RCTs), 316, 321–322, 324
Random noise, quantifying, 453
Rasnick, David, 423–424, 425, 436
Rated hardness, hierarchy of the sciences and, 81–82
Rath, Mathias, 423–424, 425, 436
Rathke, Andreas, 252, 253, 254
Raz, A., 338
Reagan, Nancy, 407
Reagan, Ronald, 398, 424
Reappraising AIDS website, 420
Redfield, Rosie, 242
Reductionism, 90
Rees, Martin, 248
Reflexology, 310, 316, 320
Registered report, 276
Regression to the mean, 365
Reiki, 311, 316, 319
Reinventing the AIDS Virus (Duesberg), 432
Remembering, 451–452
Rescorla-Wagner model, 60–61
  computer simulations of, 61–63
Research-Based Guide (NIDA), 264
"Researcher Roundup of Roundup-Ready Crops May Be Causing Animal Miscarriages and Infertility" (Huber), 118–119
Research frontier, of a discipline, 83
Research labs, for students, 411
Respectful Insolence (blog), 296
Response prevention, 180
Rethinking AIDS Society, 425, 426, 428, 434
Revue Française de Parapsychologie (journal), 386

Richards, Rodney, 436–437
Right Detox, 154–155
"Right to know" rhetoric, of anti-GM food movement, 444–445
Risk
  defined, 178
  pseudoscience and misappropriation of, 125–127
Risk compensation, 178–179
Risk-mastery skills, play and, 173
Risk perception, science of, 178–179
Risky play, 171–194
  attempts to regulate, 176–177
  benefits of, 172–173, 187–188
  categories of, 172
  complexity of play, 171–172
  evolutionary function of, 173–176, 187–188
  exaggerated emotional reactions, 184–185
  function of fear, 179
  generalized anxiety disorder, 181–182
  infantilization of young adults, 182–183
  learning to cope and, 185–186
  overparenting and, 186
  parents' worry and, 180–181
  phobias and, 180
  science of risk perception, 178–179
  student vulnerability and, 183–184
"Rochester rappings," 473
Rodin, Ervin, 303
Roosevelt, Franklin, 402
Root-Bernstein, Robert, 424
Rosa, Emily, 475
Rosenzweig, Laney, 272
Ross, Lee, 460, 461
Rough-and-tumble play, 172, 174, 175
Roulac, John, 447
Roundup. *See also* Glyphosate
  detection of, 125–127
  pseudoscientific attacks on, 113, 117, 118–122, 125–130
"Roundup Weedkiller in 75% of Air and Rain Samples, Gov't Study Finds" (GreenMedInfo), 126

Rudolf Steiner Health Center, 313
Rules-based knowledge, 227
Ruscio, J. B., 332
Ruse, Michael, 11

"Sacred Activism: Moving Beyond the Ego" (Brogan), 208
"Safe places," 183
Safety behaviors, 179
Sagan, Carl, 398, 401–402, 406, 441
Sam Houston State University, 411
Sandseter, E. B. H., 173, 174, 175–176
SAT test, 362
Scatterplot, ability to interpret, 224
Schiffrin, H. H., 186
Schizophrenia, 351
Schmidt, Casper, 424
Scholarly publishing. *See also* Science publishing
  decline in, 283–284
  dissemination of pseudoscience and, 296–297
Schopf, William, 242
Schrödinger equation, 257
Schwartz, Richard C., 273
Science
  activities of, 14
  American acceptance of, 5
  applications, 5, 6
  attributes of genuine, 89–91
  behavioral norms of, 240–241
  benefits of being within fold of, 15
  bias and, xiii
  boundary work and, 12–16, 289–290, 295–296
  canons of, 399–400
  on continuum with pseudoscience, xiv, 384
  defining for students, 396–398
  demarcation fallacy and, 3, 7–12
  emergence of, 4–5
  eroding confidence in, 5–6
  faith in scientific process, 410
  humility and, 468
  Oppenheimer on limits of, 6–7
  pseudoscience *vs.*, xii, 88–91
  psychology of, 22
  self-correction and, 242, 396–397, 468
  as social activity, 12
  as social institution, 8
  teaching limitations of, 397
  truth and, 468
  truth *vs.* warranted belief and, 17
  as way of understanding the world, 7
"Science and Pseudo-Science" *(Stanford Encyclopedia of Philosophy)*, 290
Science and Unreason (Radner & Radner), 384
Science-based medicine (SBM), 309
Science education, 21, 395–396
  active learning strategies, 411–412
  canons of science, 399–400
  case study teaching, 402–404, 412
  dealing with pseudoscience, 404–409
  defining science, 396–398
  on detecting pseudoscience, 410
  examples of pseudoscience, 407–409
  explaining harm of pseudoscience, 409–410
  on how scientists can go wrong, 400–402
  methods for, 411
  poverty of, 225
  pseudoscience vulnerability and, 233
  for public, 470
  reduction of teleological thinking and, 24
  sample case studies, 403–404
  studying distinguished scientists, 398
Science (journal), 289, 294
Science journalists, 29–30
Science literacy, pseudoscience vulnerability and, 224–225
Science Publications, 288
Science publishing
  author-pays model, 284
  boundary work and, 295–296
  open-access model, 284, 285, 291–295
  peer review and, 285–286, 289–290
  predatory journals, 284–290, 301–302
  pseudoscience and, 286–293, 296–297
  subscription publishing model, 283–284

Science Publishing Group, 286–287
Scientific community, role in evaluating scientific claims/discoveries, 397–398
Scientific discovery as dual search (SDDS) model, 23
Scientific error, 468
Scientific failure, 239–261
  BICEP2 experiment, 247–251, 256
  hydrino physics and the structure of pseudoscience, 251–254, 257
  OPERA experiment, 243–247, 255
  as opportunity for public education, 254–255
  as public good, 239–240
  structure of, 240–243
Scientific findings/discoveries
  dissemination of, 241 (see also Journals; Science publishing)
  reinterpretation of, 112–113
  replication of, 241
  role of scientific community in evaluating, 397–398
  scientists' response to, 3, 113–114, 242
  second-hand reports of, 31
Scientific Genius (Simonton), 77–78
Scientific illiteracy, 21
Scientific impotence discounting, 28
Scientific literacy, 21
  as defense against pseudoscience, 475–476
Scientific method, xiii, 8, 396
Scientific misconduct, 400–402
Scientific misinformation, 21
Scientific norms
  commandments for scientist interacting with society, 399–400
  commandments for scientist in the lab and field, 399
Scientific racism, 25
Scientific research, financing, 405
Scientific Research (publisher), 301
Scientific Research Publishing (SCIRP), 288
Scientific thinking, 22–29
  belief revision and, 28–29
  bias and, 35–36
  cognitive components, 23
  competing mental models or schemas and, 26–28
  cultural cognition and, 33–34
  cultural factors, 32–35
  cultural identity and, 32–33
  defined, 22–23
  domain-specificity of, xiv
  education about and reduction of causal illusions, 69
  essentialism and, 25–26
  evaluating expertise and, 30–32
  individual factors, 24–29
  intuitive and analytic thinking styles and, 26
  metacognitive components, 23, 35
  pseudoscientific thinking and, xv
  relationship between scientists and the public, 34–35
  social factors, 29–32
  teleological thinking and, 24–25, 35
  trust in testimony and, 30
  unnaturalness of, xiii–xiv
Scientists, relationship with public, 34–35
SCIgen, 303
Scopus (index), 293
Scott, Eugenie, 408–409
Scovill, Eliza Jane, 425, 434
Search engines, predatory journals listed in, 293–295, 297
Sears, Bob, 160, 211
Second-hand reports of scientific research, 31
Seeds of Truth (Shiva), 445–446
Segal, Z. V., 228
Segal, Zindel, 232
Segrin, C., 186
Selective reporting, in drug prevention research, 267–271
Self-correction
  OPERA experiment and, 247
  pseudoscience and lack of, 243
  science and, 242, 396–397, 468
Semmelweis, Ignaz, 404
Seneff, Stephanie, 296

Separation anxiety, risky play and, 174, 175
Seralini, Gilles Eric, 114–115, 447
Sewell, Granville, 303
Shadick, N. A., 273–274
Shakespeare, William, 468
Shapiro, T., 338
Shaw, Steve, 472
Sheehan, P. W., 338–339
Shen, F. X., 32–33
Shermer, M., 406
Shiva, Vandana, 443, 445–446, 447
Shockley, William, 361
Shtulman, A., 27, 28
Siegel, Dan, 230
Silent Spring (Carson), 5
Silove, D., 174
Simmons, J. P., 92
Simon, Henri, 353
Simon, Herbert, 63
Singh, Simon, 313
Single-study fallacy, AIDS denialism and, 421–422
Singletons, 85–86
Skeptical Inquirer (magazine), 378
Skepticism, cynicism *vs*., xv
Sleep-like state, hypnosis and, 336
Smallpox vaccine, 195, 202
Smith, P. K., 171
Smith, Tara, 422
Snyderman, Nancy, 311
Social factors, in scientific thinking, 29–32
Social media, anti-vaccine doctors and, 205, 206–207
Social phobia, treating, 184
Social physical play, 175
Social prejudice, essentialist thinking and, 25
Social psychology, pseudoscience and, 92
Social sciences
 in hierarchy of the sciences, 80–81
 parsimony and, 94
 status of, 14
Social skills, risky play and, 175

Society, norms for scientists interacting with, 399–400
Society for Integrative Oncology, 324–325
Society of Scientific Exploration, 429, 438
Sociobiology, 11
Sociology, in hierarchy of the sciences, 80, 81
Socrates, 476
Sokal hoax, 303
"Sometimes behave so strangely" illusion, 459
South Africa, AIDS denialism in, 6, 423–426
South Pole Telescope, 249
Southwest College of Naturopathic Medicine and Health Sciences, 148, 153, 157
Soybeans, genetic engineering of, 103
The Space Gods Revealed (Story), 376
Spanos, N. P., 338, 342
Spearman, Charles, 353
Special education, IQ testing and, 352
Speed, risky play and, 172, 174
Spergel, David, 249
Spiritual therapies, 319
Springer (publisher), 303
Spurgeon, C. H., xi–xii
Squaring the circle, 78–79
Stage hypnosis, 332–333
Stalker, Douglas, 310, 325
Stanford-Binet Intelligence Scales, Fifth Edition, 366
Stanford-Binet Intelligence test, online, 366
Stanford Encyclopedia of Philosophy, 290
Stanovich, Keith, 360
Steblay, N. M., 340
Steiner, Rudolf, 313
Steinhardt, Paul, 250
STEM, declining interest in, 21
Stephenson, A., 174
Stereotypes, confirmation bias and, 460–461
Stern, William, 356
Sternberg, Robert, 359
Stewart, Potter, 8
Sting operations, on publishers, 289, 294–295
Stokely, Karri, 425, 434, 438
Story, Ronald, 376

Story-telling, remembering and, 451–452
Strengthening Families Program, 267
Stressors, exposure to, 185
Stroop interference effect, hypnosis and, 338–339
Structures of intelligence, 357–359
Students. *See also* Science education
   mental health counseling and, 183–184
   trigger warnings and infantilization of, 182–183
   vulnerability to emotional distress and disorder, 183–184
"Stunning Corn Comparison," 123–124, 127
Stutz, E., 173
Subjective perception of risk, 178
Subjectivity, psychology and, 94–95
Subscription publishing model, 283–284
Substance Abuse and Mental Health Services Administration (SAMSA), National Registry of Effective Prevention Programs, 264–266, 275–276
Sugar beets, genetic engineering of, 103
Suggestibility effects, hypnosis and, 340
Suggestion effect, 339
Suleiman, Ramzi, 289
Sun, W., 269
Supersolid, 242
Superstition, confirmation bias and, 460
Superstring theory, parsimony and, 95–96
Supplements
   anti-vaccine movement and, 204
   efficacy of, 323
   naturopathic sale of, 154
   orthomolecular medicine and, 313
Sussman, S., 269–270, 271
Sutton-Smith, B., 171, 172, 173
Symptom relief, 323
System research, pseudoscience in drug prevention research, 266–267

TACT (trial to assess chelation therapy), 323–324
Tai chi, 323
Tavris, Carol, xiii
Taylor, Emery, 425
Taylor, John, 376
Teleological thinking, scientific thinking and, 24–25, 35
Telepathy, 377, 380–381
Tenpenny, Sherri, 203, 204
10 Percent Happier (Harris), 230
Terman, Lewis, 353, 356
Testing methodologies, brain training, 365–366
Thales, 89
Theistic evolution, 11
Theories-to-laws ratio, hierarchy of the sciences and, 83–84
Therapeutic touch, 316, 475
Thimerosal, 206, 312, 469–470
Thimerosal (Hyman & Kennedy), 312
Thiotepa, 147
Thom, Dickson, 154
Thomas, Paul, 203, 204
Thomson Reuters, 293
Time magazine, 5
Timm, H. W., 340
Tobacco, genetic engineering of, 110
Tomatoes, antecedents of modern, 107
"Tongue diagnosis," 311, 320
Tools, risky play with, 172
"Toxins gambit," 199
Traditional Chinese medicine (TCM), 311, 320
Trance logic, hypnosis and, 337
Traub, Michael, 146
Triarchic theory of successful intelligence, 359
Trick or Treatment (Singh & Ernst), 313
Trigger warnings, 182–183
Trilby (du Maurier), 332
Trust in experts, 35
Trust in testimony, 30
Truth, science and, 17, 467–468
The Truth about Uri Geller (Randi), 378
"Truth" (Emerson), 472–473
Truzzi, Marcello, 398
Tshabalala-Msimang, Manto, 425

Turner, John, 436
Tuskegee Institute syphylis project, 6, 404
Tversky, Amos, 63
Tversky, B., 452
Twain, Mark, xii
"200 Evidence-Based Reasons NOT to Vaccinate" (Ji), 213
Two-tailed test, 271, 276n2
Type II diabetes, 315
Tyson, Neil deGrasse, 212, 239, 398
  disproving flat-earth theory, 441–442, 443

Ukrain, 146–147
Undergraduate major, logical structure of, 82
Understanding Science website, 230, 396, 397
US Department of Education, 151, 152
US Department of Energy, 474
US National Safety Council, 171
US Public Health Service, 404
US Right to Know, 445
Universities. *See* Colleges and universities
University of California, Berkeley
  Duesberg and, 431
  Understanding Science website, 230, 396, 397
University of Colorado, 264
University of Edinburgh, Koestler Parapsychology Unit, 380–381
University of Kansas Medical Center Integrative Medicine Program, 312
University of Maryland at Baltimore, Center for Integrative Medicine, 313
University of Michigan, Holistic Family Practice Medicine Program, 313
University of Pittsburgh, 320
"The Unreported Health Benefits of Measles" (GreenMedInfo), 199
Urine, glyphosate in, 128
Us *vs.* them, confirmation bias and, 460–461
Uzick, Michael, 146, 147

Vaccine Book (Sears), 160
Vaccine neutral naturopaths, 159–160
Vaccines. *See also* Anti-vaccine movement; *individual vaccines*
  economic analysis of, 195
  facts about, 195
  naturopathic medicine and, 140, 159–161
The Vaccine Wars (Margulis), 198
VacTruth website, 214
Valcarcel, J., 27
Vanity presses, pseudoscience and, 78
Vavilov, Nikolai, 107
Vaxxed: From Cover-Up to Catastrophe (film), 203, 446, 447
Vilsack, Tom, 118–119, 121
Vioxx, 209
Virginia Polytechnic Institute and State University, Bauer and, 428, 429
Vis, 138
Visual illusions, 48–50, 54
Vitalism
  integrative medicine and, 317, 319
  naturopathy and, 137–138
von Adrian, Ulrich, 214
von Däniken, Erich, 376, 385

Wait-list control design, 231
Wakefield, Andrew, anti-vaccine movement and, 11–12, 198, 202, 239, 242, 243, 412, 447
Waldorf Schools, 313
Wallis, John, 78
Ward, 175
Warning signs of pseudoscience, xiv
Warranted belief, science and, 17
Water, risky play and desensitization to, 175
Watson, James, 86
Watt, Caroline, 381
Websites, pseudoscience and personal, 78
Wechsler, David, 353, 355, 356
Wechsler Adult Intelligence Scale, 362, 364
Wechsler-Bellevue Intelligence Scale, 356
Wechsler full-scale IQs, 363
Wechsler Intelligence Scale for Children (WISC), 354, 355, 362, 365

Wegman, Ita, 313
Wegner, Alfred, 404
Weil, Andrew, 325
Weiss, Brian, 341–342
Wellness institutes, 311, 312
Wellness packages, naturopathic, 154
Wells, Jody, 424
West, E. J., 226
What If Everything You Thought You Knew about AIDS Was Wrong? (Maggiore), 434
White, Ryan, 423
Whole medical systems, 318
Whole-person treatment, in naturopathic medicine, 156–159
Whooping cough vaccine, 209–210, 213
Whorton, James C., 13
"Why Vaccines Aren't Paleo" (Brogan), 198
Wild Chickens and Petty Tyrants (Kozak), 228
Wilde, Oscar, 380
Williams, Nushawn, 435
Wilson, Edward, 398
Wilson, Jeff, 228
Wine, glyphosate in, 129
Wired magazine, 401
Wiseman, Richard, 379, 381, 388
Wittgenstein, Ludwig, 233–234
Wolfson, Jack, 200
Wolters Kluwer, 293
Wonderlic, 362
Woodcock-Johnson test, 358, 359
Worry
  generalized anxiety disorder, 181–182
  parental, 180–181

Yoga, 315, 316, 323
Young, Fiona, 112–113
Youngren, Christina, 153
YouTube, pseudoscience presentations on, 78